Microbial Efflux Pumps

Current Research

Edited by

Edward W. Yu

Bioinformatics and Computational Biology Interdepartmental Graduate Program;
Department of Chemistry;
Department of Physics and Astronomy
Iowa State University
Ames, IA
USA

Qijing Zhang

Department of Veterinary Microbiology
Iowa State University
Ames, IA
USA

and

Melissa H. Brown

School of Biological Sciences
Flinders University
Adelaide
Australia

Caister Academic Press

Caister Academic Press
Norfolk, UK

www.caister.com

British Library Cataloguing-in-Publication Data
A catalogue record for this book is available from the British Library

ISBN: 978-1-908230-21-8

Cover image courtesy of Edward W. Yu and Chih-Chia Su

Printed and bound in Great Britain

Contents

Contributors

Stephanie Baugh
Antimicrobial Agents Research Group
College of Medical and Dental Sciences
University of Birmingham
Birmingham
UK

sxb513@bham.ac.uk

Jani Reddy Bolla
Department of Chemistry
Iowa State University
Ames, IA
USA

janreddy@iastate.edu

Melissa H. Brown
School of Biological Sciences
Flinders University
Adelaide
Australia

melissa.brown@flinders.edu.au

Silvia Buroni
Department of Biology and Biotechnology
University of Pavia
Pavia
Italy

silvia.buroni@unipv.it

Geoffrey Chang
Skaggs School of Pharmacy and Pharmaceutical Sciences and Department of Pharmacology
School of Medicine
University of California
La Jolla, CA
USA

g1chang@ucsd.edu

Jason M. Cloward
Department of Microbiology and Immunology
Emory University School of Medicine
Atlanta, GA
USA

wnv3@cdc.gov

Sylvia V. Do
Bioinformatics and Computational Biology Interdepartmental Graduate Program
Iowa State University
Ames, IA
USA

svdo@iastate.edu

Dijun Du
Department of Biochemistry
University of Cambridge
Cambridge
UK

dd339@cam.ac.uk

Bart A. Eijkelkamp
School of Biological Sciences
Flinders University
Adelaide
Australia

bart.eijkelkamp@flinders.edu.au

Maira Goytia
Department of Microbiology and Immunology
Emory University School of Medicine
Atlanta, GA
USA

mgoytia@emory.edu

Karl A. Hassan
Department of Chemistry and Biomolecular Sciences
Macquarie University
Sydney
Australia

karl.hassan@mq.edu.au

Xiao He
Skaggs School of Pharmacy and Pharmaceutical Sciences
University of California
La Jolla, CA
USA

xihe@ucsd.edu

Paul J.T. Johnson
Department of Microbiology and Immunology
Emory University School of Medicine
Atlanta, GA
USA

pjohns4@emory.edu

Justin L. Kandler
Department of Microbiology and Immunology
Emory University School of Medicine
Atlanta, GA
USA

jkandle@emory.edu

Hsiang-Ting Lei
Department of Chemistry
Iowa State University
Ames, IA
USA

hlei@iastate.edu

Feng Long
Department of Chemistry
Iowa State University
Ames, IA
USA

longfe@iastate.edu

Ben F. Luisi
Department of Biochemistry
University of Cambridge
Cambridge
UK

bfl20@cam.ac.uk

Alexandra D. Mercante
Department of Microbiology and Immunology
Emory University School of Medicine
Atlanta, GA
USA

wip2@cdc.gov

Justin D. Nguyen
Division of Biological Sciences
University of California at San Diego
La Jolla, CA
USA

jdn001@ucsd.edu

Dietrich H. Nies
Molecular Microbiology
Institute for Biology/Microbiology
Martin-Luther-University Halle-Wittenberg
Halle (Saale)
Germany

d.nies@mikrobiologie.uni-halle.de

Elizabeth A. Ohneck
Department of Microbiology and Immunology
Emory University School of Medicine
Atlanta, GA
USA

eohneck@emory.edu

Maria Rosalia Pasca
Department of Biology and Biotechnology
University of Pavia
Pavia
Italy

mariarosalia.pasca@unipv.it

Ian T. Paulsen
Department of Chemistry and Biomolecular Sciences
Macquarie University
Sydney
Australia

ian.paulsen@mq.edu.au

Laura J. V. Piddock
Antimicrobial Agents Research Group
College of Medical and Dental Sciences
University of Birmingham
Birmingham
UK

l.j.v.piddock@bham.ac.uk

Keith Poole
Department of Biomedical & Molecular Sciences
Queen's University
Kingston, ON
Canada

poolek@queensu.ca

Klaas M. Pos
Institute of Biochemistry and Cluster of Excellence Frankfurt – Macromolecular Complexes
Goethe University Frankfurt
Frankfurt
Germany

pos@em.uni-frankfurt.de

Giovanna Riccardi
Department of Biology and Biotechnology
University of Pavia
Pavia
Italy

giovanna.riccardi@unipv.it

Milton H. Saier, Jr.
Division of Biological Sciences
University of California at San Diego
La Jolla, CA
USA

msaier@ucsd.edu

William M. Shafer
Department of Microbiology and Immunology
Emory University School of Medicine
Atlanta, GA;
Laboratories of Bacterial Pathogenesis
Veterans Affairs Medical Center
Decatur, GA
USA

wshafer@emory.edu

Zhangqi Shen
Department of Veterinary Microbiology
Iowa State University
Ames, IA
USA

szq@iastate.edu

Maksim A. Shlykov
Division of Biological Sciences
University of California at San Diego
La Jolla, CA
USA

mshlykov@ucsd.edu

Chih-Chia Su
Department of Chemistry
Iowa State University
Ames, IA
USA

jacksu@iastate.edu

Paul Szewczyk
Skaggs School of Pharmacy and Pharmaceutical Sciences
University of California
La Jolla, CA
USA

pszewczy@ucsd.edu

Henrietta Venter
Division of Health Sciences
School of Pharmacy and Medical Sciences
University of South Australia
Adelaide
Australia

rietie.venter@unisa.edu.au

Eric Wang
Division of Biological Sciences
University of California at San Diego
La Jolla, CA
USA

eric.joe.wang@gmail.com

Edward W. Yu
Bioinformatics and Computational Biology Interdepartmental Graduate Program;
Department of Chemistry;
Department of Physics and Astronomy
Iowa State University
Ames, IA
USA

ewyu@iastate.edu

Yaramah M. Zalucki
Department of Microbiology and Immunology
Emory University School of Medicine
Atlanta, GA
USA

yzaluck@emory.edu

Qijing Zhang
Department of Veterinary Microbiology
Iowa State University
Ames, IA
USA

zhang123@iastate.edu

Wei Hao Zheng
Division of Biological Sciences
University of California at San Diego
La Jolla, CA
USA

whzheng87@gmail.com

Preface

Infections caused by bacteria remain a leading cause of death worldwide. Antibiotics have been key drugs for clinical therapy of bacterial infections, but development of resistance has severely compromised the effectiveness of antibiotic treatments. As a common and major resistance mechanism, bacterial efflux systems are able to extrude structurally diverse antimicrobials, facilitating their survival in toxic environments. Additionally, recent findings have revealed that antibiotic efflux systems have important physiological functions and play major roles in bacterial pathogenesis. The significance of these efflux systems is further highlighted by the fact that they are widely distributed and a single species may harbour several of such efflux systems. For example, most opportunistic nosocomial bacterial pathogens possess an extended collection of resistance strategies to circumvent the effects of continuous exposure to antimicrobial stresses. Of these, active efflux has proven to be one of the most successful detoxification mechanisms used by both Gram-positive and Gram-negative pathogens. Recent breakthroughs in the field have provided novel insights into the detailed mechanisms on how these efflux systems operate and how bacteria utilize them beyond antibiotic efflux. This book focuses on bacterial efflux systems and presents a detailed survey of the field on how microbes employ them to ensure their survival and establish/maintain successful infections. In particular, this book summarizes the spectacular discoveries taken place in the past several years by means of biochemistry and bioinformatics, together with structural and molecular biology. It is hoped that unravelling the intricacies of these mechanisms will inspire the development of new strategies to overcome antimicrobial resistance.

Edward W. Yu
Qijing Zhang
Melissa H. Brown

Transmembrane Molecular Transporters Facilitating Export of Molecules from Cells and Organelles

1

Maksim A. Shlykov, Wei Hao Zheng, Eric Wang, Justin D. Nguyen and Milton H. Saier, Jr.

Abstract

Transport systems catalyse the uptake and efflux of solutes and macromolecules from all living cells and organelles. In this chapter, we review multiple types of efflux systems responsible for the export of inorganic and organic cations and anions, metabolites, drugs, complex carbohydrates, lipids, proteins and nucleic acids. These types fall into three primary categories: simple channels, secondary carriers and primary active transporters. We select a number of well-characterized systems in order to illustrate the structures and functions of different types of transporters as well as their mechanisms of action. However, metabolite exporters, in contrast to uptake systems and drug and macromolecular exporters, are still poorly characterized. We anticipate that studies conducted in the near future will reveal a plethora of metabolite exporters while revealing relatively few novel systems for metabolite uptake.

Introduction

Transport systems found throughout the biosphere fall into a surprisingly small number of types. First, there is the class of channel- or pore-forming transmembrane proteins that constitute several subclasses as categorized in the Transporter Classification Database (TCDB, www.tcdb.org; Saier *et al.*, 2006, 2009). All such proteins belong to class 1 in TCDB. Almost without exception, they catalyse bidirectional fluxes of solutes in accordance with the directions of relevant electrochemical gradients (Saier and Paulsen, 2001; Gouaux and MacKinnon, 2005; Blanke, 2006; Karuppiah *et al.*, 2011). Second, secondary carriers (class 2 in TCDB) catalyse uniport, symport or antiport depending on the system and conditions. The uniporters, like channels, allow the free flow of solutes across the membrane in accordance with relevant electrochemical gradients, but in contrast to most channels, they exhibit much lower (~ 10,000-fold) turnover rates and exhibit stereo – and electrospecificity with respect to solute recognition at the active site. Because the membrane potential is virtually always negative inside, and because symporters and antiporters generally use cations (usually H^+ or Na^+) as the co – or anti-transported ion, the symporters catalyse solute uptake while the antiporters catalyse efflux. Finally, TC class 3, the primary active transporters, use a primary source of energy (chemical energy, light energy or electron flow) to drive the uptake or extrusion of ions and other molecules (Papa *et al.*, 1998; Hirai *et al.*, 2009). All three classes of transporters can catalyse export of small molecules and macromolecules across biological membranes, and all three can do so in a unidirectional fashion, although this is relatively rare for class 1 channels/pores.

The presence of multiple topological types within a single family has led to a better understanding of protein evolution, which involves intragenic duplications, gene fusions, deletions and sequence divergence (Saier, 2000b; Prakash *et al.*, 2003; Saier, 2003; Lolkema *et al.*, 2008). One example of a superfamily of integral membrane proteins is the ubiquitous Haem Handling Protein (HHP) family (TC# 9.B.14). Members of this family exhibit 6, 8, 10, 11, 13 or 15 putative transmembrane segments (TMS)s. Lee *et al.* (2007) showed that intragenic triplication of a two TMS element gave rise to a protein with a six TMS topology, exemplified by the CcmC protein of *Escherichia coli*, which has numerous homologues. This basic six TMS unit then gave rise to two distinct types with eight TMSs, exemplified by ResC and the archaeal CcmC, and these further underwent fusional or insertional events yielding proteins with 10, 11 and 13 TMSs (ResC homologues) as well as 15 TMSs (CcmF homologues). Specific evolutionary pathways taken were proposed (Lee *et al.*, 2007).

A second example of a superfamily of integral membrane proteins with varying topologies is the SdpI family (TC# 9.A.32) of signal transduction proteins (Povolotsky *et al.*, 2010). One member of the family in *Bacillus subtilis*, SdpI, provides immunity to cells from cannibalism in times of nutrient limitation. SdpI family members are transmembrane proteins with 3, 4, 5, 6, 7, 8, or 12 putative TMSs. The two most conserved motifs, found between TMSs 1 and 2 and TMSs 4 and 5 of the 6 TMS proteins, show significant sequence similarity, which led Povolotsky *et al.* (2010) to propose that the primordial precursor of these proteins was a three TMS-encoding genetic element that underwent intragenic duplication. Various deletional and fusional events, as well as intragenic duplications and inversions, may have yielded SdpI homologues with topologies of varying numbers and positions of TMSs. As for the HHP family, a specific evolutionary pathway that could have given rise to these distantly related bacterial immunity proteins was proposed.

Integral membrane transport protein sizes exhibit remarkable regularity, with eukaryotic members of superfamilies, on average, being 40% larger than bacterial members and archaeal members being on the average 8% smaller (Chung *et al.*, 2001).

In this chapter we provide an overview of these transporters with emphasis on exporters in classes 2 and 3 (Barabote *et al.*, 2006; Tetsch and Jung, 2009). The techniques and methods used for bioinformatic analyses of transmembrane proteins are outlined in detail by Reddy and Saier (2012) and Reddy *et al.* (2012). Details of the procedures for using these programs can be found at www.tcdb.org/discuss.

Channels

Of the ten types of channels listed in TCDB, seven of them catalyse the energy-independent flow of molecules across biological membranes. Many are non-specific and transport anything smaller than the diameter of the channel. However, others exhibit ion specificity – preferring cations to anions or vice versa. Some are even specific for a particular cation or anion, but all of them catalyse solute uptake and efflux with a specificity determined by the specific channel proteins. For example, within the Voltage-gated Ion Channel (VIC) superfamily, members are specific for Na^+, K^+ or Ca^{2+} or are non-specific – acting on any cation (Saier, 2000a).

Found ubiquitously in all organisms ranging from bacteria to higher eukaryotes, α-type channels (TC subclass 1A) usually catalyse energy-independent passage through a transmembrane aqueous pore. While these channels mainly consist of transmembrane α-helical spanners (α-TMSs), β-strands may contribute to the overall structure of the protein (Fig. 1.1A; Gulbis *et al.*, 1999; Lomize *et al.*, 2006). On the other hand, β-type porins (TC subclass 1B) usually consist of β-strands (β-TMSs) that form β-barrels (Fig. 1.1B) that are found in the outer membranes of Gram-negative bacteria, mitochondria, plastids and possibly a restricted group of Gram-positive bacteria, the Actinomycetes (Gromiha and Suwa, 2007). Pore-forming toxins (TC subclass 1C) are a special type of proteins that are secreted by one cell but insert into the membrane of another cell. The pore may allow the entry of a toxin into the cell or the free flow of electrolytes and other small molecules across the membrane. Loss of electrolytes and other solutes eventually kills the cell. These toxins range from large proteins to small ribosomally or non-ribosomally synthesized peptides (Slatin *et al.*, 2008).

Non-ribosomally synthesized channels (TC subclass 1D) typically consist mainly of L- and D-amino acids as well as other small molecular building blocks such as hydroxy acids (i.e. lactate). However, other substances completely lacking amino acids are also capable of channel formation. It has been found that voltage may induce the formation of these oligomeric transmembrane pore-forming channels (Saier, 2006; Saier *et al.*,

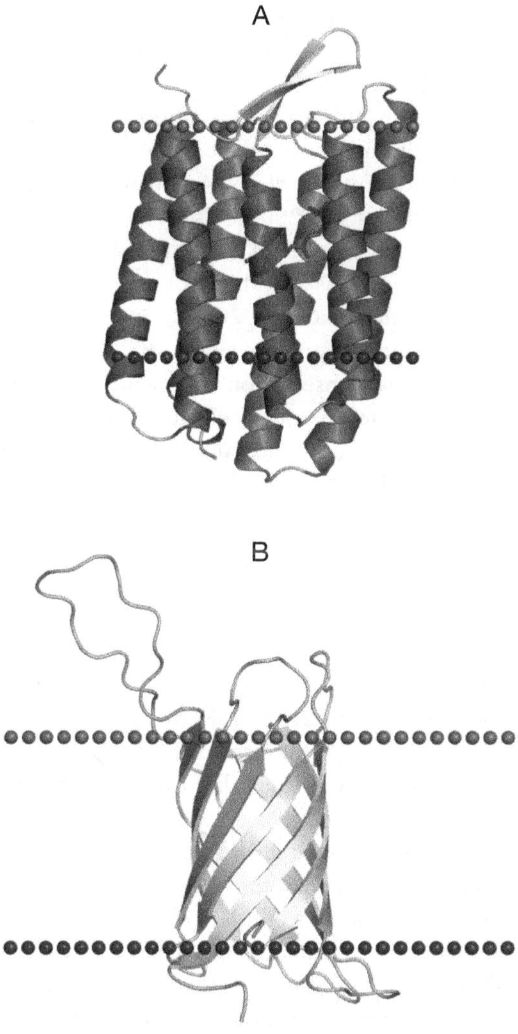

Figure 1.1 Representative structures of alpha (A) and beta (B) transmembrane proteins. The archaeal rhodopsin-1 (A; TC # 3.E.1) and beta-barrel porin (B; TC # 1.B.6) proteins are represented. Top and bottom of each figure show the extracellular and cytoplasmic sides, respectively. Both figures were obtained from the Orientations of Proteins in Membranes (OPM) database. A colour version of this figure is located in the plate section at the back of the book.

2009). Holins (TC subclass 1E) are present in the genomes of Gram-positive and Gram-negative bacteria as well as in the bacteriophages of these organisms. They act primarily as murein hydrolase exporters that, once in contact with the cell wall, hydrolyse the cell wall polymers. The holins may also facilitate the leakage of electrolytes, nutrients and nucleotides from the cytoplasm – leading to cell death. While holins are incredibly diverse, most of them are small (60–145 residues), having one, two, three or at most, four putative transmembrane α-helical spanners (TMSs) per polypeptide chain. They form homooligomeric pores of varying sizes, some of them of variable micron-diameter sizes. They can accommodate secreted proteins, regardless of their sizes, by forming large oligomeric pores (Pang *et al.*, 2009; Dewey *et al.*, 2010). In eukaryotes, vesicles sequester various substances such as neurotransmitters, proteins, and complex carbohydrates. Fusion with the plasma membrane releases their contents into the extracellular medium. Vesicle fusion is initiated by the formation of a pore complex (TC subclass 1F) of variable sizes that releases the intravesicular substances into the extracellular space at rates dependent on the pore size (He *et al.*, 2006). On the other hand, viral fusion pores (TC subclass 1G) contain different types of fusion peptides and vary in their reliance on accessory proteins. However, despite this diversity, all characterized viral fusion proteins convert from a fusion-competent state (dimers and trimers) to a membrane-embedded homotrimeric prehairpin. This structure then proceeds to a trimer-of-hairpins that brings the fusion peptides, attached to the target membrane, and the transmembrane domain, attached to the viral membrane, into close proximity. This process eventually allows the union of viral and target membranes (White *et al.*, 2008).

Recently discovered paracellular channels (TC subclass 1H) mediate the passive but selective transport of solutes across tight junctions allowing changes to transcellular electrical potentials and concentration gradients. Specific proteins, the claudins, tighten the paracellular cleft and also form the selectively permeable paracellular ion pores (Chiba *et al.*, 2008). Membrane-bounded channels (TC subclass 1.I) and virions egress pyramidal apertures represent two additional channel types not to be discussed here (see TCDB).

Secondary active efflux pumps

Secondary active transport is contingent upon an exergonic reaction carried out by primary active transport systems. The energy released from the exergonic reaction drives the creation of an electrochemical gradient for related molecules, often Na^+ or H^+ (Pasternak, 1990). Dissipation of the gradient occurs in part via secondary active efflux pumps (SAEPs), which couple the release of potential energy stored within the gradient with the movement of mono/di/trivalent cations, anions, drugs, macromolecules, metabolites and a host of other solutes across cell membranes against their gradients (Brey *et al.*, 1979; Maloney *et al.*, 1990). The proposed mechanisms of action of secondary active porters (SAPs) follow two models. The 'rocker-switch/alternating access' model suggests an inversion of the protein's configuration and the subsequent alternating accessibility of the binding site to the cytoplasm and the external medium. The competing 'gated pore' model considers the local motion of inside and outside gates, alternatively opening while flanking the binding sites (Enkavi *et al.*, 2010).

Current evidence points to both conformational modes being utilized in secondary transport (Boudker *et al.*, 2010). Further, the functioning of secondary transporters must be physically prevented without the binding of all molecules involved in the transport reaction such that substrate/cation gradients are not dissipated. Depending on the polarity of the gradients for the molecules involved, SAPs can be further subdivided into either export antiporters or uptake symporters (SAUPs), based on whether movement of the primary substrate is in the outward or inward direction, respectively.

SAEPs are present in all domains of life and probably all living cells. We aim to describe specific SAEP families related to export of most of the aforementioned molecules, their mechanisms of action, and any further properties they may confer.

Monovalent cations

The monovalent cation:proton antiporter (CPA) families 1, 2 and 3 (TC #s 2.A.36, 2.A.37 and 2.A.63), the NhaA, NhaB, NhaC and NhaD Na^+:H^+ antiporter families (TC #s 2.A.33, 2.A.34, 2.A35 and 2.A.62), and the mitochondrial inner membrane K^+/H^+ and Ca^{2+}/H^+ exchanger (LetM1) family (TC#2.A.97) are all involved in secondary active monovalent cation efflux. Most of these families display a similar topology with 10–14 TMSs.

Transport of monovalent cations is crucial for ionic homeostasis, pH control and volume control (Nakamura *et al.*, 1984; Padan *et al.*, 2001; Nowikovsky *et al.*, 2004). They are of further importance because high or conversely low concentrations of ions may be toxic and affect cell growth (Radchenko *et al.*, 2006).

A common feature of many Na^+-dependent SAEPs is the conservation of essential aspartyl residues, which interact electrostatically with sodium (Nakamura *et al.*, 2001). For example, an NhaA homologue (TC#2.A.33.1.1) has four such conserved aspartyl residues where D164 binds Na^+, D133 and D65 are crucial for pH regulation, and D163 controls alternating access to the binding site from the cytoplasm and periplasm (Fig. 1.2; Williams *et al.*, 1999; Hunte *et al.*, 2005; Olkhova *et al.*, 2006; Screpanti *et al.*, 2006; Arkin *et al.*, 2007; Padan *et al.*, 2008; Appel *et al.*, 2009; Herz *et al.*, 2010). Protonation of D163 and D164 provides access to the periplasm and releases Na^+, respectively.

Electroneutral transporters with a 1:1 stoichiometry of Na^+ to H^+ are common, but electrogenic carriers, with a Na^+:H^+ ratio of < 1, are also known. An NhaA family member (TC#2.A.33.1.1) has a ratio of 2. Stoichiometric relationships demonstrate the robustness and effectiveness of the transporter in its ability to pump against a gradient.

Although Na^+:H^+ transport is most commonly observed within all of the Nha families and the CPA1 family, functional outliers exist. CPA1 members that are able to transport other monovalent cations including Li^+, K^+, Rb^+ and Cs^+ have been identified (see TC#2.A.36.3.2 and 2.A.36.5.1; Fujisawa *et al.*, 2005). Low-affinity Na^+ SAEPs can substitute other monovalent cations for Na^+ including H^+, K^+, Cs^+, NH_4^+ and Li^+. More tightly regulated and higher affinity transporters have also been observed to exhibit promiscuity. Several homologues of the NhaA and NhaB families, which normally extrude Na^+, can mediate export of K^+ when inhibitory cellular concentrations of K^+ exist (Radchenko *et al.*, 2006). Other functional outliers belonging primarily to Na^+ SAEP

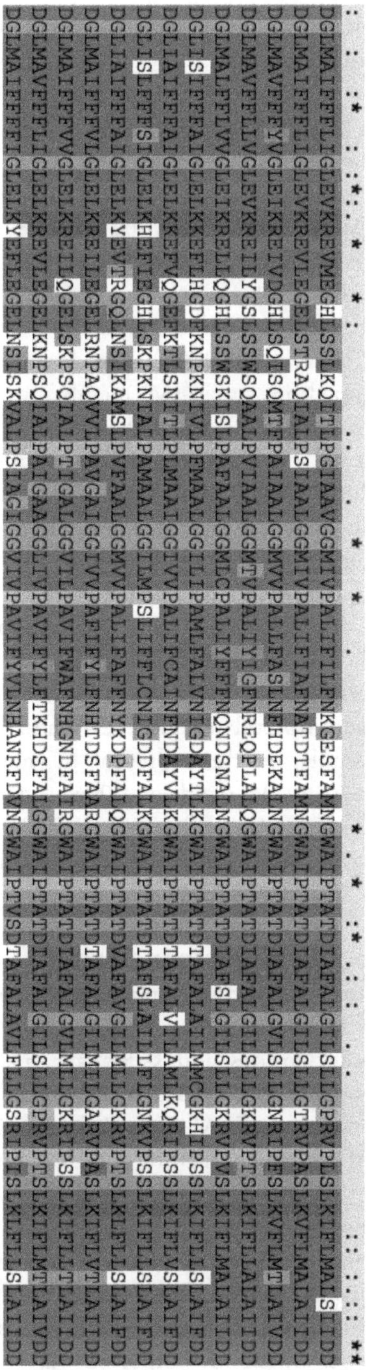

Figure 1.2 Alignment of NhaA homologues displaying high levels of conservation for the functionally important D residues in positions 65, 133, 163 and 164. Only a small portion of the multiple alignment is displayed, explaining why some residues that appear to be fully conserved do not have symbols of conservation. (*, full conservation; :, close substitutions only; ., more distantly related similarities.)

families include exporters for amino acids and exchangers for malate + H^+:lactate + Na^+.

A 1013-residue *D. melanogaster* protein containing one putative TMS in the LetM1 family has been proposed to mediate Ca^{2+}:H^+ antiport (Santo-Domingo and Demaurex, 2010). The primary role of LetM1 was proposed to be 1:1 electrogenic and voltage-dependent uptake of Ca^{2+} upon mitochondrial alkalinisation, but data indicating a role of Ca^{2+} efflux upon mitochondrial acidification suggest bifunctionality (Jiang *et al.*, 2009). Because no other secondary carrier has been reported to have less than four TMSs, the mechanism of action reported by Santo-Domingo and Demaurex (2010) must be considered tentative. A channel mechanism could account for most of the data.

A second role reported for LetM1 is K^+/H^+ antiport-mediated K^+ uptake (Froschauer *et al.*, 2005). Both the TMHMM 2.0 and HMMTOP programs predicted a conserved N-terminal TMS, but another central region was predicted to be transmembrane by HMMTOP but not TMHMM 2.0 (Sonnhammer *et al.*, 1998; Tusnády *et al.*, 1998; 2001; Krogh *et al.*, 2001). This central 18-residue putative TMS stretch was used to create a helical wheel diagram using the Helixator (http://www.tcdb.org/progs/helical_wheel.php), which suggested the presence of an amphipathic α-helix. The function of LetM1 is limited and dependent upon the pH gradient generated by the mitochondrial electron transport chain (Santo-Domingo and Demaurex, 2010).

Di/trivalent cations

Many divalent cations are essential for normal cell function and development, but high concentrations of heavy metal cations can lead to toxicity (Au *et al.*, 2008). In mammals, heavy metals can readily enter the CNS, impair the blood brain barrier, and deposit themselves in the brain (Fitsanakis and Aschner, 2005). The high levels of trivalent cations such as Al^{3+} found in the brains of Alzheimer's sufferers, as well as Mn^{2+} deposit-derived symptoms that are similar to those observed in idiopathic Parkinson's disease (iPD) patients, a disease with a different route of pathogenesis, have led some to postulate that di/trivalent cations may play a role in the pathogenesis of neurodegenerative diseases (Au *et al.*, 2008; Tomljenovic, 2011).

Di/trivalent cation SAEPs exported by members of the families mentioned below have been shown to be capable of exporting a variety of mono/di/trivalent cations including, but not limited to Ag^{+}, Al^{3+}, Cu^{+}, Cu^{2+}, Cd^{2+}, Co^{2+}, Fe^{2+}, Fe^{3+}, Hg^{2+}, Ni^{2+}, Pb^{2+} and Zn^{2+}. Most members of these families drive heavy metal efflux by a proton antiport mechanism.

All members of the ubiquitous cation diffusion facilitator (CDF) family (TC#2.A.4), the prokaryotic heavy metal efflux (HME) family (TC#2.A.6.1) within the resistance–nodulation–cell division (RND) Superfamily, the Ni^{2+}–Co^{2+} transporter (NiCoT) family (TC#2.A.52), and the nickel resistance (Nre) family (TC#2.A.1.31) of the major facilitator superfamily appear to exclusively catalyse active divalent cation efflux. A variation on this theme has been reported within the CDF family. CzcD of *B. subtilis* (TC#2.A.4.1.3) may couple H^{+}/K^{+} uptake to the efflux of Zn^{2+}, Cd^{2+}, Co^{2+}, Ni^{2+}, Cu^{2+}, or Pb^{2+} (Guffanti *et al.*, 2002). This possibly suggests electroneutrality for monovalent cation uptake coupled to divalent heavy metal ion export, but the exact stoichiometry has not been established. Surprisingly, evidence has been presented leading to the conclusion that the ancient, ubiquitous, six TMS CDF carriers gave rise through evolution to four TMS CRAC Ca^{2+} channels in animals (Matias *et al.*, 2010). This seems to be a rare instance of 'reverse evolution' where a complex protein was the precursor of a structurally and functionally simpler protein.

In addition to their usual roles in the plasma membranes of eukaryotes, some members of these families are found in eukaryotic organelles including the Golgi apparatus, mitochondria, the endoplasmic reticulum, and vacuoles. While most members of the Nramp family appear to catalyse cellular cation import via a symport mechanism, one member named Nramp1 or Slc11A1 has been reported to catalyse cation efflux (Courville *et al.*, 2004; Nevo, 2007; Czachorowski *et al.*, 2009).

HME family homologues (TC# 2.A.6.1), mostly restricted to Gram-negative bacteria, have been observed to mediate efflux of a wide range of divalent cations as well as Ag^{+} and Cu^{+}. All well-characterized HME porters function in conjunction with members of the membrane fusion protein (MFP) family (TC#8.A.1) and the outer membrane factor (OMF) family (TC#1.B.17). In such a system, an HME SAEP is situated in the cytoplasmic membrane, while the MFP protein serves as an adaptor protein to connect the HME protein to an OMF protein. The OMF family member serves as a channel in the outer membrane (Touzé *et al.*, 2004). The divalent cations are pumped out across two membranes and the periplasm in a single energy coupled step without periplasmic equilibration (Nikaido and Takatsuka, 2009). These RND-type porters can also function as 'vacuum cleaners,' removing heavy metal ions from the cytoplasmic membrane and the periplasm (Nikaido and Takatsuka, 2009).

YiiP (CDF family: TC#2.A.4.7.1) of *E. coli* functions as a homodimer, and, as for other members of the CDF family, it possesses highly conserved aspartyl residues (D49 and D147 in YiiP), similar to those present in monovalent cation SAEPs (Wei and Fu, 2006). CDF family members cluster according to their divalent cation specificities (Montanini *et al.*, 2007; Matias *et al.*, 2010).

Metabolites and other solutes

Aside from ions, drugs and macromolecules, a host of solutes including amino acids, carbohydrates, nucleosides and secondary metabolites are exported by SAEPs. The metabolite and solute exporting group of SAEPs are the most diverse in terms of function, structure, and the compounds that they export.

Secondary active transporters for all 20 common amino acids found in proteins have been identified. A search through TCDB revealed that all amino acids except glutamate and phenylalanine, which are imported via secondary active transport, are exported via SAEPs. Several families catalysing the secondary active export of lysine, arginine, threonine, serine, branched chain amino acids and several other compounds have been identified.

The LysE superfamily consists of the L-lysine exporter (LysE), the resistance to homoserine/threonine (RhtB), the cadmium resistance

(CadD), and the Ni^{2+} uptake transporter (HupE-UreJ) families (TC #s 2.A.75, 2.A.76, 2.A.77, 2.A.99, respectively). These families have similar sizes, use H^+ as the antiported cation, usually exhibit six TMS topologies and are restricted to bacteria and archaea. However, the exported substrates vary substantially.

Only two members of the LysE family have been functionally characterized, but uncharacterized homologues have been discovered in the genomes of various bacteria. The LysE family member, LysE (TC#2.A.75.1.1) of *Corynebacterium glutamicum*, was the first to be characterized and mediates the efflux of L-lysine (Vrljic *et al.*, 1999). Another LysE member, ArgO (TC#2.A.75.1.2) of *E. coli*, exports arginine and the plant derived polyamine, canavanine, in addition to lysine (Nandineni and Gowrishankar, 2004). The majority of RhtB family members remains uncharacterized, but those that have had a function assigned to them were found to possess the ability to export neutral amino acids including threonine, cysteine, leucine and other less common compounds (Aleshin *et al.*, 1999; Zakateva *et al.*, 1999; Franke *et al.*, 2003; Kutukova *et al.*, 2005). The ubiquitous Threonine/Serine Exporter (ThrE) family (TC#2.A.79), which has yet to be assigned to a superfamily, like RhtB, mediates the proton motive force-dependent export of threonine, in addition to serine (Ziegler *et al.*, 2000). A phylogenetic relationship between the ten TMS ThrE family and the RhtB family with members of the LysE superfamily has not yet been established, but can reasonably be proposed.

Whereas the LysE and RhtB families export amino acids, members of the CadD family play a role in conferring cadmium resistance and extruding quaternary amines (Crupper *et al.*, 1999). In contrast to the functions of the LysE and RhtB families, CadD family members export Cd^{2+} while Ni^{2+} uptake transporters (TC#2.A.99) within the same superfamily transport these heavy metals.

Members of the ubiquitous branched-chain amino acid exporter (LIV-E) family (TC#2.A.78) utilize H^+ as the antiported cation and function as pairs of integral membrane proteins to export branched-chain amino acids and methionine. The AzlC (six putative TMS) and AzlD (three or four TMS) pair of *B. subtilis* is reported to export leucine, isoleucine and valine whereas the homologous BrnE (four TMS) and BrnF (6 TMS) pair in *C. glutamicum* additionally extrudes methionine (Belitsky *et al.*, 1997; Kennerknecht *et al.*, 2002; Trötschel *et al.*, 2005). It has been postulated that these systems arose as a result of a rare gene triplication of a four TMS primordial peptide, which resulted in eight and four TMS-encoding elements (Kennerknecht *et al.*, 2002). Subsequently, two TMSs at the N-terminus are believed to have been lost to yield the six-TMS BrnF protein (Fig. 1.3).

Recently, a functional relationship has been observed between the ubiquitous arsenical resistance-3 (ACR3) and the auxin efflux carrier (AEC) families within the bile/arsenite/riboflavin transporter (BART) superfamily (Mansour *et al.*, 2007) and the ubiquitous arsenite–antimonite (ArsB) Efflux family within the Ion Transporter (IT) superfamily. Most members of the ACR3 and AEC families possess 10 TMSs, whereas most members of the ArsB family members possess 12 TMSs, although some 10 TMS members may exist. The evolutionary relationships between these two topological types, if verified, should prove interesting.

ArsB (TC#2.A.45.1.1), an ArsB family member, possesses the ability to export arsenite or antimonite via either a secondary active process, or by utilizing the ATP hydrolysis function of the ArsA protein to drive efflux (Rosen, 1996; Xu *et al.*, 1998; Rensing *et al.*, 1999). Bioinformatic and biochemical studies have led to the proposal that ArsA proteins can also function with ACR3 homologues, thus converting these secondary active arsenite/antimonite transporters that are non-homologous to ArsB into primary active transporters (Castillo and Saier, 2010; Fu *et al.*, 2010). It has not been demonstrated that auxin efflux pumps of the AEC family can function with an ArsA protein like its superfamily partner.

The large and diverse MFS family, catalysing the secondary active uptake or efflux of ions, drugs, metabolites and other solutes, has been reviewed extensively (Pao *et al.*, 1998; Saier *et al.*, 1999; Yen *et al.*, 2010; Reddy *et al.*, 2012) and will not be considered further in this treatise.

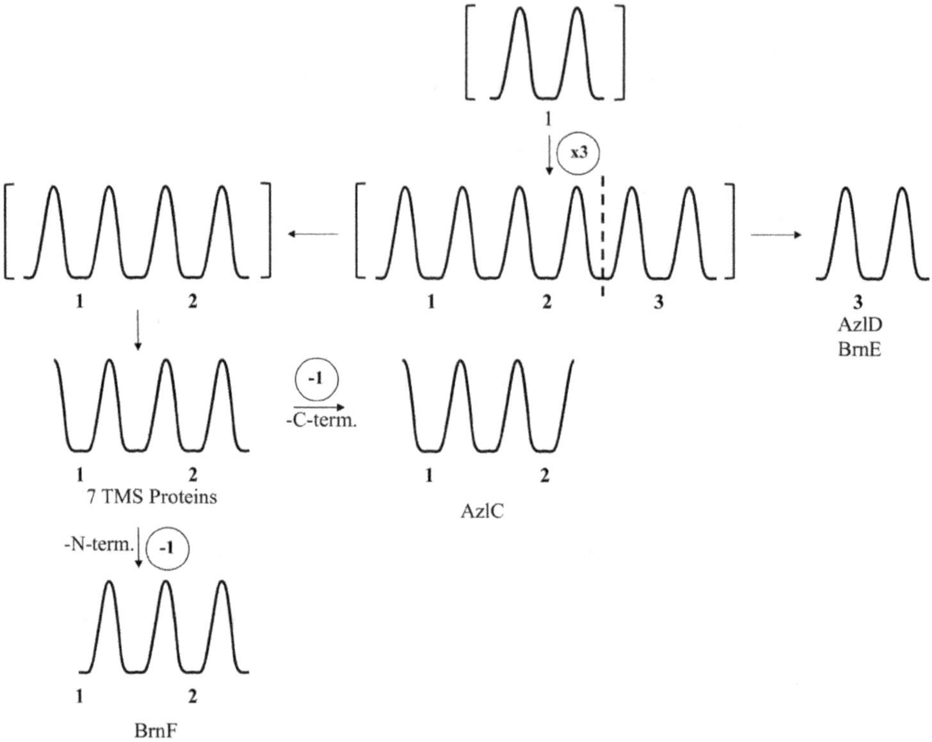

Figure 1.3 Proposed evolutionary pathway for members of the LIV-E family. A four TMS unit is believed to have triplicated, and the resultant 12 TMS product was split into eight and four TMS proteins. For the large subunits of the two-component LIV-E systems, six and seven TMS topologies are found in present-day systems due to loss of one or two TMSs from the putative primordial eight TMS protein. Brackets imply that these ancestral proteins have not yet been identified in current protein sequence databases.

Multidrug resistance

Multidrug resistance (MDR) is a characteristic of many harmless and disease-causing organisms. Terrestrial soil-dwelling bacteria have large numbers of these pumps, presumably to protect themselves against biological warfare with other microbes. All of these organisms utilize efflux systems that transport a variety of non-related chemicals out of the cell, thereby conferring resistance to them. Organisms that display such behaviour include pathogenic bacteria, archaea, eukaryotic protists, fungi, animals and plants. Thus, they can be found in all types of living organisms.

Pathogenic organisms utilize proton or sodium motive force-energized systems to export drugs. Such multidrug exporters belong to four different superfamilies: the multidrug/oligosaccharide/polysaccharide (MOP; TC# 2.A.66) superfamily, the drug metabolite transporter (DMT; TC#2.A.7) superfamily, the resistance/nodulation/division (RND; TC#2.A.6) superfamily and the major facilitator superfamily (MFS; TC#2.A.1) (Paulsen *et al.*, 1996; Yen *et al.*, 2010).

All members of the MOP superfamily (Fig. 1.4D) appear to catalyse active efflux (Hvorup *et al.*, 2003). This superfamily includes the ubiquitous multidrug and toxin extrusion (MATE) family and the bacterial mouse virulence factor (MVF) family. Only members of the MATE family are known to export drugs, but members of the characterized MVF family may do so as well.

The DMT superfamily includes members that can catalyse nutrient uptake, drug and metabolite efflux, and solute:solute exchange. The best characterized of the DMT superfamily families is the small multidrug resistance (SMR) family with proteins of about 110 amino acids with four TMSs (Fig. 1.4A). They catalyse cationic drug export

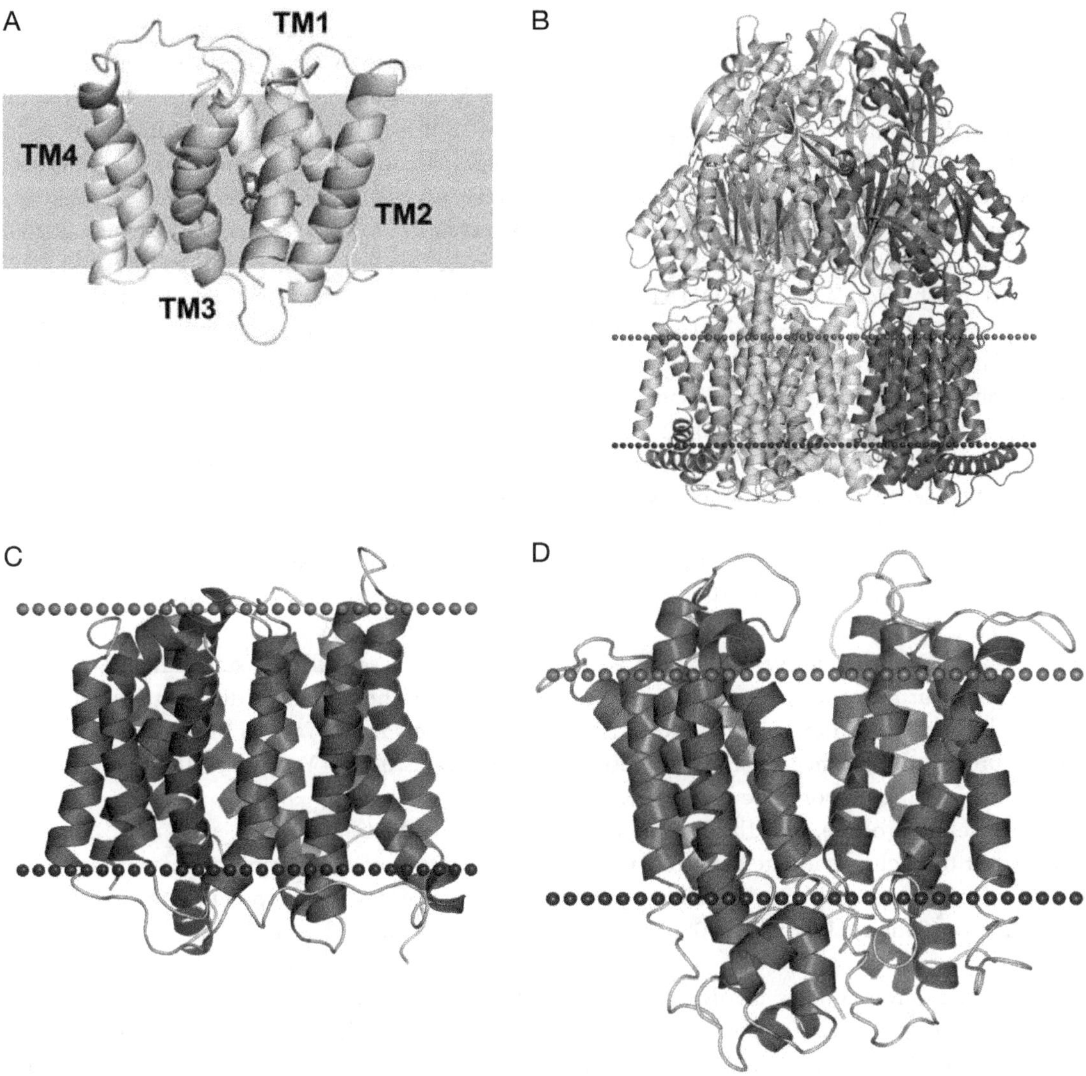

Figure 1.4 Structures of the dimeric EmrE (A; TC # 2.A.7.1.3; DMT superfamily), the trimeric AcrB (B; TC # 2.A.6.2.2; RND superfamily), EmrD (C; TC# 2.A.1.2; MFS) and NorM (D; TC # 2.A.66.1.2; MOP superfamily) proteins. Red and blue dots signify the periplasmic and cytoplasmic sides of the Gram-negative bacterial envelope, respectively. The image for EmrE was published by Chen *et al.* (2007). A–C were obtained from the OPM database. A colour version of this figure is located in the plate section at the back of the book.

in exchange for a proton. The drug/metabolite exporter (DME) family within the DMT superfamily has been implicated in metabolite uptake, but some members may catalyse drug export as well (Jack *et al.*, 2003).

RND superfamily members (Fig. 1.4B) that probably export drugs are found in the hydrophobe/amphiphile efflux-1 (HAE1) family (primarily found in Gram negative bacteria), the hydrophobe/amphiphile efflux-2 (HAE2) family (Gram positive bacteria), and the hydrophobe/amphiphile efflux-3 (HAE3) family (archaea and bacteria). Porters in these families may export drugs and other amphipathic compounds (Tseng *et al.*, 1999). While HAE1 family members are primarily drug exporters, HAE2 family members are known to export complex lipids, but many lipid exporters also catalyse drug efflux (see below). No member of the HAE3 family has yet been characterized.

The MF superfamily (Fig. 1.4C) contains several drug:H^+/antiporter families (e.g. DHA1, 2 and 3), many of which export normal cellular metabolites as well as drugs. Other members of these three families export metabolites exclusively, and several additional MFS families export both drugs and metabolites. Within the DHA1 family, the Blt porter of *B. subtilis* (2.A.1.2.8) and TPO1 of *Saccharomyces cerevisiae* (TC#2.A.1.2.16) export polyamines. VMAT1, VMAT2, and VAT1 of mammals (TC#2.A.1.2.11, 29 and 12, respectively) export monoamines such as dopamine, noradrenaline, serotonin and histamine as well as drugs (Yen *et al.*, 2010).

It has been shown that many of these export systems play important roles in medicine. Bacteria, archaea and eukaryotes have evolved these and other MDR exporters over billions of years for different purposes, long before the advent of animals. First, they protect against toxic substances produced by other organisms, second, they maintain low-levels of toxic substances that they themselves produce, and third, they allow drug export through metabolite exporters where structural characteristics of the drug and metabolite are similar (Sahl and Bierbam, 2008; Yen *et al.*, 2010).

There are many examples of organisms that use recognized metabolite exporters to pump out drugs. For instance, the MDR protein, MdtI of *E. coli* (TC#2.A.7.1.9), is a spermidine exporter, a member of the SMR family within the DMT superfamily. In the RND superfamily, the HAE1 family of drug resistance pumps includes the AcrAB pump in *E. coli* (2.A.6.2.2) which exports phospholipids. The MtrCDE pump of *Neisseria gonorrhoeae* (2.A.6.2.5), the MexAB-OprM system of *Pseudomonas aeruginosa* (TC #2.A.6.2.6) and the BpeAB-OprB system of *Burkholderia pseudomallei* (TC #2.A.6.2.23) are also known to export both lipids and drugs (Yen *et al.*, 2010).

While members of each superfamily may share characteristics between superfamilies, the differences are more pronounced. This is not surprising since these superfamilies may have evolved independently of each other. Further, proteins within each of these superfamilies have distinct three-dimensional folds (see Fig. 1.4A–D); only MFS and MOP superfamily porters have sufficiently similar three-dimensional structural features to suggest a common evolutionary origin (Fig. 1.4; Mukarami *et al.*, 2002, 2006; Lomize *et al.*, 2006; Yin *et al.*, 2006; Chen *et al.*, 2007; He *et al.*, 2010).

Functionally, members of the DMT and MFS superfamilies pump substrates from the cytoplasm via drug:H^+ antiport, but HAE1 members of the RND superfamily apparently are capable of 'extracting' hydrophobic and amphipathic substrates from the cell membrane and periplasm like a vacuum cleaner (Nikaido and Takatsuka, 2009). In contrast to the DMT and MFS superfamilies, characterized MATE family members within the MOP superfamily usually use Na^+ instead of H^+ as the antiported cation. The distinctive features of each type of pump must have evolved independently in these different superfamilies, possibly explaining the different distributions of these transporters in the various organismal types that have been studied (Yen *et al.*, 2010).

Macromolecules

The export of macromolecules such as lipids, polysaccharides and proteins into the extracytoplasmic space has been implicated in roles including energy balance maintenance, metabolism, cell signalling, cell structure and cellular organization. Although macromolecular export is a relatively common phenomenon, only a small number of macromolecular SAEPs have been identified.

The polysaccharide transport (PST) family (TC#2.A.66.2) within the MOP superfamily can be subdivided into two main functional groups (Paulsen *et al.*, 1997a). The first consists of flippases, which mediate the flipping of lipopolysaccharides and lipopolysaccharide precursors, such as the undecaprenol-pyrophosphate O-antigen, from the inner leaflet to the outer leaflet of the inner membrane in Gram-negative bacteria. This cluster is similar in function to the oligosaccharidyl-lipid flippase (OLF) family (TC#2.A.66.3), which is also a part of the MOP superfamily. The second includes members that mediate export of exopolysaccharides (EPSs) or capsular polysaccharides (CPSs) (Paulsen *et al.*, 1997a). Among other functions (Kumar *et al.*, 2007), EPSs play a role in protection against antibiotics and heavy metals, whereas CPSs assist in host immune evasion. MOP superfamily homologues range from 400 to 500 amino acyl residues in size and generally possess 12 TMSs.

In the inner membranes of Gram-negative bacteria, PST porters function along with cytoplasmic membrane-periplasmic auxiliary-1 (MPA1-C) proteins (TC#8.A.3.3.2) containing a C-terminal ATP-binding regulatory domain and an outer membrane auxiliary (OMA) protein (TC#1.B.18) in the outer membrane (Paulsen *et al.*, 1997b). In Gram-positive bacteria, PST porters function with a pair of MPA1+C (TC#8.A.3) proteins that are homologous to the MPA1-C protein found in Gram-negative bacteria (Paulsen *et al.*, 1997b). The C-terminal end of an MPA1 – C protein was shown to possess protein–tyrosine kinase activity and to have the ability to autophosphorylate its homodimeric partner in the presence of ATP (Vincent *et al.*, 1999). This suggests a regulatory role for the C-terminal domains of these transport-associated proteins. They may, however, be multifunctional.

The mode of transport for PST family members has not been established; however homology with the MATE family (TC#2.A.66.1) of the MOP superfamily, which can utilize Na^+ as the antiported cation, suggests that the PST family also consists of secondary carriers.

Families of lipid SAEPs have not yet been identified and annotated in TCDB. Certain marine bacteria grown under limited phosphorus or nitrogen conditions, but in the presence of an abundant source of carbon, have been shown to export lipids (Bredemeier *et al.*, 2003). A study was performed in which *Alcanivorax borkumensis* and *Alcanivorax jadensis* were grown under unbalanced growth conditions and were subjected to transposon-induced mutagenesis. This resulted in the generation of several mutant strains unable to export lipids such as triacylglycerols and wax esters (Manilla-Pérez *et al.*, 2010). The insertions were mapped to genes encoding a heavy metal resistance SAEP of the RND superfamily (TC#2.A.6), a multidrug efflux protein of the MATE family, a dicarboxylate/amino acid:cation (Na^+ or H^+) symporter (DAACS) family (TC#2.A.23) member, and a putative magnesium transporter of unknown family assignment (Manilla-Pérez *et al.*, 2010). Apart from the putative magnesium transporter, the three remaining proteins appear to utilize secondary active transport. The mechanisms used by and the proteins involved in extracellular export of bacterial lipids will be of great interest.

Members of the ubiquitous twin arginine translocase (Tat) family (TC#2.A.64) function as part of multicomponent systems that use the proton motive force to export fully folded proteins containing an N-terminal (S/T)RRXFLK 'twin arginine' leader motif (Müller, 2005). The transported substrates are often reductases, hydrogenases, and dehydrogenases (Saier *et al.*, 2008). However, proteins without enzymatic functions have been observed to be the substrates of Tat transport (Berks *et al.*, 2003). The Tat system, its components, the mechanism of transport, and the phylogenetic relationships between the Tat components have been described in detail (Yen *et al.*, 2002; Saier, 2006; Saier *et al.*, 2008).

Primary active efflux pumps

ABC transporters

ATP-binding cassette (ABC) transporters use ATP hydrolysis to drive the transport of ions, metabolites, drugs, macromolecules and various other types of substrates across cell membranes. The importers take up substances from the environment, while the exporters pump substances out of cells and organelles (Davidson *et al.*, 2008). Although ABC importers are found only in prokaryotes and chloroplasts, exporters are ubiquitous in all life forms. Wang *et al.* (2009) have tabulated the types of substrates transported by members of the 50 currently identified families of ABC efflux porters. The known families of ABC uptake porters are summarized in TCDB.

Two essential domains are present in all ABC transporters: the transmembrane domain (TMD) and the nucleotide-binding domain (NBD). The TMDs are composed of hydrophobic α-TMSs capable of translocating various substrates by undergoing structural changes coupled to the transport cycle. Considering the diverse range of substrates transported, it is not surprising that the sequences, orientations, and folding of the TMD regions vary tremendously (Davidson *et al.*, 2008; Rees *et al.*, 2009).

The TMDs of ABC exporters arose through three different independently occurring pathways (Fig. 1.5). ABC1 efflux porters arose by intragenic triplication of a primordial two-TMS-encoding genetic segment (Fig. 1.5A; Wang *et al.*, 2009). ABC2 porters arose by duplication of a primordial three-TMS-encoding genetic element (Fig. 1.5B; Wang *et al.*, 2009; Zheng *et al.*, 2012; submitted). ABC3 porters arose by duplication of a primordial four-TMS-encoding genetic segment (Fig. 1.5C; Khwaja *et al.*, 2004). ABC3 porters either have two TMD proteins of four TMSs or one TMD protein of eight or ten TMSs. X-ray crystallography of ABC1 exporters and ABC2 importers are available, showing no structural similarities between them. This observation is in agreement with the conclusion that ABC1 and ABC2 porters arose independently of each other (Wang *et al.*, 2009).

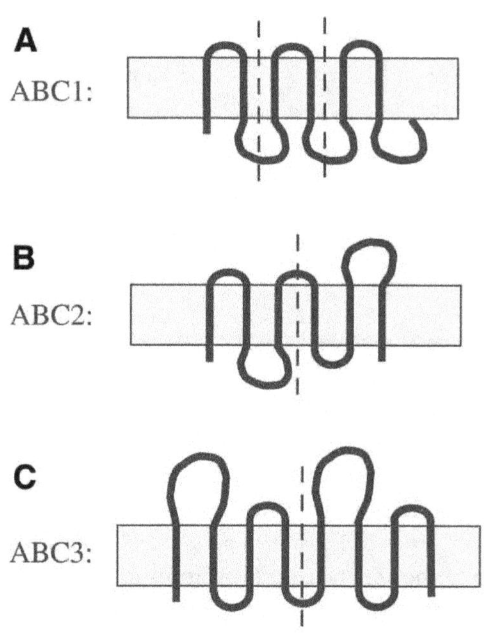

Figure 1.5 Three topological types of ABC exporters, illustrating the types of internal repeats present in each one. In all cases, *vertical dashed lines* separate the repeat units. (A) ABC1: A six-TMS topology resulting from intragenic triplication of a primordial two-TMS-encoding genetic element. The three hairpin repeats have the same orientation in the membrane. (B) ABC2: A six-TMS topology resulting from intragenic duplication of a primordial three-TMS-encoding genetic element. The two three-TMS repeats have opposite orientation in the membrane. (C) ABC3: An eight-TMS topology resulting from intragenic duplication of a primordial four-TMS-encoding genetic element (Khwaja *et al.* 2005). The two four-TMS repeats have the same orientation in the membrane. Reproduced with permission from Wang *et al.* (2009).

The transport mechanisms of ABC exporters may be as follows: in the resting state, the apo-porter opens towards the cytoplasm of the cell. Upon binding of substrate and then ATP, the porter undergoes a series of conformational changes in both the NBD and the TMD. This leads to alternate exposure of the binding site to the two sides of the membrane, resulting in transport of the substrate across the lipid bilayer. At the end of the half process, the porter opens towards the outside of the cell before returning to the inner surface to complete the cycle (Higgins *et al.*, 2004; Reyes *et al.*, 2006).

Monovalent cation porters

Out of the 50 ABC efflux families identified, only one includes members that transport a monovalent cation. The Na^+ exporter (TC#3.A.1.115.1) from *B. subtilis* can actively pump out sodium ions and produce an electrochemical gradient, negative inside. This gradient can drive secondary transport of substrates by sodium-coupled processes (Cheng *et al.*, 1997). Interestingly, high Na^+ concentrations in the cytoplasm may be toxic to bacteria. Na^+ exporters may thus be important in protecting the cell against sodium toxicity (Padan and Schuldiner, 1994).

Divalent and trivalent cation porters

ABC exporters can pump a number of divalent and trivalent ions out of the cytoplasm, contributing to cell homeostasis. An example is the putative mitochondrial iron transporter, ATM1, which transports iron from the matrix into the intermembrane space. Mitochondria with a nonfunctional ATM1 porter exhibit increased (up to 30-fold) concentrations of iron in the matrix, promoting damage of mitochondrial DNA via oxidation. Decreased levels of mitochondrial cytochromes as well as cytosolic haem proteins are also observed (Chloupková *et al.*, 2004).

Metabolites and other solutes

There are two types of metabolites, primary and secondary. Primary metabolites are the products of the primary catabolic and anabolic pathways generating energy and intermediates for macromolecular biosynthesis, while secondary metabolites do not directly affect the development and reproduction of the organism. These secondary metabolites are used in processes such as communication, biological warfare, and metal chelation (Fraenkel, 1959; Demain, 1980). ABC efflux porters are capable of transporting many of these metabolites.

Haem is an example of a metabolite transported by ABC porters. CcmA and CcmB proteins (TC#3.A.1.107.1) comprise a type of ABC haem exporter in Gram-negative bacteria where CcmA is the NBD while CcmB is the TMD. Together, CcmA and CcmB export haem across the cytoplasmic membrane into the periplasm where the haem will undergo covalent modification (Stevens *et al.*, 2004; Christensen *et al.*, 2007). Another such ABC complex of the same designation (TC#3.A.1.107.2) exists in the mitochondrial inner membrane. This complex is capable of transporting cytochrome c from the matrix to the intermembrane space where the cytochrome will be modified (Rayapuram *et al.*, 2007). A third example is the haem exporter hrtAB (TC#3.A.1.122.4) in *Staphylococcus aureus*. This porter has been shown to be important in eliminating the toxic effects of excessive amounts of haem, which can damage nucleic acids and proteins within the bacterial cell via oxidation (Stauff *et al.*, 2008).

Multidrug resistance

Novel types of drug resistance arise in tumours, fungi, bacteria and viruses every year (Royal Society, 2008). This renders drugs that are anti-tumour, anti-fungal, anti-bacterial and anti-HIV agents less effective (Lubelski *et al.*, 2007). ABC transporters contribute to MDR (Oswald *et al.*, 2007). Increased numbers of sequenced genomes indicate that ABC MDR pumps are nearly ubiquitous (Lubelski *et al.*, 2007). The efflux porters can actively pump out a large variety of drugs, influencing the drug's pharmacokinetics by altering the absorption and distribution of drugs in the human body (Hollenstein *et al.*, 2007). For instance, some of the recently developed drugs targeting the HIV virus such as Lopinavir can be transported via ABC efflux porters (Agarwal *et al.*, 2007). Yeast can pump out virtually every known anti-fungal agent commonly used in medicine and agriculture (Kolaczkowski *et al.*, 1998).

In cancer treatment, prolonged exposure of a patient to chemotherapeutic drugs such as daunorubicin and doxorubicin may cause an over-expression of P-glycoprotein (Pgp) in the tumour cell membrane (Senior *et al.*, 1995). Knowledge of its three-dimensional structure may aid in the development of new drugs and inhibitors of drug export (Kolaczkowski *et al.*, 1998; Aller *et al.*, 2009). This porter is known to export anticancer drugs, leading to increased MDR. The DrrAB protein (TC#3.A.1.105.1), a bacterial transporter, is homologous to Pgp, and DrrAB may contribute to bacterial resistance to daunorubicin and doxorubicin (Kaur, 1997). DrrA is the NBD while DrrB is the TMD, predicted to have six TMSs with both the N- and C-termini in the cytoplasm (Gandlur *et al.*, 2004). It is of the ABC2 type (Wang *et al.*, 2009; see Fig. 1.5).

Macromolecules

Numerous ABC efflux porters transport macromolecules such as polysaccharides, lipids, large peptides and proteins (Wang *et al.*, 2009). For instance, KpsMT (TC#3.A.1.101.1) of *E. coli* is an exporter responsible for the secretion of capsular polysaccharide, a capsule capable of 'hiding' antigenic proteins found on the surfaces of bacteria that can provoke host immune responses (Smith *et al.*, 1990).

In many Gram-negative bacteria, lipoproteins are anchored in the periplasmic and outer faces of the bacterial outer membrane. Synthesized as precursors in the cytoplasm, the lipoproteins are transported across the inner membrane by the actions of five different membrane proteins, LolA, LolB, LolC, LolD and LolE (Remans *et al.*, 2010). The LolCDE complex (TC#3.A.1.125) is an ABC porter. Both LolC and LolE are TM proteins while LolD is the conserved NBD. LolC and LolE, of the ABC3 type, have four TMSs each (Wang *et al.*, 2009; Nagao *et al.*, 2010).

ABC porters also play important roles in the synthesis and localization of cell surface antigens. For instance, lipopolysaccharides (LPS) in the outer membranes of Gram-negative bacteria contribute to the diversity of cell surface antigens that give the bacteria their 'fingerprint'. These macromolecules can be targeted by host immune systems (Reeves *et al.*, 1995). During the production of LPS, sugar repeat units are added to the non-reducing end of the polymer to form an O-antigen on the cytosolic face of the plasma membrane (Bronner *et al.*, 1994). The whole molecule is then transported across the plasma membrane by the Wzm (TMD)-Wzt (NBD) ABC2 complex before being anchored to the cell surface (Feng *et al.*, 2004).

The *Serratia marcescens* metalloprotease, PrtA, a type of haemolysin found in Gram-negative bacteria, is secreted by a transporter containing three different proteins. Two proteins, PrsD and PrsE, are found in the inner membrane while the third protein, PrtF, is found in the outer membrane. PrsE and PrtF also span the periplasm. PrsD alone is the ABC membrane protein with both the TMD and the NBD. PrsE is a MFP adaptor and auxiliary protein that connects the ABC transporter in the cytoplasmic membrane to the PrtF pore-forming protein in the outer membrane. By functioning together, these proteins allow export of molecules across both membranes of the Gram-negative bacterial envelope and the periplasm in a single energy coupled step (Dinh *et al.*, 1994; Saier *et al.*, 1994). This tripartite arrangement is observed for secondary carriers as well as ABC porters, and membrane fusion proteins can be found in Gram-positive bacteria where they may be essential for transport activity (Paulsen *et al.*, 1997b; Harley *et al.*, 2000; Xiao *et al.*, 2010).

P-type ATPases

P-type ATPases of eukaryotes (Thever and Saier, 2009) and of prokaryotes (Chan *et al.*, 2010) have recently been surveyed and analysed for topology, function, phylogeny and structure. In addition to nine functionally characterized families, 13 eukaryotic functionally uncharacterized P-type ATPase (FUPA) families and 11 prokaryotic FUPA families were discovered. Although the characteristics of these novel proteins were described, their biochemical and physiological functions have yet to be elucidated. These studies are currently under way. The reader is referred to the two above-mentioned papers for details regarding these pumps.

Protein secretion systems

Over 16 evolutionarily distinct protein secretion systems have been characterized in bacteria alone, and several others are known to exist (Saier, 2006; Saier *et al.*, 2008). These can use either the proton motive force or ATP/GTP to energize transport, but several use both forms of energy to activate different steps in the overall secretion process. Because these have been reviewed recently, they will not be considered further here. For the most recent treatment of these systems, see Saier *et al.* (2008).

Conclusions and perspectives

With the exception of drug and macromolecular export systems, which have been recognized for several decades, the discoveries of metabolite exporters have been relatively recent developments. While uptake systems were first characterized, with a plethora of information currently available, we recognize that exporters are, with near certainty, underrepresented in TCDB because of the paucity of studies that have been devoted to their characterization. Hundreds of transporter families of unknown function are recognized (see TCDB class 9). We anticipate that a majority of these will prove to be families of exporters rather than importers. This will undoubtedly be a fertile field of future investigation. Such studies are likely to allow us to attain new horizons with respect to our understanding of the fields of transport molecular biology, physiology and evolution.

Acknowledgements

We thank Carl Welliver for his assistance in the preparation of this book chapter. The work reported in this paper was supporter by NIH Grant 2 RO1 GM077402-05A1.

References

Agarwal, S., Pal, D., and Mitra, A.K. (2007). Both P-gp and MRP2 mediate transport of Lopinavir, a protease inhibitor. Int. J. Pharm. *339*, 139–147.

Aleshin, V.V., Zakataeva, N.P., and Livshits, V.A. (1999). A new family of amino acid efflux proteins. Trends Biochem. Sci. *24*, 133–135.

Aller, S.G., Yu, J., Ward, A., Weng, Y., Chittaboina, S., Zhuo, R., Harrell, P.M., Trinh, Y.T., Zhang, Q., Urbatsch, I.L., *et al.* (2009). Structure of P-glycoprotein reveals a molecular basis for poly-specific drug binding. Science. *323*, 1718–1722.

Appel, M., Hizlan, D., Vinothkumar, K.R., Ziegler, C., and Kühlbrandt, W. (2009). Conformations of NhaA, the Na/H exchanger from *Escherichia coli*, in the pH-activated and ion-translocating states. J. Mol. Biol. *386*, 351–365.

Arkin, I.T., Xu, H., Jensen, M.Ø., Arbely, E.R., Bowers, K.J., Chow, E., Dror, R.O., Eastwood, M.P., Flitman-Tene, R., Gregersen, B.A., *et al.* (2007). Mechanism of Na^+/H^+ antiporting. *317*, 799–803.

Au, C., Benedetto, A., and Aschner, M. (2008). Manganese transport in eukaryotes: the role of DMT1. Neurotoxicology *29*, 569–576.

Barabote, R.D., Tamang, D.G., Abeywardena, S.N., Fallah, N.S., Fu, J.Y.C., Lio, J.K., Mirhosseini, P., Pezeshk, R., Podell, S., Salampessy, M.L., *et al.* (2006). Extra domains in secondary transport carriers and channel proteins. Biochim. Biophys. Acta *1758*, 1557–1579.

Belitsky, B.R., Gustafsson, M.C.U., Sonenshein, A.L., and von Wachenfeldt, C. (1997). An Irp-like gene of *Bacillus subtilis* involved in branched-chain amino acid transport. J. Bacteriol. *179*, 5448–5457.

Berks, B.C., Palmer, T., and Sargent, F. (2005). Protein targeting by the bacterial twin – arginine translocation (Tat) pathway. Curr. Opin. Microbiol. *8*, 174–181.

Blanke, S.R. (2006). Portals and pathways: principles of bacterial toxin entry into host cells. Microbe. *1*, 26–32.

Boudker, O., and Verdon, G. (2010). Structural perspectives on secondary active transporters. Cell Press. *31*, 418–426.

Bredemeier, R., Hulsch, R., Metzger, J.O., and Berthe-Corti, L. (2003). Submersed culture production of extracellular wax esters by the marine bacterium *Fundibacter jadensis*. Mar. Biotechnol. *5*, 579–583.

Brey, R.N., Rosen, B.P., and Sorensen, E.N. (1979). Cation/proton antiport systems in *Escherichia coli*: properties of the potassium/proton antiporter. J. Biol. Chem. *255*, 39–44.

Bronner, D., Clarke, B.R., and Whitfield, C. (1994). Identification of an ATP-binding cassette transport system required for translocation of lipopolysaccharide O-antigen side-chains across the cytoplasmic membrane of *Klebsiella pneumoniae* serotype O1. Mol. Microbiol. *14*, 505–519.

Chan, H., Babayan, V., Blyumin, E., Gandhi, C., Hak, K., Harake, D., Kumar, K., Lee, P., Li, T.T., Liu, H.Y., *et al.* (2010). The P-type ATPase superfamily. J. Mol. Microbiol. Biotech. *19*, 5–104.

Chen, Y.J., Pornillos, O., Lieu, S., Ma, C., Chen, A.P., and Chang, G. (2007). X-ray structure of EmrE supports dual topology model. Proc. Natl. Acad. Sci.. *104*, 18999–19004.

Cheng, J., Guffanti, A.A., and Krulwich, T.A. (1997). A two-gene ABC-type transport system that extrudes Na+ in *Bacillus subtilis* is induced by ethanol or protonophore. Mol. Microbiol. *23*, 1107–1120.

Chiba, H., Osanai, M., Murata, M., Kojima, T., and Sawada, N. (2008). Transmembrane proteins of tight junctions. Biochim. Biophys. Acta *1778*, 588–600.

Chloupková, M., Reaves, S.K., LeBard, L.M., and Koeller, D.M. (2004). The mitochondrial ABC transporter Atm1p functions as a homodimer. FEBS Lett. *569*, 65–69.

Christensen, O., Harvat, E.M., Thöny-Meyer, L., Ferguson, S.J., and Stevens, J.M. (2007). Loss of ATP hydrolysis activity by CcmAB results in loss of c-type cytochrome synthesis and incomplete processing of CcmE. FEBS J. *274*, 2322–2332.

Chung, Y.J., Krueger, C., Metzgar, D., and Saier, M.H., Jr. (2001). Size comparisons among integral membrane transport protein homologues in *bacteria, arcahea, and eucarya*. J. Bacteriol. *183*, 1012–1021.

Courville, P., Chaloupka, R., Veyrier, F., and Cellier, M.F.M. (2004). Determination of transmembrane topology of the *Escherichia coli* natural resistance-associated macrophage protein (Nramp) ortholog. J. Biol. Chem. *279*, 3318–3326.

Crupper, S.S., Worell, V., Stewart, G.C., and Iandolo, J.J. (1999). Cloning and expression of *cadD*, a new cadmium resistance gene of *Staphylococcus aureus*. J. Bacteriol. *181*, 4071–4075.

Czachorowski, M., Lam-Yuk-Tseung, S., Cellier, M., and Gros, P. (2009). Transmembrane topology of the mammalian Slc11a2 iron transporter. Biochemistry. *48*, 8422–8434.

Davidson, A.L., Dassa, E., Orelle, C., and Chen, J. (2008). Structure, function, and evolution of bacterial ATP-binding cassette systems. Microbiol. Mol. Biol. Rev. *72*, 317–364.

Demain, A.L. (1980). Microbial production of primary metabolites. Naturwissenschaften *67*, 582–587.

Dewey, J.S., Savva, C.G., White, Vitha, R.L. , S., Holzenburg, A., and Young, R. (2010). Micron-scale holes terminate the phage infection cycle. Proc. Natl. Acad. Sci. *107*, 2219–2223.

Dinh, T., Paulsen, I.T., and Saier, Jr, M.H. (1994). A family of extracytoplasmic proteins that allow transport of large molecules across the outer membranes of gramnegative bacteria. J. Bacteriol. *176*, 3825–3831.

Enkavi, G., and Tajkhorshid, E. (2010). Simulation of spontaneous substrate binding revealing the binding pathway and mechanism and initial conformational response of GlpT. Biochem. *49*, 1105–1114.

Feng, L., Senchenkova, S.N., Yang, J., Shashkov, A.S., Tao, J., Guo, H., Cheng, J., Ren, Y., Knirel, Y.A., Reeves, P.R., *et al.* (2004). Synthesis of the heteropolysaccharide O antigen of *Escherichia coli* O52 requires an ABC transporter: structural and genetic evidence. J Bacteriol. *186*, 4510–4519.

Fitsanakis, V.A., and Aschner, M. (2005). The importance of glutamate, glycine, and gamma-aminobutyric acid transport and regulation in manganese, mercury and lead neurotoxicity. Toxicol. Appl. Pharmacol. *204*, 343–354.

Fraenkel, G.S. (1959). The raison d'Etre of secondary plant substances. Science *129*, 1466–1470.

Franke, I., Resch, A., Dassler, T., Maier, T., and Bock, A. (2003). YfiK from *Escherichia coli* promotes export of *O*-acetylserine and cysteine. J. Bacteriol. *185*, 1161–1166.

Froschauer, E., Nowkovsky, K., and Schweyen, R.J. (2005). Electroneutral K^+/H^+ exchange in mitochondrial membrane vesicles involves Yol027/Letm1 proteins. Biochim. Biophys. Acta *1711*, 41–48.

Fujisawa, M., Kusumoto, A., Wada, Y., Tsuchiya, T., and Ito, M. (2005). NhaK, a novel monovalent cation/H^+ antiporter of *Bacillus subtilis*. Arch. Microbiol. *183*, 1–10.

Gandlur, S.M., Wei, L., Levine, J., Russell, J., and Kaur, P. (2004). Membrane topology of the DrrB protein of the doxorubicin transporter of *Streptomyces peucetius*. J. Biol. Chem. *279*, 27799–27806.

Gottesman, M.M., and Pastan, I. (1993). Biochemistry of multidrug resistance mediated by the multidrug transporter. Annu. Rev. Biochem. *62*, 385–427.

Gouaux, E., and MacKinnon, R. (2005). Principles of selective ion transport in channels and pumps. Science *310*, 1461–1465.

Gromiha, M.M., and Suwa, M. (2007). Current developments on β-barrel membrane proteins: sequence and structure analysis, discrimination and prediction. Curr. Protein Pept. Sci. *8*, 580–599.

Guffanti, A.A., Wei, Y., Rood, S.V., and Krulwich, T.A. (2002). An antiport mechanism for a member of the cation diffusion facilitator family: divalent cations efflux in exchange for K^+ and H^+. Mol. Microbiol. *45*, 145–153.

Gulbis, J.M., Mann, S., and MacKinnon, R. (1999). Structure of a voltage-dependent K^+ channel beta subunit. Cell *97*, 943–952.

Harley, K.T., Djordjevic, G.M., Tseng, T.T., and Saier, M.H., Jr. (2000). Membrane-fusion protein homologues in Gram-positive bacteria. Mol. Microbiol. *36*, 516–517.

He, L., Wu, X.S., Mohan, R., and Wu, L.G. (2006). Two modes of fusion pore opening revealed by cell-attached recordings at a synapse. Nature *444*, 102–105.

Herz, K., Rimon, A., Olkhova, E., Kozachkov, L., and Padan, E. (2010). Transmembranesegment II of NhaA Na^+/H^+ antiporter lines the cation passage, and Asp65 is critical for pH activation of the antiporter. J. Biol. Chem. *285*, 2211–2220.

Higgins, C.F., and Linton, K.J. (2004). The ATP switch model for ABC transporters. Nat. Struct. Mol. Biol. *11*, 918–926.

Hirai, T., Subramaniam, S., and Lanyi, J.K. (2009). Structural snapshots of conformational changes in a seven-helix membrane protein: lessons from bacteriorhodopsin. Curr. Opin. Struct. Biol. *19*, 433–439.

Hollenstein, K., Dawson, R.J., and Locher, K.P. (2007). Structure and mechanism of ABC transporter proteins. Curr. Opin. Struct. Biol. *17*, 412–418.

Hunte, C., Screpanti, E., Venturi, M., Rimon, A., Padan, E., and Michel, H. (2005). Structure of a Na^+/H^+ antiporter and insights into mechanism of action and regulation by pH. Nature *435*, 1197–1202.

Hvorup, R.N., Winnen, B., Chang, A.B., Jiang, Y., Zhou, X.F., and Saier, M.H., Jr. (2003). The multidrug/oligosaccharidyl-lipid/polysaccharide (MOP) exporter superfamily. Eur. J. Biochem. *270*, 799–813.

Ito, Y., Kanamaru, K., Taniguchi, N., Miyamoto, S., and Tokuda, H. (2006). A novel ligand bound ABC transporter, LolCDE, provides insights into the molecular mechanisms underlying membrane detachment of bacterial lipoproteins. Mol. Microbiol. *62*, 1064–1075.

Jack, D.L., Yang, N.M., and Saier, Jr, M.H. (2001). The drug/metabolite transporter superfamily. Eur. J. Biochem. *268*, 3620–3639.

Jiang, D., Zhao, L., and Clapham, D.E. (2009). Genome-wide RNAi screen identifies LetM1 as a mitochondrial Ca^{2+}/H^+ antiporter. Science. *326*, 144–147.

Karuppiah, V., Berry, J.L., and Derrick, J.P. (2011). Outer membrane translocons: structural insights into channel formation. Trends Mirbiobiol. *19*, 40–48.

Kaur, P. (1997). Expression and characterization of DrrA and DrrB proteins of *Streptomyces peucetius* in *Escherichia coli*: DrrA is an ATP binding protein. J. Bacteriol. *179*, 569–575.

Kennerknecht, N., Sahm, H., Yen, M.R., Patek, M., Saier, Jr, M.H., and Eggeling, L. (2002). Export of L-isoleucine from *Corynebacterium glutamicum*: a two-gene-encoded member of a new translocator family. J. Bacteriol. *184*, 3947–3956.

Khwaja, M., Ma, Q., and Saier, M.H., Jr. (2004). Topological analysis of integral membrane constituents of prokaryotic ABC efflux systems. Res. Microbiol. *156*, 270–277.

Kolaczkowski, M., Kolaczowska, A., Luczynski, J., Witek, S., and Goffeau, A. (1998). *In vivo* characterization of the drug resistance profile of the major ABC transporters and other components of the yeast pleiotropic drug resistance network. Microb. Drug Resist. *4*, 143–158.

Krogh, A., Larsson, G., von Heijne, G., and Sonnhammer, E.L.L. (2001). Predicting transmembrane protein topology with a hidden Markov model: application to complete genomes. J. Mol. Biol. *305*, 567–580.

Kumar, A.S., Mody, K., and Jha, B. (2007). Bacterial exopolysaccharides – a perception. J. Basic Microbiol. *47*, 103–117.

Kutukova, E.A., Livshits, V.A., Altman, I.P., Ptitsyn, L.R., Zyiatdinov, M.H., Tokmakova, I.L., and Zakataeva, N.P. (2005). The *yeaS* (*leuE*) gene of *Escherichia coli* encodes an exporter of leucine, and the Lrp protein regulates its expression. FEBS Lett. *579*, 4629–4634.

Lee, J., Harvat, E.M., Stevens, J.M., Ferguson, S.J., and Saier, Jr, M.H. (2010). Evolutionary origins of members of a superfamily of integral membrane cytochrome *c* biogenesis proteins. Biochim. Biophys. Acta *1768*, 2164–2181.

Létoffé, S., Ghigo, J.M., and Wandersman, C. (1993). Identification of two components of the *Serratia marcescens* metalloprotease transporter: protease SM secretion in *Escherichia coli* is TolC dependent. J. Bacteriol. *175*, 7321–7328.

Lolkema, J.S., Dobrowolski, A., and Slotboom, D. (2008). Evolution of antiparallel two-domain membrane proteins: tracking multiple gene duplication events in the DUF606 family. J. Mol. Biol. *378*, 596–606.

Lomize, M.A., Lomize, A.L., Pogozheva, I.D., and Mosberg, H.I. (2006). OPM: orientations of proteins in membranes database. Bioinformatics. *22*, 623–625.

Lubelski, J., Konings, W.N., and Driessen, A.J. (2007). Distribution and physiology of ABC-type transporters contributing to multidrug resistance in bacteria. Microbiol. Mol. Biol. Rev. *71*, 463–476.

Maloney, P.C., Ambudkar, S.V., Anatharam, V., Sonna, L.A., and Varadhachary, A. (1990). Anion-exchange mechanisms in bacteria. Microbiol. Rev. *54*, 1–17.

Manilla-Pérez, E., Reers, C., Baumgart, M., Hetzler, S., Reichelt, R., Malkus, U., Kalscheuer, R., Wältermann, M., and Steinbüchel, A. (2010). Analysis of lipid export in hydrocarbonoclastic bacteria of the genus *Alcanivorax*: identification of lipid export-negative mutants of *Alcanivorax borkumensis* SK2 and *Alcanivorax jadensis* T9. J. Bacteriol. *192*, 643–656.

Mansour, N.M., Sawhney, M., Tamang, D.G., Vogl, C.C., and Saier, Jr, M.H. (2007). The bile/arsenite/riboflavin transporter (BART) superfamily. FEBS J. *274*, 612–629.

Markin, V.S. (1987). On the mechanism of amiloride-sensitive nonelectrogenic Na^+-H^+ exchange in cell membranes: Na^+/H^+ antiport or $Na^+/OH-$ symport? Gen. Physiol. Biophys. *6*, 449–468.

Matias, M.G., Gomolplitinant, K.M., and Saier, Jr, M.H. (2010). Animal Ca^{2+} release-activated Ca^{2+} (CRAC) channels appear to be homologous to and derived from the ubiquitous cation diffusion facilitators. BMC Res. Notes. *3*, 158.

Montanini, B., Blaudez, D., Jeandroz, S., Sanders, D., and Chalot, M. (2007). Phylogenetic and functional analysis of the cation diffusion facilitator (CDF) family: improved signature and prediction of substrate specificity. BMC Genomics. *8*, 107.

Mukarami, S., Nakashima, R., Yamashita, E., and Yamaguchi, A. (2002). Crystal structure of bacterial multidrug efflux transporter AcrB. Nature *419*, 587–593.

Mukarami, S., Nakashima, R., Yamashita, E., Matsumoto, T., and Yamaguchi, A. (2006).Crystal structures of a multidrug transporter reveal a functionally rotating mechanism. Nature *443*, 173–179.

Müller, M. (2005). Twin-arginine-specific protein export in *Escherichia coli*. Res. Microbiol. *156*, 131–136.

Nagao, K., Kimura, Y., Mastuo, M., and Ueda, K. (2010). Lipid outward translocation by ABC proteins. FEBS Lett. *584*, 2717–2723.

Nakamura, T., Tokuda, H., and Unemoto, T. (1984). K^+/H^+ antiporter as a regulator of cytoplasmic pH in a marine bacterium, *Vibrio alginolyticus*. Biochim. Biophys. Acta *776*, 330–336.

Nakamura, T., Fujisaki, Y., Enomoto, H., Nakayama, Y., Takabe, T., Yamaguchi, N., and Uozumi, N. (2001). Residue aspartate-147 from the third transmembrane region of Na^+/H^+ antiporter NhaB of *Vibrio alginolyticus* plays a role in its activity. J. Bacteriol. *183*, 5762–5767.

Nandineni, M.R., and Gowrishankar, J. (2004). Evidence for an arginine exporter encoded by yggA (argO) that is regulated by the LysR-type transcriptional regulator ArgP in *Escherichia coli*. J. Bacteriol. *186*, 3539–3546.

Nevo, Y. (2007). Site-directed mutagenesis investigation of coupling properties of metal ion transport by DCT1. Biochim. Biophys. Acta *1778*, 334–341.

Nikaido, H., and Takatsuka, Y. (2009). Mechanisms of RND multidrug efflux pumps. Biochim. Biophys. Acta *1794*, 769–781.

Nowikovsky, K., Froschauer, E.M., Zsurka, G., Samaj, J., Reipert, S., Kolisek, M., Wiesenberger, G., and Schweyen, R.J. (2004). The *LETM1/YOL027* gene family encodes a factor of the mitochondrial K^+ homeostasis with a potential role in the Wolf-Hirschhorn syndrome. J. Biol. Chem. *279*, 30307–30315.

Olkhova, E., Hunte, C., Screpanti, E., Padan, E., and Michel, H. (2006). Multiconformation continuum electrostatics analysis of the NhaA Na^+/H^+ antiporter of *Escherichia coli* with functional implications. Proc. Natl. Acad. Sci. U.S.A. *103*, 2629–2634.

Oswald, S., Grube, M., Siegmund, W., and Kroemer, H.K. (2007). Transporter-mediated uptake into cellular compartments. Xenobiotica *37*, 1171–1195.

Padan, E. (2008). The enlightening encounter between structure and function in the NhaA Na^+-H^+ antiporter. Trends Biochem. Sci. *33*, 435–443.

Padan, E., and Schuldiner, S. (1994). Molecular physiology of Na^+/H^+ antiporters, key transporters in circulation of Na^+ and H^+ in cells. Biochim. Biophys. Acta *1185*, 129–151.

Padan, E., Venturi, M., Gerchman, Y., and Dover, N. (2001). Na^+/H^+ antiporters. Biochim. Biophys. Acta *1505*, 144–157.

Pang, T., Savva, C.G., Fleming, K.G., Struck, D.K., and Young, R. (2009). Structure of the lethal phage pinhole. Proc. Natl. Acad. Sci. U.S.A. *106*, 18966–18971.

Pao, S.S., Paulsen, I.T., and Saier, M.H., Jr. (1998). Major facilitator superfamily. Microbiol. Mol. Biol. Rev. *62*, 1–34.

Papa, S., Capitanio, N., Villani, G., Capitanio, G., Bizzoca, A., Palese, L.L., Carlino, V., and De Nitto, E. (1998). Cooperative coupling and role of heme a in the proton pump of heme-copper oxidases. Biochimie *80*, 821–836.

Pasternak, C.A. (1990). Monovalent Cations in Biological Systems (CRC Press, Boca Raton, FL), pp. 212–219.

Paulsen, I.T., Brown, M.H., and Skurray, R.A. (1996). Proton-dependent multidrug efflux systems. Microbiol. Rev. *60*, 575–608.

Paulsen, I.T., Beness, A.M., and Saier, M.H., Jr. (1997a). Computer-based analyses of the protein constituents of transport systems catalyzing export of complex carbohydrates in bacteria. Microbiology *143*, 2685–2699.

Paulsen, I.T., Park, J.H., Choi, P.S., and Saier, M.H., Jr. (1997b). A family of Gram-negative bacterial outer membrane factors that function in the export of proteins, carbohydrates, drugs and heavy metals from Gram-negative bacteria. FEMS Microbiol. Lett. *156*, 1–8.

Povolotsky, T.L., Orlova, E., Tamang, D.G., and Saier, M.H., Jr. (2010). Defense against cannibalism: the SdpI family of bacterial immunity/signal transduction proteins. J. Membr. Biol. *235*, 145–162.

Prakash, S., Cooper, G., Singhi, S., and Saier, Jr, M.H. (2003). The ion transporter superfamily. Biochim. Biophys. Acta *1618*, 79–92.

Radchenko, M.V., Waditee, R., Fukuhara, M., Takabe, T., and Nakamura, T. (2006). Cloning, functional expression and primary characterization of *Vibrio parahaemolyticus* K^+/H^+ antiporter genes in *Escherichia coli*. Mol. Microbiol. *59*, 651–663.

Rayapuram, N., Hagenmuller, J., Grienenberger, J.M., Giegé, P., and Bonnard, G. (2007). AtCCMA interacts with AtCcmB to form a novel mitochondrial ABC transporter involved in cytochrome c maturation in Arabidopsis. J. Biol. Chem. *282*, 21015–21023.

Reddy, V., and Saier, M.H., Jr. (2012). BioV Suite – a collection of programs for the study of transport protein evolution. FEBS J. *279*, 2036–2046.

Reddy, V., Shlykov, M.A., Castillo, R., Sun, E.I., and Saier, M.H., Jr. (2012). The major facilitator superfamily (MFS) revisited. FEBS J. *279*, 2022–2035.

Rees, D.C., Johnson, E., and Lewinson, O. (2009). ABC transporters: the power to change. Nat. Rev. Mol. Cell Biol. *10*, 218–227.

Reeves, P.R. (1995). Role of O-antigen variation in the immune response. Trends Microbiol. *3*, 381–386.

Rensing, C., Ghosh, M., and Rosen, B.P. (1999). Families of soft-metal-ion transporting ATPase. J. Bacteriol. *181*, 5891–5897.

Reyes, C.L., Ward, A., Yu, J., and Chang, G. (2006). The structures of MsbA: insight into ABC transporter-mediated multidrug efflux. FEBS Lett. *580*, 1042–1048.

Rosen, B.R. (1996). Bacterial resistance to heavy metals and metalloids. J. Biol. Inorg. Chem. *1*, 273–277.

Royal Society (2008). Innovative mechanism tackling antibacterial resistance. Available at: http://royalsociety.org/document.asp?tip=0&id=7888 (accessed June 2012).

Sahl, H.G., and Bierbaum, G. (2008). Multiple activities in natural antimicrobials. Microbe *3*, 467–473.

Saier, M.H., Jr. (2000a). Families of proteins forming transmembrane channels. J. Membrane Biol. *175*, 165–180.

Saier, M.H., Jr. (2000b). Vectorial metabolism and the evolution of transport systems. J. Bacteriol. *182*, 5029–5035.

Saier, M.H., Jr. (2003). Tracing pathways of transport protein evolution. Mol. Microbiol. *48*, 1145–1156.

Saier, M.H., Jr. (2006). Protein secretion and membrane insertion systems in Gram-negative bacteria. J. Membr. Biol. *214*, 75–90.

Saier, M.H., Jr., and Paulsen, I.T. (2001). Phylogeny of multidrug transporters. Semin. Cell Dev. Biol. *12*, 205–213.

Saier, M.H., Jr., Tam, R., Reizer, A., and Reizer, J. (1994). Two novel families of bacterial membrane proteins concerned with nodulation, cell division and transport. Mol. Microbiol. *11*, 841–847.

Saier, M.H., Jr., Beatty, J.T., Goffeau, A., Harley, K.T., Heijne, W.H., Huang, S.C., Jack, D.L., Jähn, P.S., Lew, K., Liu, J., *et al.* (1999). The major facilitator superfamily. J. Mol. Microbiol. Biotechnol. *1*, 257–279.

Saier, M.H., Jr., Tran, C.V., and Barabote, R.D. (2006). TCDB: the Transporter Classification Database for membrane transport protein analyses and information. Nucl. Acids Res. *34*, 181–186.

Saier, M.H., Jr., Ma, C.H., Rodgers, L., Tamang, D.G., and Yen, M.R. (2008). Protein secretion and membrane insertion systems in bacteria and eukaryotic organelles. Adv. Appl. Microbiol. *65*, 141–197.

Saier, M.H., Jr., Yen, M.R., Noto, K., Tamang, D.G., and Elkan, C. (2009). The Transporter Classification Database: recent advances. Nucl. Acids Res. *37*, 274–278.

Santo-Domingo, J., and Demaurex, N. (2010). Calcium uptake mechanisms of mitochondria. Biochim. Biophys. Acta *1797*, 907–912.

Screpanti, E., Padan, E., Rimon, A., Michel, H., and Hunte, C. (2006). Crucial steps in the structure determination of the Na^+/H^+ antiporter NhaA in its native conformation. J. Mol. Biol. *362*, 192–202.

Senior, A.E., al-Shawi, M.K., and Urbatsch, I.L. (1995). The catalytic cycle of P-glycoprotein. FEBS Lett. *377*, 285–289.

Slatin, S.L., Finkelstein, A. , and Kienker, P.K. (2008). Anomalous proton selectivity in a large channel: colicin A. Biochem. *47*, 1778–1788.

Sonnhammer, E.L.L., von Heijne, G., and Krogh, A. (1998). A hidden Markov model for predicting transmembrane helices in protein sequences. In proceedings of the Sixth International Conference on Intelligent Systems for Molecular Biology, 28 June–1 July 1998, Montreal, Canada. Glasgow, J., Littlejohn, T., Major, F., Sankoff, D., and Sensen, C., eds. (AAAI Press, Menlo Park, CA), pp. 175–182.

Smith, A.N., Boulnois, G.J., and Roberts, I.S. (1990). Molecular analysis of the *Escherichia coli* K5 kps locus: identification and characterization of an inner-membrane capsular polysaccharide transport system. Mol. Microbiol. *4*, 1863–1869.

Stauff, D.L., Bagaley, D., Torres, V.J., Joyce, R., Anderson, K.L., Kuechenmeister, L., Dunman, P.M., and Skaar, E.P. (2008). *Staphylococcus aureus* HrtA is an ATPase required for protection against heme toxicity and prevention of a transcriptional heme stress response. J Bacteriol. *190*, 3588–3596.

Stevens, J.M., Daltrop, O., Allen, J.W., and Ferguson, S.J. (2004). C-type cytochrome formation: chemical and biological enigmas. Acc. Chem. Res. *37*, 999–1007.

Taglicht, D., Padan, E., and Schuldiner, S. (1993). Proton-sodium stoichiometry of NhaA, an electrogenic antiporter from *Escherichia coli*. J. Biol. Chem. *268*, 5382–5387.

Tetsch, L., and Jung, K. (2009). The regulatory interplay between membrane-integrated sensors and transport proteins in bacteria. Mol. Microbiol. 73, 982–991.

Thever, M.D., and Saier, Jr, M.H. (2009). Bioinformatic characterization of P-type ATPases encoded within the fully sequenced genomes of 26 eukaryotes. J. Membr. Biol. *229*, 115–130.

Tomljenovic, L. (2011). Aluminum and Alzheimer's disease: after a century of controversy, is there a plausible link? J. Alzheimers Dis. *23*, 567–598.

Touzé, T., Eswaran, J., Bokma, E., Koronakis, E., Hughes, C., and Koronakis, V. (2004).Interactions underlying assembly of the *Escherichia coli* AcrAB-TolC multidrug efflux system. Mol. Microbiol. *53*, 697–706.

Trötschel, C., Deutenberg, D., Bathe, B., Burkovski, A., and Krämer, R. (2005). Characterization of methionine export in *Corynebacterium glutamicum*. J. Bacteriol. *187*, 3786–3794.

Tseng, T.T., Gratwick, K.S., Kollman, J., Park, D., Nies, D.H., Goffeau, A., and Saier, Jr, M.H. (1999). The RND permease superfamily: an ancient, ubiquitous and diverse family that includes human disease and development proteins. J. Mol. Microbiol. Biotechnol. *1*, 107–125.

Tusnády, G.E., and Simon, I. (1998). Principles governing amino acid composition of integral membrane proteins: applications to topology prediction. J. Mol. Biol. *283*, 489–506.

Tusnády, G.E., and Simon, I. (2001). The HMMTOP transmembrane topology prediction server. Bioinformatics *17*, 849–850.

Vincent, C., Doublet, P., Grangeasse, C., Vaganay, E., Cozzone, A.J., and Duclos, B. (1999). Cells of *Escherichia coli* contain a protein-tyrosine kinase, Wzc, and a phosphotyrosine-protein phosphatase, Wzb. J. Bacteriol. *181*, 3472–3477.

Vrljic, M., Garg, J., Bellmann, A., Wachi, S., Freudl, R., Malecki, M.J., Sahm, H., Kozina, V.J., Eggeling, L., and Saier, Jr, M.H. (1999). The LysE superfamily: topology of the lysine exporter LysE of *Corynebacterium glutamicum*, a paradyme for a novel superfamily of transmembrane solute translocators. J. Mol. Microbiol. Biotechnol. *1*, 327–336.

Wang, B., Dukarevich, M., Sun, E.I., Yen, M.R., and Saier, Jr, M.H. . (2009). Membrane porters of ATP-binding cassette transport systems are polyphyletic. J. Membr. Biol. *231*, 1–10.

Wei, Y., and Fu, D. (2006). Binding transport of metal ions at the dimmer interface of the *Escherichia coli* metal transporter YiiP. J. Biol. Chem. *281*, 23492–23502.

White, J.M., Delos, S.E., Brecher, M., and Schornberg, K. (2008). Structures and mechanisms of viral membrane fusion proteins: multiple variations on a common theme. Crit. Rev. Biochem. Mol. Biol. *43*, 189–219.

Williams, A., Geldmacher-Kaufer, U., Padan, E., Schuldiner, S., and Kühlbrandt, W. (1999). Projection structure of NhaA, a secondary transporter from *Escherichia coli*, at 4.0 Å resolution. EMBO J. *18*, 3558–3563.

Xiao, A.Y., Wang, J., and Saier, M.H., Jr. (2010). Bacterial adaptor membrane fusion proteins and the structurally dissimilar outer membrane auxiliary proteins have exchanged central domains in α-proteobacteria. Int. J. Microbiol. *2010*, 1–5.

Xu, C., Zhou, T., Kuroda, M., and Rosen, B.P. (1998). Metalloid resistance mechanisms in prokaryotes. J. Biochem. *123*, 16–23.

Yen, M.R., Tseng, Y.H., Nguyen, E.H., Wu, L.F., and Saier, Jr, M.H. (2002). Sequence and phylogenetic analyses of the twins-arginine targeting (Tat) protein export system. Arch. Microbiol. *177*, 441–450.

Yen, M.R., Chen, J.S., Marquez, J.L., Sun, E.I., and Saier, Jr, M.H. . (2010). Multidrug resistance: phylogenetic characterization of superfamilies of secondary carriers that include drug exporters. Methods Mol. Biol. *637*, 47–64.

Yin, Y., He, X., Szewczyk, P., Nguyen, T., and Chang, G. (2006). Structure of the multidrug transporter EmrD from *Escherichia coli*. Science. *312*, 741–744.

Zakataeva, N.P., Aleshin, V.V., Tokmakova, I.L., Troshin, P.V., and Livshits, V.A. (1999). The novel transmembrane *Escherichia coli* proteins involved in the amino acid efflux. FEBS Lett. *452*, 228–232.

Ziegler, P., Sahm, H., and Eggeling, L. (2001). L-threonine export: use of peptides to identify a new translocator from *Corynebacterium glutamicum*. J. Bacteriol. *183*, 5317–5324.

Structures of Multidrug Efflux Pumps from the MFS, SMR, MATE and ABC Transporter Families

2

Geoffrey Chang, Paul Szewczyk and Xiao He

Abstract

Drug efflux pumps play an important role in multidrug resistance (MDR) affecting the treatment of several diseases like infection and cancer. In this chapter, we review some of the molecular structures of MDR transporters that have their substrate binding sites located in the lipid bilayer. These transporters include those from the major facilitator superfamily (MFS), small multidrug resistance (SMR) family, the multidrug and toxic-compound extrusion (MATE) family, and the ATP-binding cassette (ABC) family. From these models, common structural themes have emerged that are consistent with their function, including a hydrophobic substrate-binding pocket comprising of several aromatic residues facilitating poly-specificity, V-shapes to allow access of substrate from the lipid bilayer, and distinct inward – and outward-facing conformations facilitating transport across the cell membrane.

Importance of MDR transporters

The neutralization of chemotherapies in the treatment of infectious diseases and cancer, collectively termed multidrug resistance (MDR), is a major limitation in pharmacology and drug efficacy (Persidis, 1999; Nikaido, 2009). Possible mechanisms for innate or acquired drug resistance include enzymatic degradation, glutathione conjugation, transcriptional regulation, decreased drug permeability, enhanced drug efflux, and alternative metabolic pathways (Nikaido, 2009). However, probably one of the most ubiquitous mechanisms among all organisms is the direct multidrug efflux mediated by MDR transporters.

A large body of work has been done characterizing the function and localization of MDR transporters to the cell membrane/wall perimeter. The cumulative sum of genome sequencing data has established that five families of drug efflux pump proteins directly cause or are closely associated with MDR. These transporters include those from the major facilitator superfamily (MFS), the small multidrug resistance (SMR), the multidrug and toxic-compound extrusion (MATE) family, the ATP-binding cassette (ABC) family, and the resistance nodulation and the cell division (RND) family (Piddock, 2006; Saier *et al.*, 2006). Representative X-ray structures of each of these families have been determined (Yin *et al.*, 2006; Chen *et al.*, 2007; Ward *et al.*, 2007; Aller *et al.*, 2009; He *et al.*, 2010) (Fig. 2.1). Transporters from the MFS, SMR, and MATE families use an antiport mechanism that energetically couples substrate export with cation (H^+ or Na^+) import. In contrast, ABC transporters function by coupling nucleotide binding and hydrolysis to the drug export.

MDR transporters from the RND family are, by far, the most dominant MDR transporters in Gram-negative pathogens (Tseng *et al.*, 1999). Their molecular structure and organization are different than those transporters from the MFS, SMR, MATE, and ABC families. RND transporters form a tripartite complex that recruits components in the periplasm and the outer membrane, which then collectively spans throughout the periplasmic space from the inner to the outer membranes necessary for direct drug efflux. The X-ray structures of several members of the RND family have been well-established (Murakami *et al.*, 2002; Murakami *et al.*, 2006; Sennhauser *et al.*, 2009; Long *et al.*, 2010) and reveal that their

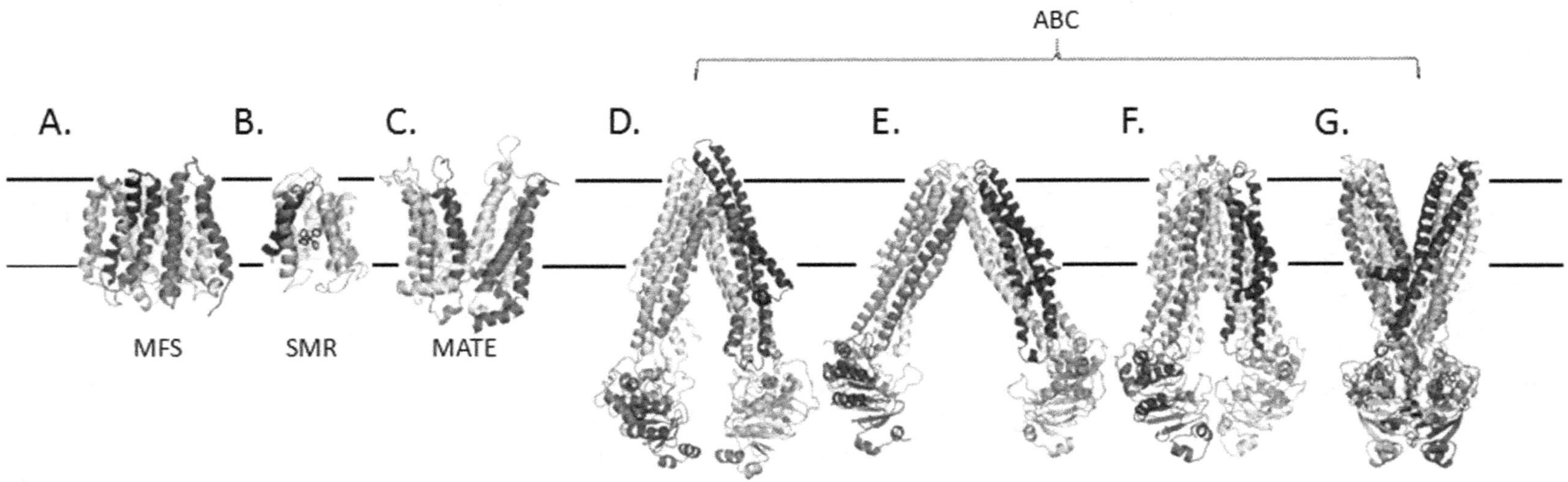

Figure 2.1 X-ray structures from the four MDR/lipid transporter families found in nature where the polyspecific binding site is located in the cell membrane (indicated by black lines). These crystal structures provide a molecular structural basis for drug/lipid transport and are from the major facilitator superfamily (MSF), small multidrug resistance (SMR), multidrug and toxin extrusion family (MATE), and ATP-binding cassette (ABC) transporter families. The X-ray structures, from left to right, are EmrD (Yin *et al.*, 2006), EmrE (Chen *et al.*, 2007), and NorM (He *et al.*, 2010), P-glycoprotein (Aller *et al.*, 2009), and the open conformation of MsbA, closed conformation of MsbA, outward-facing conformation (AMPPNP) of MsbA (Ward *et al.*, 2007). A colour version of this figure is located in the plate section at the back of the book

substrate binding sites reside within the periplasmic space instead of being located in the cell membrane.

For the sake of focus, this chapter will concentrate on the four MDR transporter families (MFS, SMR, MATE, and ABC) that have the poly-specific substrate-binding pocket imbedded in the cell membrane. Presumably the plasma membrane retains a significant fraction of the hydrophobic substrates for these MDR transporters, and the binding of these substrates from within the plasma membrane provides an efficient mechanism to decrease their concentration within the cell. Alternatively or in conjunction, substrates entering from the cytoplasmic opening may also be bound and effluxed as well. Here, we compare the structures of these transporters and draw some common trends in their architectural designs enabling them to pump hydrophobic substrates across the plasma membrane.

Structure of the MDR–MFS transporters

The MFS transporters are found among all kingdoms of life (Marger and Saier, 1993), and are one of the most functionally diverse families of secondary transporters (Pao *et al.*, 1998). These transporters participate in the symport, antiport, or uniport of various substrates, which include natural compounds and metabolites such as sugars, oligosaccharides, phosphate esters, Kreb cycle intermediates, and xenobiotics (Marger and Saier, 1993). Hydropathy analysis of the most conserved core regions of MFS drug/H^+ antiporters (DHA) divides them into two separate clusters, those with 12 (DHA12) and those with 14 (DHA14) transmembrane (TM) helices (Paulsen and Skurray, 1993). Although several MFS transporter structures have been determined by X-ray crystallography (Abramson *et al.*, 2003; Huang *et al.*, 2003; Dang *et al.*, 2010; Newstead *et al.*, 2011), only one is classified as a putative MDR transporter, the *Escherichia coli* (*E. coli*) proton-driven antiporter EmrD. EmrD was first identified as an efflux pump for uncouplers of oxidative phosphorylation (Naroditskaya *et al.*, 1993), which can rapidly arrest growth in bacteria by depleting the H^+ gradient (Krulwich *et al.*, 1990). Uncouplers can be structurally unrelated, such as meta–chloro carbonylcyanide phenylhydrazone (CCCP) and tetrachlorosalicylanilide, as long as they are amphiphilic and undergo protonation/deprotonation cycle within the physiological pH range. It was later discovered that EmrD could also transport antiseptics such as benzalkonium and detergents such as sodium dodecyl sulphate (Nishino and Yamaguchi, 2001). Sequence alignments suggested that EmrD was a close homolog to other MDR MFS transporters (Paulsen *et al.*, 1996) from the DHA12 multidrug efflux subfamily (Saier *et al.*, 2006), which is consistent with its role transporting hydrophobic substrates.

Structure

The overall structural topology of EmrD (Yin *et al.*, 2006) is reminiscent of LacY, GlpT, FucP, and the core structure of $PepT_{so}$ (Abramson *et al.*, 2003; Huang *et al.*, 2003; Dang *et al.*, 2010; Newstead *et al.*, 2011). Twelve TM helices form a compact structure that spans ~ 50 Å in the plane of the lipid bilayer and ~45 Å along the membrane norm (Fig. 2.1A). The TM helices facing away from the interior (TM3, TM6, TM9, and TM12) have an organization similar to other MFS transporters, with the remaining TM helices forming the drug-binding pocket. The structure of EmrD (Yin *et al.*, 2006) revealed a conformation representing more like an intermediate between the inward – and the outward-facing in the transport cycle. Compared to other MFS transporter structures, the periplasmic loops in EmrD are more imbedded in the cell membrane, and the loop connecting the two halves of the transporter is shorter. Like all other MFS transporter structures solved to date, an intramolecular twofold axis relates the N – and C-terminal halves of EmrD and supports the notion that the MFS likely arose from a genomic duplication event of an ancestral gene encoding a six-helix protein domain (Pao *et al.*, 1998; Saier, 2003).

A hydrophobic substrate-binding pocket

The most notable difference between the structure of EmrD and those of other MFS transporters (Abramson *et al.*, 2003; Huang *et al.*, 2003; Dang *et al.*, 2010; Newstead *et al.*, 2011) is the composition of residues in the substrate-binding pocket . Whereas LacY, GlpT, FucP, and $PepT_{so}$ have more hydrophilic interiors, reflecting their functions to bind water soluble substrates, the internal cavity of EmrD comprises mostly hydrophobic residues, consistent with its role as hydrophobic compounds efflux pump. Several of these residues are bulky aliphatic or aromatics (Ile28, Ile217, Ile253, Tyr52, Tyr56, Trp300, and Phe249) and forms a similar hydrophobic core observed in some other families of MDR transporters (Chen *et al.*, 2007; Ward *et al.*, 2007; Aller *et al.*, 2009; He *et al.*, 2010) and proteins. These residues likely play a role dictating drug specificity either through steric or aromatic interactions. The internal cavity of EmrD also has some uncharged polar residues, such as glutamines (Gln21, Gln24, Gln46, and Gln60), and a basic arginine residue (Arg118) located closer to the opening towards the cytoplasm. On the periplasmic side, a threonine (Thr25), an aspartate (Asp33), and a glutamate residue (Glu227) could readily reorient into the cavity during the transport cycle and may participate in substrate translocation. The corresponding residues in other closely related orthologues of EmrD have suggested that they may play a role in substrate binding and transport (Yin *et al.*, 2006). EmrD also has two pairs of stacked aromatic groups (Tyr52/Tyr56 and Trp300/Phe249) that may play an important role in poly-specific binding, similar to that of a pair of phenylalanine residues in the substrate recognition in the MDR MFS transporter Bmr (Klyachko *et al.*, 1997). The energetic cost of transporting charged amphipathic compounds may then generally be compensated for by these types of hydrophobic interactions (Paulsen and Lewis, 2002). The hydrophobic interior of EmrD suggests a generalized pathway and mechanism for transporting a variety of structurally distinct substrates in drug efflux systems (Paulsen and Lewis, 2002).

Structure of SMR transporters

EmrE from *E. coli* represents another type of secondary transporter that directly couples the efflux of drugs to the influx of protons across the cell membrane (Chung and Saier, 2001; Schuldiner *et al.*, 2001). EmrE is a prototypical member of the SMR family and is one of the smallest known transporters in nature, composed of merely 110 amino acid residues. However, as shown by oligomerization assays, substrate binding experiments, negative dominance studies, and crosslinking analyses (Yerushalmi *et al.*, 1996; Rotem *et al.*, 2001; Soskine *et al.*, 2002; Tate *et al.*, 2003; Butler *et al.*, 2004; Ninio *et al.*, 2004), the basic functional unit of EmrE is a homodimer formed by a pair of monomers of four TM helixes apiece. This conclusion is further supported by the existence of naturally occurring heterodimeric SMR transporter proteins, such as YdgE/YdgF of *E. coli*, and EbrA/EbrB and YkkC/YkkD of *Bacillus subtilis*. The functionality of these transporter systems requires the co-expression of the two component polypeptides for proper drug efflux activity and presumably the formation of heterodimers analogous to the EmrE homodimer (Jack *et al.*, 2000; Masaoka *et al.*, 2000; Nishino and Yamaguchi, 2001; Kikukawa *et al.*, 2007; Zhang *et al.*, 2007). Electron microscopy (EM) studies of EmrE in complex with a transport substrate, tetraphenylphosphonium (TPP), reconstituted in lipid bilayers also suggested an asymmetrical homo-dimer (Ubarretxena-Belandia *et al.*, 2003).

Structure

The X-ray structure of EmrE bound with TPP reveals the minimal functional unit, a homodimer composed of two monomers each having four TM helices (Chen *et al.*, 2007). The two most highly conserved helices, TM1 and TM3, are flanked by the less conserved TM2 and TM4. In this structure of EmrE, the bound substrate TPP is located in an internal chamber surrounded by six helices, TM1, TM2 and TM3 from each monomer, and is in contact with Glu 14, which has been shown to be very important for substrate recognition and transport (Chen *et al.*, 2007). Because of the modest diffraction resolution, the TPP position defined in this structure is based on the anomalous signal from the arsenic analogue of TPP, TPA. The two TM4 helices do not constitute the binding site and only participate in the dimerization interactions, forming a four-helix bundle with the TM3 helices.

EmrE is an asymmetrical dimer with anti-parallel topology

EM and X-ray crystallography studies, and functional structural data from NMR/FRET, demonstrate that the two monomers constituting the EmrE dimer are not structurally identical but rather form an asymmetrical tertiary structure. The first three TM helices of one monomer are superimposed on the corresponding helices of the other monomer with an average root mean square deviation (rmsd) of 1.2 Å over equivalent helical C_α atoms, whereas superposition of all four TM helices overlap less, giving an average rmsd of 2.2 Å. The relative positions of TM1–3 are more conserved than that of TM4 among SMR transporters.

The X-ray structure of the TPP-bound EmrE dimer also supports the conclusion that the EmrE protein is inserted into the cell membrane in two orientations (termed anti-parallel, or dual topology) in the *E. coli* membrane. Previous studies suggested that EmrE has a unique N_{in}/C_{in} topology (Ninio *et al.*, 2004); however, biochemical studies (Daley *et al.*, 2005; Granseth *et al.*, 2005; Kikukawa *et al.*, 2007), recent NMR and FRET work (Morrison *et al.*, 2012), and the X-ray structure itself corroborate the dual topology model. Selenomethionine marking of the corresponding residues on each monomer also validated the topology of the EmrE X-ray structure, relating the two dimers by a twofold rotation axis parallel to the cell membrane (Chen *et al.*, 2007). Evidence that later members of the SMR family in the DMT superfamily likely have anti-parallel configurations is also compelling, suggesting that anti-parallel topology is not only possible but likely for all members of this transporter family (Jack *et al.*, 2001).

Substrate transport mechanism

The structural asymmetry of EmrE yields some insight into its function as a unidirectional, poly-specific substrate transporter. EmrE binds to its substrate TPP as an asymmetrical dimer, in a chamber that appears open to one side of the bilayer, similar to other MDR transporters. The binding of H^+ through the same residue (Glu14) may displace the bound TPP in the outward-facing conformation and reset the functional unit back to the inward-facing conformation where the proton is released into the cytosol. Here the dimer is competent for binding substrate again and re-initiating the transport cycle. Further studies, including computational simulations, probing this aspect are on-going and should reveal more about the transport mechanics of this small transport system.

Structure of MATE transporters

Members of MATE family of transporters represent the last class of MDR transporters to be structurally characterized (Saier *et al.*, 2006). They are involved in a wide variety of biological functions across all kingdoms of life. In plants, MATE transporters are highly prevalent, with 58 paralogues found in *Arabidopsis thaliana* alone (Hvorup *et al.*, 2003), where they play important roles in metabolite transport (Omote *et al.*, 2006; Morita *et al.*, 2009), the secretion of secondary metabolites as a defence mechanism against herbivores and microbial pathogens (Omote *et al.*, 2006; Morita *et al.*, 2009), and conferring tolerance towards phytotoxic aluminium in acidic soils, a major limitation of crop production in 50% of the world's arable land (Wood, 2000). Mammalian and bacterial MATE transporters primarily mediate MDR (Omote *et al.*, 2006), modulating the efficacy of many pharmaceutical drugs used in the treatment of a variety of diseases (Kaatz *et al.*, 2005; McAleese *et al.*, 2005; Becker *et al.*, 2009; Tsuda *et al.*, 2009). In mammals, MATE transporters export a structurally diverse array of xenobiotic cations in the liver and kidney, a process that influences the plasma concentrations of many pharmaceutical drugs including metformin, a widely prescribed Type 2 diabetes drug, mitigating its therapeutic efficacy (Kaatz *et al.*, 2005; McAleese *et al.*, 2005; Becker *et al.*, 2009; Tsuda *et al.*, 2009). Bacterial MATE transporters function primarily as xenobiotic efflux pumps and have been reported to confer resistance to tigecycline, a new glycylcycline class antibiotic developed to overcome methicillin-resistant (MR) and vancomycin-resistant *Staphylococcus aureus* (SA) (Becker *et al.*, 2009; Tsuda *et al.*, 2009). MATE transporters use either H^+ or Na^+ gradients across the membrane to drive the substrate export though the coupling mechanism is not well understood. All MATE transporters share ~40% protein sequence similarity (Omote *et al.*, 2006), suggesting an overall conservation of structure and transport function.

Structure

The X-ray structure of NorM from *Vibrio cholerae* (NorM-VC) spans ~50 Å in the plane of the lipid bilayer and represents an outward-facing conformation, where two bundles of six TM helices (TMs 1–6 and TMs 7–12) form a large internal cavity open to the extracellular space (Fig. 2.1C) (He *et al.*, 2010). The two halves are related by an intra-molecular twofold symmetry, which is likely a result of a gene duplication event as suggested by the sequence homology between the two halves. Like the transporters from the MFS family, these halves are connected by a cytoplasmic loop between TMs 6 and 7, while the initial helix of each half (TM 1 and TM7) are preceded by a helical extension embedded in the inner membrane leaflet side. An additional short helix after TM12 is nestled under the cytoplasmic side of TM11.

The overall structure of NorM reveals a novel topology distinct from all other MDR transporter families (Murakami *et al.*, 2002; Yin *et al.*, 2006; Chen *et al.*, 2007; Ward *et al.*, 2007; Aller *et al.*, 2009) (Fig. 2.1C). Unlike the ABC, MFS, and SMR drug efflux pumps, no TM helices cross over one another. Two non-equivalent portals open within the lipid bilayer, allowing the expulsion of substrates directly to the outer membrane leaflet or the extracellular space. These portals are formed by TMs 1 and 8 on one side and TMs 2 and 7 on the other. At the widest point within the lipid bilayer, the portals are ~12 Å wide and can accommodate the passage of substrates into the outer membrane leaflet. The volume of the internal cavity embedded within the lipid bilayer in this structure of NorM is estimated ~4300 $Å^3$, approximately four times larger than the substrate-binding pocket s of the transcription MDR regulators xPXR and BmrR (Klyachko *et al.*, 1997; Zheleznova *et al.*, 1999; Watkins *et al.*, 2001), and nearly two thirds that of the ABC exporters P-glycoprotein (Pgp) or MsbA (Ward *et al.*, 2007; Aller *et al.*, 2009). Residues facing the internal cavity are relatively conserved among MDR MATE transporters and contributed by all TM helices except for TMs 3 and 9, which are located at the periphery in this conformation. Of the 91 residues facing the cavity in this conformation of NorM-VC, 57 are hydrophobic, 14 aromatic, 20 polar and five charged. Nine of the 14 aromatic residues are clustered in the C-terminal half of the internal cavity.

This outward-facing conformation of NorM-VC binds to a rubidium (Rb^+) or a caesium (Cs^+) ion, which are heavier alkali metal analogues of the sodium ion and more easily detectable by X-ray crystallography, at a structurally discrete site in the C-terminal half of the internal cavity (He *et al.*, 2010). This conformation likely represent a state of the transporter where substrate is released (low substrate affinity). The binding of Na^+ at the cation-binding site in this outward-facing conformation is thought to trigger the conformational change of the transporter to the inward-facing conformation capable of binding substrates with higher affinity (He *et al.*, 2010). This cation-binding site is bounded by residues from TMs 7, 8, and 10–12, which have a configuration of different topology than found in members of the MFS family.

Structure of MDR-ABC Transporters

The ABC is a highly conserved domain that couples ATP hydrolysis to biological work, and can be found in a variety of prokaryotic and eukaryotic proteins. ABC transporters are integral membrane proteins that use the binding and hydrolysis of ATP to power the transport of diverse substrates such as ions, lipids, peptides, metabolites, chemotherapeutic drugs, and antibiotics across cellular membranes.

ABC transporters play a wide role not only in MDR (Gottesman *et al.*, 1996), but also in homeostasis of the cell. Mutations in these proteins have been shown to be cause several human diseases (Gottesman and Ambudkar, 2001; Dean, 2005; Dean and Annilo, 2005). A functional ABC transporter unit is minimally composed of two nucleotide binding domains (NBDs), which contain the ABC signature motifs and two TM domains (TMDs). The NBDs are the sites of ATP hydrolysis and are highly conserved among the ABC transporter superfamily. Like the MFS, MATE, and SMR, the TMDs form the transport pathway that mediates the passage of substrates across the membrane and dictate functional diversity. Despite differences in both genetic organization of the cognate domains (Biemans-Oldehinkel *et al.*, 2006) and substrate specificity, the known structures of ABC transporters and isolated NBDs exhibit an internal twofold or pseudo-twofold symmetry between the two interacting halves (Hung *et al.*, 1998; Karpowich *et al.*, 2001; Yuan *et al.*, 2001; Locher *et al.*, 2002; Smith *et al.*, 2002; Chen *et al.*, 2003; Schmitt *et al.*, 2003; Karcher *et al.*, 2005; Lu *et al.*, 2005; Zaitseva *et al.*, 2005; Dawson and Locher, 2006; Pinkett *et al.*, 2007).

Structure of P-glycoprotein

Owing to its high clinical relevance, the ABC family transporter Pgp (ABCB1) is perhaps the most studied MDR transporter in the modern literature. Commensurate with the high general interest in this family, the structures of several bacterial homologues determined by X-ray crystallography (Locher *et al.*, 2002; Dawson and Locher, 2006; Hollenstein *et al.*, 2007; Oldham *et al.*, 2007; Pinkett *et al.*, 2007; Ward *et al.*, 2007) and that of Pgp and other ABC transporters characterized at low resolution by electron microscopy (Rosenberg *et al.*, 2003; Lee *et al.*, 2008). The structure of mouse Pgp (Aller *et al.*, 2009) likely is similar to that of human Pgp, which shares an 87% sequence identity with only two residues different in the drug-binding pocket. Two co-crystal structures of mouse Pgp in complex with two different stereoisomers of the cyclic hexapeptide inhibitors, cyclic-tris-(R)-valineselenazole (QZ59-RRR) and cyclic-tris-(S)-valineselenazole (QZ59-SSS) revealed a molecular basis for its poly-specificity (Aller *et al.*, 2009).

The X-ray structure of mouse Pgp (Fig. 2.1D) (Aller *et al.*, 2009) represents a nucleotide-free inward-facing conformation, where the TM helices arranged as two 'halves' of six helices each (TMs 1–3, 6, 10, 11 and TMs 4, 5, 7–9, 12), forming in a large internal cavity open to both the cytoplasm and the inner leaflet. Akin to previous ABC transporter structures, the two halves are related by a pseudo two-fold intramolecular symmetry. This structure spans ~136 Å perpendicular to and ~70 Å in the plane of the bilayer, with the two nucleotide binding domains (NBDs) separated by ~30 Å. Like the other MDR transporter families already discussed, two portals open to the inner leaflet in this conformation, allowing access/entry of hydrophobic molecules directly from the membrane. For Pgp, these portals are formed by TMs 4/6 and 10/12, each of which are composed of smaller side chains that could allow tighter packing during NBD dimerization. At the widest point within the bilayer, the portals are ~9 Å wide and are formed by an intertwined interface where TMs 4/5 for one portal, and TMs 10/11 the other cross over to make extensive contacts with the opposite α-helical bundle. The intertwined interface buries ~2800 $Å^2$, contributing to the stability of the interface between the dimers, and is a conserved motif among all exporters in the ABCB subfamily solved to date. The volume of the internal cavity within the lipid bilayer is substantial (~6000 $Å^3$) and could easily accommodate at least two compounds simultaneously (Loo *et al.*, 2003). The composition of Pgp's substrate-binding pocket consists mostly of hydrophobic and aromatic residues. Of the 73 solvent accessible residues in the substrate-binding pocket of Pgp in the inward-facing conformation, 15 are polar and only two (His60 and Glu871), located in the N-terminal half of the TMD, are charged or potentially charged. The outer membrane leaflet half of Pgp's drug-binding pocket contains predominantly hydrophobic/aromatic residues while the inner membrane leaflet half has more polar and charged residues.

Polyspecificity and Pgp

The co-crystal structures of Pgp with QZ59 compounds demonstrate that it has the capacity to distinguish between stereo-isomers of cyclic peptides, resulting in different binding locations, orientation, and stoichiometry (Aller *et al.*, 2009). These co-crystal structures also suggest that the inward-facing conformation is competent to bind drugs/substrates and likely represents the high-affinity drug-binding state of Pgp. Previous studies identified several residues that interact with Pgp substrates and are highly conserved among its close mammalian (Aller *et al.*, 2009). In the inward-facing conformation, many of these residues are facing the drug-binding pocket, suggesting a common mechanism of poly-specific drug recognition. The binding positions of several QZ59 compounds, including those published, have revealed that there are functionally discrete binding sites within Pgp's large substrate-binding pocket . For each substrate, the specific roles of each of these residues contacting the ligand are different.

Structures of MsbA: insights into conformational changes and structural flexibility of transporters of the ABCB family

Bacterial ABC transporters such as MsbA and LmrA, function as homodimers or 'half-transporters', and share a similar topology and function to Pgp. The *msbA* gene encodes a TMD of six TM helices, followed by a single NBD. When translated and inserted into the membrane, two of these polypeptides assemble to form a functional homodimeric unit. MsbA is found in Gram-negative bacteria and transports lipid A and lipopolysaccharide (LPS) from the cytoplasmic leaflet (inward-facing) to the periplasmic leaflet (outward-facing) of the plasma membrane (Karow and Georgopoulos, 1993; Polissi and Georgopoulos, 1996; Zhou *et al.*, 1998; Doerrler *et al.*, 2001; Doerrler and Raetz, 2002; Wang *et al.*, 2004; Doerrler *et al.*, 2004). Functional studies have also shown that MsbA and, more significantly, Pgp are both poly-specific transporters capable of recognizing and transporting a wide spectrum of drug molecules (Reuter *et al.*, 2003; Woebking *et al.*, 2005; Shilling *et al.*, 2006). Additionally, MsbA shares significant sequence identity with other bacterial ABC 'half-transporters' that have been implicated in MDR, such as LmrA from Gram-positive *Lactococcus lactis* (van Veen *et al.*, 1998; Dean and Annilo, 2005).

Structures of both the inward – and the outward-facing conformations of MsbA from different bacterial species have already been solved (Ward *et al.*, 2007) and homology models made from available structures suggest a plausible transport pathway for hydrophobic substrates shared by all MDR ABC transporters in the ABCB subfamily (Fig. 2.1E–G). In some of the inward-facing conformations, the structures are quite wide open with the NBDs ~50 Å apart, and the two halves of the ATP-binding sites still face each other (Fig. 2.1E and F). Like Pgp, the most striking feature of the MsbA models is that TM4/TM5 crosses over and associates with the opposing monomer and is important to maintain the structural integrity of the transporter throughout the transport cycle. This cross-over motif is visibly accentuated in the open apo-structure and provides the only means of interaction between the two monomers. For the inward-facing conformations of MsbA, the cross-over interaction buries ~2300 Å^2 surface area per monomer and likely holds the dimer together during the open inverted V (inward-facing) conformation (Fig. 2.1E–F). When nucleotide is bound, MsbA undergoes a structural change to the outward-facing conformation (Fig. 2.1G) where the main interactions between the two halves are between the nucleotide binding domains (Ward *et al.*, 2007).

Pgp and MsbA share the same structural scaffold but have very different substrate specificities. Pgp can bind to a broad range of cationic hydrophobic compounds while transporters like MsbA have a much higher affinity for its native substrate, Kdo$_2$-lipid A (MW 2.3 kDa in *E. coli*), yet still retain some degree of drug polyspecificity (Reuter *et al.*, 2003). In comparison, the substrate-binding pocket s of both MsbA and Pgp posses hydrophobic residues (Fig. 2.2). However, the residues in the binding pocket of MsbA are relatively smaller and hydrophobic, while those in Pgp's binding pocket are mostly aromatic, which could greatly facilitate (favour) a wider range of substrate specificity (poly-specificity). Aromatic residues are known to play key roles in protein–substrate interactions, particularly among MDR transporters and transcriptional regulators. Presumably some of these interactions are mediated through pi stacking between the substrate(s) and the aromatic residues in the drug-binding pocket.

General trends observed among MDR transporter structures

With a complete collection of structural representatives now available for MDR transporters

A.

B.

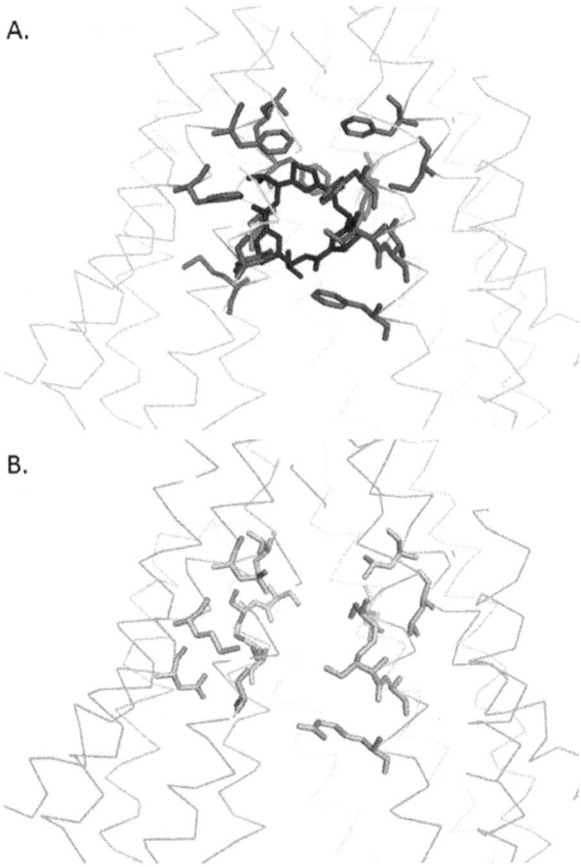

C.

Comparison of Pgp and MsbA

TM	Mouse Pgp	EC-MsbA	TM	Mouse Pgp	EC-MsbA
1	His 60	Asn 37	7	Asn 717	Asn 37
	Gly 61	Ala 38		Gln 721	Asp 41
	Ala 63	Ser 40		Phe 724	Met 44
	Leu 64	Asp 41		Phe 728	Leu 48
	Pro 65	Thr 42	8	Leu 758	Leu 74
	Met 67	Met 44		Ser 762	Arg 78
	Met 68	Leu 45		Phe 766	Ser 82
	Phe 71	Leu 48	9	Ala 830	Ile 144
2	Tyr 110	Met 67		Phe 833	Val 147
	Tyr 113	Val 70	10	Ile 860	Leu 174
	Val 121	Arg 78		Val 861	Ala 175
3	Gly 187	Ile 144		Ile 864	Val 178
	Phe 190	Val 147		Ala 865	Ser 179
	Gln 191	Arg 148		Ala 867	Ala 181
	Ala 194	Ala 151		Gly 868	Ile 182
4	Gly 222	Ser 179		Glu 871	Val 185
	Ala 225	Ile 182	11	Phe 938	Asp 252
	Ala 229	Ser 186		Thr 941	Ile 255
	Ser 233	Arg 190		Gln 942	Gln 256
5	Gly 296	Pro 253		Met 945	Ala 259
	Phe 299	Gln 256		Ser 948	Ala 262
	Leu 300	Leu 257		Tyr 949	Leu 263
	Tyr 303	Ser 260		Ala 952	Phe 265
	Tyr 306	Leu 263		Phe 953	Leu 267
	Phe 310	Leu 267	12	Val 970	Ile 284
	Trp 311	Tyr 268		Leu 971	Thr 285
6	Leu 328	Thr 285		Phe 974	Phe 288
	Thr 329	Val 286		Ser 975	Ser 289
	Phe 332	Ser 289		Val 978	Ile 292
	Ser 333	Ser 290		Gly 980	Leu 294
	Leu 335	Ile 292		Ala 981	Met 295
	Ile 336	Ala 293		Met 982	Arg 296
	Ala 338	Met 295		Gly 985	Lys 299
	Phe 339	Arg 296		Gln 986	Ser 300
	Gly 342	Lys 299		Ser 988	Thr 302
	Gln 343	Ser 300		Ser 989	Asn 303
	Pro 346	Asn 303			

Internal Cavity Residues

Figure 2.2 Comparison of residues in the substrate-binding pocket between mouse Pgp (Aller *et al.*, 2009) and MsbA from *E. coli* (Ward *et al.*, 2007). (A) Pgp structure with QZ59-RRR (Aller *et al.*, 2009) bound. (B) Residues corresponding to MsbA mapped onto the Pgp structure as a reference. (C) Comparison of residues lining the substrate-binding pocket of mouse Pgp and MsbA from *E. coli*. Aromatic residues in the binding pocket of Pgp and MsbA are highlighted yellow and cyan, respectively. A colour version of this figure is located in the plate section at the back of the book

from the MFS, MATE, SMR, and ABC families, some general trends are starting to emerge with regards to their structures and functions. First, the drug-binding pockets for each of these MDR transporters do indeed reside in the cell membrane and, interestingly but not surprisingly, are formed by residues of TM helices. Access of hydrophobic substrates to the binding sites directly from the cytoplasm and/or the lipid bilayer is another common theme and is a feature highly consistent with the function of these transporters – moving substrates from the inner to the outer membrane leaflet and/or from cytosol to extracellular space. Many of the substrates of these transporters readily partition into the plasma membrane and can be 'intercepted' by these pumps before entering the cytoplasm. In the case of MDR ABC transporters like Pgp and MsbA, phospholipids themselves may also be substrates (Callaghan *et al.*, 1997), and transported in a unidirectional way (Sharom, 2008) between the inner and the outer leaflets of the bilayer (Bosch *et al.*, 1997).

Poly-specificity among MDR transporters may also be greatly enhanced by breaking the internal symmetry of the substrate-binding pocket . The symmetric features of MsbA's homo-dimeric structure largely limit the possibilities of re-orienting residues to accommodate the binding of asymmetrical substrates. In contrast, (monomeric) Pgp's binding pocket is inherently asymmetrical.

This combination of structural asymmetry and a substrate-binding pocket enriched with aromatic hydrophobic residues has also been observed in EmrD, NorM, and EmrE, suggest a general structural basis better suited for poly-specific substrate recognition and binding (Yin *et al.*, 2006; Chen *et al.*, 2007; He *et al.*, 2010). This certainly opens up the possibilities and potentials of structure-based design of the substrate-binding pockets and bioengineering of efflux pumps with desired specificity to facilitate the efflux of certain substrates.

Another structural feature among all available MDR transporter structures in the four families discussed is a V-shape (or inverted V-shape), allowing access to the aforedescribed poly-specific binding pocket directly from the lipid bilayer. In all the MDR transporter structures solved so far, access to the substrate-binding pocket can be granted through the two portals that open wide enough to accommodate hydrophobic molecules and phospholipids (Fig. 2.1). The presence of these portals allows the internal cavity of the transporters to form a contiguous space at the entrance of the drug-binding pocket that can encompass the widest axis of the substrate molecules. This open structure then enables these transporters to 'scan' the inner leaflet to select and bind specific lipids and/or hydrophobic drugs for efflux. When the portals are large enough, lipids and substrates might remain together during initial entry into the internal cavity. In the case of MDR-ABC transporters, this could explain the requirement for, and the ATPase activity-stimulating effect of certain lipids. Some of these V-shape conformations can be quite wide and may be necessary for MDR ABC transporters to accommodate very large substrates, such as Kdo$_2$-lipid A (MW 2.3 kDa) (Ward *et al.*, 2007) for MsbA and β-amyloid (MW ~ 4 kDa) in the case of Pgp (Lam *et al.*, 2001). The idea of such astonishing degree of structural flexibility to accommodate functional diversity is also supported by EPR data of MsbA in the lipid membrane (Dong *et al.*, 2005), as well as structures of the inward-facing conformations of other ABC transporters that also have NBDs far apart from one another (Chami *et al.*, 2002; Hofacker *et al.*, 2007).

For MDR transporters, the V-shapes of these molecules are always closed on one side, suggesting an 'alternating access' model in all of these unidirectional transport systems. This type of structural mechanism is fundamentally different than that of a channel, which functions more like a selective pore allowing continuous passage of the substrate. For the MDR transporters from the SMR, MATE, and MFS families, we hypothesize that the alternating access model is mediated by a rocker-switch mechanism where the relative rigid body movement between the two halves ensure the alternative access to the substrate-binding pocket . However, in the case of MDR ABC transporters in the ABCB1 family (such as Pgp and MsbA), the structural changes throughout the transport cycle is more of a twisting motion, where the two bundles of six TM helices each slide against each other (Gutmann *et al.*, 2010). We hypothesize that this twisting motion disrupts the interactions of key residues in the drug-binding pocket that exist only in the high-affinity inward-facing conformation (Gutmann *et al.*, 2010). In the case of MsbA and Pgp, the directions of the portal openings in the inward – and the outward-facing conformations are nearly orthogonal and the results of substantial repositioning of the TM helices. In these systems, the inward-facing conformation probably represents the molecule in a pre-transport state where drug binding occurs (Aller *et al.*, 2009). During the catalytic cycle, the binding of ATP, stimulated/facilitated by substrate binding, likely causes/stabilizes the dimerization of the NBDs, which in turn produces large structural changes and leads to an outward-facing conformation similar to the nucleotide-bound structures of MsbA or Sav1866 (Dawson and Locher, 2006; Ward *et al.*, 2007). Substrates are then released as a consequence of decreased binding affinity due to the loss of specific residue contacts between the transporter and drug. ATP hydrolysis then likely disrupts NBD dimerization and resets the system back to the inward-facing confirmation, and reinitiates the transport cycle (Tombline *et al.*, 2005).

References

Abramson, J., Smirnova, I., Kasho, V., Verner, G., Iwata, S., and Kaback, H.R. (2003). The lactose permease of *Escherichia coli*: overall structure, the sugar-binding site and the alternating access model for transport. FEBS Lett. 555, 96–101.

Aller, S.G., Yu, J., Ward, A., Weng, Y., Chittaboina, S., Zhuo, R., Harrell, P.M., Trinh, Y.T., Zhang, Q., Urbatsch, I.L., *et al.* (2009). Structure of P-glycoprotein reveals a molecular basis for poly-specific drug binding. Science *323*, 1718–1722.

Becker, M.L., Visser, L.E., Van Schaik, R.H., Hofman, A., Uitterlinden, A.G., and Stricker, B.H. (2009). Genetic variation in the multidrug and toxin extrusion 1 transporter protein influences the glucose-lowering effect of metformin in patients with diabetes: a preliminary study. Diabetes, *58*, 745–749.

Biemans-Oldehinkel, E., Doeven, M.K., and Poolman, B. (2006). ABC transporter architecture and regulatory roles of accessory domains. FEBS Lett. *580*, 1023–1035.

Bosch, I., Dunussi-Joannopoulos, K., Wu, R.L., Furlong, S.T., and Croop, J. (1997). Phosphatidylcholine and phosphatidylethanolamine behave as substrates of the human MDR1 P-glycoprotein. Biochemistry *36*, 5685–5694.

Butler, P.J., Ubarretxena-Belandia, I., Warne, T., and Tate, C.G. (2004). The *Escherichia coli* multidrug transporter EmrE is a dimer in the detergent-solubilised state. J. Mol. Biol. *340*, 797–808.

Callaghan, R., Berridge, G., Ferry, D.R., and Higgins, C.F. (1997). The functional purification of P-glycoprotein is dependent on maintenance of a lipid–protein interface. Biochim. Biophys. Acta, *1328*, 109–124.

Chami, M., Steinfels, E., Orelle, C., Jault, J.M., di Pietro, A., Rigaud, J.L., and Marco, S. (2002). Three-dimensional structure by cryo-electron microscopy of YvcC, an homodimeric ATP-binding cassette transporter from Bacillus subtilis. J. Mol. Biol. *315*, 1075–1085.

Chen, J., Lu, G., Lin, J., Davidson, A.L., and Quiocho, F.A. (2003). A tweezers-like motion of the ATP-binding cassette dimer in an ABC transport cycle. Mol. Cell *12*, 651–661.

Chen, Y.J., Pornillos, O., Lieu, S., M.A., C., Chen, A.P., and Chang, G. (2007). X-ray structure of EmrE supports dual topology model. Proc. Natl. Acad. Sci. U.S.A. *104*, 18999–19004.

Chung, Y.J., and Saier, M.H., J.R. (2001). SMR-type multidrug resistance pumps. Curr. Opin. Drug Discov. Devel. *4*, 237–245.

Daley, D.O., Rapp, M., Granseth, E., Melen, K., Drew, D., and Von Heijne, G. (2005). Global topology analysis of the *Escherichia coli* inner membrane proteome. Science *308*, 1321–1323.

Dang, S., Sun, L., Huang, Y., Lu, F., Liu, Y., Gong, H., Wang, J., and Yan, N. (2010). Structure of a fucose transporter in an outward-open conformation. Nature *467*, 734–738.

Dawson, R.J., and Locher, K.P. (2006). Structure of a bacterial multidrug ABC transporter. Nature *443*, 180–185.

Dean, M. (2005). The genetics of ATP-binding cassette transporters. Methods Enzymol., *400*, 409–429.

Dean, M., and Annilo, T. (2005). Evolution of the ATP-binding cassette (ABC) transporter superfamily in vertebrates. Annu. Rev. Genomics Hum. Genet. *6*, 123–142.

Doerrler, W.T., and Raetz, C.R. (2002). ATPase activity of the MsbA lipid flippase of *Escherichia coli*. J. Biol. Chem. *277*, 36697–36705.

Doerrler, W.T., Reedy, M.C., and Raetz, C.R. (2001). An *Escherichia coli* mutant defective in lipid export. J. Biol. Chem. *276*, 11461–11464.

Doerrler, W.T., Gibbons, H.S., and Raetz, C.R. (2004). MsbA-dependent translocation of lipids across the inner membrane of *Escherichia coli*. J. Biol. Chem. *279*, 45102–45109.

Dong, J., Yang, G., and Mchaourab, H.S. (2005). Structural basis of energy transduction in the transport cycle of MsbA. Science *308*, 1023–1028.

Gottesman, M.M., and Ambudkar, S.V. (2001). Overview: ABC transporters and human disease. J. Bioenerg. Biomembr. *33*, 453–458.

Gottesman, M.M., Pastan, I., and Ambudkar, S.V. (1996). P-glycoprotein and multidrug resistance. Curr. Opin. Genet. Dev. *6*, 610–617.

Granseth, E., Daley, D.O., Rapp, M., Melen, K., and Von Heijne, G. (2005). Experimentally constrained topology models for 51,208 bacterial inner membrane proteins. J. Mol. Biol. *352*, 489–494.

Gutmann, D.A., Ward, A., Urbatsch, I.L., Chang, G., and Van Veen, H.W. (2010). Understanding polyspecificity of multidrug ABC transporters: closing in on the gaps in ABCB1. Trends Biochem. Sci. *35*, 36–42.

He, X., Szewczyk, P., Karyakin, A., Evin, M., Hong, W.X., and Zhang, Q.H. (2010). Structure of a cation-bound multidrug and toxin compound extrusion (MATE) transporter. Nature *467*, 991–994.

Hofacker, M., Gompf, S., Zutz, A., Presenti, C., Haase, W., van der Does, C., Model, K., and Tampe, R. (2007). Structural and functional fingerprint of the mitochondrial ATP-binding cassette transporter Mdl1 from Saccharomyces cerevisiae. J. Biol. Chem. *282*, 3951–3961.

Hollenstein, K., Frei, D.C., and Locher, K.P. (2007). Structure of an ABC transporter in complex with its binding protein. Nature *446*, 213–216.

Huang, Y., Lemieux, M.J., Song, J., Auer, M., and Wang, D.N. (2003). Structure and mechanism of the glycerol-3-phosphate transporter from *Escherichia coli*. Science *301*, 616–620.

Hung, L.W., Wang, I.X., Nikaido, K., Liu, P.Q., Ames, G.F., and Kim, S.H. (1998). Crystal structure of the ATP-binding subunit of an ABC transporter. Nature *396*, 703–707.

Hvorup, R.N., Winnen, B., Chang, A.B., Jiang, Y., Zhou, X.F., and Saier, M.H. (2003). The multidrug/oligosaccharidyl-lipid/polysaccharide (MOP) exporter superfamily. Eur. J. Biochem. *270*, 799–813.

Jack, D.L., Storms, M.L., Tchieu, J.H., Paulsen, I.T., and Saier, M.H., J.R. (2000). A broad-specificity multidrug efflux pump requiring a pair of homologous SMR-type proteins. J. Bacteriol. *182*, 2311–2313.

Jack, D.L., Yang, N.M., and Saier, M.H., J.R. (2001). The drug/metabolite transporter superfamily. Eur. J. Biochem. *268*, 3620–3639.

Kaatz, G.W., Mcaleese, F., and Seo, S.M. (2005). Multidrug resistance in *Staphylococcus aureus* due to overexpression of a novel multidrug and toxin extrusion (MATE) transport protein. Antimicrob. Agents Chemother. *49*, 1857–1864.

Karcher, A., Buttner, K., Martens, B., Jansen, R.P., and Hopfner, K.P. (2005). X-ray structure of RLI, an essential twin cassette ABC ATPase involved in ribosome biogenesis and HIV capsid assembly. Structure, *13*, 649–659.

Karow, M., and Georgopoulos, C. (1993). The essential *Escherichia coli* msbA gene, a multicopy suppressor of null mutations in the htrB gene, is related to the universally conserved family of ATP-dependent translocators. Mol. Microbiol. *7*, 69–79.

Karpowich, N., Martsinkevich, O., Millen, L., Yuan, Y.R., Dai, P.L., Macvey, K., Thomas, P.J., and Hunt, J.F. (2001). Crystal structures of the MJ1267 ATP binding cassette reveal an induced-fit effect at the ATPase active site of an ABC transporter. Structure, *9*, 571–586.

Kikukawa, T., Miyauchi, S., Araiso, T., Kamo, N., and Nara, T. (2007). Anti-parallel membrane topology of two components of EbrAB, a multidrug transporter. Biochem. Biophys. Res. Commun. *358*, 1071–1075.

Klyachko, K.A., Schuldiner, S., and Neyfakh, A.A. (1997). Mutations affecting substrate specificity of the *Bacillus subtilis* multidrug transporter Bmr. J. Bacteriol. *179*, 2189–2193.

Krulwich, T.A., Quirk, P.G., and Guffanti, A.A. (1990). Uncoupler-resistant mutants of bacteria. Microbiol. Rev. *54*, 52–65.

Lam, F.C., Liu, R., Lu, P., Shapiro, A.B., Renoir, J.M., Sharom, F.J., and Reiner, P.B. (2001). beta-Amyloid efflux mediated by p-glycoprotein. J. Neurochem. *76*, 1121–1128.

Lee, J.Y., Urbatsch, I.L., Senior, A.E., and Wilkens, S. (2008). Nucleotide-induced structural changes in P-glycoprotein observed by electron microscopy. J. Biol. Chem. *283*, 5769–5779.

Locher, K.P., Lee, A.T., and Rees, D.C. (2002). The *E. coli* BtuCD structure: a framework for ABC transporter architecture and mechanism. Science *296*, 1091–1098.

Long, F., Su, C.C., Zimmermann, M.T., Boyken, S.E., Rajashankar, K.R., Jernigan, R.L., and Yu, E.W. (2010). Crystal structures of the CusA efflux pump suggest methionine-mediated metal transport. Nature *467*, 484–488.

Loo, T.W., Bartlett, M.C., and Clarke, D.M. (2003). Simultaneous binding of two different drugs in the binding pocket of the human multidrug resistance P-glycoprotein. J. Biol. Chem. *278*, 39706–39710.

Lu, G., Westbrooks, J.M., Davidson, A.L., and Chen, J. (2005). ATP hydrolysis is required to reset the ATP-binding cassette dimer into the resting-state conformation. Proc. Natl. Acad. Sci. U.S.A. *102*, 17969–17974.

Marger, M.D., and Saier, M.H., J.R. (1993). A major superfamily of transmembrane facilitators that catalyse uniport, symport and antiport. Trends Biochem. Sci. *18*, 13–20.

Masaoka, Y., Ueno, Y., Morita, Y., Kuroda, T., Mizushima, T., and Tsuchiya, T. (2000). A two-component multidrug efflux pump, EbrAB, in Bacillus subtilis. J. Bacteriol. *182*, 2307–2310.

McAleese, F., Petersen, P., Ruzin, A., Dunman, P.M., Murphy, E., Projan, S.J., and Bradford, P.A. (2005). A novel MATE family efflux pump contributes to the reduced susceptibility of laboratory-derived *Staphylococcus aureus* mutants to tigecycline. Antimicrob. Agents Chemother. *49*, 1865–1871.

Morita, M., Shitan, N., Sawada, K., van Montagu, M.C., Inze, D., Rischer, H., Goossens, A., Oksman-Caldentey, K.M., Moriyama, Y., and Yazaki, K. (2009). Vacuolar transport of nicotine is mediated by a multidrug and toxic compound extrusion (MATE) transporter in Nicotiana tabacum. Proc. Natl. Acad. Sci. U.S.A. *106*, 2447–2452.

Morrison, E.A., Dekoster, G.T., Dutta, S., Vafabakhsh, R., Clarkson, M.W., Bahl, A., Kern, D., Ha, T., and Henzler-Wildman, K.A. (2012). Antiparallel EmrE exports drugs by exchanging between asymmetric structures. Nature *481*, 45–50.

Murakami, S., Nakashima, R., Yamashita, E., and Yamaguchi, A. (2002). Crystal structure of bacterial multidrug efflux transporter AcrB. Nature *419*, 587–593.

Murakami, S., Nakashima, R., Yamashita, E., Matsumoto, T., and Yamaguchi, A. (2006). Crystal structures of a multidrug transporter reveal a functionally rotating mechanism. Nature *443*, 173–179.

Naroditskaya, V., Schlosser, M.J., Fang, N.Y., and Lewis, K. (1993). An *E. coli* gene emrD is involved in adaptation to low energy shock. Biochem. Biophys. Res. Commun. *196*, 803–809.

Newstead, S., Drew, D., Cameron, A.D., Postis, V.L., Xia, X., Fowler, P.W., Ingram, J.C., Carpenter, E.P., Sansom, M.S., Mcpherson, M.J., *et al.* (2011). Crystal structure of a prokaryotic homologue of the mammalian oligopeptide-proton symporters, PepT1 and PepT2. EMBO J. *30*, 417–426.

Nikaido, H. (2009). Multidrug resistance in bacteria. Annu. Rev. Biochem. *78*, 119–146.

Ninio, S., Elbaz, Y., and Schuldiner, S. (2004). The membrane topology of EmrE – a small multidrug transporter from *Escherichia coli*. FEBS Lett. *562*, 193–196.

Nishino, K., and Yamaguchi, A. (2001). Analysis of a complete library of putative drug transporter genes in *Escherichia coli*. J. Bacteriol. *183*, 5803–5812.

Oldham, M.L., Khare, D., Quiocho, F.A., Davidson, A.L., and Chen, J. (2007). Crystal structure of a catalytic intermediate of the maltose transporter. Nature *450*, 515–521.

Omote, H., Hiasa, M., Matsumoto, T., Otsuka, M., and Moriyama, Y. (2006). The MATE proteins as fundamental transporters of metabolic and xenobiotic organic cations. Trends Pharma. Sci. *27*, 587–593.

Pao, S.S., Paulsen, I.T., and Saier, M.H., J.R. (1998). Major facilitator superfamily. Microbiol. Mol. Biol. Rev. *62*, 1–34.

Paulsen, I.T., and Lewis, K. (2002). Microbial Multidrug Efflux (Horizon Scientific Press, Norfolk).

Paulsen, I.T., and Skurray, R.A. (1993). Topology, structure and evolution of two families of proteins involved in antibiotic and antiseptic resistance in eukaryotes and prokaryotes – an analysis. Gene *124*, 1–11.

Paulsen, I.T., Brown, M.H., and Skurray, R.A. (1996). Proton-dependent multidrug efflux systems. Microbiol. Rev. *60*, 575–608.

Persidis, A. (1999). Cancer multidrug resistance. Nat. Biotechnol. *17*, 94–95.

Piddock, L.J. (2006). Multidrug-resistance efflux pumps – not just for resistance. Nat. Rev. Microbiol. *4*, 629–636.

Pinkett, H.W., Lee, A.T., Lum, P., Locher, K.P., and Rees, D.C. (2007). An inward-facing conformation of a putative metal-chelate-type ABC transporter. Science *315*, 373–377.

Polissi, A., and Georgopoulos, C. (1996). Mutational analysis and properties of the msbA gene of *Escherichia coli*, coding for an essential ABC family transporter. Mol. Microbiol. *20*, 1221–1233.

Reuter, G., Janvilisri, T., Venter, H., Shahi, S., Balakrishnan, L., and van Veen, H.W. (2003). The ATP binding cassette multidrug transporter LmrA and lipid transporter MsbA have overlapping substrate specificities. J. Biol. Chem. *278*, 35193–35198.

Rosenberg, M.F., Kamis, A.B., Callaghan, R., Higgins, C.F., and Ford, R.C. (2003). Three-dimensional structures of the mammalian multidrug resistance P-glycoprotein demonstrate major conformational changes in the transmembrane domains upon nucleotide binding. J. Biol. Chem. *278*, 8294–8299.

Rotem, D., Sal-Man, N., and Schuldiner, S. (2001). *In vitro* monomer swapping in EmrE, a multidrug transporter from *Escherichia coli*, reveals that the oligomer is the functional unit. J. Biol. Chem. *276*, 48243–48249.

Saier, M.H., Jr. (2003). Tracing pathways of transport protein evolution. Mol. Microbiol. *48*, 1145–1156.

Saier, M.H., Jr, Tran, C.V., and Barabote, R.D. (2006). TCDB: the Transporter Classification Database for membrane transport protein analyses and information. *Nucl. Acids Res.*, *34*, D181–186.

Schmitt, L., Benabdelhak, H., Blight, M.A., Holland, I.B., and Stubbs, M.T. (2003). Crystal structure of the nucleotide-binding domain of the ABC-transporter haemolysin B: identification of a variable region within ABC helical domains. J. Mol. Biol. *330*, 333–342.

Schuldiner, S., Granot, D., Mordoch, S.S., Ninio, S., Rotem, D., Soskin, M., Tate, C.G., and Yerushalmi, H. (2001). Small is mighty: EmrE, a multidrug transporter as an experimental paradigm. News Physiol. Sci. *16*, 130–134.

Sennhauser, G., Bukowska, M.A., Briand, C., and Grutter, M.G. (2009). Crystal structure of the multidrug exporter MexB from *Pseudomonas aeruginosa*. J. Mol. Biol. *389*, 134–145.

Sharom, F.J. (2008). ABC multidrug transporters: structure, function and role in chemoresistance. Pharmacogenomics *9*, 105–127.

Shilling, R.A., Venter, H., Velamakanni, S., Bapna, A., Woebking, B., Shahi, S., and van Veen, H.W. (2006). New light on multidrug binding by an ATP-binding-cassette transporter. Trends Pharmacol. Sci. *27*, 195–203.

Smith, P.C., Karpowich, N., Millen, L., Moody, J.E., Rosen, J., Thomas, P.J., and Hunt, J.F. (2002). ATP binding to the motor domain from an ABC transporter drives formation of a nucleotide sandwich dimer. Mol. Cell *10*, 139–149.

Soskine, M., Steiner-Mordoch, S., and Schuldiner, S. (2002). Crosslinking of membrane-embedded cysteines reveals contact points in the EmrE oligomer. Proc. Natl. Acad. Sci. U.S.A. *99*, 12043–12048.

Tate, C.G., Ubarretxena-Belandia, I., and Baldwin, J.M. (2003). Conformational changes in the multidrug transporter EmrE associated with substrate binding. J. Mol. Biol. *332*, 229–242.

Tombline, G., Muharemagic, A., White, L.B., and Senior, A.E. (2005). Involvement of the "occluded nucleotide conformation" of P-glycoprotein in the catalytic pathway. Biochemistry, *44*, 12879–12886.

Tseng, T.T., Gratwick, K.S., Kollman, J., Park, D., Nies, D.H., Goffeau, A., and Saier, M.H., J.R. (1999). The RND permease superfamily: an ancient, ubiquitous and diverse family that includes human disease and development proteins. J. Mol. Microbiol. Biotechnol. *1*, 107–125.

Tsuda, M., Terada, T., Mizuno, T., Katsura, T., Shimakura, J., and Inui, K. (2009). Targeted disruption of the multidrug and toxin extrusion 1 (Mate1) gene in mice reduces renal secretion of metformin. Mol. Pharmacol. *75*, 1280–1286.

Ubarretxena-Belandia, I., Baldwin, J.M., Schuldiner, S., and Tate, C.G. (2003). Three-dimensional structure of the bacterial multidrug transporter EmrE shows it is an asymmetric homodimer. EMBO J. *22*, 6175–6181.

van Veen, H.W., Callaghan, R., Soceneantu, L., Sardini, A., Konings, W.N., and Higgins, C.F. (1998). A bacterial antibiotic-resistance gene that complements the human multidrug-resistance P-glycoprotein gene. Nature *391*, 291–295.

Wang, X., Karbarz, M.J., Mcgrath, S.C., Cotter, R.J., and Raetz, C.R. (2004). MsbA transporter-dependent lipid A 1-dephosphorylation on the periplasmic surface of the inner membrane: topography of francisella novicida LpxE expressed in *Escherichia coli*. J. Biol. Chem. *279*, 49470–49478.

Ward, A., Reyes, C.L., Yu, J., Roth, C.B., and Chang, G. (2007). Flexibility in the ABC transporter MsbA: Alternating access with a twist. Proc. Natl. Acad. Sci. U.S.A. *104*, 19005–19010.

Watkins, R.E., Wisely, G.B., Moore, L.B., Collins, J.L., Lambert, M.H., Williams, S.P., Willson, T.M., Kliewer, S.A., and Redinbo, M.R. (2001). The human nuclear xenobiotic receptor PXR: Structural determinants of directed promiscuity. Science *292*, 2329–2333.

Woebking, B., Reuter, G., Shilling, R.A., Velamakanni, S., Shahi, S., Venter, H., Balakrishnan, L., and van Veen, H.W. (2005). Drug-lipid A interactions on the *Escherichia coli* ABC transporter MsbA. J. Bacteriol. *187*, 6363–6369.

Wood, S., Sebastian, K., and Scherr, S.J. (2000). Pilot Analysis of Global Ecosystems: Agroecosystems (World Resources Institute, Washington, D.C.).

Yerushalmi, H., Lebendiker, M., and Schuldiner, S. (1996). Negative dominance studies demonstrate the oligomeric structure of EmrE, a multidrug antiporter from *Escherichia coli*. J. Biol. Chem. *271*, 31044–31048.

Yin, Y., He, X., Szewczyk, P., Nguyen, T., and Chang, G. (2006). Structure of the multidrug transporter EmrD from *Escherichia coli*. Science *312*, 741–744.

Yuan, Y.R., Blecker, S., Martsinkevich, O., Millen, L., Thomas, P.J., and Hunt, J.F. (2001). The crystal structure of the MJ0796 ATP-binding cassette. Implications for the structural consequences of ATP hydrolysis in the active site of an ABC transporter. J. Biol. Chem. *276*, 32313–32321.

Zaitseva, J., Jenewein, S., Jumpertz, T., Holland, I.B., and Schmitt, L. (2005). H662 is the linchpin of ATP hydrolysis in the nucleotide-binding domain of the ABC transporter HlyB. EMBO J. *24*, 1901–1910.

Zhang, Z., Ma, C., Pornillos, O., Xiu, X., Chang, G., and Saier, M.H., J.R. (2007). Functional characterization of the heterooligomeric EbrAB multidrug efflux transporter of *Bacillus subtilis*. Biochemistry *46*, 5218–5225.

Zheleznova, E.E., Markham, P.N., Neyfakh, A.A., and Brennan, R.G. (1999). Structural basis of multidrug recognition by BmrR, a transcription activator of a multidrug transporter. Cell *96*, 353–362.

Zhou, Z., White, K.A., Polissi, A., Georgopoulos, C., and Raetz, C.R. (1998). Function of *Escherichia coli* MsbA, an essential ABC family transporter, in lipid A and phospholipid biosynthesis. J. Biol. Chem. *273*, 12466–12475.

The Machinery and Mechanism of Multidrug Efflux in Gram-negative Bacteria

3

Dijun Du, Henrietta Venter, Klaas M. Pos and Ben F. Luisi

Abstract

The Gram-negative bacteria are enveloped with a protective double layer of lipid membranes, across which they expel noxious compounds by active transmembrane transport. Several classes of membrane-spanning proteins mediate the transport and thereby confer the bacteria with the capacity to occupy hazardous ecological niches or to evade the cytotoxic effects of antimicrobial compounds. The proteins can be grouped into diverse structural families, but all drive transport uphill energetically, against concentration gradients, using the free energy of the electrochemical gradient or of ATP hydrolysis. One group of transmembrane proteins forms an assembly that spans the inner and outer membranes and the periplasm. The inner membrane component of the assembly recognizes substrates for transport and harnesses energy for the processes, whereas a major role of the periplasmic adapter is to connect the inner membrane component to the outer membrane channel to safeguard the transport of the captured substrate across the outer membrane. The outer membrane component has an adjustable aperture that may be opened through interaction with its partner pump proteins. Using the AcrA/AcrB/TolC assembly of *Escherichia coli* as a paradigm efflux machine, we describe the current knowledge of the pump components and speculate on how they might interact in an allosteric manner during the transport process.

Introduction

When considering the function of multidrug efflux pumps, it may be useful to reflect on the remarkable evolutionary history of the prokaryotes. Over the course of several billion years, there has been ceaseless competition within and between bacterial species, and relentless pressure to cope with infiltration and subversion by phages. Over the eons, many species have evolved synthetic pathways that generate toxic compounds that could be used to defend established ecological niches or as offensive weapons for invading and dominating new ones. It seems likely that the process has generated, cumulatively, astronomical numbers of compounds. Consider, for example, the modular polyketide synthase assemblies: these comprise distinct modules organized like a conveyer belt and are capable of generating staggering combinatorial diversity through simple shuffling of chemical modules, and the chemical space explored by this modular design in the course of evolution is likely to have been immense. The continuous presence of such toxic compounds is envisaged to sustain great selective pressure to counter their effects, for instance by inactivating modifications or simply by driving the compounds from intracellular locations (D'Costa *et al.*, 2011). The sheer diversity of the compounds encountered means that the protective systems must have some capacity for broad recognition. Additionally, the potentially small size of many of these compounds means that recognition must be achieved with an economy of interactions.

With the advent of clinically applied antibiotics over the last 70 years, the well-tested and proven mechanisms for evading environmental toxins have found their way to the transport of drugs, which after all resemble compounds like those encountered in the course of bacterial evolution. Five families of transporters have been identified that contribute to multidrug resistance in prokaryotes. These can be categorized into primary transporters, which use the energy of ATP binding and hydrolysis to drive transport, and secondary transporters that harness the energy of electrochemical potential. Examples of primary transporters involved in multidrug efflux are members of the ATP-binding cassette (ABC) superfamily, and the secondary transporters are the major facilitator superfamily (MFS) and the (super)families known as 'small multidrug resistance' (SMR), 'resistance/nodulation/cell division' (RND), and 'multi antimicrobial extrusion' (MATE). Homologues of these proteins have been identified also in eukaryotic organisms.

The envelope of Gram-negative bacteria contains two membrane bilayers of distinct lipid composition, and while this barrier can retard the influx of many toxic compounds, it cannot completely prevent some from entering. The multidrug transporters counter this breach by forming a complex that actively transports antibiotics back across the two membrane layers and the intervening periplasm (Nikaido, 2001). Generally these transport systems consist of an outer membrane protein (OMP), such as TolC (Koronakis *et al.*, 2000), a membrane fusion protein (MFP), such as AcrA (Mikolosko *et al.*, 2006), or MexA (Akama *et al.*, 2004; Higgins *et al.*, 2004), and an inner-membrane protein (IMP), usually a MFS (Furukawa *et al.*, 1993; Lee *et al.*, 1999), RND (Mokhonov *et al.*, 2005) or ABC (Rouquette-Loughlin *et al.*, 2005) family transporter. These proteins form an energy-transducing tripartite multidrug efflux pump. The pumps endow the bacteria with the capacity to endure a wide range of toxic compounds, including detergents, bile salts, complex aromatic dyes, organic solvents and chemically diverse antibiotics.

Here, we will describe the structure and function of the AcrA-AcrB-TolC assembly of *Escherichia coli* as a paradigm system of the tripartite pumps. We will explain the structure of the individual components of the pump assembly; discuss ideas about how is it assembled, and comment on how it might work. Since research on this topic is still at an early stage (but rapidly progressing), hypotheses on any of the pump systems are still under severe but lively debate among the efflux pump research community.

Structure and function of the components

The transmembrane protein AcrB: a member of the RND family

As mentioned, the inner membrane component of the efflux pump, AcrB, is a member of the 'resistance/nodulation/cell division' (RND) family. This extensive group takes its descriptive name after the diverse phenotypes associated with mutations in some of its members. Family members include MexB from the pathogen *Pseudomonas aeruginosa* and the copper transporter, CusA, which has recently been characterized structurally (Long *et al.*, 2010).

AcrB is a large protein by *E. coli* standards, comprising 1049 amino acids. Numerous crystallographic studies at high resolution have revealed in stereochemical detail the quaternary and tertiary structure, identified the boundary of the transmembrane domain, and discovered pockets for ligand binding. AcrB has 12 transmembrane helices and forms a trimer, with extensive extra-membrane domains. Roughly half the mass of the protein resides in the periplasm (Fig. 3.1). This periplasmic portion forms a compact, multidomain fold made from a region linking transmembrane helices TM1 and TM2 and the pairs TM7 and TM8, including a remarkable two-stranded extension from each protomer that interacts with the adjacent protomer and resemble the feeding of a belt into a buckle. The subdomains in the periplasm are commonly referred to as PN1, PN2, PC1, PC2, DN and DC. The AcrB protomer has an internal structural repeat, suggesting that the protein most likely originated from a gene duplication and fusion at some stage of its evolutionary history. AcrB extends roughly 70 angstroms into the periplasm from the cytoplasmic membrane, but this is not sufficiently long to reach the outer membrane, and it probably meets its TolC partner somewhere near the middle of the periplasm. The DN and DC subdomains form a funnel shape that is the likely docking site

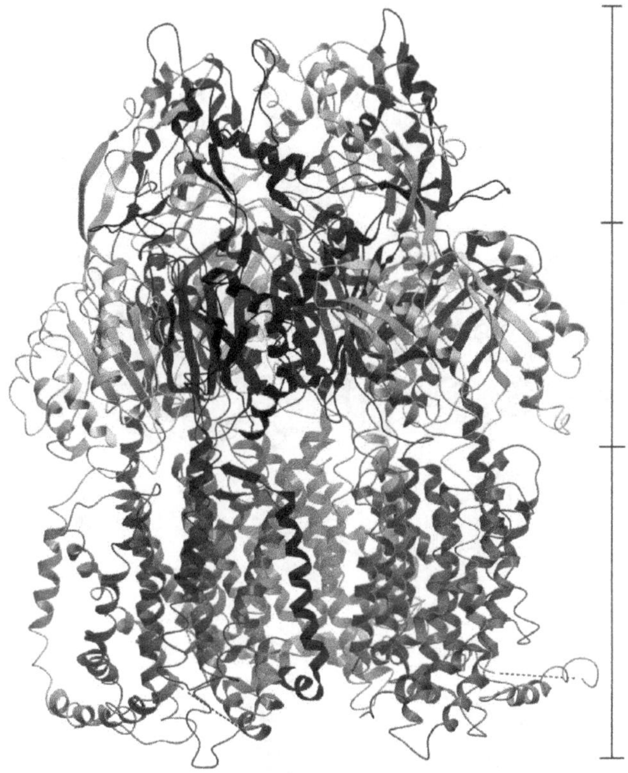

Figure 3.1 The inner membrane efflux pump component AcrB. The protein is a homotrimer, and the 3-fold axis is oriented vertically. This symmetry is lost during the transport process (see Figure 3.2).

for TolC, described further below. CusA shares a very similar fold and quaternary subunit interfaces with AcrB, and it is likely that these features are common to many other RND proteins.

In the absence of ligand, the homotrimeric AcrB has three-fold rotational symmetry. Although the subunits are chemically identical, in the sense that they all have the same amino acid sequence, they can become conformationally non-equivalent with ligand binding (Fig. 3.2). Crystal structures of AcrB reveal an asymmetry, indicating three states, referred to as Loose (L), Tight (T) and Open (O) (Murakami *et al.*, 2006; Seeger *et al.*, 2006; Sennhauser *et al.*, 2007). All protomers face onto a funnel-like opening on the central trimeric pseudo-axis; however, only one protomer, the Open protomer, has access to the funnel at a time. In the L monomer, there is access to a proximal pocket, which can hold a ligand, and a flexible loop that gates drug entry to a distal pocket and is present only in the T state. The distal pocket is located near the funnel but barred by the PN1 subdomain of the neighbouring L monomer. Drug release to the funnel occurs by simultaneous closure of the distal pocket and unbarring the exit as is the case in the Open state. In this way, there is sequentially movement of large mass ligands from the proximal pocket, to the distal pocket, and then into the funnel through successive stages of the transport process. The subunits switch states with obligatory coupling: the T → O transition requires entry of a second drug molecule into the proximal pocket of the neighbouring L monomer. Evidence suggests that these states correspond to substrate access, binding and extrusion for the L, T and O protomers. The T state subunit has a deep hydrophobic pocket and the L state has an access pocket, with a switch-loop that seems to be a gate that controls passage between these pockets (Nakashima *et al.*, 2011; Eicher *et al.*, 2012).

Evidence supporting this model comes from studies of a covalently linked AcrB trimer that carries two active and one inactive site (Takatsuka *et al.*, 2009). Furthermore, the transport activity

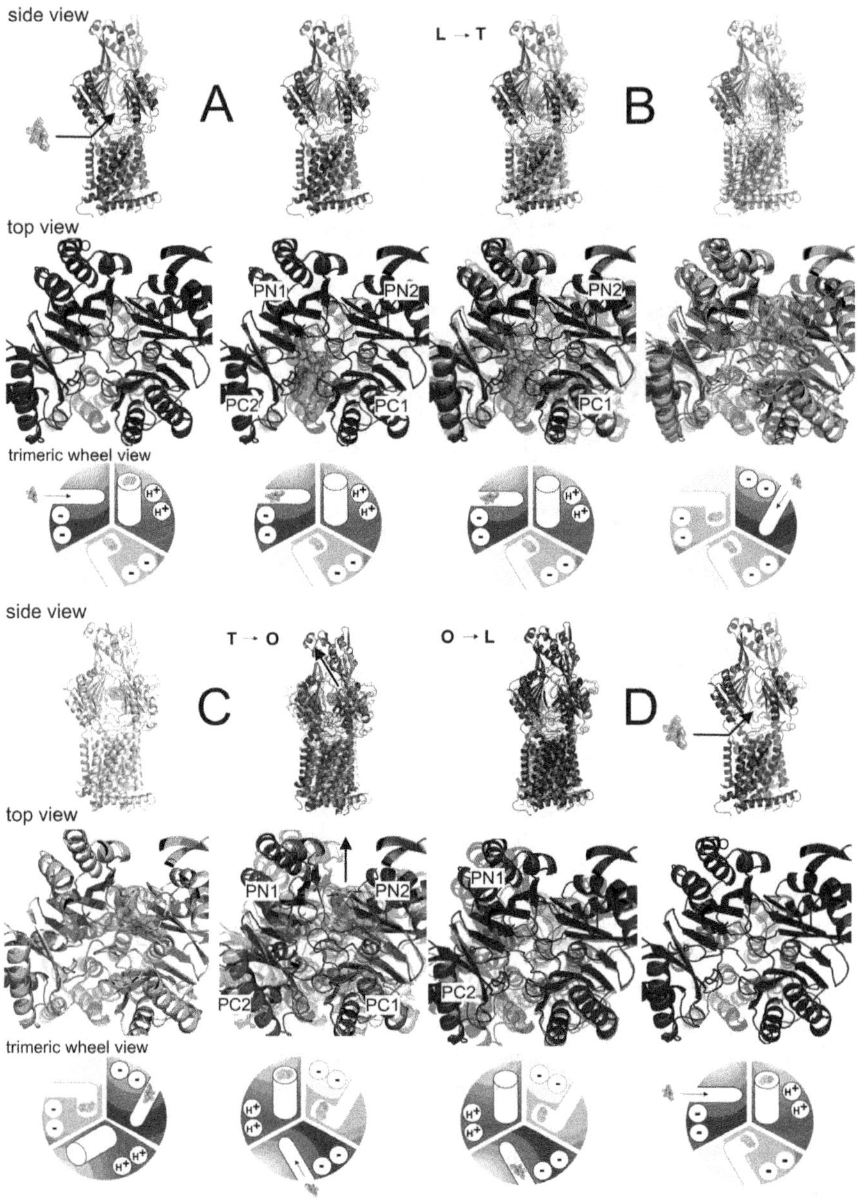

Figure 3.2 Stepwise AcrB-catalysed drug transport mediated by cyclic conformations loose (L, blue), tight (T, yellow) and open (O, red). Shown are four panels (A–D) indicating the sequential drug transport steps and conformational cycling of one AcrB monomer within the trimer. The panels show in the upper row a side view of the monomer orientated such that the membrane plane is horizontal, and the view is towards the PC1 and PC2 subdomains and the access pocket region. The second row is a view on the porter domain perpendicular to the membrane plane. The third row displays in a trimeric wheel view the state of the monomer from rows one and two (upper left monomer in the wheel, L in blue, T in yellow and O in red) shown in (A–D) in relation to the other monomers in the AcrB trimer. Substrate (doxorubicin) is depicted in green. The putative protonation state of D407 and D408 inside the transmembrane domain is indicated (L and T are deprotonated, O is protonated). (A) Initial drug uptake via the access pocket in the L monomer. (B) Transfer of the drug from the low affinity access pocket to the high affinity deep binding pocket during the L to T transition. (C) Directional drug transport during the T to O transition. Subdomains PC1 and PC2 close the lateral access and prevent backsliding of the drug, which is extruded through tunnel 3. (D) Resetting the monomer by the O to L transition to reinitialize drug uptake via the access pocket (Taken from Eicher *et al.*, 2012). A colour version of this figure is located in the plate section at the back of the book.

is nearly abolished when disulphides are introduced that restrict movement of the periplasmic subdomains, and the activity can be restored with reducing agent that breaks this constraining tie (Takatsuka *et al.*, 2007; Seeger *et al.*, 2008). Additional support for the asymmetry comes from observations of the binding to AcrB of DARPins, which are engineered protein-binding modules. Binding is limited to two at saturation, not three, suggesting that there is some conformation adjustment propagated in the trimer that prevents binding of the third DARPin (Sennhauser *et al.*, 2007; Brandstaetter *et al.*, 2011). This system has striking analogy with F_1F_0 type ATPase, hexameric ring shaped helicases and other systems in which the rotational symmetry becomes broken and the subunit interfaces become non-equivalent. The obligate coupling of subunits in AcrB accounts for the noted cooperativity in the efflux rates of beta-lactam antibiotics (Nagano *et al.*, 2009; Lim *et al.*, 2010).

One puzzling feature of AcrB is its capacity to transport a broad spectrum of compounds (Fig. 3.3). These include molecules such as erythromycin, a macrocyclic lactone belonging to the class of natural products synthesized by the modular polyketide synthases. With masses of 734 daltons and 823 daltons, respectively, erythromycin and rifampicin have the bulk and size of a compact tripeptide or tetrapeptide. These larger ligands (Fig. 3.4) appear to be recognized within the proximal access pocket, while smaller molecules, such as doxorubicin or minocycline (Fig. 3.5), are not, but are readily adapted in the distal pocket (Nakashima *et al.*, 2011; Eicher *et al.*, 2012).

Polyspecificity is an intriguing facet of drug efflux proteins, and having two pockets contributes to the ability of the pump to deal with a wide spectrum of transport substrates. Transported substrates seem to have little in common other than containing planar hydrophobic and often polar/charged moieties. They also have vast differences in substrate sizes from smallest (hexane) to largest (rifampicin). There appears to be plasticity in the binding sites that can mould to the ligands, and there is a redundancy in aromatic (specifically Phe) residues. As a result many amino acid residues within the binding pocket and even ones that are directly interacting with substrates can be mutated without affecting transport at all

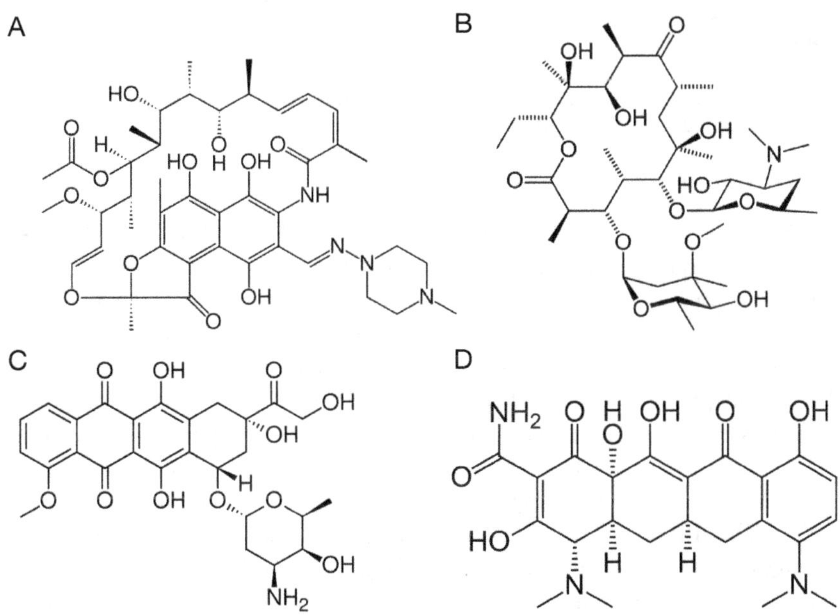

Figure 3.3 Ligands bound to the AcrB pocket, large and small. (A) The chiral rifampicin ~823 daltons, a popular reagent for terminating transcription in bacteria. (B) The chiral polyketide erythromycin, ~734 daltons. The smaller ligands: (C) Doxorubicin, a chiral anthracycline with mass ~534 daltons and (D) minocycline, a chiral tetracycline with mass ~457 daltons.

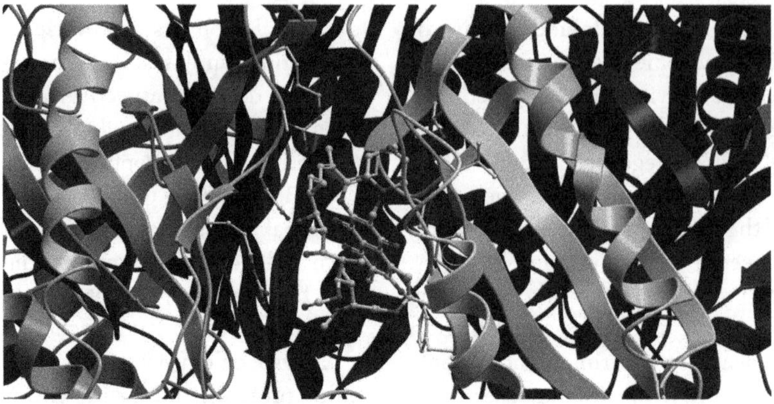

Figure 3.4 Rifampicin bound to the access pocket in the periplasmic domain of the AcrB L conformer (cartoon representation). Rifampicin is displayed as a ball-and-stick model.

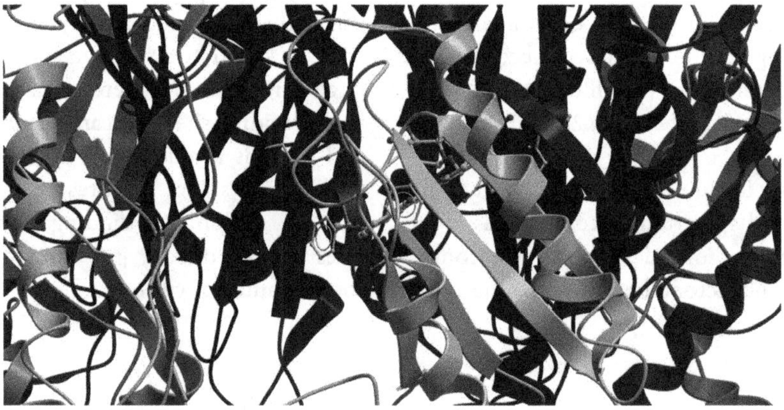

Figure 3.5 Doxorubicin bound to the deep binding pocket of the AcrB T conformer (in cartoon representation). Doxorubicin is displayed in ball-and-stick representation.

(Bohnert *et al.*, 2008; Vargiu *et al.*, 2011). This has obvious therapeutic consequences which we will return to later in the chapter.

Recently, a small protein has been identified in a co-precipitation assay as a potential partner of the AcrB/AcrA/TolC efflux pump that may enhance the activity of the assembly for particular substrates (Hobbs *et al.*, 2012). Potential substrates include the antibiotics puromycin, chloramphenicol and tetracycline. The effector protein, AcrZ, is 49 residues in length and is predicted to interact with AcrB. As AcrZ has a strongly predicted transmembrane spanning region, it may be associated with AcrB in the membrane and possibly allosterically modulate the protein.

AcrA: a multimodular periplasmic protein

Like its partner AcrB, the periplasmic AcrA is representative of a large family of homologous proteins. Crystal structures are available at good resolution of AcrA and its homologues. The AcrA monomer is composed of four linearly arranged subdomains with flexible tethers: these are the membrane proximal domain, a β-barrel domain, a lipoyl domain and an α-helical hairpin domain (Fig. 3.6). The flexibility of the intradomain linkers enables the domains to manoeuvre so as to dock and mould to the surfaces of cognate partners. We will return to this issue of AcrA 'conformability' in the section to follow on subassemblies. One astonishing feature of the structure is that the

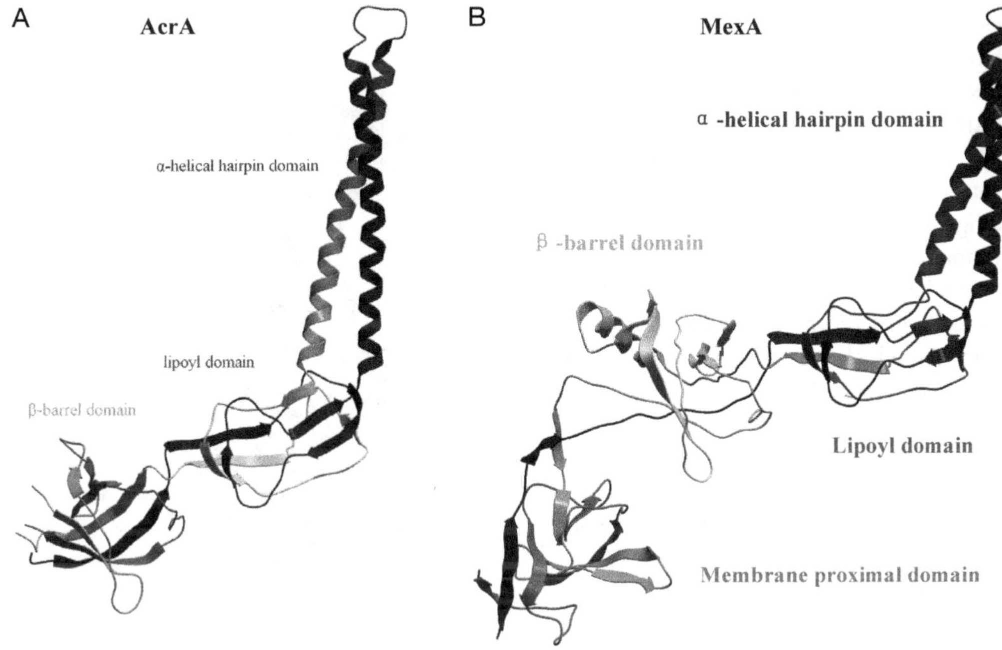

Figure 3.6 (A) The periplasmic efflux pump component AcrA. Residues 45–312 are shown, and the membrane proximal domain is missing from this structure. (B) The membrane proximal domain portion of the structure has been resolved in the structure of the AcrA homologue MexA from *Pseudomonas aeruginosa*.

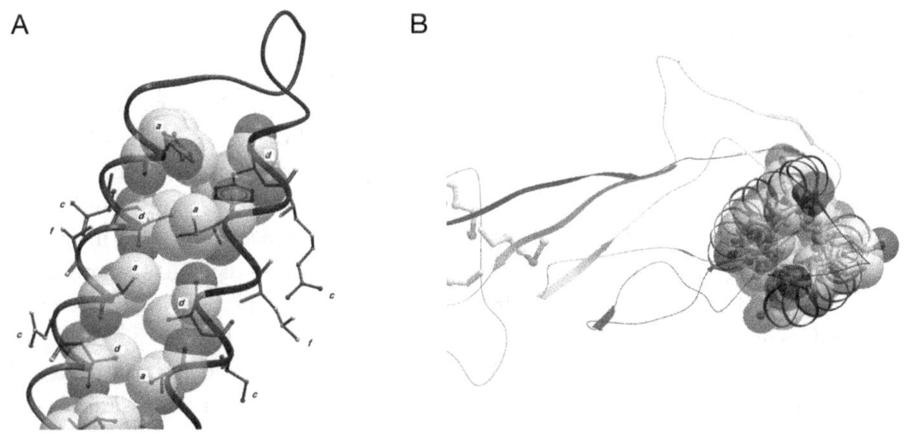

Figure 3.7 The helical hairpin of AcrA, showing the knobs-into-holes interactions of the coiled-coil. (A) View of the hairpin with the coiled-coil axis inclined to the vertical axis. Side chains at the *a* and *d* positions of the heptadic repeat are labelled. (B) A view along the coiled-coil axis.

termini are close in space, despite the elongated shape and modular organization of AcrA. The consequence is that the modules are folded from discontinuous portions of the linear sequence, with the exception of the α-helical hairpin domain that lies at the mid-point of the protein.

The α-helical hairpin domain forms a conventional anti-parallel coiled-coil, with five heptad repeats in each of its helices (Fig. 3.7). Throughout the family of periplasmic adaptors, there is a conserved pattern of large, hydrophobic side chains in the ***a*** and ***d*** positions of the heptadic repeat, Ala residues in the *f* position and hydrophilic residues such as Ser or Glu in the *c* position. The ***a*** and ***d*** positions form knobs and holes to pack together in a hydrophobic interface that is maintained along the long co-axial length of the coiled-coil (Johnson *et al.*, 1999).

A hinge between the α-helical hairpin and lipoyl domain is composed of residues 99–106 in α-helix 1 and residues 169–173 in α-helix 2. The helical winding of the hinge provides flexibility of the protein, which may play a functional role in coupling between AcrA conformations and TolC channel opening (Mikolosko *et al.*, 2006). Furthermore, the coiled-coil is expected to be able to adjust its twist a little to affect a superhelical trajectory, and thus would also permit some shape-moulding to dock with a partner.

The structure of the membrane proximal domain of AcrA is not available yet; however, this domain in MexA, a homologue of AcrA, has been solved recently, and assembles a β-roll structure contiguously from the β-barrel domain on a β-ribbon linker (Fig. 3.6) (Symmons *et al.*, 2009).

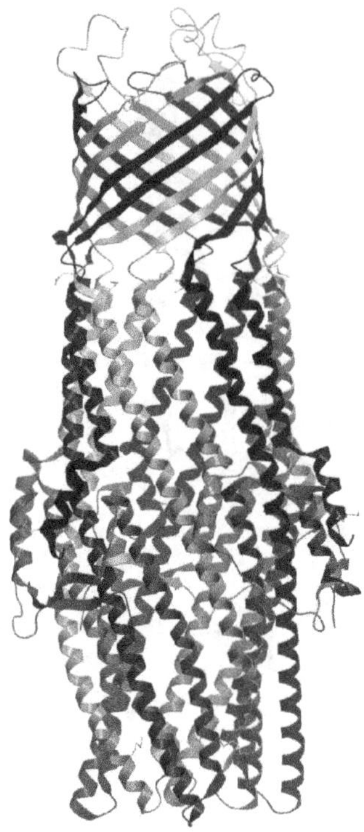

Figure 3.8 The outer membrane efflux pump component TolC.

Transmembrane protein TolC: a channel with an adjustable aperture

The outer membrane component of the pump is TolC, and, like AcrA and AcrB, there are many homologues identified amongst the Gram-negative bacteria; however, none have been found thus far in the Gram-positive bacteria, so its origin is likely to have originated close to the time of divergence of these lineages. As far as we are aware, no homologues have been identified in mitochondria. TolC is not only a partner of AcrA/AcrB, but of several other pumps with an RND inner membrane component, as well as MFS and ABC transporters. TolC thus experiences a wide diversity of transported species, ranging from metal ions to 100 kDa protein toxins, as part of the Type I secretion pathway (this uses an ABC partner).

Like AcrB, TolC is also a homotrimer and has an internal structural repeat (Fig. 3.8). It too has therefore arisen from a gene duplication event. The membrane-spanning portion of TolC has 12 strands that run anti-parallel. The strands form a beta-barrel with high inclination with respect to the axis perpendicular to the membrane plane. Because the strands are so inclined, they must twist and curve to form the barrel shape (the best way to see this is to imagine placing a straight piece of tape on the surface of a cylinder at an incline). The neighbouring strands twist with a right-handed sense, as they do in the beta sheets of globular proteins. This geometrical requirement for twisting places constraints on the pattern of amino acids on the 'inside' of the barrel and its 'outside' – i.e. lipid facing surface. There will also be constraints on the residues on the lipid-exposed surface, with preferences for aromatics near the periphery.

The portion outside the membrane is almost entirely α-helical. Notably, there is a putative anion-binding pocket at the junction of the β-barrel and α-helical domain which may contribute to stabilization of the structure (Bavro *et al.*, 2008). One portion of the α-helical domain forms a near-perfect cylinder, while the portion most distal to the outer membrane has six conventional coiled coils, constructed from heptadic repeats similar to those mentioned in the subsection on AcrA. These conventional coiled-coils taper nearly to a close, which is in agreement with the low ionic conductance of the channel (Koronakis *et al.*, 2000). The constriction is supported by buried salt bridges and van der Waals interactions between the side chains of neighbouring coiled-coils, both within and between the protomers (Bavro *et al.*, 2008). As the aperture is restrictive even for ions, it is clear that the isolated TolC structure represents a closed state that must be opened during the transport process.

Like the strands in the transmembrane β-barrel, the helices in the cylindrical portion of TolC must also twist and curve to maintain a cylindrical shape. One important consequence of this requirement is that the helices in the cylindrical portion also pack with heptadic repeats intermeshed, knobs-into-holes style, but on two faces simultaneously. Thus, this portion forms a coiled-coil, but with two heptadic repeats. This pattern has been proven very useful for elaborating geometry-based rules for the structural determinants of the fold (Calladine *et al.*, 2001) and for a pattern that has been useful for identifying remote homologues (Sharff *et al.*, 2001). Comparison of the crystal structures of diverse TolC homologues shows that the fold of the protomer and quaternary structure is well maintained (*Vibrio cholerae* VceC, *Pseudomonas aeruginosa* OprM, *E. coli* TolC).

Insight into how the channel might open has come from crystal structures of mutants in which the salt bridges have been disrupted by mutation (Bavro *et al.*, 2008; Pei *et al.*, 2010). These structures reveal a lateral displacement of the coiled-coils away from the central molecular axis, although not to the extent that is required to pass small compounds. Additional structural changes are likely to be trigged by interactions with AcrB and/or AcrA.

Models of the pump subassemblies

Biochemical evidence indicates that the three components of tripartite multidrug efflux pumps form a stable inter–membrane complex even in the absence of substrates and proton-motive force (Tikhonova *et al.* 2004; Touze *et al.* 2004; Gerken *et al.* 2004). Although the crystal structures of all three individual components of tripartite multidrug efflux pump are available, the question remains as to how the components assemble to form a functional pump.

Numerous experimental results have provided clues as to how the TolC outer membrane component might interact with AcrA. The hybrid system of *E. coli* AcrA/AcrB with the *Pseudomonas aeruginosa* TolC homologue, OprM, is not functional; however, exchange of the hairpin domain of AcrA with that of MexA restores the function of the pump, indicating that the MexA hairpin domain is involved in a key interaction with OprM. Bifunctional cross-linking experiments map an interaction face between the inner and outer helices H7/H8/H3 of TolC and the N-terminal α-helix 1 of AcrA (Symmons *et al.*, 2009). To map the interactions of AcrB and TolC, cysteine residue pairs were introduced in the two proteins and examined using *in vivo* disulphide cross-linking. The formation of spontaneous disulphide bonds between positions 255, 256, 257 in AcrB and position 147 in TolC suggests that the hairpins of AcrB and the coiled-coils of TolC are in close proximity, supporting a model for direct docking (Tamura *et al.*, 2005). The membrane proximal domain of AcrA plays an important role in the assembly and function of AcrA/AcrB/TolC efflux pump. Evidence comes from the behaviour of an $AcrA_{G363C}$ mutant in the membrane proximal domain, which interacts with both AcrB and TolC but fails to properly assemble into a functional complex, and therefore dramatically impairs the multidrug efflux activity of AcrA/AcrB/TolC (Ge *et al.*, 2009).

A model of sequential assembly of AcrA/AcrB/TolC pump has been proposed, which holds that AcrB docking of TolC contributes to the transition from its closed to the open state (Fig. 3.9) (Bavro *et al.*, 2008). AcrB docking results in the outward movement of TolC's aperture-restricting helical pairs, causing a deepening of the intra-protomer grooves of the TolC helices, into which the coiled-coil region of AcrA can be accommodated. Support for this model comes from finding that the vancomycin sensitivity of a TolC mutant carrying substitutions in the periplasmic turn region between helices was suppressed by AcrA. The interpretation is

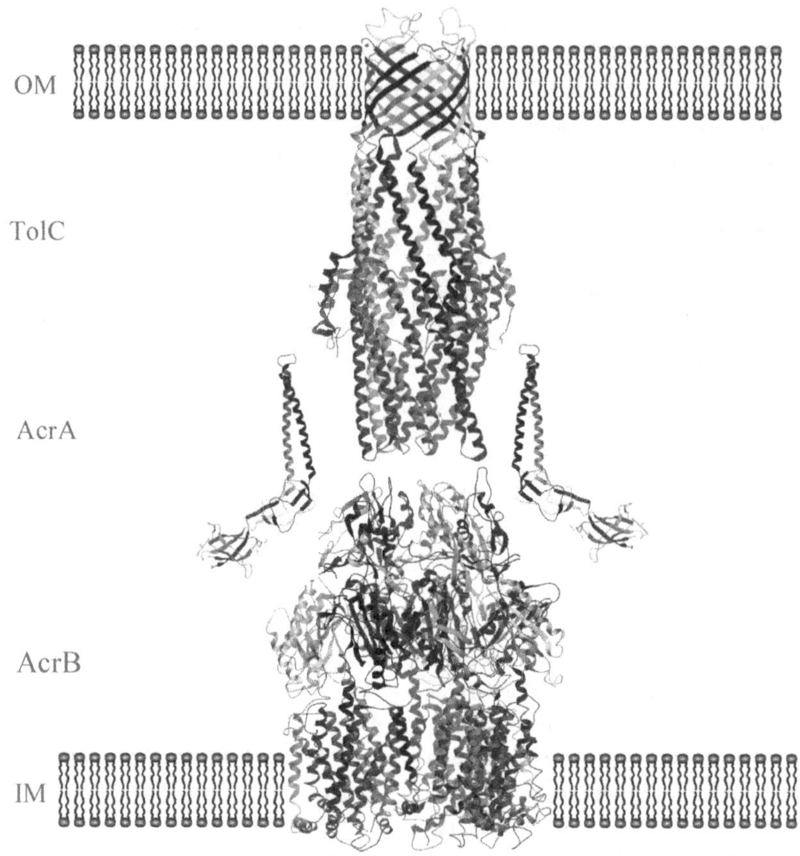

Figure 3.9 Schematic of the assembly. The subunit stoichiometry of the assembly has not been established unequivocally. The membrane-proximal domain is not included in the structure of the AcrA. A colour version of this figure is located in the plate section at the back of the book.

that AcrA dilates the TolC aperture/channel in an AcrB-dependent manner, and it underscores the importance of the TolC turn in the functional assembly of the tripartite efflux complex and AcrA in transitioning TolC from its closed to open state (Weeks *et al.*, 2010). The assembly of the AcrA/AcrB/TolC complex has been characterized by surface plasmon resonance, and these data indicate that AcrA interacts AcrB with nanomolar affinity and that TolC binds AcrB independently from AcrA (Tikhonova *et al.*, 2011). The preassembly of AcrA-TolC in solution interferes with AcrA binding to AcrB, and the conformation of the membrane proximal domain of AcrA is critical for the formation of the AcrA/AcrB/TolC assembly. The interactions between the components are pH dependent. These results are also consistent with the model of sequential assembly of the AcrA/AcrB/TolC pump proposed by Bavro *et al.* (2008).

Different models have been proposed for the AcrA/AcrB/TolC assembly based on the crystal structures and biochemical evidence. An IMP: MFP: OMP ratio of 3:3:3 has been proposed based on bifunctional chemical crosslinking data (Symmons *et al.*, 2009). Supporting this model, cross-linking of *in vivo* complexes has identified MFP trimers in both drug efflux and protein export pumps (Thanabalu *et al.*, 1998; Zgurskaya *et al.*, 2000).

Evidence also supports an IMP: MFP: OMP ratio of 3:6:3 (Janganan *et al.* 2011). The crystal structure of the *E. coli* CusBA heavy metal efflux complex supports a 3:6 complex of IMP:MFP (Fig. 3.10). Mutagenesis and crosslinking data indicate that each TolC protomer interacts with

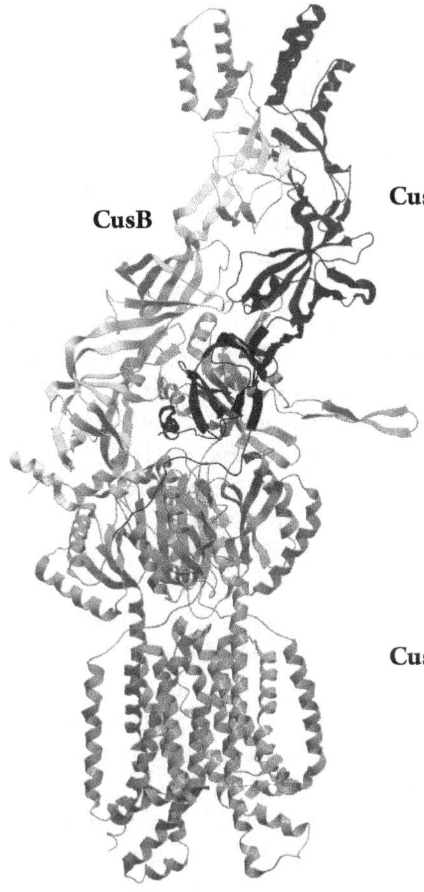

Figure 3.10 The crystal structure of the CusB/CusA complex. The CusB:CusA stoichiometry is 2:1. Three CusB/CusA (2:1) monomers form a complex with trimeric CusC to an active complex *in vivo*. The two CusB molecules are shown with gray and black tracing.

two AcrA molecules, suggesting a 3:6:3 model for the AcrB/AcrA/TolC pump (Stegmeier *et al.*, 2006).

In *Pseudomonas aeruginosa,* an efflux pump analogous to the *E. coli* AcrA/AcrB/TolC system is composed of the proteins MexA, MexB and OprM. Whole cell quantitative immunoblotting assays indicated that wild type *Pseudomonas aeruginosa* contains roughly 2500, 1000 and 1200 copies of MexA, MexB and OprM, respectively, suggesting a stoichiometry of 3:6:3 for the MexB/MexA/OprM efflux pump (Narita *et al.*, 2003). In the TriABC-OpmH triclosan extrusion system in *Pseudomonas aeruginosa,* functional activity of the pump requires two different MFPs, also suggesting a 3:6:3 stoichiometry where the two different MFPs occupy two non-equivalent surface grooves on the OpmH (Mima *et al.*, 2007; Misra and Bavro, 2009).

Akama *et al.* (2004) devised homology models for MexB and OprM based upon the structures of AcrB and TolC. In the proposed model, the helical hairpins of three pairs of MexA dimers are inclined with the surface of OprM with an angle of about 20°. The lipoyl domain fits within the gap between MexB-OprM contact regions. The β-barrel domain and the membrane proximal domain may have interacted with the MexB trimer. The IMP: MFP: OMP ratio was predicted to be 3:6:3 in this model.

The transport process

The pathway

The drug efflux pathway through the periplasmic domains of AcrB has now been very well established and characterized. Substrates are either directly captured from the periplasm/outer leaflet by AcrA/AcrB/TolC or first moved from the cytoplasm/inner membrane to the periplasm/outer leaflet by multidrug transporters of the MFS, SMR or MATE family. The question remains if all drugs are displaced from the periplasm or if substrates could also be removed directly from the cytoplasm/inner cytoplasmic membrane by the AcrA/AcrB/TolC complex. If drugs are captured solely from the periplasm/outer leaflet, the membrane domain of AcrB would only act as an energy transducer which utilize the proton motive force to cause the molecular changes in the periplasmic domain, which are then relayed to AcrA and TolC in a way similar to the workings of the TonB energy-dependent uptake system. So far, there is no evidence of amino acids in the transmembrane regions of AcrB and related MDR transporters that interact with drugs. However, recently two conserved phenylalanine residues that are aligned around the cytoplasmic side of the central cavity of MexB, were shown to be involved in substrate recognition from the cytoplasm (Ohene-Agyei *et al.*, 2012). In addition, MexB reconstituted into proteoliposomes can transport the dye Hoechst 33342 in the absence of MexA and OprM, and

MexB was active when expressed in a Gram-positive host (Welch *et al.*, 2010) indicating that MexB might have the ability to transport substrates from the cytoplasm. Independent substrate transport by other members of the RND family when reconstituted in proteoliposomes has also been observed; i.e. for the cation/proton antiporter CzcA and for the Cu^+ and Ag^+ transporter CusA (Goldberg *et al.*, 1999; Long *et al.*, 2010). Metal ion efflux by CusA occurs from the cytoplasm through the central pore (Long *et al.*, 2010). It is not clear yet if a similar pathway exists for AcrB and a more thorough investigation of this putative pathway is needed.

Energy transduction by AcrB

The energy for driving the mechanical movements required for the drug efflux originates from the electrochemical gradient of protons, which flow from the periplasm to the cytoplasm. How are protons shuttled, and how is this transduced into the mechanical movements associated with sequential movement of drug from one pocket to the next? The protons are likely to be conducted along titratable groups, and the available evidence indicates that two adjacent aspartates, D407 and D408, located within the transmembrane domain, are part of the pathway, as well as lysine K940 (Su *et al.*, 2006; Seeger *et al.* 2009). Intriguingly, a fourth residue located closer towards the cytoplasmic border of the transmembrane domain, R971, has been shown to be absolutely essential for the activity of the efflux pump as well. The four residues constitute a network remarkably similar to that of bacteriorhodopsin, where three aspartates (Asp85, Asp96 and Asp212), a retinal-coupled lysine side chain Schiff base (Lys216) and an arginine side chain residue (Arg82) form the basis of the proton pumping mechanism upon energization by light.

Some indications on the proton translocation events and its coupling to drug efflux can be deduced from the crystal structures of AcrB in its asymmetrical state (Murakami *et al.*, 2006; Seeger *et al.*, 2006; Sennhauser *et al.* 2007; Nakashima *et al.*, 2011; Eicher *et al.*, 2012). The three conformation states of the monomers in the asymmetrical AcrB structure are assumed to represent consecutive steps in the transport cycle of every single monomer (analogous to the catalytic 'binding change' cycle of the α/β subunits of the F_1F_o ATP synthase). In AcrB, the L conformer represents the access state of initial (high molecular mass) drug binding to the access pocket, the T conformation represents the tight binding of the drug into the deep binding pocket and the O conformer represents the extrusion state, where the bound drug has been expelled towards TolC and to the outside of the Gram-negative bacterial cell. Observation of the side chain configuration of the above mentioned essential charged/titratable residues in L, T and O states, show a remarkable similarity in the L and T states, and a remarkable dissimilarity between the T and O states (or between the O and L states). The current hypothesis suggests that the differences between the side chain geometries of the O conformation compared to the L and T conformations are due to the different protonation states of the Asp407/Asp408 side chains, i.e. protonated in the O conformation and deprotonated in the L and T states (Eicher *et al.*, 2012). The events of drug binding (in the access and deep binding pockets) and release (towards TolC) in the AcrB periplasmic domain must be sensed and transduced to the transmembrane domain and subsequent proton uptake and release likewise be transduced to the periplasmic domain.

Within the proposed directional conformational cycle from L (initial drug binding), T (tight drug binding in the deep binding pocket) and O (release of the drug towards TolC), it is postulated that the energy (proton motive force, i.e. H^+ transport)-dependent step is from the T to O transition. Exactly here we find the major conformational changes in both the periplasmic domain and the transmembrane domain. In the T state, protons might access the Asp407 and Asp408 side chains and protonate them causing all four residues (Asp407, Asp408, Lys940 and Arg971) to seek alternative electrostatic environments as the ionic network becomes disrupted. This causes the transmembrane helices to adopt new conformations accordingly which is transduced to the periplasmic domain most prominently indicated by the large coil to helix transition of transmembrane helix 8, which is coupled to the periplasmic PC2 (and PN1) subdomains (Seeger *et al.*, 2006; Pos *et al.*, 2009). Subsequent release of the drug in the O state is signalled back to the transmembrane domain causing deprotonation of the Asp407 and Asp408 side chains and return of the monomer to the L conformation (Eicher *et al.*, 2012). As stated above, it appears that cycling of the monomers through the different conformational changes is highly cooperative and that conformational changes are not only dependent on the signal transduction of conformational changes within one monomer, but that changes in adjacent monomers are most likely relevant to completing the L, T, O, L cycle. Moreover, there is likely to be a mechanical linkage leading from the protein–protein interactions at the AcrA–AcrB interfaces with protein movements causing signal transduction even to the TolC tunnel.

Substrate preferences

The RND pumps operate on diverse compounds, with a range of molecular mass, and the broad preferences originate from interactions with the binding pockets in AcrB. How are such a broad range of transported ligands recognized? What is the origin of specificity and broad preferences? The broad specificity of the pumps raises the important question of how the pump avoids the fate of inadvertently transporting molecules that are needed, for instance, metabolites, co-factors such as NAD and FAD, aromatic amino acids such as tryptophan, tyrosine and phenylalanine. Discrimination could originate in part from the substrate entry pathway via the periplasm and/or the outer leaflet of the inner-membrane which in effect, acts as a key filter for transport species. Tal and Schuldiner (2009) postulated the synergistic network of single component membrane transporters like EmrE and MdfA, with the tripartite AcrA/AcrB/TolC pump. Accordingly, for some compounds chromosomal deletion of both *emrE* and *mdfA* genes lead to the same drug-sensitive phenotype as the single chromosomal deletion of *acrB*. Notably, the single component inner membrane transporters display much more restricted substrate specificity so as to filter toxic drugs from essential (hydrophobic) metabolites during transport from cytoplasm to the periplasm. The action will result in the accumulation of a plethora of toxic compounds in the periplasm, which necessitates the broader specificity displayed by the AcrA/AcrB/TolC system. Perhaps the latter system gains some specificity if recognition occurs out of equilibrium during the compression stage of a pocket, when the channel might be a selective pore for shape or physiochemical properties. Allosteric control may also affect specificity of the system, as seen for instance with the effects of the newly discovered AcrZ protein (Hobbs *et al.*, 2012).

Conclusion and perspectives

The pump might seem a logical target for therapeutic intervention to treat bacterial infections. It must be borne in mind that using a drug to target a drug transporter is somewhat of a logical paradox; after all, the system has evolved over billions of years to avoid such targeting. The drug-binding pockets within AcrB are too promiscuous and would most likely not be good choices. On the other hand, targeting protein–protein interfaces is an interesting possibility for inhibitor studies, and this has become a developing theme in current drug design for other systems. Drugs that are tight binders and hence 'lock' the transporter in a certain conformation might still be an option for exploration.

Acknowledgements

Dijun Du and Ben Luisi are supported by the Wellcome Trust. Henrietta Venter is supported by the Royal Society. Research in the laboratory of Klaas M. Pos is supported by the German Research Foundation: DFG-EXC115 (Cluster of Excellence Macromolecular Complexes at the Goethe-University Frankfurt) and SFB 807 (Transport and Communication across Biological Membranes).

References

Akama *et al.* (2004). Crystal structure of the membrane fusion protein, MexA, of the multidrug transporter in *Pseudomonas aeruginosa*. J. Biol. Chem. *279*, 25939–25942.

Bavro *et al.* (2008). Assembly and channel opening in a bacterial drug efflux machine. Mol. Cell. *30*, 114–121.

Bohnert *et al.* (2008). Site-directed mutagenesis reveals putative substrate binding residues in the *Escherichia coli* RND efflux pump AcrB. J. Bacteriol. *190,* 8225–8229.

Brandstätter *et al.* (2011). Analysis of AcrB and AcrB/DARPin ligand complexes by LILBID MS. Biochim. Biophys. Acta *1808,* 2189–2196.

D'Costa, V.M., King, C.E., Kalan, L., Morar, M., Sung, W.W., Schwarz, C., Froese, D., Zazula, G., Calmels, F., Debruyne, R., *et al.* (2011). Antibiotic resistance is ancient. Nature *477,* 457–461.

Eicher *et al.* (2012). Transport of drugs by the multidrug transporter AcrB involves an access and a deep binding pocket that are separated by a switch-loop Proc. Natl. Acad. Sci. U.S.A. *109,* 5687–5692.

Furukawa *et al.* (1993). Thiolactomycin resistance in *Escherichia coli* is associated with the multidrug resistance efflux pump encoded by emrAB. J. Bacteriol. *175,* 3723–3729.

Ge *et al.* (2009). The C-terminal domain of AcrA is essential for the assembly and function of the multidrug efflux pump AcrAB-TolC. J. Bacteriol. *191,* 4365–4371.

Gerken *et al.* (2004). Genetic evidence for functional interactions between TolC and AcrA proteins of a major antibiotic efflux pump of *Escherichia coli.* Mol. Microbiol. *54,* 620–631.

Goldberg *et al.* (1999). Energetics and topology of CzcA, a cation/proton antiporter of the resistance-nodulation-cell division protein family. J. Biol. Chem. *274,* 26065–26070.

Higgins *et al.* (2004). Structure of the periplasmic component of a bacterial drug efflux pump. Proc. Natl. Acad. Sci. U.S.A. *101,* 9994–9999.

Hobbs, E.C., Yin, X., Paul, B.J., Astarita, J.L., and Storz, G. (2012). Conserved small protein associates with the multidrug efflux pump AcrB and differentially affects antibiotic resistance. Proc. Natl. Acad. Sci. U.S.A. *109,* 16696–16701.

Janganan *et al.* (2011). Evidence for the assembly of a bacterial tripartite multidrug pump with a stoichiometry of 3:6:3. J. Biol. Chem. *286,* 26900–26912.

Johnson *et al.* (1999). Alignment and structure prediction of divergent protein families: periplasmic and outer membrane proteins of bacterial efflux pumps. J. Mol. Biol. *287,* 695–715.

Koronakis *et al.* (2000). Crystal structure of the bacterial membrane protein TolC central to multidrug efflux and protein export. Nature *405,* 914–919.

Li *et al.* (2009). Efflux-mediated drug resistance in bacteria: an update. Drugs *69,* 1555–1623.

Lim *et al.* (2010). Kinetic parameters of efflux of penicillins by the multidrug efflux transporter AcrAB-TolC of *Escherichia coli.* Antimicrob. Agents Chemother. *54,* 1800–1806.

Long *et al.* (2010). Crystal structures of the CusA efflux pump suggest methionine mediated metal transport. Nature *467,* 484–488.

Mikolosko *et al.* (2006). Conformational flexibility in the multidrug efflux system protein AcrA. Structure *14,* 577–587.

Misra, R., and Bavro, V.N. (2009). Assembly and transport mechanism of tripartite drug efflux systems. Biochim. Biophys. Acta. *1794,* 817–825.

Murakami *et al.* (2006). Crystal structures of a multidrug transporter reveal a functionally rotating mechanism. Nature *443,* 173–179.

Nagano *et al.* (2009). Kinetic behavior of the major multidrug efflux pump AcrB of *Escherichia coli.* Proc. Natl. Acad. Sci. U.S.A. *106,* 5854–5858.

Nakashima *et al.* (2011). Structures of the multidrug exporter AcrB reveal a proximal multisite drug-binding pocket. Nature *480,* 565–569.

Narita *et al.* (2003). Linkage of the efflux-pump expression level with substrate extrusion rate in the MexAB-OprM efflux pump of *Pseudomonas aeruginosa.* Biochem. Biophys. Res. Commun. *308,* 922–926.

Nikaido (2001). Preventing drug access to targets: cell surface permeability barriers and active efflux in bacteria. Semin. Cell Dev. Biol. *12,* 215–223.

Ohene-Agyei, T., Lea, J.D., and Venter, H. (2012). Mutations in MexB that affect the efflux of antibiotics with cytoplasmic targets. FEMS Microb. Lett. *333,* 20–27.

Seeger *et al.* (2006). Structural asymmetry of AcrB trimer suggests a peristaltic pump mechanism. Science *313* 1295–1298.

Seeger *et al.* (2008). Engineered disulfide bonds support the functional rotation mechanism of multidrug efflux pump AcrB. Nat. Struct. Mol. Biol. *15,* 199–205.

Seeger *et al.* (2009). Crucial role of Asp408 in the proton translocation pathway of multidrug transporter AcrB: evidence from site directed mutagenesis and carbodiimide labeling. Biochemistry *48,* 5801–5812.

Sennhauser *et al.* (2007). Drug export pathway of multidrug exporter AcrB revealed by DARPin inhibitors. PLoS Biol. *5,* e7.

Sharff, A., Fanutti, C., Shi, J., Calladine, C., and Luisi, B. (2001). The role of the TolC family in protein transport and multidrug efflux. From stereochemical certainty to mechanistic hypothesis. Eur J. Biochem. *268,* 5011–5026.

Stegmeier *et al.* (2006). Importance of the adaptor (membrane fusion) protein hairpin domain for the functionality of multidrug efflux pumps. Biochemistry *45,* 10303–10312.

Su *et al.* (2006). Conformation of the AcrB multidrug efflux pump in mutants of the putative proton relay pathway. J. Bacteriol. *188,* 7290–7296.

Su *et al.* (2011). Crystal structure of the CusBA heavy metal efflux complex of *Escherichia coli.* Nature *470,* 558–562.

Symmons *et al.* (2009). The assembled structure of a complete tripartite bacterial multidrug efflux pump. Proc. Natl. Acad. Sci. U.S.A. *106,* 7173–7178.

Takatsuka *et al.* (2007). Site-directed disulfide cross-linking shows that cleft flexibility in the periplasmic domain is needed for the multidrug efflux pump AcrB of *Escherichia coli.* J. Bacteriol. *189,* 8677–8684.

Takatsuka *et al.* (2009). Covalently linked trimer of the AcrB multidrug efflux pump provides support for the functional rotating mechanism. J. Bacteriol. *191,* 1729–1737.

Takatsuka *et al.* (2010). Mechanism of recognition of compounds of diverse structures by the multidrug efflux pump AcrB of *Escherichia coli*. Proc. Natl. Acad. Sci. U.S.A. *107*, 6559–6565.

Tamura *et al.* (2005). Direct interaction of multidrug efflux transporter AcrB and outer membrane channel TolC detected via site-directed disulfide cross-linking. Biochemistry *44*, 11115–11121.

Thanabalu *et al.* (1998). Substrate – induced assembly of a contiguous channel for protein export from *E.coli*: reversible bridging of an inner-membrane translocase to an outer membrane exit pore. EMBO J. *17*, 6487–6496.

Tikhonova *et al.* (2004). AcrA, AcrB, and TolC of *Escherichia coli* form a stable intermembrane multidrug efflux complex. J. Biol. Chem. *279*, 32116–32124.

Tikhonova *et al.* (2011). Sequential mechanism of assembly of multidrug efflux pump AcrAB-TolC. Chem. Biol. *18*, 454–463.

Touze *et al.* (2004). Interactions underlying assembly of the *Escherichia coli* AcrAB-TolC multidrug efflux system. Mol. Microbiol. *53*, 697–706.

Vargiu *et al.* (2011). Effect of the F610A mutation on substrate extrusion in the AcrB transporter: explanation and rationale by molecular dynamics simulations. J. Am. Chem. Soc. *133*, 10704–10707.

Welch *et al.* (2010). Promiscuous partnering and independent activity of MexB, the multidrug transporter protein from *Pseudomonas aeruginosa*. Biochem. J. *430*, 355–364.

Weeks *et al.* (2010). AcrA suppressor alterations reverse the drug hypersensitivity phenotype of a TolC mutant by inducing TolC aperture opening. Mol. Microbiol. *75*, 1468–1483.

Zgurskaya *et al.* (2000). Cross-linked complex between oligomeric periplasmic lipoprotein AcrA and the inner-membrane-associated multidrug efflux pump AcrB from *Escherichia coli*. J. Bacteriol. *182*, 4264–4267.

Structure, Mechanism and Assembly of the Tripartite CusCBA Heavy Metal Efflux Complex

4

Sylvia V. Do, Chih-Chia Su, Feng Long, Hsiang-Ting Lei, Jani Reddy Bolla and Edward W. Yu

Abstract

Gram-negative bacteria expel various toxic chemicals via tripartite efflux pumps belonging to the resistance–nodulation–cell division (RND) superfamily. These pumps span both the inner and outer membranes of the cell. The three components of these tripartite systems are an inner membrane, substrate-binding transporter (or pump); a periplasmic membrane fusion protein (or adaptor); and an outer membrane-anchored channel. These three efflux proteins interact in the periplasmic space to form the three-part complex. One such efflux system, CusCBA, is responsible for extruding Cu(I) and Ag(I) ions, which are biocides. We previously presented the crystal structures of both the inner membrane transporter CusA and membrane fusion protein CusB of the CusCBA tripartite efflux system from *Escherichia coli*. We also described the co-crystal structure of the CusBA adaptor-transporter, revealing that the trimeric CusA efflux pump assembles with six CusB protein molecules to form the complex $CusB_6$-$CusA_3$. Recently, we reported three different conformers of the crystal structures of CusBA-Cu(I), suggesting a mechanism on how Cu(I) binding initiates a sequence of conformational transitions in the transport cycle. Herein, we summarize the structural information of these efflux proteins, and present the accumulated evidence that this efflux system utilizes conserved methionine as well as charged amino acids to bind and export Cu(I) and Ag(I) ions.

Introduction

Efflux pumps of the RND superfamily are integral membrane proteins that derive energy from the proton motive force (PMF). These efflux systems are widely distributed and found ubiquitously in bacteria, archaea and eukaryotes. In Gram-negative bacteria, which contain two membrane layers in the cell envelope, these export systems play major roles in the intrinsic and acquired tolerance to antibiotics and toxic compounds. They assemble as tripartite efflux pumps, spanning both the inner and outer membranes, to extrude antimicrobial and toxic compounds out of the cell (Fig. 4.1).

The Gram-negative bacterium *E. coli* harbours seven different RND efflux transporters (Table 4.1). These seven transporters can be categorized into two distinct subfamilies, the hydrophobic and amphiphilic efflux RND (HAE-RND) and heavy metal efflux RND (HME-RND) families (Tseng *et al.*, 1999; Nies, 2003). Six of these transporters, such as AcrB (Zgurskaya and Nikaido, 1999; Nishino and Yamaguchi, 2001; Murakami *et al.*, 2002; Yu *et al.*, 2003; Yu *et al.*, 2005; Murakami *et al.*, 2006; Seeger *et al.*, 2006; Su *et al.*, 2006; Sennhauser *et al.*, 2007; Das *et al.*, 2007; Törnroth-Horsefield *et al.*, 2007), AcrD (Rosenberg *et al.*, 2000; Nishino and Yamaguchi, 2001; Aires and Nikaido, 2005), AcrF (Ma *et al.*, 1993; Nishino and Yamaguchi, 2001; Lau and Zgurskaya, 2005), MdtB (Nishino and Yamaguchi, 2001; Baranova and Nikaido, 2002; Nagakubo *et al.*, 2002), MdtC (Nishino and Yamaguchi, 2001; Baranova and Nikaido, 2002; Nagakubo *et al.*, 2002) and YhiV (Nishino and Yamaguchi, 2001; Kobayashi *et al.*, 2001; Bohnert *et al.*, 2007), are multidrug

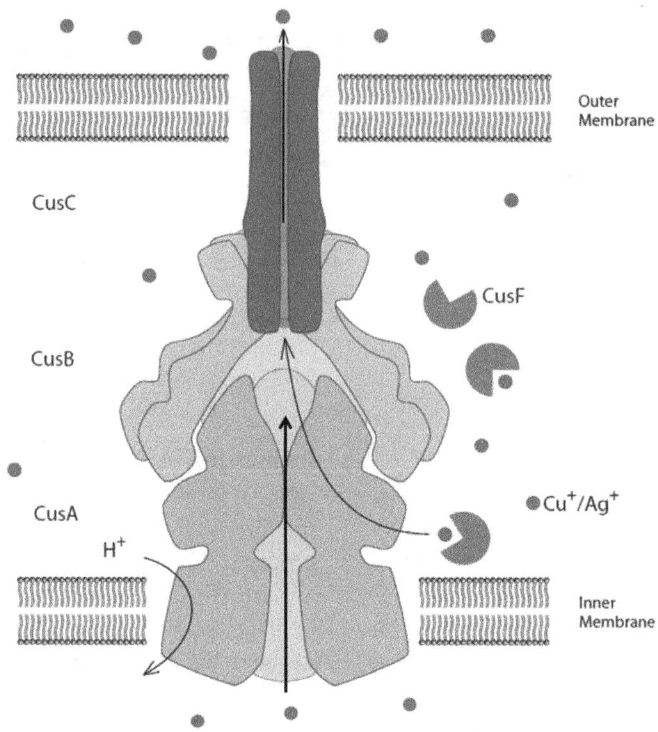

Figure 4.1 Model of the CusCBA tripartite efflux pump. The three components of this HME-RND system form a CusCBA efflux complex, which spans both the inner and outer membranes of *E. coli* to export Cu^+ and Ag^+ directly out of the cell. The periplasmic chaperone CusF is involved in delivering the metal ions to the CusCBA complex. Heavy metal efflux is driven by the proton-motive force.

efflux pumps, which belong to the HAE-RND protein family (Tseng *et al.*, 1999). In addition to these multidrug efflux pumps, *E. coli* contains one HME-RND efflux transporter, CusA, which specifically recognizes and confers resistance to Ag(I) and Cu(I) ions (Franke *et al.*, 2001, 2003).

The CusA transporter operates with a periplasmic membrane fusion protein (MFP or adaptor), CusB, and an outer membrane channel (OMC), CusC, to form a functional protein complex. CusA is a large PMF-dependent inner membrane RND efflux pump comprising 1047 amino acids (Franke *et al.*, 2001, 2003). CusC is a 440-residue protein that forms an OMC (Franke *et al.*, 2001, 2003). The MFP CusB is a 379-amino-acid polypeptide, contacting both the inner membrane CusA and outer membrane CusC efflux proteins (Franke *et al.*, 2001, 2003). The resulting CusCBA tripartite efflux system spans the entire cell envelope and confers resistance to Cu(I) and Ag(I) by exporting these metal ions directly out of the cell. Heavy metal efflux by the CusCBA complex is driven by proton import, and this process is catalysed through the inner membrane transporter CusA.

Among those transporters belonging to the RND superfamily, the structural and functional relationships of the HAE-RND pumps have been elucidated. Currently, two crystal structures of the HAE-RND-type efflux pumps, the *E. coli* AcrB (Murakami *et al.*, 2002; Yu *et al.*, 2003, 2005; Murakami *et al.*, 2006; Seeger *et al.*, 2006; Su *et al.*, 2006; Sennhauser *et al.*, 2007; Das *et al.*, 2007; Törnroth-Horsefield *et al.*, 2007) and *Pseudomonas aeruginosa* MexB (Sennhauser *et al.*, 2009) multidrug transporters, have been determined. Their structures suggest that both AcrB and MexB span the entire width of the inner membrane and protrude approximately 70 Å into the periplasm. Along with the models of these two HAE-RND transporters, the crystal structures of the other components of these tripartite complex systems have also been resolved. These include

Table 4.1 The RND efflux pumps of *Escherichia coli*

RND protein	Subfamily	Function	MFP	OMC	References
AcrB	HAE	Multidrug efflux pump (conferring resistance to chloramphenicol, tetracycline, minocycline, erythromycin, nalidixic acid, norfloxacin, enoxacin, doxorubicin, novobiocin, rifampin, trimethoprim, acriflavine, crystal violet, ethidium bromide, rhodamine 6G, tetraphenylphosphonium bromide, benzalkonium, Triton X-100, sodium dodecyl sulphate, deoxycholate, taurocholate, etc.)	AcrA	TolC	Zgurskaya and Nikaido (1999), Nishino and Yamaguchi (2001)
AcrD	HAE	Multidrug efflux pump (conferring resistance to aminoglycosides, novobiocin, sodium dodecyl sulphate and deoxycholate)	AcrA	TolC	Rosenberg *et al.* (2000), Nishino and Yamaguchi (2001), Aires and Nikaido (2005)
AcrF	HAE	Multidrug efflux pump (conferring resistance to doxorubicin, acriflavine, ethidium bromide, rhodamine 6G, sodium dodecyl sulphate, deoxycholate; important in the maintenance of cell division)	AcrE	TolC	Ma *et al.* (1993), Nishino and Yamaguchi (2001), Lau and Zgurskaya (2005)
MdtB	HAE	Multidrug efflux pump (conferring resistance to nalidixic acid, norfloxacin, fosfomycin, novobiocin, benzalkonium, sodium dodecyl sulphate, and deoxycholate)	MdtA	TolC	Nishino and Yamaguchi (2001), Baranova and Nikaido (2002), Nagakubo *et al.* (2002)
MdtC	HAE	Multidrug efflux pump (conferring resistance to nalidixic acid, norfloxacin, fosfomycin, novobiocin, benzalkonium, sodium dodecyl sulphate and deoxycholate)	MdtA	TolC	Nishino and Yamaguchi (2001), Baranova and Nikaido (2002), Nagakubo *et al.* (2002)
YhiV	HAE	Multidrug efflux pump (conferring resistance to erythromycin, doxorubicin, crystal violet, ethidium bromide, rhodamine 6G, tetraphenylphosphonium bromide, benzalkonium, sodium dodecyl sulphate and deoxycholate)	YhiU	TolC	Nishino and Yamaguchi (2001), Kobayashi (2001), Bohnert *et al.* (2007)
CusA	HME	Heavy metal efflux pump (conferring resistance to Cu(I) and Ag(I) ions)	CusB	CusC	Nishino and Yamaguchi (2001), Franke *et al.* (2001), Franke *et al.* (2003)

the OMCs, *E. coli* TolC (Koronakis *et al.*, 2000) and *P. aeruginosa* OprM (Akama *et al.*, 2004a), as well as the MFPs, *E. coli* AcrA (Mikolosko *et al.*, 2006) and *P. aeruginosa* MexA (Higgins *et al.*, 2004; Akama *et al.*, 2004b; Symmons *et al.*, 2009).

Unlike the HAE-RND family, members of the HME-RND family are highly substrate-specific. To date, *E. coli* CusA is the only HME-RND transporter that has available crystal structures. These structures immediately suggested a mechanism on how this efflux transporter specifically recognizes monovalent Cu(I) and Ag(I) ions. In addition, the crystal structures of individual components, the MFP CusB (Su *et al.*, 2009) and OMC CusC (Kulathila *et al.*, 2011), within the CusCBA tripartite efflux system have been determined. Recently, the X-ray structure of the CusBA efflux complex has also been described (Su *et al.*, 2011). This is the first co-crystal structure of an adaptor-transporter efflux complex in any efflux pump family. The structure indicates that the complete assembly of the CusCBA tripartite efflux complex should be in the form of $CusC_3$-$CusB_6$-$CusA_3$ that spans both the inner and outer membranes of *E. coli* and extrudes Cu(I) and Ag(I) ions (Su *et al.*, 2011). In this chapter, we summarize the structural information of these efflux proteins. Genetic analysis and transport assay strongly indicate that the conserved charged residues as well as the methionine pairs and clusters of CusA are essential for extruding metal ions out of the bacterial cell. Based on the current findings of the CusCBA efflux system, we put forward our opinion on how these protein machines facilitate metal transport.

The CusB membrane fusion protein

Crystal structure of the CusB membrane fusion protein

The crystal structure of CusB (pdb code: 3H9I), constituting approximately 80% of the protein (residues 89–385), was determined to a resolution of 3.40 Å (Su *et al.*, 2009). This structure has allowed us to reveal novel structural features that are not found in other known MFP structures. The crystal structure indicated that the asymmetrical unit of the crystal consists of two protomers. These two CusB molecules were found to interact with each other. However, the functional oligomerization state of this protein was still not clear at that time. Each CusB molecule is folded into an elongated polypeptide, ~ 120 Å long and ~ 40 Å wide, and can be divided into four different domains (Fig. 4.2). The first three domains of the protein are mostly β-strands. The fourth domain, however, forms an all α-helical domain, which is folded into a three-helix bundle secondary structure.

The first β-domain (Domain 1) of CusB is formed by the N and C-terminal ends of the polypeptide (residues 89–102 and 324–385). This domain is located directly above the outer-leaflet of the inner membrane. By coupling cross-linking with mass spectrometry, it has been suggested that this domain specifically interacts with the periplasmic domain of the CusA efflux pump (Su *et al.*, 2009). Overall, Domain 1 is a β-barrel domain, consisting of six β-strands, with the N-terminal end forming one of the β-strands while the C-terminus of the protein constitutes the remaining five strands (Fig. 4.2).

The second β-domain (Domain 2) of CusB is formed by residues 105–115 and 243–320. This domain comprises six β-strands and one short α-helix. Again, the N-terminal residues form one of the β-strands that is incorporated into this domain. The C-terminal residues contribute a β-strand, an α-helix and four anti-parallel β-sheets. In an asymmetrical unit of the crystal, the two molecules of CusB contact one another in an anti-parallel fashion, with Domains 1 and 2 of one molecule interact closely with Domains 2 and 1 of the other molecule, respectively (Su *et al.*, 2009). It was later found that these two domains are important for the oligomerization of this CusB adaptor.

Domain 3 is another globular β-domain adjacent to the second domain of CusB. This domain consists of residues 121–154 and 207–239, with a majority of these residues folding into eight β-strands.

Perhaps the most interesting motif appears to be in the fourth domain (Domain 4) of CusB. This region, comprising residues 156–205, forms an all-helical domain. Surprisingly, this domain is folded into an anti-parallel, three-helix bundle (Su *et al.*, 2009). This structural feature, not found in other known protein structures in the MFP family, highlights the uniqueness of the CusB protein. The helix bundle creates a ~ 27-Å-long helix–turn–helix–turn–helix secondary structure, making it at least 20 Å shorter than the two-helical hairpin domains of MexA (Akama *et al.*, 2004b; Higgins *et al.*, 2004; Symmons *et al.*, 2009) and AcrA (Mikolosko *et al.*, 2006). To date, CusB is the only periplasmic protein in the MFP family that possesses this three-helical domain instead of a two-helical hairpin motif.

The crystal structure of CusB represents the first structure of any MFP that is associated with an HME-RND-type transporter. Among all known structures of the MFP family, including AcrA (Mikolosko *et al.*, 2006) and MacA (Yum *et al.*, 2009), the CusB (Su *et al.*, 2009), MexA (Symmons *et al.*, 2009) and ZneB (De Angelis *et al.*, 2010) proteins exhibit the most complete three-dimensional structures. Like MexA and ZneB, analysis of the structure of CusB revealed that this MFP consists of four major domains, including three β-strand domains and one α-helical domain. However, CusB is folded into a distinct secondary structure when compared with the other available MFP structures. This distinct structural feature may underscore the unique functionality of CusB in the MFP family.

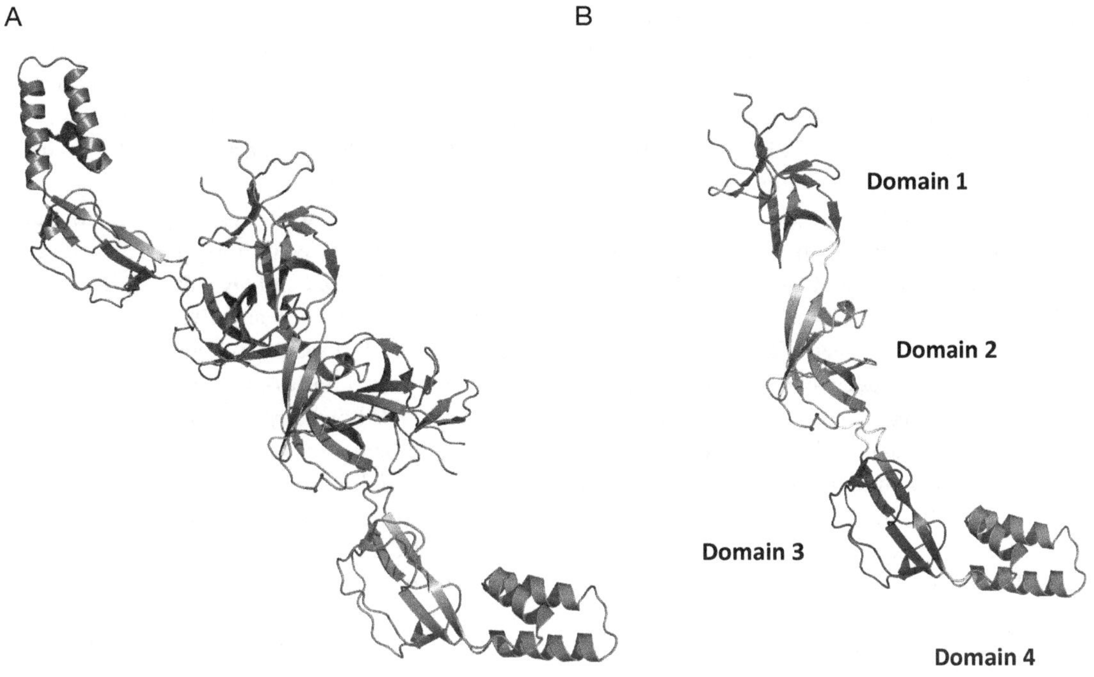

Figure 4.2 Crystal structure of the CusB membrane fusion protein. (A) Ribbon diagram of the two protomers in the asymmetrical unit of the crystal. (B) The four domains of each protomer in ribbon diagram. Domain 1 is formed by the N and C-terminal ends of the polypeptide (residues 89–102 and 324–385) and interacts with the periplasmic domain of the CusA efflux pump. Domain 2 is formed by residues 105–115 and 243–320 and comprises six β-strands and one short-helix. Domain 3 consists of residues 121–154 and 207–239, with a majority of these residues folding into eight β-strands. Domain 4 comprises residues 156–205, forming an all-helical domain. It is folded into an anti-parallel, three-helix bundle that is not found in other known protein structures in the MFP family.

Cross-linking and mass spectrometry

To determine how CusB interacts with the CusA efflux pump and the relative orientation of CusB in the efflux complex, the purified CusA and CusB proteins were cross-linked using the lysine–lysine cross-linker disuccinimidyl suberate. The resulting product was then digested with trypsin and examined using LC-MS/MS. Analysis of the mass spectral data suggests that the lysine residue of the polypeptide IDPTQTQNLGVKTATVTR originating from the N-terminal residues (84–101) of CusB directly interacts with the lysine residue of the peptide SGKHDLADLR (residues 148–157 of CusA), which is supposed to form part of the N-terminal periplasmic loop of the CusA efflux pump (Su *et al.*, 2009). Together with the crystal structure of CusB and the mass spectrometric data, it was suggested that Domain 1, formed by the N and C-terminal ends, of CusB should directly interact with the periplasmic domain of the CusA transporter. This result is indeed in good agreement with the cross-linking study that the N and C-termini of AcrA specifically interact with the periplasmic domain of AcrB (Symmons *et al.*, 2009).

The CusA heavy metal efflux pump

Crystal structure of the CusA heavy metal efflux pump

The crystal structure of the CusA efflux pump (Long *et al.*, 2010) was resolved to a resolution of 3.52 Å, with 98% of the residues (residues 5–504 and 516–1040) included in the final model (pdb code: 3K07) (Fig. 4.3). Overall, the crystal structure of CusA is quite distinct from those of the HAE-RND pumps AcrB and MexB. Superimposition of the structure of CusA with the structure of AcrB (Murakami *et al.*, 2002) results in a high RMSD of 11.4 Å for 1003 Cα atoms, suggesting highly significant differences between these two transporters. CusA exists as a homotrimer, and each subunit of CusA consists of 12 transmembrane helices (TM1-TM12) and a large periplasmic domain formed by two periplasmic loops between TM1 and TM2, and TM7 and TM8, respectively (Fig. 4.3A). In the transmembrane region, the relative locations of TM1-TM6 are related to those of TM7-TM12 by pseudo-twofold symmetry. These TM helices are arranged in such a way that TM4 and TM10 form the centre of the core and are surrounded by the other TM helices. Unlike AcrB and MexB, four helices, TM4, TM5, TM10 and TM11, extend into the cytoplasm to form the cytoplasmic domain of the pump. Two other helices, TM2 and TM8, protrude into the periplasm and contribute to part of the periplasmic domain. It is important to note that TM2, TM4 and TM5 of the N-terminal half correspond to TM8, TM10 and TM11 of the C-terminal half, respectively, in the pseudo-twofold symmetry.

Like AcrB and MexB, the periplasmic domain of CusA can be divided into six subdomains: PN1, PN2, PC1, PC2, DN and DC (Fig. 4.3B). Subdomains PN1, PN2, PC1 and PC2 form the pore domain, with PN1 making up the central pore and stabilizing the trimeric organization. Subdomains DN and DC, however, contribute to form the docking domain, presumed to be interacting with the outer membrane channel CusC. The trimeric CusA structure suggests that subdomains PN2, PC1 and PC2 are located at the outermost core of the periplasmic domain, facing the periplasm. In AcrB, subdomains PC1 and PC2 form a large external cleft and create the entrance for drug molecules from the periplasm. However, the

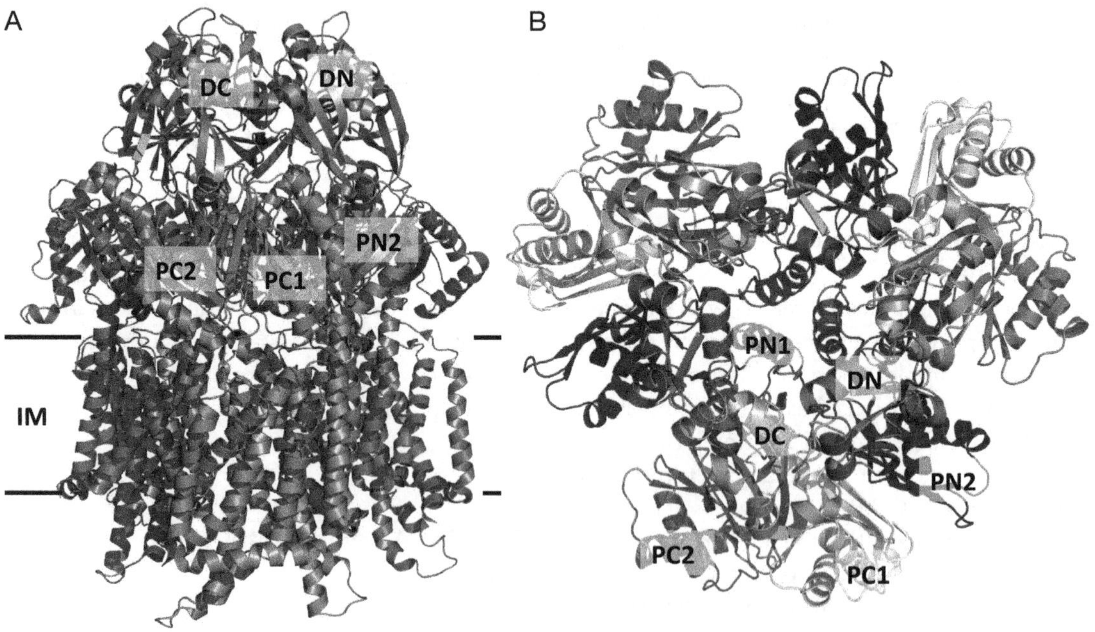

Figure 4.3 Crystal structures of the CusA efflux pump. (A) Ribbon diagram of the CusA homotrimer viewed in the membrane plane. Each molecule is labelled with a different colour: molecule 1 is purple, molecule 2 is orange, and molecule 3 is green. Each subunit of CusA consists of 12 transmembrane helices (TM1-TM12) and a large periplasmic domain formed by two periplasmic loops between TM1 and TM2, and TM7 and TM8, respectively. (B) Ribbon diagram of the CusA homotrimer subdomains as viewed from the top. The six subdomains are PN1 (red), PN2 (blue), PC1 (yellow), PC2 (teal), DN (green) and DC (pink). A colour version of this figure is located in the plate section at the back of the book.

apo-CusA structure shows that the gap between PC1 and PC2 is completely closed (Fig. 4.3B). As mentioned above, *in vitro* cross-linking coupled with mass spectrometry suggested that the polypeptide SGKHDLADLR (residues 148–157 of CusA) should interact with the N-terminus of the MFP CusB. This polypeptide was found to be located at subdomains PN2 of CusA in the crystal structure.

Perhaps the most interesting secondary structural feature appears in the cleft region of the periplasmic domain. Residues located on the left side of the wall, formed by one α-helix (residues 690–706) and three β-sheets (residues 681–687, 711–716 and 821–827), are found to tilt into the cleft to close the opening. Surprisingly, residues 665–675, located at the bottom of the cleft, form an α-helix. This structural feature, not found in AcrB and MexB, likely governs the specificity of the CusA pump. The α-helix orients horizontally and roughly divides the transmembrane and periplasmic domains into two compartments. Located just above this horizontal helix are three proximal methionine residues, M573, M623 and M672, presumably creating a three-methionine specific binding site for Cu(I) and Ag(I) ions (Zhou and Thiele, 2001; Jiang *et al.*, 2005). Notably, M672 is also one of the residues within the horizontal helix. Site-directed mutagenesis had suggested that these three methionine residues are essential for mediating copper resistance (Franke *et al.*, 2003). Indeed, a strong peak at the copper edge was observed at the centre of these three methionines in our CusA-Cu(I) crystal derivative, indicating the binding of a Cu(I) ion in a three-sulphur binding site.

Crystal structures of the CusA-Cu(I) and CusA-Ag(I) complexes

The copper-edge wavelength (1.3779 Å) was used to collect the X-ray diffraction data of the CusA-Cu(I) crystal. These data have allowed us to identify the location of the bound Cu(I). Intriguingly, the overall structure of the CusA-Cu(I) complex (pdb code: 3KSS) (Long *et al.*, 2010) is quite distinct from that of apo-CusA (Fig. 4.4). Superimposition of these two structures gives an overall RMSD of 3.9 Å (for 1006 Cα atoms). As mentioned above, the bound Cu(I) ion was found to coordinate residues M573, M623 and M672. These three methionines specifically form a typical three-methionine coordination site for binding Cu(I)/Ag(I) (Zhou and Thiele, 2001; Jiang *et al.*, 2005). Binding of Cu(I) initiates significant conformational changes in the periplasmic as well as the transmembrane domains of CusA. Perhaps, the most noticeable difference between the apo and ion-bound structures appears in the PC2 region (Fig. 4.4A and B). Cu(I) binding leads to a 30° swing of the entire PC2 subdomain. This motion shifts PC2 away from the PC1 subdomain. The hinge for this rotational movement appears to be at the junction between subdomains PC2 and DC with residues G721 and P810 forming the hinge. As a consequence, the gap between PC1 and PC2 appears to open up after binding this metal ion (Fig. 4.4A,B). This gap presumably creates an entrance for metal ions from the periplasmic space. The horizontal helix, residues 665–675, located inside the cleft also makes a substantial movement. The C-terminal end of this helix is found to tilt upwards by 21° in the Cu(I)-bound structure with respect to the apo form (Fig. 4.4C). This tilting motion allows M672 to move closer to M573 and M623 to complete the three-methionine coordination site. Coupled with this movement, TM8 also shifts in position to a more vertical orientation while retaining its α-helical structure. Overall, the N-terminal end of TM8 is found to shift away from the core by 10 Å after Cu(I) binding.

For the CusA–Ag(I) complex (pdb code: 3KSO) (Long *et al.*, 2010), an anomalous difference Fourier peak was found at the centre of residues M573, M623 and M672, indicating that the bound Ag^+ ion is coordinated by these three methionines (Fig. 4.4D). The overall conformational changes triggered by Cu(I) and by Ag(I) binding are nearly identical. Superimposition of the CusA–Cu(I) and CusA–Ag(I) structures gives an overall RMSD of 1.0 Å (for 1021 Cαs). Based on the crystal structures, the horizontal helix in the cleft directly interacts with the N-terminal end of TM8. The movement of TM8 may relate directly to transmembrane signalling and could initiate the translocation of a proton across the membrane. Indeed, in the AcrB pump, there is evidence that proton translocation is coupled

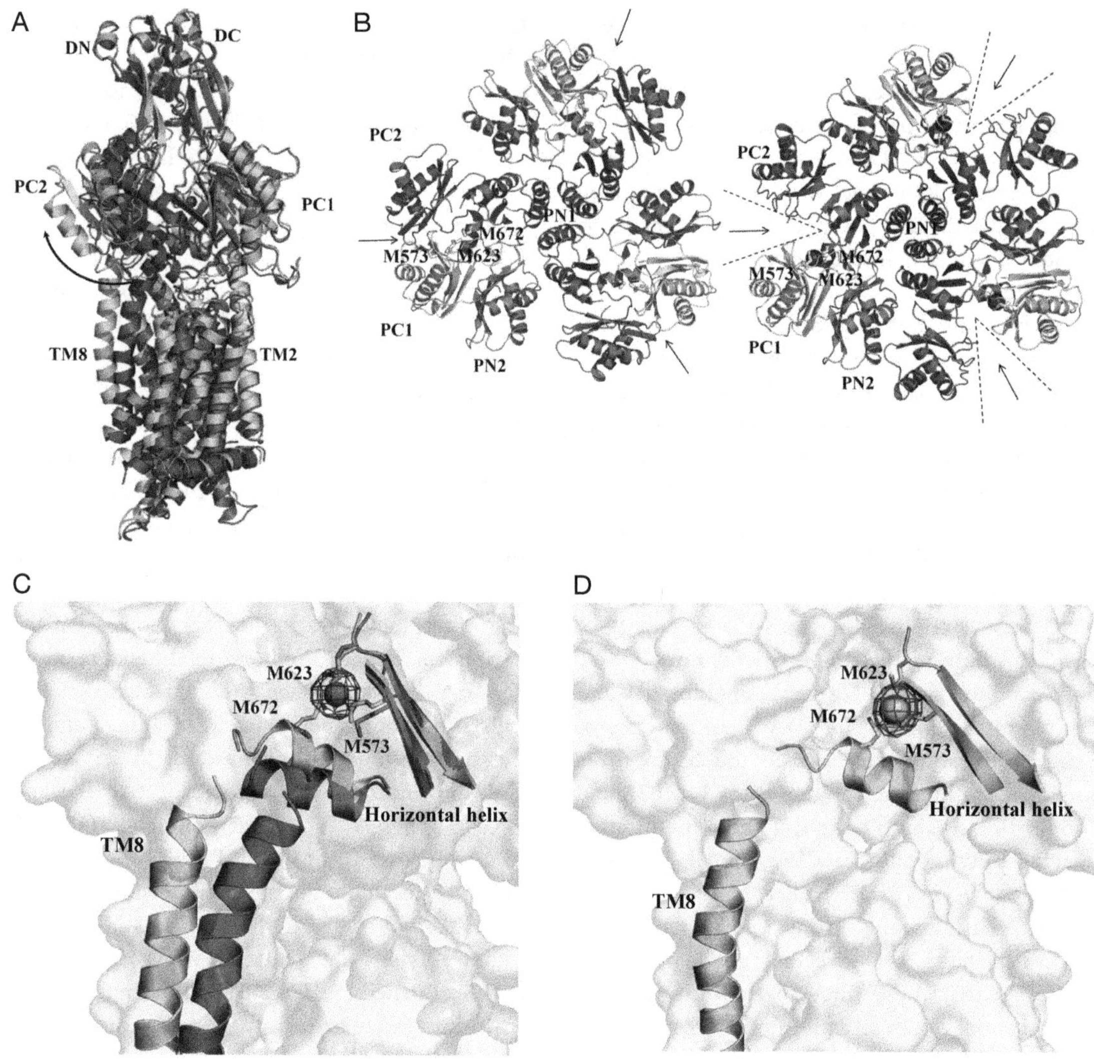

Figure 4.4 Comparison of the apo and metal bound structures of CusA. (A) Superposition of a monomer of apo-CusA onto a monomer of Cu(I) bound-CusA. The bound Cu(I) is shown as a sphere. The arrow represents a major swing of the PC2 subdomain initiated by Cu(I) binding. (B) Conformational changes of the periplasmic domain of CusA. The conformation of each subdomain of CusA before (left) and after (right) Cu(I) binding. The periplasmic cleft formed between PC1 and PC2 is opened after Cu(I) binding. The bound coppers in the CusA-Cu(I) structure are blue. M573, M623 and M672 forming a metal binding site at the periplasmic cleft are shown in stick form. (C) The changes in conformation of the horizontal helix and TM8, are shown in a superimposition of the structures of apo and Cu(I)-bound CusA. The bound Cu(I) is shown as a sphere. Anomalous map of the bound Cu(I) is contoured at 8σ. M573, M623 and M672 are shown as sticks. (D) The Ag(I) binding site. The bound Ag(I) is shown as a sphere. Anomalous map of the bound Ag(I) is contoured at 10σ.

to the conformational change in TM8 (Su *et al.*, 2006; Sennhauser *et al.*, 2007). The change in conformation takes place in TM8 by reeling in some random coil residues to extend the α helix (Sennhauser *et al.*, 2007). In addition, when individual residues of the proton-relay network was changed to alanine, disrupting the hydrogen bonds in the system, it was reported that TM8 of AcrB becomes longer and extends into the periplasmic domain (Su *et al.*, 2006). In contrast to AcrB, the helical structure of the N-terminus of TM8 in CusA is retained after metal binding. Instead, the

N-terminal end of TM8 moves outward, together with the movement of the PC2 subdomain.

Intriguingly, it appears that the binding of Cu(I) or Ag(I) also triggers significant conformational changes in the other transmembrane helices of the pump. In addition to the movement of TM8, all other transmembrane helices except TM2 shift horizontally by as much as 4 Å, mimicking the motion of TM8. Further, TM1, TM3 and TM6 readjust in an approximately 3 Å upwards shift with respect to the inner membrane surface. The net result is that all three of these transmembrane helices move towards the periplasm by one turn. Mutagenesis studies of the transmembrane domain of CusA indicated the conserved charged residue D405 of TM4 to be essential for transporter function (Franke *et al.*, 2003). Based on the crystal structure, this acidic residue interacts with E939 and K984. It is possible that these three charged residues participate in forming the proton-relay network in the transmembrane region of the pump.

Coupled with the above conformational changes, the subdomain PN1 at the periplasmic domain was also found to undergo substantial movement. Overall, PN1 also shifts upwards by 3 Å. This moves the central pore helix upwards by one turn upon metal binding. It should be noted that the binding of metal ions does not significantly affect the conformation of the subdomains DN, DC, PN2 or the transmembrane helix TM2.

The methionine-residue ion relay network

The full-length CusA includes 34 methionine residues with 18 of them located in the transmembrane domain. Of the 18 methionines, six are paired up to form three distinct methionine pairs. These methionine pairs are M410 of TM4 and M501 of TM6; M403 of TM4 and M486 of TM6; and M391 of TM4 and M1009 of TM12 (Fig. 4.5). It is known that copper tolerant proteins, such as CusF (Loftin *et al.*, 2007; Xue *et al.*, 2008), CueR (Changela *et al.*, 2003) and Atx1 (Arnesano, 2001), frequently utilize two-methionine or two-cysteine binding pockets to carry their Cu(I)/Ag(I) cargoes. Thus, these methionine pairs could potentially form binding sites for Ag(I) and Cu(I) in the transmembrane region of the pump. If this is the case, then CusA could transport these metal ions from the cytoplasm along these methionine pairs. In the periplasmic domain of CusA, we have found another methionine pair (M271-M755) in addition to the three-methionine metal binding site (Fig. 4.5). In view of the crystal structures, these five methionine pairs/clusters are deemed to be important for binding and transport of metal ions.

We then determined whether the CusA protomer forms a channel by using the program CAVER (http://loschmidt.chemi.muni.cz/caver) and found that each protomer of CusA forms a channel spanning the entire transmembrane region up to the bottom of the periplasmic funnel (Fig. 4.5). Intriguingly, the channel includes four methionine pairs, three (M410-M501, M403-M486, M391-M1009) from the transmembrane region and one (M271-M755) from the periplasmic domain, as well as the three-methionine binding site formed by M573, M623 and M672 (Fig. 4.5). Taken together, these five methionine pairs/clusters are likely to form a relay network facilitating metal ion transport. Remarkably, this channel spans almost the entire length of each protomer, from the transmembrane domain through to the bottom of the periplasmic funnel region, and is likely to represent a real path for transporting the metal ion from both the cytoplasm and periplasm to the periplasmic funnel for extrusion.

In vivo metal susceptibility assay

We made an *E. coli* knockout strain BL21(DE3)*ΔcueOΔcusA* that lacks both *cueO* and *cusA*. The *cueO* gene encodes a putative multicopper oxidase, CueO, which is also involved in copper homeostasis (Grass and Rensing, 2001). We mutated M573, M623 and M672, which are members of the binding site triad inside the cleft of the periplasmic domain, into isoleucines (M573I, M623I and M672I). We then expressed these three mutant transporters in BL21(DE3)*ΔcueOΔcusA*, and tested their ability to confer copper and silver resistance *in vivo*. We found that the CusA mutants, M573I, M623I and M672I are unable to relieve the copper or silver sensitivity of strain BL21(DE3)*ΔcueOΔcusA*, thus agreeing well with the work of Franke *et al.* (2003)

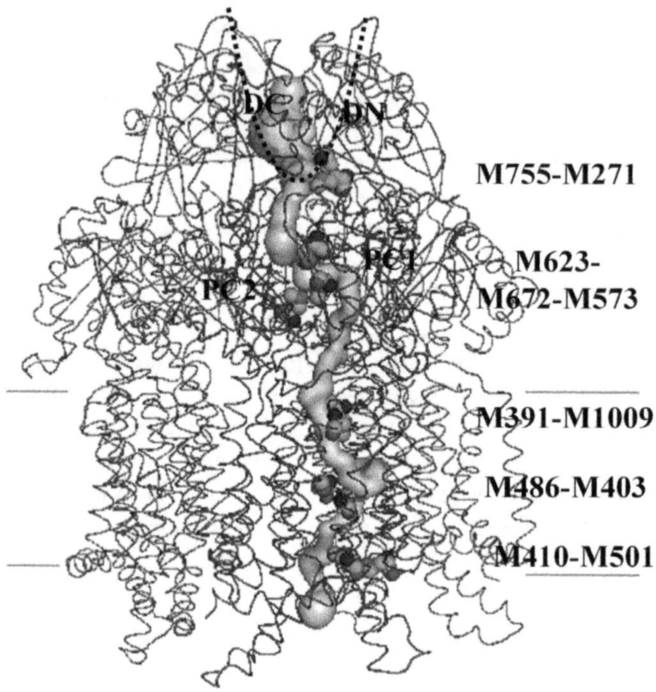

Figure 4.5 Channel in the CusA pump. The channel formed by the front protomer of apo-CusA leading through the transmembrane and periplasmic domains is shown. This channel was calculated using the position of the sulphur atom of residue M672 as a starting point. The 11 methionines forming the relay network are in spheres. Two other CusA protomers behind the front protomer are also shown as wires.

in which M573, M623 and M672 were shown to be essential for the function of CusA.

We mutated M410 into an isoleucine (M410I) to disrupt the pair formed by M410 and M501 at the bottom of the transmembrane. We also replaced M486 and M391, which pair with M403 and M1009, with isoleucines (M486I and M391I). Expression of these three mutants in BL21(DE3)*ΔcueOΔcusA* showed a significant decrease in the level of copper and silver tolerances when compared with cells expressing wild-type CusA. In addition, we introduced M755I in the periplasmic domain. Again, when expressed in BL21(DE3)*ΔcueOΔcusA*, this mutant transporter (M755I) showed a decrease in copper/silver tolerance in comparison with cells harbouring the wild-type transporter. Together, these results strongly support the hypothesis that these methionine pairs/clusters engage in the methionine-residue ion relay channel.

Based on the crystal structure of CusA, it is expected that the charged residues D405, E939 and K984 are important for the proton-relay network of the pump, so they were replaced with alanines (D405A, E939A and K984A) to disrupt the hydrogen-bonded network. Indeed, cells expressing this mutant were unable to tolerate copper and silver, demonstrating that these three charged residues are essential for the transporter's function.

In vitro metal transport assay

To investigate whether CusA can transport metal ions from the cytoplasm, we reconstituted the purified CusA protein into liposomes containing the fluorescent indicator Phen Green SK in the intravesicular space. The intravesicular and extravesicular pHs of these proteoliposomes were 6.6 and 7.0, respectively. We then determined whether these proteoliposomes can capture metal ions from the extravesicular medium with a stopped-flow transport assay. (The charged metal ions are unlikely to diffuse spontaneously across membrane bilayers.) When Ag^+ ions were added into the extravesicular medium, we detected the quenching of the fluorescence signal, suggesting the uptake of Ag^+ into the intravesicular space (Long *et al.*, 2010). The uptake into proteoliposomes is presumably due to the CusA active transport activity.

In addition to the above experiment, we investigated the methionine residues that were shown to be important for copper and silver tolerances with the above transport assay. We expressed, purified and reconstituted the CusA mutants, M573I, M623I and M672I, into liposomes encapsulated with the same fluorescence indicator. As expected, these mutant transporters do not take up Ag^+ from the extravesicular medium of the proteoliposomes (Long *et al.*, 2010). It is clear that even single point mutations M573I, M623I or M672I abolish the process of metal transport across the membrane. Thus, it is expected that this three-methionine site is requisite for both metal binding and export.

We then purified the mutants, M391I, M486I and M755I, and reconstituted them into liposomes. In these three cases, the fluorescence signals did not attenuate as indicated by the stopped-flow assay (Long *et al.*, 2010). When compared with the result from the wild-type CusA, this suggests that the M391I, M486I and M755I mutant transporters are unable to actively transport Ag^+ across the membrane. Our collective experiments provide direct compelling evidence that CusA is capable of taking up Ag^+ from the cytoplasm.

The crystal structure of CusA suggests that the charged residues D405, E939 and K984 in the transmembrane domain may be important for proton translocation. When reconstituted into liposomes, mutant transporters (D405A, E939A and K984A) do not take up Ag^+ into the intravesicular space (Long *et al.*, 2010), thus confirming the importance of D405, E939 and K984 for the function of the pump.

Dynamics simulation

It is likely that CusA operates through an alternating-access mechanism (Murakami *et al.*, 2006; Seeger *et al.*, 2006; Sennhauser *et al.*, 2007). The dynamics simulation of the trimeric CusA pump using elastic network model (Atilgan *et al.*, 2001) was then performed. The calculation indeed suggests that CusA functions through three coupled motions in which the periplasmic cleft formed by subdomains PC1 and PC2 alternately open and close (Long *et al.*, 2010).

The CusBA adaptor–transporter efflux complex

Co-crystal structure of the CusBA adaptor–transporter complex

The co-crystal structure of CusBA was determined to a resolution of 2.90 Å, with the final structure including 93% of the total amino acids (pdb code: 3NE5) (Su *et al.*, 2011). The structure suggests that each protomer of CusA specifically interacts with two elongated molecules of CusB (molecules 1 and 2) at the upper half portion of the periplasmic domain (Fig. 4.6A). The two CusB adaptors are tilted at an angle of ~ 50° with respect to the membrane surface, and establish a close fit with the transporter at the concave surface formed by Domains 1 and 2 of the adaptor. It is found that molecule 1 of CusB contacts mainly the upper regions of PN2 and PC1, and the DN subdomain of CusA. Molecule 2 of CusB, however, predominantly bridges to the upper regions of PC1 and PC2, and also the subdomain DC of the pump. These two adaptor molecules are also seen to specifically contact one another, primarily through Domains 1, 2 and 3 of these two elongated molecules. The trimeric CusA pump therefore directly contacts six CusB adaptor molecules, which form a channel at the top of the CusA trimer (Fig. 4.6B).

The overall conformation of the CusA protomer in the complex resembles the apo-CusA structure (Long *et al.*, 2010). Superimposition of these two structures results in an RMSD of only 1.0 Å for 1019 Cα atoms. For the CusB protomers, two distinct conformations of this adaptor are captured in the complex. When we superimpose these two elongated protein structures high RMSD of 7.5 Å is observed for the 321 Cα atoms, suggesting that the conformations of these two protomers are quite different. Interestingly, a comparison of these two molecules with the two previously determined structures of CusB (Su *et al.*, 2009) indicates that the four molecules display

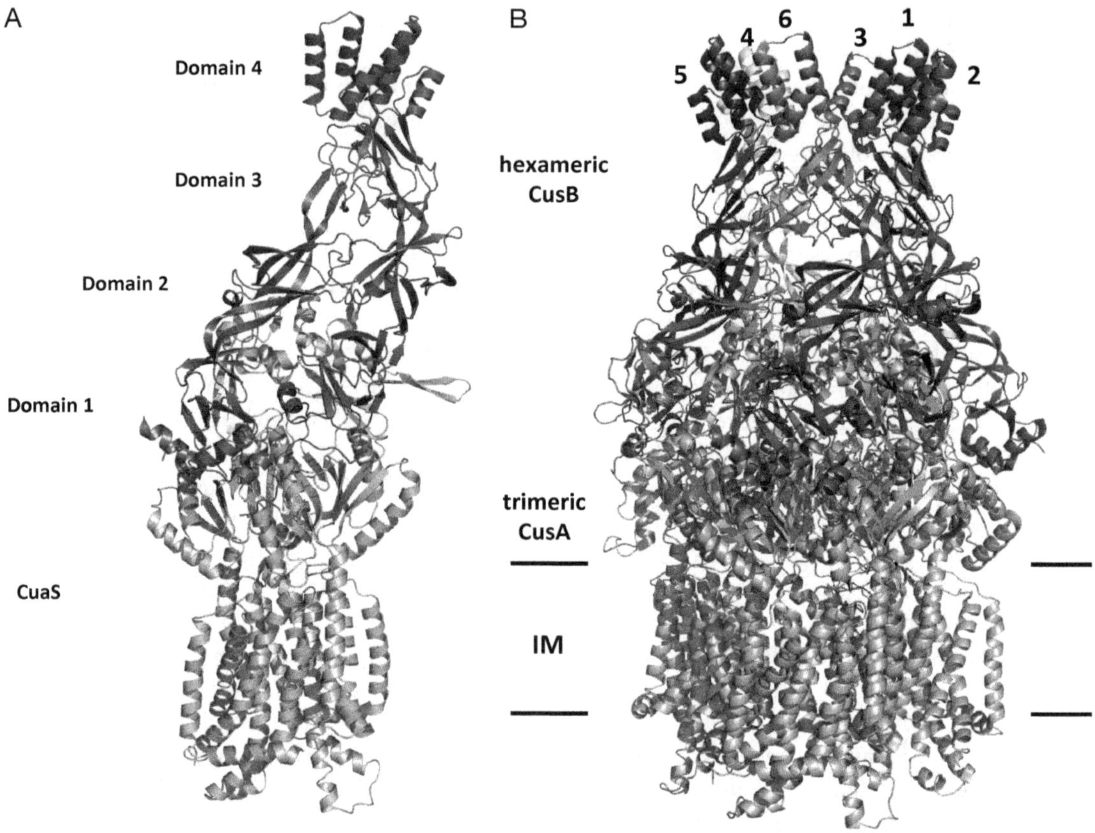

Figure 4.6 Crystal structures of the CusBA adaptor-transporter efflux complex. (A) The asymmetrical unit shows two CusB molecules (colours green and red) interacting with one CusA molecule (light orange). Domains 1–4 of CusB are labelled. (B) The full structure showing the hexameric CusB adaptor structure interacting with the trimeric CusA pump. Each CusB molecule is indicated in a different colour: green, red, light blue, yellow, purple, and orange. Each CusA molecule is indicated in different, lighter colours: light orange, light red, and light green. A colour version of this figure is located in the plate section at the back of the book.

four distinct conformations, presumably representing different transient states of this membrane fusion protein. This observation is consistent with the previous finding that these membrane fusion proteins are highly flexible (Mikolosko *et al.*, 2006; Symmons *et al.*, 2009; Su *et al.*, 2009), with the ability to change their conformations upon ligand binding (Bagai *et al.*, 2007; De Angelis *et al.*, 2010). Indeed, it has been demonstrated that binding of metal ions trigger significant conformational changes in the CusB (Bagai *et al.*, 2007) and ZneB (De Angelis *et al.*, 2010) adaptor proteins.

Previously, it was found that the N and C-terminal ends (residue 29–88 and 386–407) of the CusB adaptor are intrinsically disordered and cannot be identified in the electron density maps of the CusB crystals (Su *et al.*, 2009). Here, the co-crystal structure suggests that this region forms several short α-helices. In molecule 1 of CusB, residues 392–399 at the C-terminal end form a short α-helix. However, residues 79–95 of the N-terminus feature a long random coil and these amino acids are located immediately outside the cleft formed between subdomains PC1 and PC2 of the CusA pump. For molecule 2 of CusB, the N-terminal residues 79–85 and 86–92 participate to form a random coil and a short α-helix, respectively. However, the C-terminal residues 382–392 and 394–400 appear to create two short α-helices. Like molecule 1 of CusB, the N-terminus of molecule 2 of CusB is near the periplasmic cleft of the pump (Fig. 4.6).

It has been proposed that the N-terminal residues M49, M64 and M66 of CusB form a three-methionine metal-binding site (Bagai *et al.*, 2007). Although these three methionine residues cannot be identified in the electron density maps of our co-crystal, the co-crystal structure does show that the N-terminal tails of both molecule 1 and 2 of the CusB adaptors are located outside the cleft formed between PC1 and PC2 of the CusA pump. On the basis of this structure, it is possible that CusB may help to transfer the metal ions via the N-terminal three-methionine binding site into the periplasmic cleft of CusA.

CusA and CusB interactions

Intriguingly, molecule 1 of CusB interacts predominantly with CusA through charge–charge interactions. Residues K95, D386, E388 and R397 of this CusB molecule form four salt bridges with D155, R771, R777 and E584 of CusA, respectively (Fig. 4.7A). In addition, T89, the backbone oxygen of N91, and R292 of molecule 1 of CusB form hydrogen bonds with K594, R147, and the backbone oxygen of Q198 of CusA to secure the interaction (Su *et al.*, 2011). However, the interaction between molecule 2 of CusB and CusA appears to be governed principally by charge–dipole and dipole–dipole interactions. Specifically, Q108, S109, S253 and N312 of CusB (molecule 2) form hydrogen bonds with Q785, Q194, D800 and Q198 of CusA, respectively (Fig. 4.7B). The backbone oxygens of L92 and T335 of this CusB molecule also contribute two additional hydrogen bonds with the side chains of K591 and T808 of the CusA pump to anchor the proteins (Su *et al.*, 2011).

CusB and CusB interactions

For CusB–CusB interactions, it appears that molecule 1 of CusB makes a close contact with molecule 2 of CusB. Domains 1–3 of these two molecules are involved in the binding (Fig. 4.8A). Residues E118, Y119, R186, E252 and R292 of molecule 1 of CusB participate to form hydrogen bonds with residues T139, D142, T206, N312 and N113 of molecule 2 of CusB, respectively (Su *et al.*, 2011). Further, molecule 1 of CusB also contributes to contact molecule 6 of CusB, which is anchored to the next subunit of CusA (Fig. 4.8B). The majority of the interactions come from Domains 2 and 3 of these two molecules. Particularly, residues N113, N228, and N312 of molecule 1 of CusB pair up with R292, the backbone oxygen of A126, and E252 of molecule 6 of CusB to form three hydrogen bonds. In addition, D142 of molecule 1 of CusB participates to form two hydrogen bonds with Y119 and R297 of molecule 6 of CusB to secure the binding (Su *et al.*, 2011).

The hexameric CusB channel

The co-crystal structure suggests that CusB forms a hexamer above the trimeric CusA pump (Fig. 4.9). This hexameric arrangement allows the membrane fusion proteins to create a funnel-like structure with the central channel formed along the crystallographic 3-fold symmetry axis. Six of these CusB protomers utilize a side-by-side fashion to pack against one another to form this funnel. In view of the structure, it appears that Domain 1 and the lower half of Domain 2 of CusB primarily create a cap-like structure whereas the upper half of Domain 2, Domain 3 and Domain 4 of the adaptor contribute to the central channel of the funnel (Fig. 4.9A). The inner surface of the cap fits closely with the outer surface of the upper portion of the periplasmic domain of the CusA trimer. The channel formed above the cap of the adaptor is ~62 Å long with an average internal diameter of ~37 Å. Thus, the interior of the channel gives rise to a large elongated cavity with a volume of ~65,000 $Å^3$. The lower half of the channel is primarily created by β-barrels whereas the upper half is an all α-helical tunnel. Presumably, this channel is filled with solvent molecules. The diameter of the channel is gradually constricted and then dilated as it approaches the outer membrane. Thus, the α-helices of Domain 4 create an inverted conical structure. The narrowest section of the central channel is located at residue D232, which is close to the hinge region between Domains 3 and 4 of the membrane fusion protein (Fig. 4.9B). The widest section of the channel appears to form at the top edge with its inner diameter approximately equals 56 Å. The inner surface of the channel is predominantly negatively charged, as indicated by the electrostatic surface

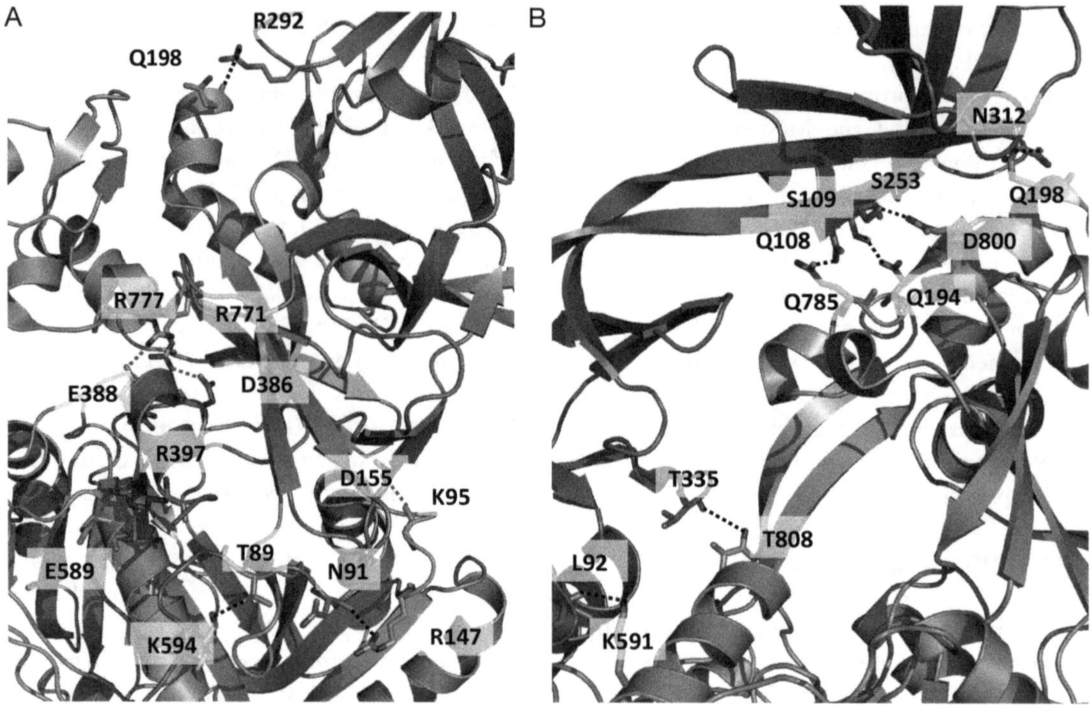

Figure 4.7 Interactions between CusA and CusB in ribbon format. Important residues involved in interactions between the molecules are shown as sticks. (A) Interaction between molecule 1 of CusB and CusA, with each molecule. Residues K95, D386, E388 and R397 of CusB form four salt bridges with D155, R771, R777 and E584 of CusA, respectively. In addition, T89, the backbone oxygen of N91, and R292 of molecule 1 of CusB form hydrogen bonds with K594, R147, and the backbone oxygen of Q198 of CusA. (B) Interaction between molecule 2 of CusB and CusA. Residues Q108, S109, S253 and N312 of CusB form hydrogen bonds with Q785, Q194, D800 and Q198 of CusA, respectively. The backbone oxygens of L92 and T335 of this CusB molecule also contribute two additional hydrogen bonds, with the side chains of K591 and T808 of CusA, anchoring the proteins.

diagram (Fig. 4.10), which suggests that the interior surface of the channel may have the capacity to bind positively charged metal ions.

The co-crystal structure of CusBA here highlights the structural importance of the periplasmic adaptor protein. Given the fact that six CusB molecules assemble to form a channel at the CusA funnel top, this suggests that the adaptor is likely to be involved in the active efflux of metal ions. The hexameric arrangement of these membrane fusion proteins has also been observed in the crystal structure of MacA, in which six protomers assemble to create a large channel (Yum *et al.*, 2009). Thus, in addition to their role as adaptors to bridge the inner and outer membrane efflux components, we can anticipate that these membrane fusion proteins are likely to actively participate in extrusion of their substrates.

Isothermal titration calorimetry for CusA and CusB binding

We used isothermal titration calorimetry (ITC) to determine the binding affinity of CusB to the CusA pump. The ITC data indicates an equilibrium dissociation constant of $5.1 \pm 0.3\,\mu$M (Su *et al.*, 2011). The titration is characterized by a negative enthalpic contribution ($\Delta H = -22.4 \pm 0.3$ kcal/mol), which yields a hyperbolic binding curve. The entropic contribution (ΔS) of this binding reaction was found to be -50.9 cal mol/deg. Interestingly, the molecular ratio for this binding reaction based on ITC is one CusA monomer for every two CusB protomers (Su *et al.*, 2011). This result is indeed confirmed by the crystal structure where each protomer of CusA binds two molecules of CusB.

Recently, artificial peptides called DARPins

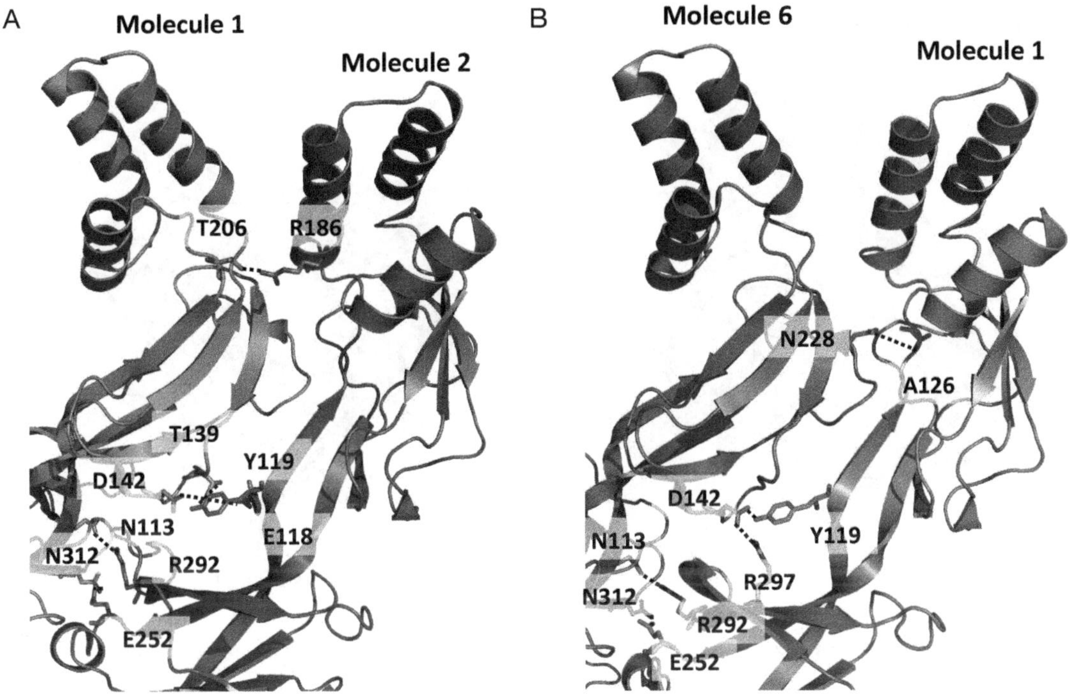

Figure 4.8 Interactions between the CusB molecules in forming the hexameric structure in ribbon format. Important residues involved in interactions between the molecules are shown as sticks. (A) Interactions between molecule 1 and molecule 2 of CusB. Residues E118, Y119, R186, E252 and R292 of molecule 1 of CusB participate to form hydrogen bonds with residues T139, D142, T206, N312 and N113 of molecule 2 of CusB, respectively. (B) Interactions between molecule 1 and molecule 6 of CusB. Residues N113, N228 and N312 of molecule 1 of CusB pair with R292, the backbone oxygen of A126, and E252 of molecule 6 of CusB, respectively, to form three hydrogen bonds. D142 of molecule 1 of CusB also participates to form two hydrogen bonds, with Y119 and R297 of molecule 6 of CusB.

have been introduced to bind and effectively inhibit the AcrB transporter (Sennhauser *et al.*, 2007). The inhibitor-binding site is found right above the cleft formed between PC1 and PC2 of AcrB (Sennhauser *et al.*, 2007). Interestingly, the location of this inhibitor-binding site corresponds to the binding site for molecule 2 of CusB in the CusBA complex. Thus, it is likely that the mechanism of action for these inhibitors may be the disruption of the adaptor–transporter interaction by competition. If this is the case, then a proposed $AcrB_3$-$AcrA_3$-$TolC_3$ model (Symmons *et al.*, 2009) containing only one protomer of AcrA per AcrB protomer would not be sufficient to form a functional pump. Indeed, it has been recently demonstrated that AcrA and AcrB interact with the stoichiometry of 2:1 adaptor-to-transporter molar ratio (Tikhonova *et al.*, 2011).

The CusBA-Cu(I) complexes

Crystal structures of the CusBA-Cu(I) complexes

In the absence of the CusB adaptor, two distinct structures of CusA were obtained by X-ray crystallography (Long *et al.*, 2010). These structures most likely capture two different conformational states of the pump in the transport cycle. The apo-CusA conformation should represent the 'resting' state where the external periplasmic cleft is closed. However, the Cu(I) and Ag(I)-bound CusA structures should correspond to the 'binding' state where the periplasmic cleft is open. Upon metal ion transport, the pump goes through other transient states to actively remove the metal ions. Recently, the co-crystal structures of the CusBA-Cu(I) complexes have been determined. These structures should represent different

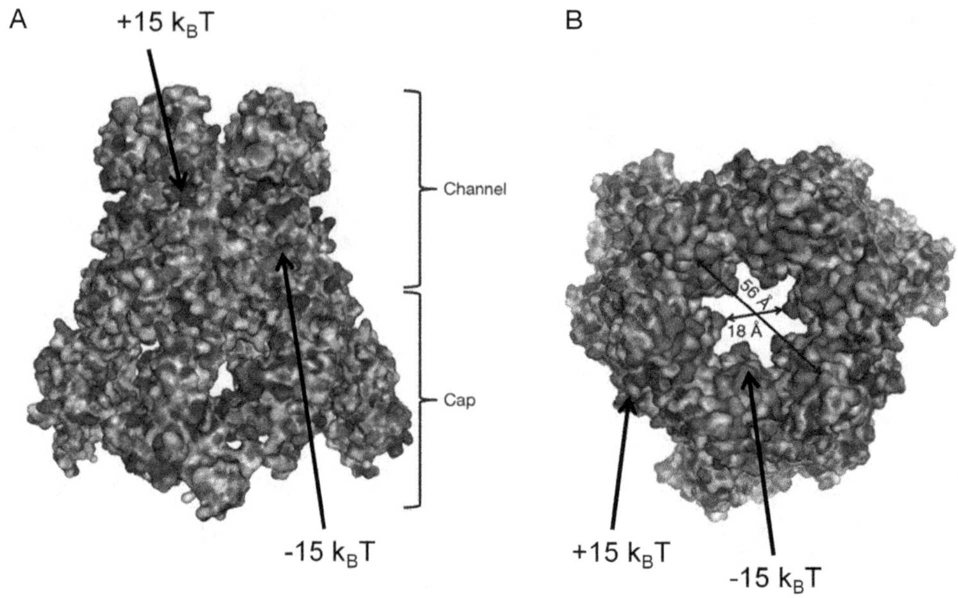

Figure 4.9 Structure of the hexameric CusB channel. (A) Side view of the CusB channel. The six molecules forming the hexameric channel are labelled. Domain 1 and the lower half of domain 2 of the molecules of the CusB hexamer form a cap, whereas the upper half of domain 2, together with domains 3 and 4, of these molecules creates a channel. (B) Top view of the CusB channel with the narrowest point (D232) shown as a stick. The six molecules are the same as depicted in (A). The internal diameter in the narrowest region is ~18 Å.

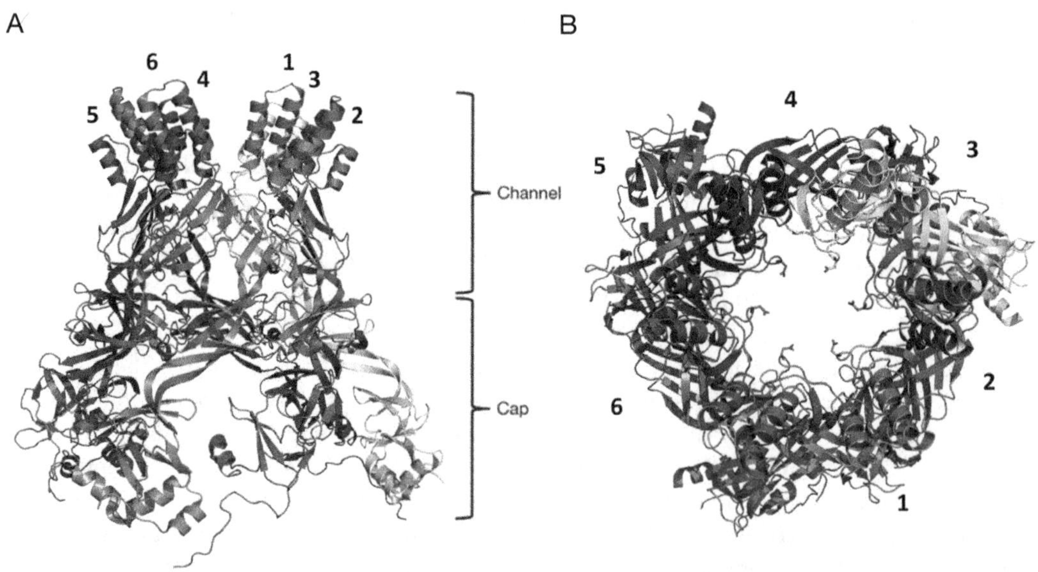

Figure 4.10 Electrostatic surface potential of the hexameric CusB channel. (A) Side view of the electrostatic surface potential of the hexameric CusB channel. This view shows the cap and channel regions formed by the CusB hexamer. (B) Top view of the electrostatic surface potential of the hexameric CusB channel. The widest section of the hexameric channel appears to form at the top edge with its internal diameter of ~56 Å.

intermediates transitioning between the 'binding' and 'resting' states in the transport cycle.

There are three distinct conformations of CusBA-Cu(I) based on the crystal structures (pdb codes: 3T56, 3T51 and 3T53) (Su *et al.*, 2012). These conformations were designated as 'pre-extrusion 1', 'pre-extrusion 2' and 'extrusion' states (Fig. 4.11). The conformation of the 'pre-extrusion 1' state appears to be similar to but is also different from the 'binding' structure, where the periplasmic cleft is open. However, the conformation of the 'pre-extrusion 2' state seems to be closer to but is distinct from the 'resting' structure of the pump. In this conformation, the periplasmic cleft is closed. In the 'pre-extrusion 1' state, a single bound Cu(I) is found to coordinate residues M573, M623 and M672, which forms the distinct three-methionine ion binding site inside the cleft of the periplasmic domain of CusA. Surprisingly, the nearby conserved charged residue E625 seems to be involved in the binding (Fig. 4.12). In comparison with the structure of the 'binding' state, the horizontal helix, formed by residues 665–675, inside the cleft of the structure of 'pre-extrusion 1' state is lower in position (Fig. 4.13). The change can be interpreted as a 3° downwards tilting motion of the C-terminal end of the horizontal helix in this structure when compared with the CusA-Cu(I) form. This motion also shifts M672 away from M573 and M623, seemingly to disassemble the three-sulphur coordination site and to release the bound Cu(I) ion from this site. In contrast to the 'binding' and 'resting' structures, no continuous channel is found in each CusA protomer of 'pre-extrusion 1' state as indicated by the program CAVER (Fig. 4.12A).

In the 'pre-extrusion 2' state, the channel formed by the methionine relay network of each protomer of CusA has been closed up (Su *et al.*, 2012). The only cavity that can be identified is the space nearby the copper signal (Fig. 4.12B,C). This copper signal is found in the familiar three-methionine binding site formed by M573, M623 and M672. However, its binding mode is quite distinct from that found in the 'binding' state. First, the C-terminal end of the horizontal helix (residues 665–675) significantly tilts downwards by ~ 10° in this 'pre-extrusion 2' state in comparison with the 'binding' form. This motion shifts the position of M672 away from M573 and M623, probably weakening the binding for Cu(I) within the methionine triad. Secondly, coupled with this movement, the side chain of a conserved anionic charged residue E625, located deep inside the periplasmic cleft, is found to flip towards the bound Cu(I) and interact with this ion by electrostatic interaction. In addition, a cluster of conserved charged residues, including R83, E567, D617, E625, R669 and K678, is found nearby this copper signal. Interestingly, these conserved charged residues line along the metal-ion transport channel formed by the methionine relay network (Fig. 4.14), suggesting that these charged residues may be important for the transporter's functioning.

In addition to the above movements, the transmembrane helices 5 and 6 (residues 447–495) are found to shift downwards by 5 Å with respect to the membrane plane in both 'pre-extrusion 1' and 'pre-extrusion 2' states, mimicking the change in position of the horizontal helix, when compared with the conformation of the 'binding' state (Fig. 4.13).

The conformation of the 'extrusion' state is nearly identical to that of apo-CusBA, and one bound Cu(I) is found to coordinate M573, M623, M672 and E625 (Fig. 4.12D). The conformation of CusA in this complex structure is also very similar to that of apo-CusA at the 'resting' state. However, a closer inspection of these structures suggests that the CusA protomers in the two CusBA complexes are in a different conformational state when compared with the apo-CusA form. Binding of the CusB adaptor triggers a subtler but significant conformational change at the upper portion of subdomain PC1 of the CusA transporter. It is observed that a short C-terminal helix (residues 391–400) of molecule 1 of CusB directly interacts with and pushes a helix (residues 582–589) located at the upper half of PC1 of CusA (Fig. 4.13). The consequence is that the N-terminal end of this PC1 helix is found to tilt downwards by 8° in the CusBA form in comparison with the apo-CusA structure. This tilting motion in turn further pushes a loop (residues 609–626) located right below the PC1 helix downwards. On the basis of these structures, CusB may be involved

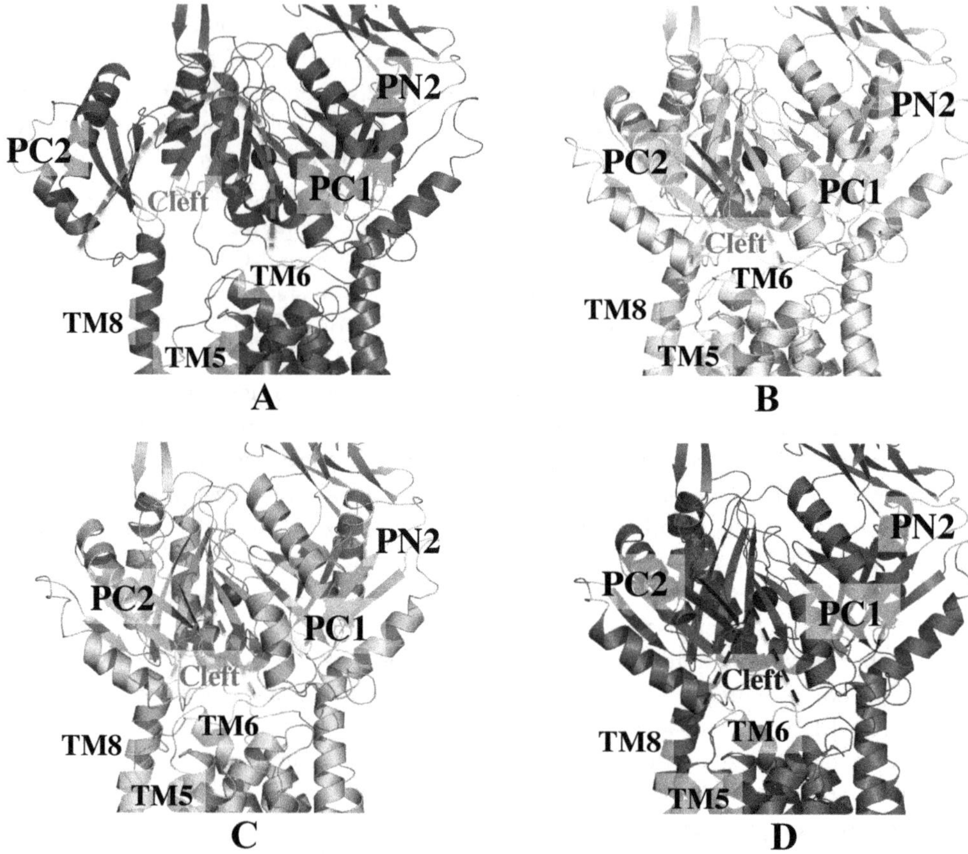

Figure 4.11 Structures of the CusBA-Cu(I) efflux complexes. (A) Ribbon diagram of the structure of form Ia viewed in the membrane plane. The bound copper is shown as a sphere. This structure represents the 'pre-extrusion 1' state of the pump. (B) Ribbon diagram of the structure of form Ib viewed in the membrane plane. The bound copper is shown as a sphere. This structure represents the 'pre-extrusion 2' state of the pump. (C) Ribbon diagram of the structure of form II viewed in the membrane plane. The bound copper is shown as a sphere. This structure represents the 'pre-extrusion 2' state of the pump. (D) Ribbon diagram of the structure of form III viewed in the membrane plane. The bound copper is shown as a sphere. This structure represents the 'extrusion' state of the pump. For clarity, only the periplasmic domain (subdomains PN2, PC1 and PC2) and part of the transmembrane region of CusA are shown in A–D. The molecules of CusB in each complex are not included.

in tuning the width of the channel formed by the methionine metal-ion relay network to its optimal level through the above mechanism.

In the 'pre-extrusion 2' structures, the channel formed by the methionine metal-ion relay network is occluded (Fig. 4.12B,C). It is observed that the region nearby E622 and M812 form the narrowest region immediately after the three-methionine binding site. Interestingly, M812 was found to be significant in relieving copper sensitivity of *E. coli* (Franke *et al.*, 2003). Thus, an empirical measurement defined as the Cα-Cα distance between E622 and M812 was used to identify the degree of opening of the channel. These distances are 13 Å, 12 Å, 10 Å, 10 Å and 11 Å for the 'resting', 'binding', 'pre-extrusion 1', 'pre-extrusion 2' and 'extrusion' states, respectively. In the 'resting' state, the channel is fully open. This channel is also open and is believed to be at its optimal degree of opening in the 'extrusion' form (Su *et al.*, 2012).

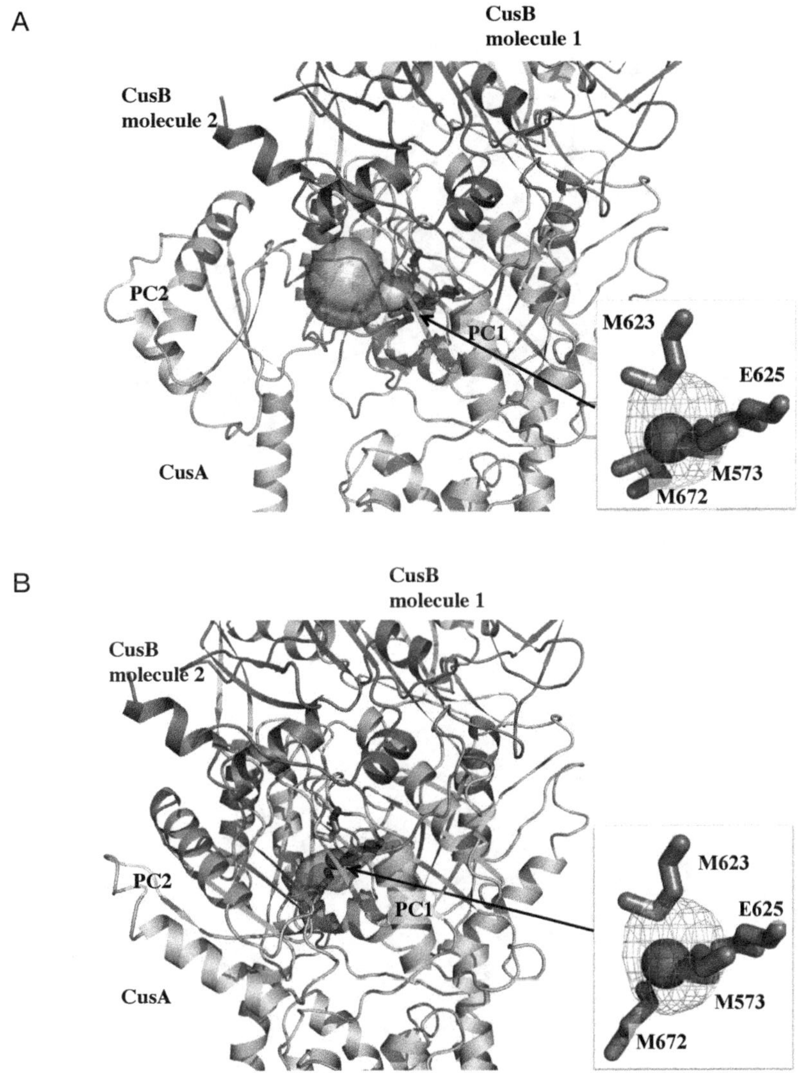

Figure 4.12 Formation of channels at different conformational states. (A) The channel formed by a protomer of the form Ia structure of CusBA at the periplasmic domain is shown. The bound Cu(I) is shown as a sphere. M573, M623, M672 and E625, which form a Cu(I) binding site, are shown as sticks. The CusA and two CusB protomers are ribbons. (B) The channel formed by a protomer of the form Ib structure of CusBA at the periplasmic domain is shown. The bound Cu(I) is shown as a sphere. Anomalous map of the bound Cu(I) is contoured at 3s. M573, M623, M672 and E625, which form a Cu(I) binding site of the form Ib conformation, are shown as sticks. The CusA and two CusB protomers are in ribbons. (C) The channel formed by a protomer of the form II structure of CusBA at the periplasmic domain is shown. The bound Cu(I) ion is shown as a sphere. Anomalous map of bound Cu(I) ion is contoured at 3s. Residues forming the Cu(I) binding sites M573, M623, E625 and M672 are shown as sticks. The CusA and two CusB protomers are in ribbons. (D) The channel formed by a protomer of the form III structure of CusBA at the periplasmic domain is shown. The bound Cu(I) is shown as a sphere. Anomalous map of the bound Cu(I) is contoured at 3s. M573, M623, E625 and M672, which form the Cu(I) binding site, are shown as sticks. The CusA and two CusB protomers are in ribbons.

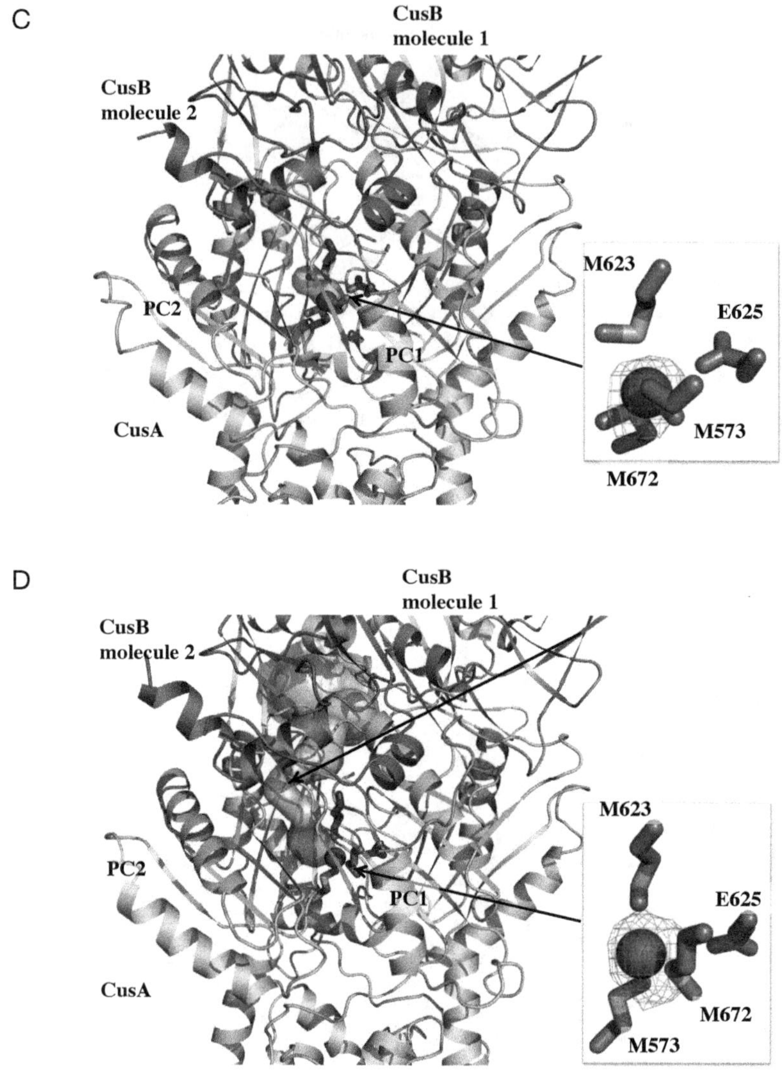

Figure 4.12 (Continued)

Functional studies of the conserved charged residues

The structures of CusBA-Cu(I) highlight the importance of charged residues for the transporter's functioning. These conserved charged residues (R83, E567, D617, E625, R669 and K678) seem to form a network and line along the wall of the methionine relay tunnel in the periplasmic domain of CusA. These residues were mutated into alanines (R83A, E567A, D617A, E625A, R669A, K678A) or aspartate (E625D), and the corresponding mutant transporters were expressed in BL21(DE3)*ΔcueOΔcusA* which lacks both the *cueO* and *cusA* genes. The ability of these mutant transporters to confer copper resistance was then tested *in vivo* using susceptibility assay. Similar to the methionine mutations mentioned earlier, the results indicate that all these CusA mutants are unable to relieve the copper sensitivity of strain BL21(DE3)*ΔcueOΔcusA*, suggesting that these residues are critical for the function of the pump (Su *et al.*, 2012).

The purified R83A, E567A, D617A, E625A, R669A and K678A mutant proteins, which are shown to abolish the function of the pump, were then reconstituted into liposomes containing the

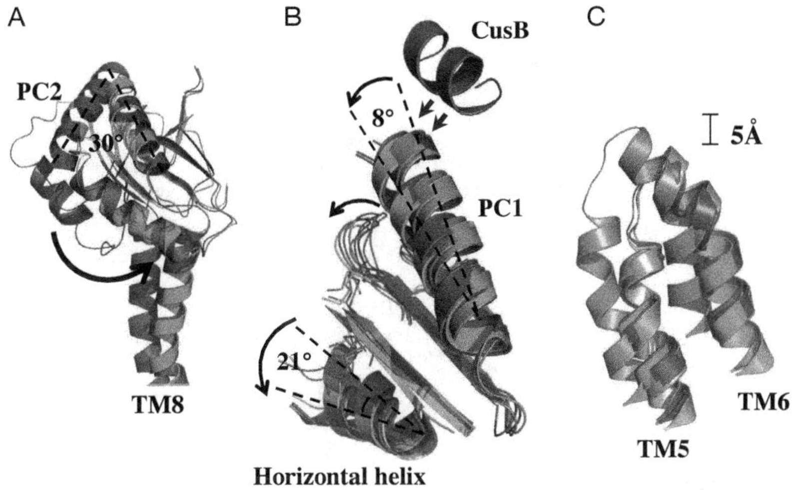

Figure 4.13 Different conformational states of CusBA-Cu(I). (A) Superimposition of subdomain PC2 and TM8 of each state of the pump. The arrow indicates a 30o swing of PC2 upon conformational transition from the 'pre-extrusion 1' to 'pre-extrusion 2' and 'extrusion' states. The structures of forms Ia, II and III are used to represent the 'pre-extrusion 1', 'pre-extrusion 2' and 'extrusion' states, respectively. (B) Superimposition of subdomain PC1 and the horizontal helix of each state of the pump. The figure illustrates the change in conformations of the PC1 helix (residues 582–589) and the flexible loop (residues 609–626) upon CusB binding. For clarity, only the short C-terminal helix (residues 391–400) of molecule 1 of CusB is included. The change in conformation of the horizontal helix at different states of the pump is also shown in this superimposition. (C) Superimposition of the transmembrane helices 5 (TM5) and 6 (TM6) of each state of the pump. The figure illustrates the change in positions of TM5 and TM6 (residues 447–495) within the transport cycle.

fluorescent indicator PGSK in the intravesicular space, respectively. A stopped-flow transport assay was used to determine whether these proteoliposomes can capture metal ions from the extravesicular medium. Unlike the wild-type CusA pump, the results suggest that these mutant transporters are incapable of taking up Ag^+ from the extravesicular medium into the intravesicular space of the proteoliposomes (Su *et al.*, 2012). These collective experiments provide compelling evidence that these conserved charged residues are critical for the transporter's functioning.

The stopped-flow assay was further used to study the transport activity of the reconstituted CusBA proteoliposomes. The results suggest that the CusBA proteoliposomes are more active than those liposomes containing CusA only for metal transport as indicated by the position of the 50% attenuation of the fluorescent signal (Su *et al.*, 2012). No transport activity has been observed in liposomes containing CusB only, indicating that these CusB proteoliposomes cannot uptake silver ions (Su *et al.*, 2012).

Sequential transitions of the CusA efflux pump

Four lines of evidence suggest that the structures identified can be understood in terms of sequential transition of conformations leading to the extrusion of metal ions from the CusA pump (Fig. 4.15). First is the sequential decrease in distance between subdomains PC1 and PC2 from the 'binding' to 'resting' states. This change highlights the sequential movements of CusA upon transiting from one state to the other. Second is the subtler shift in position of the horizontal helix leading to the departure of M672 from the methionine triad and a release of the bound Cu(I) from these methionines. In addition, transmembrane helices 5 and 6 are found to shift in position at different conformational states (Fig. 4.15), mimicking the change of the horizontal helix. Last is a smaller change in conformation of subdomain

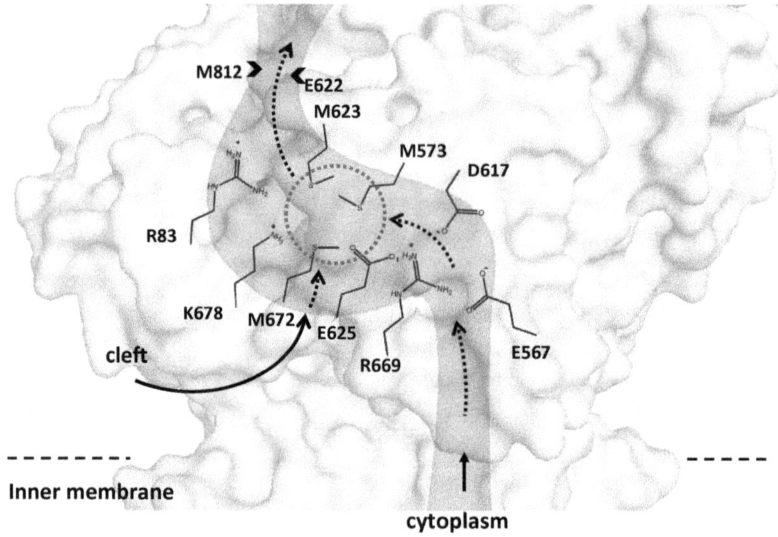

Figure 4.14 Cu(I) binding site and conserved charged residues. This is a schematic representation of the CusA channel. The conserved residues R83, E567, D617, E625, R669 and K678, lining the channel at the periplasmic domain are indicated. The dotted circle marks the location of the Cu(I) binding site formed by the methionine triad M573, M623 and M672. The paths for metal transport through the periplasmic cleft and transmembrane region are illustrated with black curves.

PC1 of CusA upon CusB binding, which in turn controls the degree of opening of the channel formed by the methionine relay network. These structures also suggest that the CusB adaptor may play an essential role in tuning the extent of channel opening upon conformational transition through different states.

The co-crystal structures of CusBA-Cu(I) extend our knowledge on how successive motions in the CusBA efflux pump are coordinated and how CusBA extrudes the bound metal ions. This protein complex undergoes cycles of conformational changes that drive the passage of the metal ions. In the absence of both CusB and Cu(I) or Ag(I), the CusA pump prefers the 'resting' conformation where the periplasmic cleft between subdomains PC1 and PC2 are closed. Cu(I) or Ag(I) binding triggers a large conformational change where subdomain PC2 performs a 30° swing to open this cleft. In the presence of CusB, the subsequent steps may be a transitions to the 'pre-extrusion 1' and then 'pre-extrusion 2' states in which PC2 is found to swing back to close the cleft (Fig. 4.16). The majority of the channel formed by the methionine relay network becomes occluded at the 'pre-extrusion 2' state. Subsequently, the last step is a transition to the 'extrusion' form where CusB is responsible for stabilizing the conformation of this state, and for tuning the channel at the optimal degree of opening to conduct metal ion export. As mentioned earlier, it is very likely that the CusCBA pump operates through an alternating-access mechanism (Murakami *et al.*, 2006; Seeger *et al.*, 2006; Sennhauser *et al.*, 2007). Indeed, functional dynamics simulation has already suggested that the periplasmic cleft of CusA may alternately open and close upon metal transport (Long *et al.*, 2010). As the crystal structures of the CusBA adaptor–transporter complex depict that there are several intermediates participating in the transport cycle, it is difficult to correlate all these conformational states to those of AcrB. Nonetheless, the 'binding' state of the CusBA system should correspond to the 'T' state of the AcrB pump (Murakami *et al.*, 2006; Seeger *et al.*, 2006; Sennhauser *et al.*, 2007).

Unresolved problem

Clearly, the CusA, CusB and CusC proteins need to assemble in order to form the tripartite CusCBA complex to export metal ions. Although the co-crystal structure of the CusBA adaptor–transporter complex has been determined, it is unclear how CusC interacts with CusBA to function. The crystal structure of CusC (Kulathila *et al.*, 2011) suggests that this channel protein resembles the architectures of *E. coli* TolC (Koronakis *et al.*, 2000) and *P. aeruginosa* OprM (Akama *et al.*, 2004a). To understand how the CusBA complex interacts with CusC, molecular dynamics simulation was used to rigidly dock the trimeric CusC channel to the crystal form of the hexameric CusB channel bound to the trimeric CusA pump. The final fitting CusCBA model, which is in the form of $CusC_3$-$CusB_6$-$CusA_3$, was then constructed. This model represents a 750 kDa tripartite efflux complex spanning both the inner and outer membranes of *E. coli* to extrude Cu(I)/Ag(I) (Fig. 4.17). Nonetheless, it is necessary to obtain the co-crystal structure of the entire CusCBA tripartite efflux complex to understand how these three efflux proteins assemble as a functional unit.

Proposed mechanism for metal-ion export

The metal ions can enter the CusA pump via the periplasmic cleft as well as the cytoplasm (Fig. 4.14). Presumably, the methionine pairs/clusters are responsible for metal ion transport. In the transmembrane regions, these methionine pairs are ~15 Å apart. Thus, it is possible that a direct transfer of metal ions from one methionine pairs to the other contributes to deliver the ions through the inner membrane. (Fig. 4.18). It is believed that CusA predominately carries out metal ion efflux through the periplasm (Kim *et al.*, 2011). Thus, the major path for taking up metals should be across the cleft in the periplasmic space. We propose that the CusA transporter utilizes methionine pairs/clusters to export Cu(I) and Ag(I). The periplasmic cleft of CusA presumably remains closed when there is no Cu(I)/Ag(I). In the presence of Cu(I) or Ag(I) ions, the periplasmic cleft opens. Metal ions could enter the three-methionine binding site, formed by M573, M623 and M672, inside the cleft between subdomains PN2 and PC1 on the periplasmic portion of CusA or via the methionine pairs within the transmembrane domain of the pump. It has been demonstrated that the chaperone protein CusF can directly transfer its bound Cu(I) to the CusB membrane fusion protein (Bagai *et al.*, 2008). Thus, it is likely that CusF is responsible for delivering metal ions to the CusCBA tripartite efflux system within the periplasm. There is a chance that the initial step of metal transport through the periplasmic cleft of the CusA pump may involve a direct transfer from CusF to the previously proposed three-methionine metal binding site (Bagai *et al.*, 2007) at the long N-terminal tail of CusB. Based on the co-crystal structure of CusBA, each flexible tail of CusB is located near the PN2/PC1

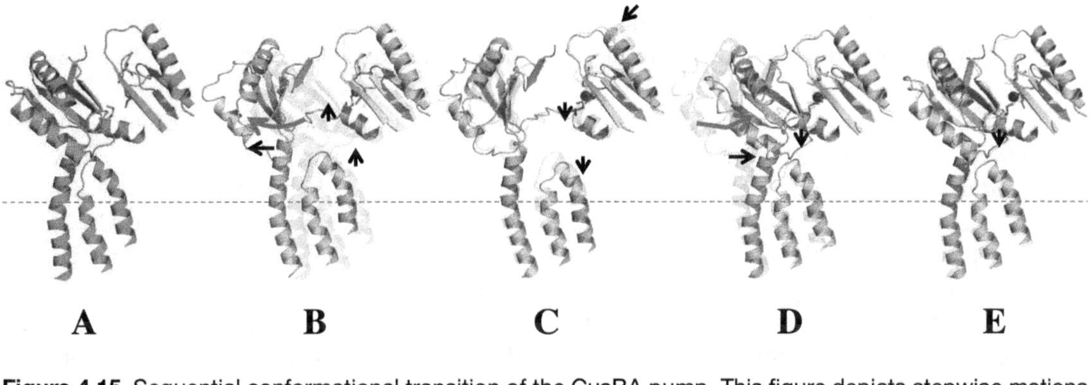

Figure 4.15 Sequential conformational transition of the CusBA pump. This figure depicts stepwise motions of CusA in the adaptor–transporter complex transitioning between the 'resting' and 'extrusion' states (A, 'resting' state; B, 'binding' state; C, 'pre-extrusion 1' state; D, 'pre-extrusion 2' state; E, 'extrusion' state). For clarity, only subdomains PC1 and PC2 in the periplasmic domain and transmembrane helices 5, 6 and 8 in the transmembrane region of CusA are shown. The bound copper is in sphere. The arrows indicate the change in conformation of the protein within the transport cycle.

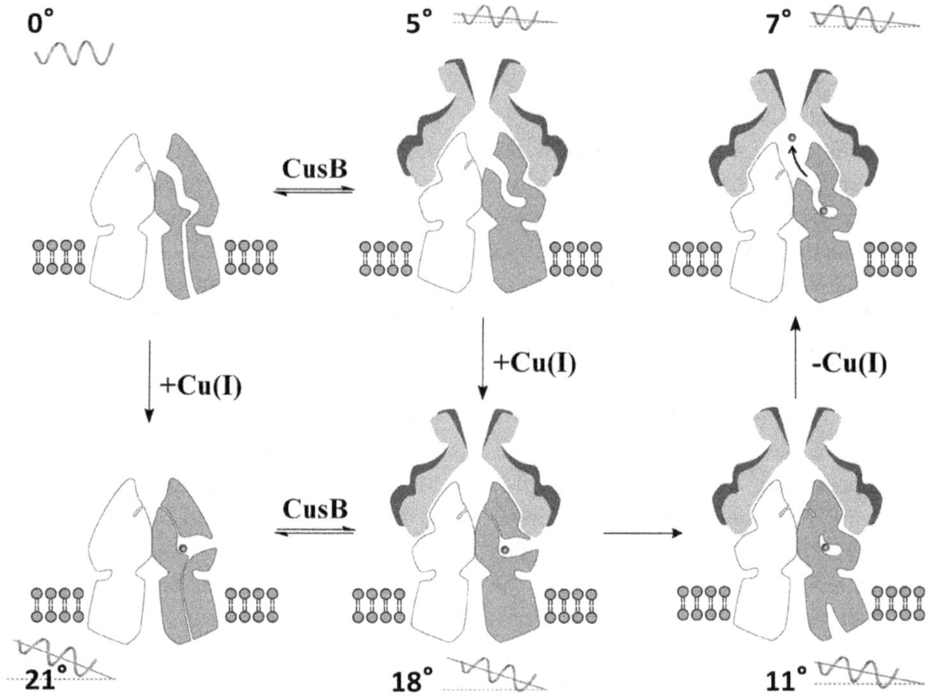

Figure 4.16 Proposed model of metal ion export within the transport cycle. The model is based on different structures obtained by X-ray crystallography. The angle of inclination of the horizontal helix at each state is shown in the figure. For clarity, the front protomers (one CusA and two CusB molecules) are not included.

cleft of the CusA pump. Thus, the second step could well be a deliverance of the bound metal ion from CusB to the three-methionine cluster (M573, M623 and M672) inside the periplasmic cleft of CusA (Long *et al.*, 2010). The bound metal ion could then be released to the nearest methionine pair (M271–M755) located directly above the three-methionine metal binding site (Fig. 4.5), from which the ion could be released into the central funnel of CusA and eventually to the CusC channel for final extrusion. Genetic data and transport assay have shown that the conserved charged residues lining the methionine relay channel are essential for the transporter's functioning (Fig. 4.14). Therefore, it is possible that these conserved charged amino acids are responsible for passing the transported metal ion along from one residue to another in the methionine relay channel.

During the transport cycle, the CusBA efflux complex must go through different transient states, including the 'resting', 'binding', 'pre-extrusion 1', 'pre-extrusion 2' and 'extrusion' conformations, to complete the transport cycle (Figs. 4.15 and 4.16). Thus, in coordination with the motion of the periplasmic cleft, the transmembrane portion of the methionine relay channel should also be synchronized to open and close to accommodate for the export of metal ions.

It is not yet known whether the interior of the hexameric CusB channel forms part of the extrusion pathway for Cu(I) and Ag(I). Based on the crystal structures of CusBA (Su *et al.*, 2011) and CusC (Kulathila *et al.*, 2011), the interior surface of both the hexameric CusB and trimeric CusC channels are predominantly negatively charged and that the interior surface of CusC is more electronegative. Thus, it is likely that the last step of transport of cationic Cu(I) and Ag(I) is facilitated by the negative potential gradient created by the CusB and CusC channels. During metal-ion extrusion, a large conformational change of CusC may occur to open this outer membrane channel, similar to the cases of TolC (Vaccaro *et al.*, 2008) and OprM (Phan *et al.*, 2010). Exactly what this tripartite efflux system's mechanism is must await

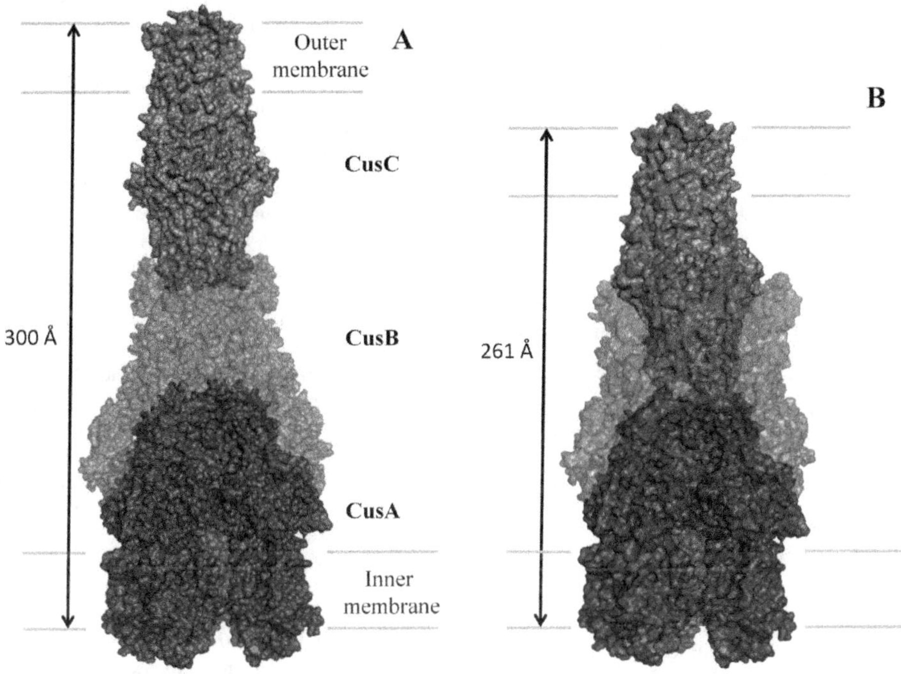

Figure 4.17 Docking of CusC to CusBA. (A) The a-helical end of CusC interacts with the α-helices (Domain 4) of CusB in the CusBA complex. (B) The funnel top of CusA, α-helical end of CusC and Domain 2 of CusB contact one another when the CusC channel slides further down the hexameric CusB channel.

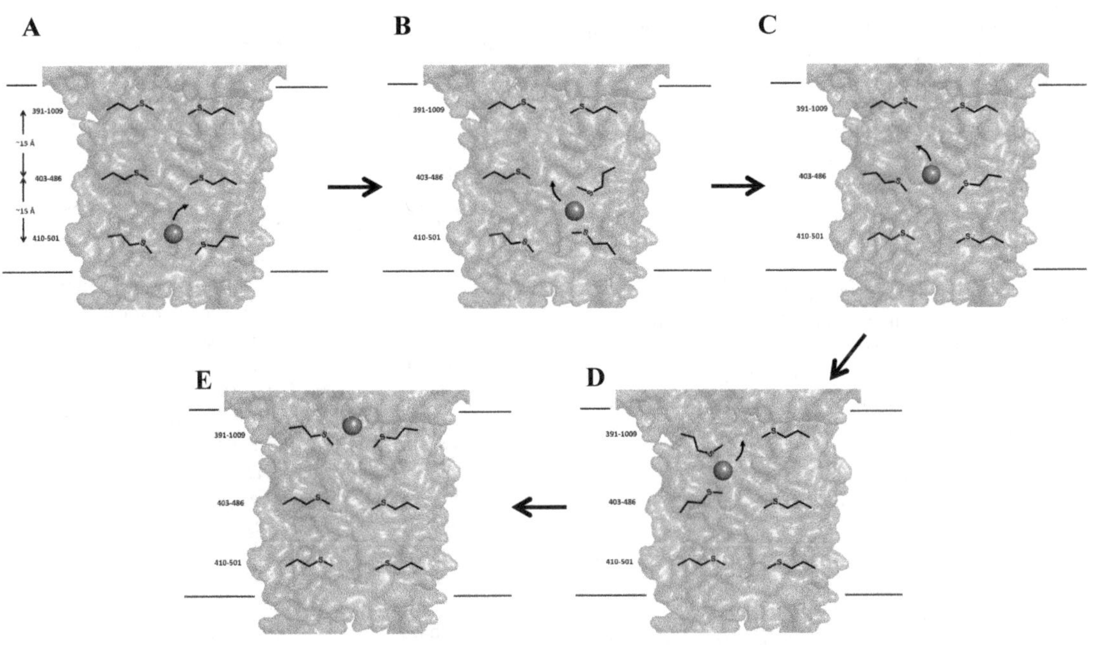

Figure 4.18 Schematic diagram of the methionine pairs in the transmembrane regions of CusA in transport of a Cu(I) ion. After interacting with the ion (A), one of the methionines in the pair moves the ion towards an outreached neighbouring methionine on the next level (B, D). The ion then interacts with the second methionine of the new pair (C, E) before being handed off to a neighbouring methionine on the next level.

confirmation by elucidation of additional crystal structures of the CusCBA tripartite complex.

Acknowledgements

This work is supported by an NIH Grant R01GM086431 (E.W.Y.).

References

Aires, J.R., and Nikaido, H. (2005). Aminoglycosides are captured from both periplasm and cytoplasm by the AcrD multidrug efflux transporter of *Escherichia coli*. J. Bacteriol. *187*, 1923–1929.

Akama, H., Kanemaki, M., Yoshimura, M., Tsukihara, T., Kashiwag, T., Yoneyama, H., Narita, S., Nakagawa, A., and Nakae, T. (2004a). Crystal Structure of the Drug Discharge Outer Membrane Protein, OprM, of *Pseudomonas aeruginosa*. J. Biol. Chem. *279*, 52816–52819.

Akama, H., Matsuura, T., Kashiwag, S., Yoneyama, H., Narita, S., Tsukihara, T., Nakagawa, A., and Nakae, T. (2004b). Crystal Structure of the Membrane Fusion Protein, MexA, of the Multidrug Transporter in *Pseudomonas aeruginosa*. J. Biol. Chem. *279*, 25939–25942.

Arnesano, F., Banci, L., Bertini, I., Huffman, D.L., and O'Halloran, T.V. (2001). Solution Structure of the Cu(I) and Apo-forms of the yeast metallochaperone, Atx1. Biochemistry. *40*, 1528–1539.

Atilgan, A.R., Durell, S.R., Jernigan, R.L., Demirel, M C., Keskin, O., and Bahar, I. (2001). Anisotropy of fluctuation dynamics of proteins with an elastic network model. Biophys. J. *80*, 505–515.

Bagai, I., Liu, W., Rensing, C., Blackburn, N.J., and McEvoy, M.M. (2007). Substrate-linked conformational change in the periplasmic component of a Cu(I)/Ag(I) efflux system. J. Biol. Chem. *282*, 35695–35702.

Bagai, I., Rensing, C., Blackburn, N.J., and McEvoy, M.M. (2008). direct metal transfer between periplasmic proteins identifies a bacterial copper chaperone. Biochemistry. *47*, 11408–11414.

Baranova, N., and Nikaido, H. (2002). The BaeSR two-component regulatory system activates transcription of *yegMNOB* (*mdtABCD*) transporter gene cluster in *Escherichia coli* and increases its resistance to novobiocin and deoxycholate. J. Bacteriol. *184*, 4168–4176.

Bohnert, J.A., Schuster, S., Fähnrich, E., Trittler, R., and Kern, W.V. (2007). Altered spectrum of multidrug resistance associated with a single point mutation in the *Escherichia coli* RND-type MDR efflux pump YhiV (MdtF). J. Antimicrob. Chemother. *6*, 1216–1222.

Changela, A., Chen, K., Xue, Y., Holschen, J., Outten, C.E., O'Halloran, T.V., and Mondragón, A. (2003). Molecular basis of metal-ion selectivity and zeptomolar sensitivity by CueR. Science *301*, 1383–1387.

Das, D., Xu, Q.S., Lee, J.Y., Ankoudinova, I., Huang, C., Lou, Y., Degiovanni, A., Kim, R., and Kim, S.H. (2007). Crystal structure of the multidrug efflux transporter AcrB at 3.1 Å resolution reveals the N-terminal region with conserved amino acids. J. Struct. Biol. *158*, 494–502.

De Angelis, F., Lee, J.K., O'Connell I.I.I., J.D., Miercke, L.J.W., Verschueren, K.H., Srinivasan, V., Bauvois, C., Govaerts, C., Robbins, R.A., Ruysschaert, J.-M., *et al.* (2010). Metal-induced conformational changes in ZneB suggest an active role of membrane fusion proteins in efflux resistance systems. Proc. Natl. Acad. Sci. U.S.A. *107*, 11038–11043.

Franke, S., Grass, G., and Nies, D.H. (2001). The Product of the *ybdE* Gene of the *Escherichia coli* chromosome is involved in detoxification of silver ions. Microbiology *147*, 965–972.

Franke, S., Grass, G., Rensing, C., and Nies, D.H. (2003). Molecular analysis of the copper-transporting efflux system CusCFBA of *Escherichia coli*. J. Bacteriol. *185*, 3804–3812.

Grass, G., and Rensing, C. (2001). Genes involved in copper homeostasis in *Escherichia coli*. J. Bacteriol. *183*, 2145–2147.

Higgins, M.K., Bokma, E., Koronakis, E., Hughes, C., and Koronakis, V. (2004). Structure of the periplasmic component of a bacterial drug efflux pump. Proc. Natl. Acad. Sci. U.S.A. *101*, 9994–9999.

Jiang, J., Nadas, I.A., Kim, M.A., and Franz, K.J. (2005). A mets motif peptide found in copper transport proteins selectively binds Cu(I) with methionine-only coordination. Inorg. Chem. *44*, 9787–9794.

Kim, E.-H., Nies, D.H., McEvoy, M.M., and Rensing, C. (2011). Switch or funnel: how RND-type transport systems control periplasmic metal homeostasis. J. Bacteriol. *193*, 2391–2387.

Kobayashi, A., Hirakawa, H., Hirata, T., Nishino, K., and Yamaguchi, A., (2001). Growth phase-dependent expression of drug exporters in *Escherichia coli* and its contribution to drug tolerance. J. Bacteriol. *16*, 5693–5703.

Koronakis, V., Sharff, A., Koronakis, E., Luisi, B., and Hughes, C. (2000). Crystal structure of the bacterial membrane protein TolC central to multidrug efflux and protein export. Nature *405*, 914–919.

Kulathila, R., Kulathila, R., Indic, M., and van den Berg, B. (2011). Crystal structure of *Escherichia coli* CusC, the outer membrane component of a heavy metal efflux pump. PLos One. *6*, e15610.

Lau, S.Y., and Zgurskaya, H.I. (2005). Cell division defects in *Escherichia coli* deficient in the multidrug efflux transporter AcrEF–TolC. J. Bacteriol. *187*, 7815–7825.

Loftin, I.R., Franke, S., Blackburn, N.J., and McEvoy, M.M. (2007). Unusual Cu(I)/Ag(I) coordination of *Escherichia coli* CusF as revealed by atomic resolution crystallography and X-ray absorption spectroscopy. Protein Sci. *16*, 2287–2293.

Long, F., Su, C.-C., Zimmermann, M.T., Boyken, S.E., Rajashankar, K.R., Jernigan, R.L., and Yu, E.W. (2010). Crystal structures of the CusA efflux pump suggest methionine-mediated metal transport. Nature *467*, 484–488.

Ma, D., Cook, D.N., Alberti, M., Pon, N.G., Nikaido, H., and Hearst, E. (1993). Molecular cloning of *acrA* and *acrE* genes of *Escherichia coli*. J. Bacteriol. *175*, 6299–6313.

Mikolosko, J., Bobyk, K., Zgurskaya, H.I., and Ghosh, P. (2006). Conformational flexibility in the multidrug efflux system protein AcrA. structure. *14*, 577–587.

Murakami, S., Nakashima, R., Yamashita, E., and Yamaguchi, A. (2002). Crystal structure of bacterial multidrug efflux transporter AcrB. Nature *419*, 587–593.

Murakami, S., Nakashima, R., Yamashita, E., Matsumoto, T., and Yamaguchi, A. (2006). Crystal structures of a multidrug transporter reveal a functionally rotating mechanism. Nature *443*, 173–179.

Nagakubo, S., Nishino, K., Hirata, T., and Yamaguchi, A. (2002). The putative response regulator BaeR stimulates multidrug resistance of *Escherichia coli* via a novel multidrug exporter system, MdtABC. J. Bacteriol. *184*, 4161–4167.

Nies, D.H. (2003). Efflux-mediated heavy metal resistance in prokaryotes. FEMS Microbiol. Rev. *27*, 313–339.

Nishino, K., and Yamaguchi, A. (2001). Analysis of a complete library of putative drug transporter genes in *Escherichia coli*. J. Bacteriol. *183*, 5803–5812.

Percival, S.L., Bowler, P.G., and Russell, D. (2005). Bacterial resistance to silver in wound care. J. Hosp. Infect. *60*, 1–7.

Phan, G., Benabdelhak, H., Lascombe, M.-B., Benas, P., Rety, S., Picard, M., Ducruix, A., Etchebest, C., and Broutin, I. (2010). Structural and dynamical insights into the opening mechanism of *P. aeruginosa* OprM channel. Structure *18*, 507–517.

Rosenberg, E.Y., Ma, D., and Nikaido, H. (2000). AcrD of *Escherichia coli* is an aminoglycoside efflux pump. J. Bacteriol. *182*, 1754–1756.

Seeger, M.A., Schiefner, A., Eicher, T., Verrey, F., Dietrichs, K., and Pos, K.M. (2006). Structural asymmetry of AcrB trimer suggests a peristaltic pump mechanism. Science. *313*, 1295–1298.

Sennhauser, G., Amstutz, P., Briand, C., Storchenegger, O., and Grütter, M.G. (2007). Drug Export Pathway of Multidrug Exporter AcrB Revealed by DARPin Inhibitors. PLoS Biol. *5*, e7.

Sennhauser, G., Bukowska, M.A., Briand, C., and Grütter, M.G. (2009). Crystal structure of the multidrug exporter MexB from *Pseudomonas aeruginosa*. J. Mol. Biol. *389*, 134–145.

Su, C.-C., Li, M., Gu, R., Takatsuka, Y., McDermott, G., Nikaido, H., and Yu, E.W. (2006). Conformation of the AcrB multidrug efflux pump in mutants of the putative proton relay pathway. J. Bacteriol. *188*, 7290–7296.

Su, C.-C., Yang, F., Long, F., Reyon, D., Routh, M.D., Kuo, D.W., Mokhtari, A.K., Van Ornam, J.D., Rabe, K.L., Hoy, J.A., *et al.* (2009). Crystal structure of the membrane fusion protein CusB from *Escherichia coli*. J. Mol. Biol. *393*, 342–355.

Su, C.-C., Long, F., Zimmermann, M.T., Rajashankar, K.R., Jernigan, R.L., and Yu, E.W. (2011). Crystal structure of the CusBA heavy metal efflux complex of *Escherichia coli*. Nature *470*, 558–562.

Su, C.-C., Long, F., Lei, H.T., Bolla, J.R., Do, S.V., Rajashankar, K.R., and Yu, E.W. (2012). Charged amino acids (R83, E567, D617, E625, R669, and K678) of CusA are required for metal ion transport in the Cus efflux system. J. Mol. Biol. *422*, 429–441.

Symmons, M., Bokma, E., Koronakis, E., Hughes, C., and Koronakis, V. (2009). The assembled structure of a complete tripartite bacterial multidrug efflux pump. Proc. Natl. Acad. Sci. U.S.A. *106*, 7173–7178.

Tikhonova, E.B., Yamada, Y., and Zgurskaya, H.I. (2011). Sequential mechanism of assembly of multidrug efflux pump AcrAB-TolC. Chem. Biol. *18*, 454–463.

Törnroth-Horsefield, S., Gourdon, P., Horsefield, R., Brive, L., Yamamoto, N., Mori, H., Snijder, A., and Neutze, R. (2007). Crystal structure of AcrB in complex with a single transmembrane subunit reveals another twist. Structure *15*, 1663–1673.

Tseng, T.T., Gratwick, K.S., Kollman, J., Park, D., Nies, D.H., Goffeau, A., and Saier, M.H., Jr. (1999). The RND permease superfamily: an ancient, ubiquitous and diverse family that includes human disease and development protein. J. Mol. Microbiol. Biotechnol. *1*, 107–125.

Vaccaro, L., Scott, K.A., and Sansom, M.S.P. (2008). Gating at both ends and breathing in the middle: conformational dynamics of TolC. Biophys. J. *95*, 5681–5691.

Xue, Y., Davis, A.V., Balakrishnan, G., Stasser, J.P., Staehlin, B.M., Focia, P., Spiro, T.G., Penner-Hahn, J.E., and O'Halloran, T.V. (2008). Cu(I) recognition via cation-π and methionine interactions in CusF. Nat. Chem. Biol. *4*, 107–109.

Yu, E.W., McDermott, G., Zgruskaya, H.I., Nikaido, H., and Koshland, D.E., Jr. (2003). Structural basis of multiple drug binding capacity of the AcrB multidrug efflux pump. Science *300*, 976–980.

Yu, E.W., Aires, J.R., McDermott, G., and Nikaido, H. (2005). A periplasmic drug-binding site of the AcrB multidrug efflux pump: a crystallographic and site-directed mutagenesis study. J. Bacteriol. *187*, 6804–6815.

Yum, S., Xu, Y., Piao, S., Sim, S.-H., Kim, H.-M., Jo, W.-S., Kim, K.-J., Kweon, H.-S., Jeong, M.-H., Lee, K., *et al.* (2009). Crystal structure of the periplasmic component of a tripartite macrolide-specific efflux pump. J. Mol. Biol. 387, 1286–1297.

Zgurskaya, H., and Nikaido, H. (1999). Bypassing the periplasm: reconstitution of the AcrAB multidrug efflux pump of *Escherichia coli*. Proc. Natl. Acad. Sci. U.S.A. *96*, 7190–7195.

Zhou, H., and Thiele, D.J. (2001). Identification of a novel high affinity copper transport complex in the fission yeast *Schizosaccharomyces pombe*. J. Biol. Chem. *276*, 20529–20535.

RND Efflux Pumps for Metal Cations

5

Dietrich H. Nies

Abstract

This review will concentrate on metal-transporting RND (resistance–nodulation–cell division) proteins in bacteria since such proteins have not been found in archaea or eucarya. Metal cations play an important role in cellular biochemistry. They are counter-ions of the negative charges in proteins, nucleic acids and phospholipids, and bridge these negative charges. Transitions metals serve as stability cores of proteins, catalyse redox reactions and rearrange small molecules. However, essential transition metal cations and those not needed by cells are toxic due to their interference with other metal cations, binding to wrong sites and dangerous redox reactions. It is therefore an interesting question how cells manage to allocate the correct metal to the right site and prevent others from doing harm. The first step into such a distribution process is a kinetic flow equilibrium of metal uptake and efflux processes that adjusts the cellular metal cation bouquet in just the right composition. Since metals may change their oxidation state or not but they cannot be degraded as organic molecules, metals are excellent subjects to understand basic homeostatic mechanisms of a living cell.

RND proteins of the HME (heavy metal efflux) protein family are central to the process of metal cation homeostasis in bacteria that are adapted to high concentrations of Zn^{2+}, Cd^{2+}, Ni^{2+}, Co^{2+}, Cu^{+} or Ag^{+}. Such a process can be studied in the beta-protobacterial model bacterium *Cupriavidus metallidurans* strain CH34. Together with the MFP (membrane fusion) and OMF (outer membrane factors), RND proteins form huge transenvelope protein complexes in Gram-negative bacteria, which export metal cations from the cell directly to the outside. This reaction exports periplasmic substrates (PPE –periplasmic efflux) predominantly, while transport from the cytoplasmic to the outside (TEE –transenvelope efflux) may also be possible. RND-driven efflux pumps cooperate with exporters of the inner membrane such as CDF (cation diffusion facilitators) proteins, P-type ATPases, and with periplasmic metal binding proteins. Expression of genes encoding RND proteins is mostly controlled by regulators that should be able to sense the periplasmic metal cation concentrations, e.g. two-component regulatory systems and extracytoplasmic (ECF) sigma factors.

This chapter will introduce the HME family of the RND superfamily. The biochemical problems that are being solved with the help of these proteins will be defined: metal cation homeostasis in environments containing high concentrations of these metals. To set the stage that HME-RND proteins are playing on, we will start with the description of the problems of cellular metal homeostasis and proceed with the introduction of the other proteins involved in handling transition metals such as uptake, regulatory and efflux proteins. These are the interaction partners of HME-RND proteins in the cell. By integrating the structure of HME-RND proteins, the available *in vitro* and *in vivo* data, and the distribution of members of the HME-RND subfamilies in the bacterial kingdom, we will finally try to paint a picture describing what these proteins might do and how they have evolved.

Introduction and screenplay: how to adjust the cellular metal bouquet

Organism in general, and bacteria more specifically, have 92% of their cellular dry matter consisting of the non-metals, C, H, O, and N, which constitute proteins, nucleic acids, carbohydrates including the murein sacculus, and lipids such as phospholipids, and lipopolysaccharides. Additionally, about 4% of the dry mass is P and

S, mainly present as phosphate in nucleic acids and phospholipids, or in thiol groups in proteins, respectively. The remaining 4% of the cellular dry mass are mostly metals plus a little Se, Cl and other halogens. Monovalent alkali metal cations K^+ and Na^+ regulate the osmotic pressure of the cytoplasm, counteract the negative charges of the non-metals, adjust the cytoplasmic pH value and form gradients across the cytoplasmic membrane. Divalent earth alkali metal cations Mg^{2+} and Ca^{2+} bridge negative charges in the cytoplasm and cytoplasmic membrane, respectively (Nies, 2004a).

Transition metals are below 1% of the cellular dry mass but essential for all kind of living organisms. Concerning cations, those with a beneficial effect can usually be found in the first transition period of the periodic system of the elements: Mn^{2+}, Fe^{3+}, Co^{2+}, Ni^{2+}, Cu^{2+} and Zn^{2+}. In contrast, metal cations from the second and third transition period have no beneficial function with one described example for Cd^{2+} being the exception (Xu *et al.*, 2008). Nevertheless, Mo and W belong to these two periods. Together with V from the first period, these elements make up the three beneficial transition metal oxyanions (Schwarz *et al.*, 2007), and listed together with the six cations and Se mentioned above as 'trace elements' (Nies, 2004a). Oxyanions, however, will not be discussed here, because there is currently no known example of a metal oxyanion-transporting HME-RND protein.

Transition metals usually have two valence electrons in an outer s-orbital and incompletely filled d-orbitals located more closely to the atomic nucleus (Housecroft and Constable, 2006). This enables most of these cations to form complex compounds, which may be redox active as in case of Mn, Fe and Cu. Mn serves as 'electron buffer' in the water-splitting complex of the cyanobacterial photosynthetic reaction centre II (Umena *et al.*, 2011) or may substitute many other cations including Mg^{2+} as bridging cations for negative charges. Iron mediates one-electron transfer reaction such as in haem groups or iron–sulphur centres. Copper is used for reactions involving molecular oxygen as substrate. Redox activity of Co and Ni is limited. These metals are compounds of cofactor B_{12} and enzymes re-arranging small molecules such as hydrogenases and ureases, respectively. Zinc has no redox activity at all and Zn^{2+} nucleates wobbly polypeptide chains into a more rigid conformation or serves in Lewis acid in biochemical reactions. Finally, molybdate performs oxyanion catalysis as part of the molybdate cofactors, for instance, in nitrate reductase or in an even more complicated cofactor in nitrogenases. W or V may do a similar job in enzymes not so widely distributed as Mo-dependent enzymes (Nies, 2004a; Nies and Grass, 2009).

This illustrates the essential function of transition metals in cellular biochemistry. To perform these functions, the metals must be imported into the cells. Since diffusion of charged oxyanions or cations across the hydrophobic inner core of biological membranes would be an extremely slow process, import is mediated by transport proteins (Fig. 5.1). These proteins may transport the metals in their ionic form. Alternatively, metals may be complexed by organic compounds that were excreted by bacteria or other organism to chelate insoluble metal ions and increase their availability. The best examples of such substances are the siderophores, which bind Fe^{3+} that otherwise would form insoluble hydroxide complexes (Braun *et al.*, 1998; Braun and Hantke, 2001; Braun and Braun, 2002). A similar 'chalkophore' designated methanobactin acquires copper for methanotrophic bacteria that need this metal for an effective initial oxidation of the C1 substrate methane (Kim *et al.*, 2004; Choi *et al.*, 2006a,b). To some extent, vitamin B_{12} and its derivatives may also be addressed as an exchange form of Co between bacteria (Nies and Grass, 2009). The nickel-containing cofactor F_{430}, needed for methane synthesis, is probably not able to play a similar role as a 'nickel exchange form': Co(III) in B_{12} and its derivatives forms a kinetically stable octahedral low spin complex with three completely filled non-binding d-orbitals. This leaves only anti-binding d-orbitals with a much higher energy available for any d-electrons of Co(III) or an additional electron in Co(II). This is the 'secret' of B_{12} stability. In contrast, Ni(II) already contains one electron in each of the two anti-binding d-orbitals, rendering Ni(II)-haem complexes much more unstable than those containing Co(III).

Transport of these metal complexes across

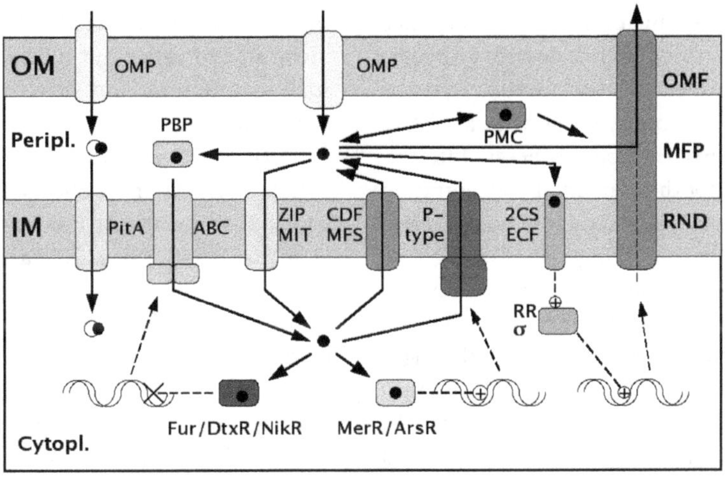

Figure 5.1 Basic components of transition metal cation homeostasis on bacteria: defining the physiological role of RND systems. Metal cations (black dots) cross the outer membrane (OM) via outer membrane porins (OMP, yellow). They are imported further on by fast and unspecific secondary uptake systems, e.g. from the ZIP or MIT protein families (green). If this is not sufficient, genes repressed by regulatory proteins of the Fur, DtxR, or NikR family (dark blue) are being de-repressed, which encode primary uptake systems like ABC transporters (medium green) that interact with periplasmic metal-binding proteins (PBP, medium green). If, on the other hand, a metal is in surplus, regulatory proteins of the MerR or ArsR protein families activate expression of genes for primary efflux pumps such as P-type ATPases (red) that cooperate with secondary efflux such as CDF and MFS members (orange). There might be also uptake of metal:phosphate complexes by OMPs and the PitA transporter (on the left hand). Periplasmic cations are being sensed by two-component or ECF sigma factor regulatory systems (light blue), which control expression of the genes for RND-driven systems and maybe periplasmic metal chaperones (purple). For the RND protein, PPE is indicated as thick line, TEE as thin and dashed line. A colour version of this figure is located in the plate section at the back of the book.

the outer membrane (OM) in Gram-negative bacteria is a complicated active transport process. The transenvelope protein complex TonB–ExbB–ExbD uses the proton-motive force across the cytoplasmic membrane to change into a conformation of high energy. Specific OM proteins bind the siderophore:Fe complexes or B_{12} at the outside of the outer membrane, and transport into the periplasm is coupled to a release of the conformational energy of the TonB–ExbB–ExbD complex (Braun, 1995; Braun and Braun, 2002). The RND proteins may be involved in the export of synthesized or 'recycled' empty metal chelators (Bleuel *et al.*, 2005).

Transport of non-complexed metals is likely by facilitated diffusion through other outer membrane proteins such as OmpC or OmpF in *Escherichia coli* (Fig. 5.1). Further uptake of ionic metals into the cytoplasm is by primary or secondary transport systems. While primary transport systems form gradients across biological membrane by using other forms of energy such as light, redox or other exergonic chemical reactions, secondary transport systems use one gradient as the driving force to build up another one. The known primary transport systems for metals are mostly driven by ATP hydrolysis. One exception might be the GTP-hydrolysing FeoB protein of the Feo system involved in iron import. However, the situation is not clear here and Feo-mediated iron transport driven by GTP-hydrolysis has not been shown *in vitro* (Braun and Hantke, 2007; Nies and Grass, 2009). Usually, production of the highly specific primary uptake systems, which need more energy for the transport process than the secondary transport systems, is strictly regulated by metal availability (Fig. 5.1).

The secondary transport systems are uniporters or metal-proton symporters, and belong to several protein families. Usually, these importers have a rather low substrate specificity but a high import rate, thus providing a broad range of metal ions to the cell with similar ionic diameters. Since no time is wasted to bind and identify a specific substrate, these rather unspecific uptake systems have a high catalytic turnover rate with a relative low energy cost per transport process. Interestingly, the ranking of the elemental concentration in environments such as seawater is not so much different from that of a living cell (Fig. 5.2), so that these unspecific uptake systems seem to provide the required metals cheaply to the cell, and in most environments roughly in the right bouquet composition (Nies, 2007a). It is interesting to note that the major bioelements Na, K, Mg, Ca and S cluster as one group but P and the transition metal cations in another one (Fig. 5.2). This agrees with the stability of transition metal cation phosphate complexes and the resulting importance of the metal:phosphate importer PitA in bacteria (Kirsten *et al.*, 2011). Thus, metal phosphate complexes may be an important source of both phosphate and metal cations, including those of transition metals.

This combination of secondary, unspecific uptake systems with inducible, primary and specific ones manages an efficient supply of metals to the cell (Fig. 5.1). However, as a disadvantage, high concentrations of an essential metal ion in the environment leads to too high amounts of this metal in the cells, and unwanted metal cations such as cadmium and silver may also enter the cytoplasm.

Both situations lead to metal toxicity, which is mainly based on three effects. First, there is competition of metals for binding sites (Irving and Williams, 1948; Nies, 2007a). The primary site for nickel toxicity in *E. coli* is the fructose-1,6-bisphosphate aldolase FbaA and nickel binds here to the non-catalytic zinc-site to exert the toxic action (Macomber *et al.*, 2011). Co and probably also Ni compete with Fe during the formation of iron–sulphur complexes, making it important for the cell to diminish use of cobalt and nickel (Wu *et al.*, 1994; Ranquet *et al.*, 2007; Thorgersen and Downs, 2007). This may be the reason why Co is mainly used in form of Co(III) in the kinetically inert B_{12} complex and derivatives thereof, which are also exchanged between different cells in the environment (Nies and Grass, 2009). The number of enzymes using nickel is low and specific nickel chaperones sequester the metal, providing it only to the insertion machinery of the respective enzyme such as hydrogenase, urease or anaerobic carbon monoxide reductase (Hausinger and Zamble, 2007).

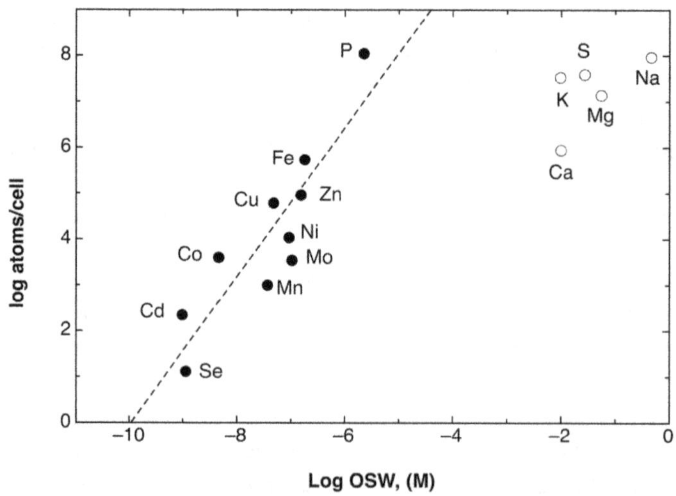

Figure 5.2 The number of atoms per cell follows the concentration of these metals in the 'standard' environment sea water (Nies, 2012). The numbers of atoms per cell was taken from *C. metallidurans* strain CH34 wild type grown in a Tris-buffered mineral salts medium (Kirsten *et al.*, 2011) and the OSW (occurrence in sea water) from the Handbook of Chemistry and Physics (Weast, 1984). Please note that P groups with the transition metals.

Secondly, there is a toxic effect due to the binding of Cd, Cu and probably all transition metal cations of the second and third transition periods to thiol residues and sulphides. This leads to numerous effects as analysed with cadmium (Helbig *et al.*, 2008b): inhibition of cysteine biosynthesis, leading to glutathione starvation followed by impairment of the repair processes for oxidized proteins, sequestration of glutathione, production of reactive oxygen species and oxidized glutathione by metal-bis-glutathionato complexes reacting with molecular oxygen, protein damage in general, inhibition of formation of iron–sulphur clusters leading to disturbance of the cellular iron homeostasis leading to redox stress, and damage of sensitive iron complexes such as that of aconitase. Third, some metal cations, e.g. iron and copper, produce toxic reactive oxygen species in the Fenton reaction as described by Haber and Weiss (Haber and Weiss, 1932). Toxic action of a metal may occur in the cytoplasm and in the periplasm of a Gram-negative bacterium.

So, superfluous essential and non-beneficial metals have to be eliminated from the cell, if their uptake cannot be limited. However, import cannot be turned down if an unspecific uptake system imports the superfluous metal. In this case, metal efflux systems have to export the metal from the cytoplasm. This is not a futile cycle because the energy spent on im – and export is used to adjust the exact composition of the cellular metal ion bouquet. Maintenance of this bouquet seems to be the prerequisite for allocating the correct metal to the right protein, which is not a trivial problem since about 40% of all enzymes contain metals, ranking from Mg (16%) > Zn (9%) > Fe (8%) > Mn (6%) > Ca (2%) > Co, Cu (1%) down to K, Na, Ni, V, Mo, W and the one example for Cd (Tottey *et al.*, 2007; Waldron *et al.*, 2009).

Overview: metal-transporting proteins and protein families in bacteria

Organisms, starting on the species level, are sorted into taxonomic groups such as genera, families, orders and classes. When the taxonomic system used is a natural one, it mirrors evolution, meaning that the taxonomic groups are 'generalization units': a result obtained with one member of the unit can be transferred as hypothesis onto a second unit with a probability of a verification that decreases with increasing taxonomic order. What this means is that a result generated with *Escherichia coli* has a high probability to be also true for *Shigella,* a smaller one for *Salmonella,* an even smaller one for *Bacillus* but also some residual probability even for an elephant (Silver, 2011).

Proteins can be sorted at least into families and superfamilies too, which also resemble a common heritage. For transport proteins higher 'taxonomic' units are artificial and based on energetics and transport mode. Nevertheless, the 'transporter categorization', which is based on a mixture of a natural and artificial taxonomic system, is an important tool for molecular biologists, comparable to the enzyme categorization, the EC classes (Busch and Saier, 2002; Saier *et al.*, 2006). The transporter categorization and TC classes of efflux systems is covered by Milton Saier in this book.

Some of the transporter protein families are big ones, occurring in most living organisms, and perhaps even with more than one member in a particular organism. Members of the same protein family are usually homologues stemming from a common ancestor. A pair of paralogues would be two homologous proteins that originated during speciation of the organism as consequence of a gene duplication event. Genes for paralogues may be expressed or silent, and may evolve into different functions during evolution. In contrast to paralogues, orthologues are homologues in two different organisms that perform more or less the same function (Tatusov *et al.*, 1997). Nevertheless, two proteins belonging to the same protein family are not always paralogues or orthologues, and these expressions have to be used carefully (von Rozycki and Nies, 2009).

To further sort families of transport protein families, the criteria that may be used are (i) mode of energetic coupling, e.g. primary transporter driven by ATP hydrolysis or secondary transporter driven by the proton motive force, (ii) membership of protein families that transport cations or oxyanions, (iii) family members occurring as monomeric or homo-multimeric proteins or as parts of heteromultimeric protein complexes, and (iv) membership in mostly efflux or uptake proteins (Table 5.1).

Efflux and uptake proteins may be members of the same protein family (Table 5.1) since transport proteins perform reaction possible in either direction, similar to 'non-transporting' enzymes. In general, there are no differences in the mechanism of uptake and efflux reactions. If for instance an efflux protein functions as a cation/proton antiporter, a typical transport cycle would be composed of (i) binding of a cytoplasmic cation to a substrate-binding site in the transmembrane part of the transport protein, followed by (ii) release of the cation to the outside (the periplasm in Gram-negative bacteria) due to protonation of the binding site by periplasmic protons, and (iii) closing of the catalytic cycle by import of the protons to the cytoplasm. In such a case, a related uptake protein could be simply mediate steps 2 and 1 in the inverse orientation, i.e. binding of a periplasmic cation to the substrate-binding site followed by import into the cytoplasm. This illustrates how both kinds of transport reactions are connected and how easily an efflux system may be turned into an uptake system or vice versa by evolution. Which reaction occurs also depends on thermodynamics, the differences in free energy and the concentrations of substrates and products, in this case the concentration in either compartment. It may even be possible that a protein serves as importer and as exporter, with the mode of action depending on the actual concentrations of substrates and products on either side of the membrane.

Thus, uptake and efflux proteins act together and adjust the concentration of a single metal in the cytoplasm or the periplasm, and they fine-tune the composition of the cellular metal bouquet.

Table 5.1 Important protein families of that contain transition metal cation transportersa

Name	TC#	Energy	Typical substrates	Configuration	Direction	Examples
Primary transport systems						
	3. Primary active transporters; 3.A P-P-bond-driven transporters					
ABC	3.A.1	ATP	many[b]	Heteromultimeric	Mainly uptake	ZnuABC
P1-type ATPase	3.A.3	ATP + pmf	Ca^{2+}, Mg^{2+}, Cu^{2+}, Zn^{2+}	Monomeric	Mainly efflux	CopA, ZntA
	9. Incompletely characterized transporters; 9.A. unknown biochemical mechanism					
FeoB	9.A.8	GTP	Fe^{2+}	Heteromultimeric	Uptake	FeoAB
Secondary transport systems						
	1. Channel/Pores; 1.A α-type channels					
MIT	1.A.35	Pmf, $\Delta\Psi$	Mg^{2+}, Zn^{2+}	Homopentameric	Mainly uptake	CorA, ZntB
Mer	1.A.72	Pmf, $\Delta\Psi$	Hg^{2+}	Unknown	Uptake	MerT
	2. Electrochemical Potential-driven Transporters; 2.A Porters					
MFS	2.A.1	Pmf	Ni^{2+}	Unknown	efflux	NreB
CDF	2.A.4	Pmf, ΔpH	Zn^{2+}, Co^{2+}, Cd^{2+}	Dimers	Mainly efflux	CzcD
ZIP	2.A.5	Pmf, $\Delta\Psi$	Zn^{2+}, Fe^{2+}	Unknown	Uptake	ZupT
DMT	2.A.7	Pmf, ΔpH	Ni^{2+}	Unknown	efflux	CnrT
RND-HME	2.A.6.1	Pmf, ΔpH	Zn^{2+}, Ni^{2+}, Cu^{2+}	Heteromultimeric	Efflux	CusA, CzcA
PiT	2.A.20	Pmf, ΔpH	Metal phosphates	Unknown	Uptake	PitA
NiCoT	2.A.52	Pmf, $\Delta\Psi$	Ni^{2+}, Co^{2+}	Unknown	Mainly uptake	HoxN, RcnA
NRAMP	2.A.55	Pmf	Mn^{2+}, Fe^{2+}, Cd^{2+}	Unknown	Uptake	MntH

[a]Classification according to the transport classification database (Saier *et al.*, 2006) at www.tcdb.org. Protein family names are explained in the TCDB or in the text. Other abbreviations: pmf, proton-motive force, portions indicated when available.

[b]Cations, anions, complexes and organic molecules, depending on the individual protein complex.

Therefore, to understand the function of metal-transporting RND proteins, contribution of all of these transport proteins to the cellular metal homeostasis has to be defined. Moreover, the activity of a transport protein may be regulated on the flux control level (Nies, 2007b) and by transcriptional control. Therefore, the regulators of gene expression involved in metal homeostasis also need to be introduced. When this is done the stage is prepared to address metal-transporting RND proteins.

The other side of the coin: metal uptake systems and their energetics

In bacteria, there is usually a proton gradient across the cytoplasmic or inner membrane, the pmf (proton motive force). In respiring bacteria for instance, electron transport in a respiratory chain is coupled to proton extrusion, which builds up the pmf. The electron transfer is from components with a negative or low E_0' value such as NADH (−320 mV) to those with a positive E_0' value such as molecular oxygen (+816 mV) or other electron acceptors. The pmf may be used to drive transport processes or the formation of ATP by the F_1F_0 ATPase. This protein complex functions in either direction, synthesis of ATP driven by proton influx or proton efflux driven by ATP hydrolysis. The ratio of ATP (synthesized or split) to the proton being transported strictly depends on the number of c-subunits of the F_0 membrane part of the protein complex with respect to the three ATP-binding sites in its F_1 part. The *E. coli* protein contains 10 c-subunits. So, the movement of 10 protons would allow for the synthesis of three ATP. Since the synthesis of one ATP under physiological conditions needs +50 kJ/mol of free energy, three ATP would need +150 kJ/mol, so each proton import has to release at least −15 kJ/mol. Since $\Delta G = -nF\Delta E$ (ΔG, difference in free Gibbs energy; n, charge transferred, F, Faraday constant that is about 96,485 Cb/mol; ΔE potential difference), the pmf that drives this process has to be larger than 150 mV. In fact, the pmf in most bacteria is about 200 mV because some entropy needs to be produced during ATP synthesis. If the pmf decreases below 150 mV, there is danger that the F_1F_0 complex functions in the opposite direction and results in hydrolysing the cytoplasmic ATP pool to maintain the pmf (Junge *et al.*, 2009, Watt *et al.*, 2010).

The pmf is composed of two parts – the charge gradient across the cytoplasmic membrane $\Delta\Psi$ and the potential resulting from the different proton concentrations on both sides, $\text{pmf} = \Delta\Psi + Z\Delta\text{pH}$ ($Z = 59$ mV at 30°C). Since the cytoplasmic pH_i in mesophilic bacteria is usually around pH 7.6 (Padan *et al.*, 2001; Hunte *et al.*, 2005) and the pH_a outside depends largely on the growth medium, ΔpH in a mesophilic growth medium at neutral pH value is $\Delta\text{pH} = 0.6$ and therefore $Z{\cdot}\Delta\text{pH} = 35$ mV. In this case, a pmf of 200 mV is predominantly composed of $\Delta\Psi = 165$ mV. In contrast, a fermenting *E. coli* culture may decrease the pH_a value below pH 5, leading to a $\Delta\text{pH} \geq 2.6$, a $Z{\cdot}\Delta\text{pH} \geq 153$ mV and a $\Delta\Psi$ of only 47 mV (Lengeler *et al.*, 1999).

This consideration is important for the understanding of metal cation import reactions, and later on to discuss the mode of action of metal-transporting RND proteins. Since many uptake systems such as MIT (metal inorganic transport), NiCoT (nickel cobalt transporters) or ZIP (zinc iron permeases) family members are uniporters that are driven by the charge gradient $\Delta\Psi$, the maximum value of the gradient inside/outside they are able to build up depends on $\Delta\Psi$, and therefore on the external pH values. Since $\Delta G = -RT\ln K$ (*R*, general gas constant 8.3145 J/mol/K; *T*, absolute temperature) and $\Delta G = -nF\Delta E$, $-RT\ln K = -nF\Delta E$ and therefore $K = e^{(nF\Delta E/R/T)}$, a divalent metal cation ($n = 2$) may be accumulated 308,000-fold at pH_a 7.0 but only 37-fold at pH_a 5.0.

The MIT family contains mostly magnesium uptake systems that also import some transition metal cations such as cobalt. A typical MIT protein is CorA from *Salmonella*. MIT proteins form homopentameric complexes that seem to be under flux control (Smith *et al.*, 1993; Lunin *et al.*, 2006). If in surplus, cytoplasmic magnesium cations bind to the protein, thereby preventing uptake of more magnesium or decreasing the transport rate. Such a process has also been considered for another family of magnesium importers, the MgtE family not introduced here (Townsend *et al.*, 1995; Hattori *et al.*, 2007) and, the other way around, also for efflux systems (Lu and Fu, 2007; Nies, 2007b). One MIT protein, ZntB from *Salmonella*, may be a zinc efflux system (Wan *et al.*, 2011; Worlock and Smith, 2002).

ZIP proteins are mostly zinc and iron importers, but they are also unspecific in their substrate range and most divalent metal cations of the first transition period may serve as substrate (Grass *et al.*, 2002; Grass *et al.*, 2005a; Taudte and Grass, 2010). The first ZIP member in bacteria was ZupT from *E. coli*. NiCoT proteins, with HoxN from *Cupriavidus eutrophus* as first member, supply nickel and cobalt to the synthesis of enzymes such as hydrogenases that need these metals (Eitinger and Friedrich, 1991; Wolfram *et al.*, 1995; Eitinger *et al.*, 2005). Again, some NiCoT proteins are efflux systems (Rodrigue *et al.*, 2005; Iwig *et al.*, 2006; Koch *et al.*, 2007), while a ZIP exporter has not been found yet.

NRAMP (natural resistance-associated macrophage proteins) systems for manganese and iron are special because they are cation/proton symporters, leading to a much higher maximum accumulation and independence of the accumulation reached from the pH_a outside (Kehres *et al.*, 2000; Makui *et al.*, 2000; Agranoff *et al.*, 2005). NRAMP systems also transport cadmium. An example for an NRMAP protein from bacteria is MntH from *E. coli*. The metal-resistant bacterium *C. metallidurans* seems to lack an NRAMP system to decrease import of Cd^{2+}, and may have sacrificed usage of a Mn-containing superoxide dismutase for this purpose (Kirsten *et al.*, 2011). The PiT protein family contains metal phosphate uptake systems such as PitA from *E. coli* and *C. metallidurans*, which seem to be important contributors of cytoplasmic metal cations (Jackson *et al.*, 2008; Kirsten *et al.*, 2011).

It may be a surprise that bacteria contain specific uptake systems (e.g. MerT) for the toxic mercury cation. However, Hg^{2+} is detoxified by reduction to the metallic, volatile form Hg^{o} by the NADPH-depending protein MerA. Since there is only NADPH in the cytoplasm, Hg^{2+} has to be imported for detoxification. A specific periplasmic mercury-binding metal-chaperone, MerP, and the uptake reaction are also necessary to protect the proteins in the periplasm of Gram-negative bacteria. It is possible that Hg^{2+} is never released during transfer from MerP to MerT and finally MerA, because it is always bound to thiol groups and transferred from one thiol pair to the next one (Silver and Phung, 1996; Silver, 2003, 2011).

Thus, in general, uptake of metal cations is mediated by secondary transport systems with a rather low substrate specificity, which are not so strongly controlled on the level of transcription but maybe rather by flux control (Kirsten *et al.*, 2011). Examples are MIT, ZIP, PiT and NRAMP proteins. These importers provide the basic metal ion bouquet to the cell, which resembles many environments such as sea water in its composition (Fig. 5.2). Additionally, some secondary transport systems serve specialized needs such as NiCoT and Mer (mercury resistance) proteins. If, however, metal import by these systems does not supply amounts of a metal to the cell, primary uptake systems are being induced.

Most important among those are the ABC (ATP-binding cassette) transport systems that are composed of a periplasmic metal-binding protein, two transmembrane proteins or protein domains, and two ATP-binding proteins or protein domains. Examples exist for phosphate, sulphate, and all essential transition metals except copper: ZnuABC for zinc (Patzer and Hantke, 1998; Li and Jogl, 2007), NikABCDE for nickel (Wu *et al.*, 1991; Navarro *et al.*, 1993; Depina *et al.*, 1995; Allan *et al.*, 1998; Rowe *et al.*, 2005; Shepherd *et al.*, 2007), SitABCD for manganese and iron (Ikeda *et al.*, 2005; Sabri *et al.*, 2006; Davies and Walker, 2007; Sabri *et al.*, 2008), CbiMNQO for cobalt, which lacks the periplasmic binding protein (Rodionov *et al.*, 2006), and ABC importers for molybdate and tungstate (Schwarz *et al.*, 2007). Uptake by these primary systems needs ATP-hydrolysis, which is more expensive than using the pmf. This might be the reason why primary uptake systems are only synthesized in times of special need or metal starvation.

So, the secondary transport systems provide a metal cation bouquet to the cytoplasm. If one 'flower' in this bouquet is too short, specialized secondary or primary uptake systems are being synthesized and fill in. If, on the other hand, a 'flower' is too long, efflux systems remove these surplus cations.

Inner membrane export: cation diffusion facilitators and P-type ATPases

The most prominent protein families containing efflux systems for transition metal cations are the P-type ATPases and the CDF (cation diffusion facilitators) proteins (Fig. 5.1). CDF proteins (Nies and Silver, 1995; Paulsen and Saier, 1997) are homodimers of the cytoplasmic membrane and are driven by the pmf (Chao and Fu, 2004b; Wei and Fu, 2005; Lu and Fu, 2007). The proteins function as proton–cation antiporters with a ratio of one to two protons/cation. Protonation of the metal-binding site is a prerequisite for release of the metal cation to the other side of the inner or cytoplasmic membrane (Chao and Fu, 2004a,b). So, the transport cycle is a typical transport cycle being composed of (i) binding of a cytoplasmic metal cation to a central metal-binding site in the transmembrane part of the protein, (ii) release of the metal to the outside by protonation using protons from the outside or periplasm, and (iii) closing the cycle by proton release into the cytoplasm. However, and as described above, a ratio of protons imported to cation exported of 1:1 is likely an *in vitro* artefact because import of one protons at pH = 7 is driven by the 35 mV Z·ΔpH portion of the pmf while export of one positive charge (divalent cation charge minus that of one proton) is against ΔΨ and −165 mV, yielding −130 mV or about +13 kJ/mol as endergonic thermodynamical balance of the transport process. In contrast, a 2:1 ratio would be an electroneutral export of the cation driven by 2 35 mV = 70 mV or about −7 kJ/mol.

At low pH values such as those reached by fermenting *E. coli* cultures on the other hand, one proton import yield 153 mV at pH_a 5 and export is against a small ΔΨ = −47 mV, yielding 106 mV or −11 kJ/mol. However, it is unlikely that CDF proteins shift their mode of action depending on the outside pH value and switch between protonation of the central metal-binding site by one or two protons.

The first bacterial CDF protein described was CzcD from *C. metallidurans* (Nies *et al.*, 1989; Nies, 1992a). CDF proteins have been found in all domains of life with half a dozen paralogues in humans (Nies, 2003). The typical CDF substrate is zinc but all cations of the first transition period (Mn^{2+}, Fe^{2+}, Co^{2+}, Ni^{2+}, Cu^{2+}, Zn^{2+}) plus Cd^{2+} may be substrate of individual CDF proteins. It may be the function of bacterial CDF proteins to remove surplus metal cations, which have just been imported by secondary uptake systems, rapidly and before they can reach 'wrong' binding sites (Fig. 5.1).

If a metal cation reaches one of these 'wrong' binding sites, P-type ATPases may be needed to extract and export them. The catalytic cycle of P-type ATPases was examined with the calcium-transporting protein of the sarcoplasmic membrane of muscles.

1. The calcium substrate binds to a central metal-binding site in the transmembrane part of the protein.
2. The ATP-binding domain of the protein opens up and binds ATP.
3. ATP – hydrolysis leads to a phosphorylated intermediate that gave the name to this protein family, and to protonation of the central metal-binding site by protons on the other side of the membrane, followed by metal release into this compartment.
4. Dephosphorylation, proton import and relaxation of the conformation back to the initial state (Toyoshima *et al.*, 2004; Echarte *et al.*, 2007).

It is interesting to note that if the ATP part is being subtracted, the catalytic cycle of the P-type ATPases becomes similar to that of a CDF protein or any other 'standard' proton/cation antiporter.

P-type ATPases are a large superfamily that can be divided into single protein families, e.g. the P_{IB}-type family transporters of transition metal cations. This family can be further divided into several subfamilies. P_{IB1}-type ATPases transport monovalent cations such as those of copper and silver, an example is CopA from *E. coli*, P_{IB2}-type divalent metal cations of zinc, cadmium and lead (Argüello *et al.*, 2007; Raimunda *et al.*, 2011). *C. metallidurans* contains four P_{IB1} proteins and three P_{IB2} proteins (von Rozycki *et al.*, 2005; von Rozycki and Nies, 2009), the latter exporting zinc (ZntA), cadmium (CadA) or lead (PbrA). Additionally, the bacterium possesses a member of yet another P_{IB} subfamily, the P_{IB4}-type protein CzcP. Compared to the three P_{IB2}-type proteins, CzcP has a much higher transport rate, but is not able to mediate much metal resistance when being the only P-type ATPase in the cell for zinc or cadmium (Scherer and Nies, 2009). Reminiscent to CDF proteins, the function of CzcP could be rapid export of just arriving surplus zinc and cadmium cations, before these are able to reach 'wrong sites'.

There are also efflux proteins for transition metal cations from other protein families, such as CnrT from the MFS (major facilitator superfamily) that exports nickel. As stated above, some protein families that are mainly composed of metal importers also contain a few efflux systems such as ZntB from the MIT family and RcnA from the NiCoT family. Sometimes, an efficient combination of a secondary uptake system and a secondary efflux system is enough to mediate an extraordinary level of metal resistance (Marrero *et al.*, 2007), probably by using the differences in the relationship of the $\Delta\Psi$ and the $Z{\cdot}\Delta pH$-portions of the pmf as explained above to decrease metal uptake but increase metal efflux rate at low pH_a values. There are also protein families containing efflux proteins for metal oxyanions such as chromate or arsenate but these are not covered in this book chapter because their physiological function is not connected to that of RND-HME proteins.

The regulators of gene expression

This topic will be lightly covered because there are excellent reviews and book chapters in this field (Helmann *et al.*, 2007; Hobman *et al.*, 2007; Reyes-Caballero *et al.*, 2011). Members of many protein families of regulatory proteins sense the cytoplasmic metal concentration and regulate transcription initiation of the genes for transport and other proteins. Members of the Fur family (Fur, Zur, Mur, Nur), the DtxR (DtxR, IdeR, MntR, ScaR, SirR) and the NikR (NikR) family sense the availability of cytoplasmic manganese, iron, nickel and zinc cations and mostly regulated synthesis of uptake systems. On the other hand, MerR (MerR, ZntR, CueR, PbrR) and ArsR (ArsR, SmB, CadC, CzrA, NmtR) sense surplus cytoplasmic metal cations and regulated synthesis of efflux systems such as P-type ATPases (Fig. 5.1).

Other regulators seem to sense the periplasmic metal concentration, two-component regulatory systems and ECF (extracytoplasmic functions) sigma factors. Two-component regulatory systems are composed of at least two 'core' proteins, typically a membrane-bound histidine kinase sensor protein and a soluble response regulator. Upon receiving a signal, the histidine kinase uses ATP to phosphorylate a histidine residue of the sensor protein, then the phosphate residue is transferred from there to an aspartate residue of the response regulator, which is usually now able to activate transcription initiation of a target promoter (van der Lelie *et al.*, 1997; Nies and Brown, 1998; Große *et al.*, 1999). Some of these two-component regulatory systems contain additional components such as signal input domains for the sensor protein, more than one domain able to accept and transfer the phosphate residue, or phosphatases that reset the signal.

ECF sigma factors (Lonetto *et al.*, 1994) are a family of proteins that are usually regulated by anti-sigma factors and sense signals from outside the cytoplasm (Nies, 2004b). If a signal is received by the anti-sigma factor or one or more additional signal input proteins on one side of the membrane, this leads to a release of an otherwise sequestered ECF sigma factor on the cytoplasmic side of the membrane. The ECF sigma factor is now able to bind to the core RNA polymerase (RNAP) protein or is even being loaded into RNAP core by the anti-sigma factor. The resulting RNAP holoenzyme is now able to initiate transcription. Examples for ECF sigma factors are RpoE in *E. coli* (Egler *et al.*, 2005; Rouviere *et al.*, 1995) controlling envelope stability, FecI in *E. coli* for iron citrate uptake (Mahren and Braun, 2003) and CnrH in *C. metallidurans* for nickel resistance (Liesegang, 1993; Grass *et al.*, 2000; Tibazarwa *et al.*, 2000; Grass *et al.*, 2005b).

The RND proteins: structure, function, families and subfamilies

The RND superfamily

RND proteins (Saier *et al.*, 1994) are a large superfamily of proteins with at least seven subfamilies involved in many processes such as export of proteins, organic substances or transition metals. Besides the RND-components SecDF of the Sec general secretion pore (Tseng *et al.*, 1999), Gram-negative bacteria may contain many members of the important RND families RND-HAE (hydrophobic amphoteric efflux) or HME-RND (Table 5.1). RND-HAE proteins export antibiotics, organic solvents, the messenger molecule indole or bile acid. The best studied examples for HAE-RND proteins are as AcrB from *E. coli* and MexB from *Pseudomonas aeruginosa.* HME-RND proteins mediate efflux of transition metal cations including Cu^{+}, Ag^{+}, Ni^{2+}, Co^{2+}, Zn^{2+} and Cd^{2+}. The main models of HME-RND proteins are CusA from *E. coli* and CzcA from *C. metallidurans.* More details on the RND protein superfamily are provided in Chapter 1 of this book by Miton Saier.

The RND-driven transenvelope protein complex: an efflux gun

RND proteins are part of heteromultimeric protein complexes composed of 12 subunits. This protein complex spans the complete cell wall of a Gram-negative bacterium from the cytoplasmic face of the inner membrane to the outside face of the outer membrane. The actual RND protein is a homotrimer located in the inner membrane and extending into the periplasm. Here, the RND trimer connects to a outer membrane factor OMF (Paulsen *et al.*, 1997), a tube-like trimer located as a beta-barrel in the outer membrane and extending also into the periplasm. The RND trimer and OMF trimer are connected by the third component, the membrane fusion or adaptor protein MFP (Saier *et al.*, 1994), which forms a hexameric ring around the contact side of RND and OMF, stabilizing and allowing contact (Koronakis *et al.*, 2000; Murakami *et al.*, 2002; Yu *et al.*, 2003; Murakami *et al.*, 2004; Yum *et al.*, 2009; Long *et al.*, 2010; Su *et al.*, 2011). In total, these RND-driven protein complexes resemble huge 'guns' traversing the cell wall of Gram-negative bacteria.

Metal-transporting RND proteins may sustain the periplasmic metal bouquet by export of surplus cations from the periplasm to the outside (periplasmic efflux, PPE) or, alternatively, from the cytoplasm to the outside (transenvelope efflux, TEE). In case of PPE, RND proteins counteract diffusion of metal cations across the outer membrane though outer membrane proteins and they compete with metal uptake systems for their substrates. In case of TEE, RND proteins counteract metal uptake proteins and they compete with efflux proteins such as P-type ATPase or CDF proteins for their substrate. Before the question can be addressed if the ammunition of the efflux gun comes from the cytoplasm or the periplasm, the structure and functional mode of metal transporting RND proteins is being discussed.

RND primary structures and early evolution

HAE-RND and HME-RND proteins are huge, composed of more than 1000 amino acid residues. They contain 12 transmembrane alpha helices (TMHs) and two large periplasmic portions. TMH number I is located at or close to the N-terminus of the proteins, TMHs II to VII in the middle of the protein and THMs VIII to XII at or close to the C-terminus. The large periplasmic domains are located between TMHs I and II, and between TMHs VII and VIII.

The halves of the HME or HAE monomers are related to each other, indicating that these proteins originated from an ancient gene-duplication and – fusion event (Saier *et al.*, 1994). Thus, the ancestral 'RND half protein' may have contained a periplasmic domain and a transmembrane domain composed of six transmembrane alpha helices. SecD and SecF resemble this ancestral protein. They are still separate proteins, each containing six TMHs. Similar to the ancestral 'RND half protein', both start with an N-terminal TMH I, contain a large periplasmic part, and end with five additional TMHs (II to VI). After the fusion, helix I of the protein now comprising the

C-terminal domain of the merged polypeptide became helix VII. So, I/VII, II/VIII, III/IX, IV/X, V/XI and VI/XII are paralogous TMHs.

One of the first identified RND proteins was the NolGHI protein complex from *Rhizobium*, which is required for nodulation. In this case, the three polypeptides, NodG, NodH and NodI, assembled head-to-tail and aligned with big RND proteins such as CzcA (Saier *et al.*, 1994). This protein complex might resemble an even earlier ancestor of RND proteins, if it is not the product of a gene-splitting event. NolH forms the central part of the RND-like protein complex and is composed of about five or six transmembrane alpha-helices, resembling many transport proteins with six transmembrane spans. NolG plays the part of the N-terminus of RND proteins, containing the initial helix I and the first periplasmic domain. NolI starts with the second periplasmic domain and finishes with the five C-terminal transmembrane domains.

Gene duplication event seems to be not uncommon in bacteria. A chromate efflux protein, ChrA, was obviously also the result of such a process and 'ChrA half proteins' exist in many genomes such as *C. metallidurans* and *Bacillus subtilis*. Since SecDF proteins occur in all three domains of life but HAE – and HME-RND proteins only in bacteria, the ancestral RND protein may have been present even in the last universal ancestor (LUCA) of all life forms on this planet.

RND structures: AcrB, MexB, CusA and SecDF

Three-dimensional structures are available for HAE-RND proteins AcrB (Murakami *et al.*, 2002), MexB (Sennhauser *et al.*, 2009), the HME-RND protein CusA (Long *et al.*, 2010), and recently SecDF from the SecDF–RND family (Tsukazaki *et al.*, 2011). The CusA structure will be discussed in detail by Edward W. Yu in Chapter 4 of this book.

Both HAE-RND proteins and the HME-RND protein form homo-trimeric 'jellyfish' structures composed of a membrane domain and a large periplasmic head portion. The 12 TMHs per monomer form one transmembrane domain per monomer. TMHs II, III, V, VI, and their paralogues VIII, IX, XI, XII build the shell of a hydrophobic transmembrane 'block' and serve as border to the surrounding phospholipids. TMHs IV and its paralogue X are sitting inside this domain, indicating a special function. TMHs I and VII, the two TMHs at the N-terminus of the possible ancestral 'RND half protein', are outside of the big transmembrane domain with TMH I between adjacent monomers (Murakami *et al.*, 2002). This orients the transmembrane 'block' of RND proteins in two halves, as has been predicted 20 years ago (Nies, 1992b).

While the periplasmic parts of AcrB and SecDF were different, the transmembrane regions were similar (Tsukazaki *et al.*, 2011). The six TMHs of SecD and SecF are again forming a solid transmembrane domain, with TMH I of SecD and TMH VII (first TMH of SecF) slightly outside of this domain and TMH IV and X in its middle (Tsukazaki *et al.*, 2011).

Each of the two periplasmic parts of HME – or HAE-RND proteins contains three subdomains designated PN1, PN2, and DN in case of the first one and PC1, PC2, DC in case of the second one (Table 5.2). The three DNs and DCs per trimer build a docking domain to contact the OMF, MFP and the neighbouring RND monomers, and are located on top of the 'jellyfish'. The four other periplasmic subdomains per monomer are sitting between the transmembrane part and the docking domain with PN2, PC1, and PC2 forming the outside shell of the trimer with a large cleft between the PC2 and PN2 subdomains of adjacent monomers. They have similar structures, each is composed of four beta sheet structures and two alpha helices, and they are responsible for the function of the RND protein. The orientation of the total RND monomer and possible functions of the domains are listed as an overview in Table 5.2.

The RND trimer contains a central pore formed by the three PN1 subdomains of the three monomers (Long *et al.*, 2010). Domain DN contains a large arm in the middle that connects to the adjacent monomer. The paralogous arm region of DC points to the interior of the docking domain and may contact the arm from the respective adjacent monomer (Murakami *et al.*, 2002)

A central substrate-binding site of CusA is formed by three methionine residues, M573,

Table 5.2 Structural components of RND monomers, their location and possible function[a]

Component[b]	Location	Possible function
TMH I	Between 'membrane blocks'	Monomer-monomer contact
PN1	Inner wall of RND trimer	Substrate transfer to central pore
PN2a	Outer wall of RND trimer	
DN	Top of monomer	Contact to OMF and next monomer
PN2b	Outer wall of RND trimer	
TMH II tp VI	Half 'membrane block'	TMH IV: proton transport
TMH VII	Outside of 'membrane block'	
PC1	Outer wall of RND trimer	Substrate binding
PC2a	Outer wall of RND trimer	Substrate binding
DC	Top of monomer	Contact to OMF and next monomer
PN2b	Outer wall of RND trimer	
TMHs VIII to XII	Half 'membrane block'	TMH XI: proton transport

[a]The RND monomer is organized in three levels, a 'membrane block' in the cytoplasmic membrane composed of eight transmembrane alpha-helices (TMHs) with TMHs IV/X inside the 'block' TMHs I/VII outside, a first periplasmic level and the top level.
[b]The components are listed from the amino terminus of the RND polypeptide chain to the carboxy terminus.

M623 and M672, all located in PC1 and the following domain PC2a. Binding of a metal here opens another cleft between PC1 and PC2 (Long *et al.*, 2010). In AcrB, these periplasmic domains are also responsible for the substrate specificity (Elkins and Nikaido, 2002). The alpha helices of PN1 directly lining the central pore of the AcrB trimer needs to change its conformation during the transport reaction mediated by this protein (Murakami *et al.*, 2004), indicating that the substrates may be moved from the central substrate binding site into the central pore. A similar movement was also observed for an AcrB orthologue from *Haemophilus influenza* (Dastidar *et al.*, 2007). The amino acyl residues Phe664, Phe666 and Glu673 might be involved in substrate-binding in AcrB (Yu *et al.*, 2005). These residues are also located in the PC1-PC2a region, not too far away from the positions of the CusA-Met residues mentioned above, indicating that the position of the substrate-binding sites of RND proteins may be conserved between HAE and HME proteins. The C-terminal 'RND-half protein' seems to play the role of the substrate-specific part while its N-terminal paralogue may be responsible for proton transport, as has been suggested 20 years ago (Nies, 1992b).

Since the conformations of the individual monomers of the trimer differ, consecutive rotation of the monomeric conformations between (i) a loose conformation L open to the periplasm, (ii) a closed or tight conformation T and (iii) an conformation O open to the central pore was suggested (Seeger *et al.*, 2006; Murakami, 2008). In conformation L, a tunnel for substrate diffusion starts at the PC1/PC2 cleft leading to a possible hydrophobic substrate-binding site. Preventing this cleft opening and closing inhibits function of AcrB (Takatsuka and Nikaido, 2007). In conformation O, another tunnel leads from here to the central pore (Seeger *et al.*, 2006). The suggested 'peristaltic rotation mechanism' was supported by linking the AcrB monomers covalently (Seeger *et al.*, 2008; Takatsuka and Nikaido, 2009).

RND proteins are molecular engines driven by the proton-motive force

The HME-RND protein CzcA transports Zn^{2+}, Co^{2+} and Cd^{2+} across the membrane of proteoliposomes and is driven *in vivo* and *in vitro* by the proton-motive force (Nies, 1995; Goldberg *et al.*, 1999) or more exactly its ΔpH portion and not the charge portion ΔΨ. This was also confirmed for AcrB (Zgurskaya and Nikaido, 1999). TMH IV of RND proteins, which is located in the middle of the monomeric transmembrane 'block' formed by most of the other TMHs, contains amino acid residues conserved in most RND proteins (Tseng *et al.*, 1999). The conserved motif in CzcA is DFGX_3DGAX_3VEN in TMH IV while that in CusA is AVGX_3DGAX_3VEN. Specifically, the Asp residues in the middle of this TMH (D408 in CzcA, D405 in CusA, D340 in SecD from *Thermus thermophilus*) and the Glu-Asn at the cytoplasmic face of TMH IV are conserved in most RND proteins and essential for function of CzcA, CusA, SecDF and AcrB (Goldberg *et al.*, 1999; Su *et al.*, 2006; Long *et al.*, 2010; Tsukazaki *et al.*, 2011). Therefore, these residues might be involved in proton transport and release to the cytoplasm. To prevent the uncoupling of the pmf this proton transport should exclusively be allowed to occur if the respective RND monomer is in the closed 'T' conformation.

In HAE-RND proteins, the aspartate residue in the middle of TMH IV is duplicated in a highly conserved 'DD' motif, and both Asp residues are essential (Su *et al.*, 2006). In AcrB, both aspartate residues form a hydrogen bond with K940 of TM XI, adjacent to the paralogous TMH X of TMH IV.

Vibrio alginolyticus contains a SecDF complex driven by a sodium gradient instead of the proton-motive force. With this protein complex, a clear sodium dependence of protein export could be shown in patch-clamp experiments (Tsukazaki *et al.*, 2011). This gave clear evidence that RND proteins are indeed driven by protons or Na^+ cations that were previously expelled into the periplasm by respiration, the F_1F_0 ATPase or other processes. So, if they are not driven by a sodium gradient, RND proteins are driven by the ΔpH portion of the pmf.

SecDF are part of the general protein secretion apparatus Sec in bacteria. During protein export by Sec, a N-terminal leader sequence of the nascent polypeptide chain mediates sequestration of this polypeptide by SecG or other chaperones, preventing folding of the protein. The sequestered polypeptide migrates to the Sec pore in the cytoplasmic membrane and is injected here into the SecYEG pore by the SecA ATPase, which pushes 20 amino acyl residue into SecA per ATP hydrolysed. The function of SecDF is to release the polypeptide from the Sec apparatus after movement through SecY. In this case, the proton (or Na^+) movement leads to a conformational change of the periplasmic P1 head of SecD, which helps to pull the preprotein through and release it from SecYEG (Tsukazaki *et al.*, 2011). Thus, in SecDF, proton movement drives the conformational change of a periplasmic domain of this protein complex.

In HAE-RND and HME-RND proteins, proton movement drives the 'peristaltic' rotation of the three monomers between three different conformations (Yu *et al.*, 2005; Seeger *et al.*, 2006; Takatsuka and Nikaido, 2006, 2009). Therefore, a general motif for RND protein seems to be that these proteins are molecular engines in which a conformational change is driven by the proton (or sodium) motive force, or more specifically, release of individual protons from the periplasm to the cytoplasm. Furthermore, conserved amino acid residues in or downstream of TMH IV are involved in proton transport.

The HME–RND subfamilies and representative proteins

C. metallidurans contains 12 members of the HME-RND protein family but does not express all 12 genes (von Rozycki *et al.*, 2005; Nies *et al.*, 2006; von Rozycki and Nies, 2009). Based on the sequence of the conserved TMH IV and sequence similarities in 63 sequenced prokaryotic genomes those days (Nies, 2003), the HME-RND proteins were sorted into five subfamilies HME1 to HME5 (Table 5.3 and Fig. 5.3). *C. metallidurans* does not contain a gene encoding a HME5 protein (Fig. 5.3).

CzcA from HME1 is the archetype of the RND proteins and encoded as part of the *czc* metal resistance determinant of the native plasmid pMOL30 of the beta-proteobacterium *C. metallidurans* strain CH34 (Mergeay *et al.*, 1985). Strain CH34, which contains a second native plasmid (pMOL28) in addition to pMOL30, had many names in the past, e.g. *Wautersia metallidurans*, *Ralstonia metallidurans*, *Ralstonia* spec., *Alcaligenes eutrophus* or *Pseudomonas* (Mergeay *et al.*, 1978, 1985, 2003; Taghavi *et al.*, 1997; Mergeay, 2000; Goris *et al.*, 2001; von Rozycki *et al.*, 2005; von Rozycki and Nies, 2009; Janssen *et al.*, 2010).

The *czc* metal resistance determinant mediates

Table 5.3 HME-RND subfamilies and their members in *C. metallidurans*

	Name		Subfamily	Status	Environment
Gene bank	Old	New			
YP_145595	CzcA	CzcA	HME1	430@Zn(II)	pMOL30, MFP, OMF, CDF, MFS, PMCs, 2CS
YP_586602	Hme465	HmuA	HME1	Silent	MFP.
YP_161705	CnrA	CnrA	HME2	18@Ni(II)	pMOL28, MFP, OMF, DMT, ECF
YP_145656	NccA	NccA	HME2	Silent	pMOL30, MFP, OMF, NiCoT
YP_587447	HME647a	ZniA	HME3a	5@Zn(II)	MFP, OMF
ZP_00273751	HME593	HmvA	HME3a	constitutive	nonsense@RND, MFP, OMF.
YP_587457	HME647b	ZneA	HME3a	Silent	MFP, OMF, 2CS
YP_586259	HME695	HmyA	HME3c	Silent	
ZP_00274918	HME711	$NimA_{1/2}$	HME3b	3@Co(II)	Tn@RND, MFP
YP_585153	HME696	3011[a]	HME3b	Silent	MFP, OMF, 2CS
YP_587164	HME619	CusA	HME4	20@Ag(I)	MFP, OMF
YP_145664	HME709	SilA	HME4	constitutive	pMOL30, MFP, OMF

The table lists the current gene bank number, the old (Nies, 2003, Nies *et al.*, 2006) and new name (Janssen *et al.*, 2010) of each protein. Status indicates which metal induces expression of the genes and how many fold. Environment says plasmid or not, and which proteins are encoded adjacent to the RND protein. The interrupted *nimA* gene and [a]Rmet_3011 were annotated as pseudo-genes for whatever reason. Abbreviations not used in the text: 2CS, two-component regulatory system, DMT, divalent metal transporter family, NiCoT, nickel-cobalt transporter family; PMC, periplasmic metal chaperone.

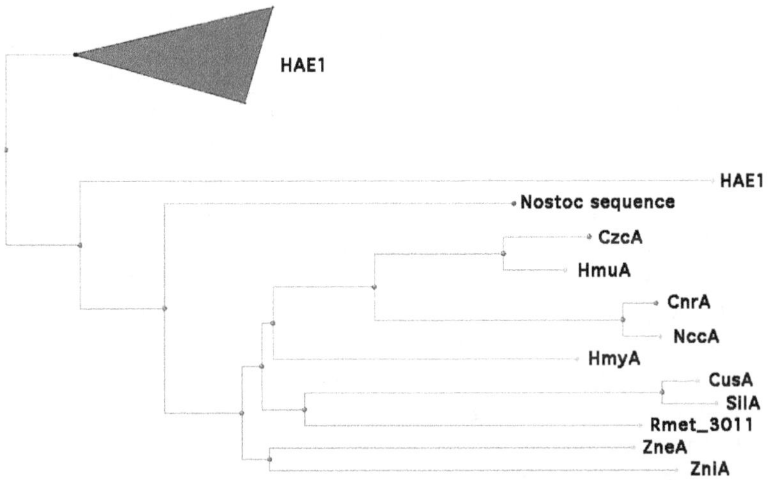

Figure 5.3 Relationship of RND proteins from *C. metallidurans* with sequence All7631 from *Nostoc* sp. The figure shows the relationship of the HME-RND subgroups and the extract of BLAST searches with more than 1000 bacterial genomes in May 2011. The close relationship of the HME1 proteins CzcA/HmuA, the HME2 proteins CnrA/NccA, and the HME4 proteins CusA/SilA is shown. The *Nostoc* sequence represents HME5, which does not occur in *C. metallidurans*. Please note also the fact that HME proteins may have originated from HAE1-RND proteins, indicated by a closely related HAE1 protein and the large cluster of these proteins represented by the triangle. NimA1/A2 and HmvA are not shown here.

resistance to Co^{2+}, Zn^{2+} and Cd^{2+} (= Czc). These metal cations are being imported into the *C. metallidurans* cytoplasm by fast and unspecific uptake systems (Kirsten *et al.*, 2011) and exported back into the periplasm by CDF proteins and P-type ATPases (Scherer and Nies, 2009). The *czc* region was the first RND-encoding gene region cloned (Nies *et al.*, 1987), connected to a physiological function (Nies and Silver, 1989), sequenced, and characterized in detail (Nies *et al.*, 1989; Nies, 1995; Rensing *et al.*, 1997). CzcA was the first RND protein purified and characterized (Goldberg *et al.*, 1999), showing that RND proteins are cation-proton antiporters. CzcA interacts with the MFP CzcB and the OMF CzcC in the ratio CzcC:B:A = 1:2:1 (Rensing *et al.*, 1997). Unfortunately, later authors did not follow the Czc nomenclature so that AcrB and MexB are the RND proteins of their respective systems, AcrA and MexA the MFPs, with the OMFs having totally unrelated names. This situation may lead to confusion.

Another HME1 protein, HmuA, is a close relative of CzcA and its gene is encoded by a disrupted operon, which is also present in the closest relatives of *C. metallidurans*. This *czcICBA*-like *hmuICB//B'A* operon is located on the second chromosome of the bacterium and contains the genes for a putative small periplasmic CzcI-like HmuI protein in addition to the genes for a CzcCBA-like HmuCBA efflux protein complex. However, the *hmuA* gene was not expressed under several conditions tested (Nies *et al.*, 2006). *C. metallidurans* may have inherited a *czcICBA*-like operon from its ancestor, copied it to plasmid pMOL30, framed the new copy with additional genes, and disrupted in the *hmuB* gene the old *hmuICBA* copy because *hmu* was no longer needed. In addition to the genes for the CDF protein CzcD, the P_{IB4}-type ATPase CzcP, the two-component regulatory system CzcRS and the protein with unknown function CzcN, a number of genes for small periplasmic metal chaperones became also part of the *czc* determinant during evolution.

A similar situation presents itself concerning the *cnr* operon on plasmid pMOL28. CnrA and NccA are the two HME2 members in *C. metallidurans*. The CnrCBA system mediates nickel and cobalt resistance in *C. metallidurans* and is encoded by *cnrYXHCBAT* on plasmid pMOL28 (Liesegang, 1993). Expression of *cnr* is regulated by the ECF sigma factor CnrH and the membrane-bound anti-sigma factor complex CnrYX (Grass *et al.*, 2000, 2005b; Tibazarwa *et al.*, 2000; Trepreau *et al.*, 2011). Plasmid pMOL30 contains a highly homologous *nccCBA* operon that, however, does not contain genes for regulators, is not expressed and a 'silent' operon (Nies *et al.*, 2006).

Two operons encode Cus-like systems with signatures of monovalent cation exporters: CusA and SilA of the HME4 protein subfamily. The *cusA* gene is located on plasmid pMOL30 and is constitutively expressed while *silA* is harboured by chromosome 2 and heavily induced by silver.

The remaining six RND-encoding systems may be three putative cobalt and three putative zinc exporters according to their induction profiles. They belong to the HME3 subfamily that may be deeply separated into at least two clusters, HME3a and HME3b. Of the HME3a proteins, ZniA is part of a complete *zniCBA* operon and its production is 5-fold upregulated by the presence of Zn^{2+}. HmvA is constitutively synthesized. Again, the *hmvCBA* operon is complete and encodes a RND, MFP and OMF protein. However, the *hmvA* gene contains a nonsense mutation so that this RND protein probably cannot be expressed. Although the *zneCBA* operon is again complete and contains the genes for a two-component regulatory system in the vicinity, *zneA* was silent under the conditions tested (Nies *et al.*, 2006).

The three HME3b proteins were probably also not expressed (Nies *et al.*, 2006). The *hmyA* gene was silent and contains no gene for a MFP or OMF in its vicinity. The *nimA* gene was interrupted by a transposon insertion and this annotated a pseudogene. Rmet_3011 was not even given a name although the gene contains the genes for a MFP, OMF and a two-component regulatory system at its side. Nevertheless, Rmet_3011 was also silent.

When the amino acid residues conserved among the HME-RND proteins from *C. metallidurans* and CusA from *E. coli* were analysed (Table 5.4), most conserved residues had structural functions in CusA. Exceptions were those in TMH IV, which are probably involved in proton transport.

Table 5.4 Conserved amino acid residues among HME-RND proteins. The numbering follows that of CusA from *E. coli*. Possible functions are predicted, proton transport underlined. Residues in bold letters are even conserved among the two halves of the protein

Residue	Position	Predicted function	Remark	Corresp.
R13	TMH I, cytopl. face	Anchoring at phospolipids		
G27	TMH I, 2/3 to peripl.	Keeps TMH I straight		
P40	After TMH1	Turn to head portion		
E62	PN1	?		
P68	PN1	?		
G78	PN1	Turn to beta sheet		
R83	PN1	?	Close to K684	
Y104	PN1, close to IM	Anchoring at membrane		
P122	PN1	Turn to alpha helix		
W162	PN2a, top	Contact to DC		
L168	PN2a	Hydrophobic core of PN2		
V174	PN2a	Hydrophobic core of PN2		
G181	PN2a, top	Turn		
G182	PN2a, top	Turn		
A209	DN, close to arm	Monomer-monomer inter.		
N214	DN, beginning of arm	Monomer-monomer inter.		
G218	DN, arm	Keeps arm straight		
R232	DN, arm	Monomer-monomer inter.		
R272	DN/PN transition	?		
G274	DN/PN transition	Turn		
P316	PN2b	Turn		
L342	TMH II	Entry into membrane		
V348	TMH II	Middle of membrane		
L357	TMH II	**Begin. of membrane span**		L887
R361	TMH III, cytopl. face	Anchoring at phospholipids		
S392	TMH IV, peripl. Face	?		
G394	TMH IV	**Keeps TMH straight**		G935
A397	TMH IV	Keeps TMH in line		
G401	TMH IV	Keeps TMH straight		
D405	TMH IV	Proton transport		
E412	TMH IV, cytopl face	Proton transport		
N413	TMH IV, cytopl face	Proton transport		
P463	TMH V	**Keeps TMH straight**		P997
P477	TMH VI, peripl face	Slight curve leading to IM		
P498	TMH VI, cytopl. face	**Turn**		P1032
G558	TMH VII, peripl. face	Turn to head portion	Like P40	
F561	TMH VII, peripl. face	Anchoring at membrane	Like Y104	
P563	TMH VII, peripl. Face	Turn to head portion		
E567	PC1	Beginning of PC1		
P578	PC1 top	Turn to beta sheet	Like G78	
G607	PC1	Keeps beta sheet straight		

Table 5.4 (Continued)

Residue	Position	Predicted function	Remark	Corresp.
D617	PC1, inner cavity	Substrate-binding		
W636	PC1 bottom	Anchoring at membrane	Like Y104	
G657	PC1, top	Turn to alpha helix	Like P122	
I666	PC1 bottom	Anchoring at membrane	Like Y104	
G676	PC1	Turn		
K684	PC1	?	Close to R63	
G687	PC1	Turn		
G709	PC2a, bottom	Turn to beta sheet	Like G78	
R735	DC, top	docking		
G737	DC, top	Turn back down		
G753	DC	Turn to beta sheet	Like G78	
R765	DC, bottom	Monomer contact		
R771	DC	Docking/monomer cont.		
R777	DC	Docking/monomer cont.		
L789	DC	Hydrophobic core/docking		
G809	DC/PC2	Turn		
R831	PC2b, bottom	Contact to MFP		
D832	PC2b, bottom	Contact to MFP		
G860	PC2b/TMH8	Turn to TMH VIII		
E863	TMH VIII, peripl. Face	Contact to MFP		
A868	TMH VIII, peripl. Face	Helix packaging		
L887	TMH VIII	**Begin. of membrane span**		L357
P905	TMH IX	Keeps TMH straight		
G910	TMH IX	Keeps TMH straight		
G911	TMH IX	Keeps TMH straight		
L915	TMH IX	Begin. of membrane span		
G929	TMH X	Keeps TMH straight		
G935	TMH X	**Keeps TMH straight**		G394
A972	Cytoplasm	?		
R980	TMH XI cytopl. face	Anchoring at phospholipids		
R982	TMH XI cytopl. face	Proton transport	Close to E412	
P983	TMH XI	Keeps TMH straight		
M986	TMH XI	?		
T987	TMH XI	Proton transport	Close to D405	
G994	TMH XI	Keeps TMH straight		
P997	TMH XI	**Keeps TMH straight**		P463
G1003	TMH XI/XII	Turn		
G1005	TMH XI/XII	Turn		
V1008	TMH XII	Begin. Membrane span		
A1013	TMH XII, peripl. Face	Helix packaging		
G1018	TMH XII	Keeps TMH straight		
G1019	TMH XII	Keeps TMH straight		
P1032	TMH XII/cytoplasm	Turn to cytopl. C-term.		P498

Gold – always the exception!

There is one RND-driven protein complex that seems to be involved in detoxification of gold in *Salmonella enterica*. Au(I) and Au(III) are formed from metallic gold deposits by bacteria and fungi. Although both ions would be immediately reduced to metallic gold again when free, both ions may be stable as part of gold complexes. So gold complexes may occur in nature (Reith *et al.*, 2007; Southam *et al.*, 2009; Williams-Jones *et al.*, 2009).

Nobody knows up to now why this pathogenic gut bacterium has the need to detoxify gold complexes but it contains *gol* genes that are specifically being induced by the precious metal. The Gol system is composed of the gold-specific regulatory protein GolS of the MerR protein family, the Au(I)-exporting P-type ATPase GolT, the periplasmic gold-binding protein GolB and the possible Au(I)-exporting RND-driven efflux system GesCBA (Pontel *et al.*, 2007). Interestingly, the RND protein does not show the 'monovalent cation' nor the 'divalent cation' sequence signature but rather, the 'DD' motif of an HAE-RND protein and belongs into this family. Indeed, GesCBA mediates export of fluoroquinones, chloramphenicol and biocides (Conroy *et al.*, 2010). It may be that not ionic Au(I) but Au(I)-containing complexes are the real substrates of this RND-driven exporter.

The burning question: transenvelope efflux or periplasmic efflux?

The problem

When it became clear that RND-containing protein complexes are spanning the complete cell wall of a Gram-negative bacterium the question arose which substrates are being transported by these huge efflux guns, those located in the periplasm or the cytoplasm or both compartments? From the cytoplasm to the outside would be transenvelope efflux (TEE) and from the periplasm this would be periplasmic efflux (PPE).

This question cannot be so easily solved as it may seem. Physiological data may indicate PPE or TEE but these effects may be the result of the combined action of an RND-driven efflux 'gun' and another transport protein (Fig. 5.4). So, metal cation uptake into the cytoplasm followed by TEE would appear as PPE since the periplasmic metal concentration is decreased by this process. Alternatively, cytoplasmic metal cations may be exported into the periplasmic membrane by P-type ATPases or CDF proteins performing inner membrane efflux (IME), and IME followed by PPE would appear in total as TEE. These processes may be addressed as indirect PPE or indirect TEE, respectively. Moreover, individual transition metal cations may have different toxicities in the cytoplasm and periplasm.

To learn about the process, the metal concentration may be determined in each compartment independently, for instance by measuring upregulation of genes by regulators with known localization in the cell. MerR-type proteins are usually bound at their operator structure on the DNA and thus react exclusively to cytoplasmic signals. Two-component regulatory systems, on the other hand, should respond to periplasmic signals. Studying regulation of gene expression with either system would allow measurement of decreasing metal concentrations in the periplasm or cytoplasm, respectively. Using such an experimental approach may reveal if net TEE or PPE is happening, however, not the difference between a direct or an indirect process.

To address this question further, deletions of the genes of interfering uptake or IM efflux systems in combination with gene expression studies could reveal which process is being catalysed by a HME-RND-driven protein complex. In the case of HAE-RND proteins, the problem might be even more complicated because hydrophobic substrates may have a higher solubility in the cytoplasmic membrane than in the periplasm or cytoplasm, leading to diffusion of these substrates from one compartment into the other. On the other hand, the PPE/TEE problem may not even exist for HAE-RND proteins: if they bind substrates from out of the cytoplasmic membrane, these RND systems may use the inner membrane as a trap for the hydrophobic substrates, elegantly removing them from here to the outside.

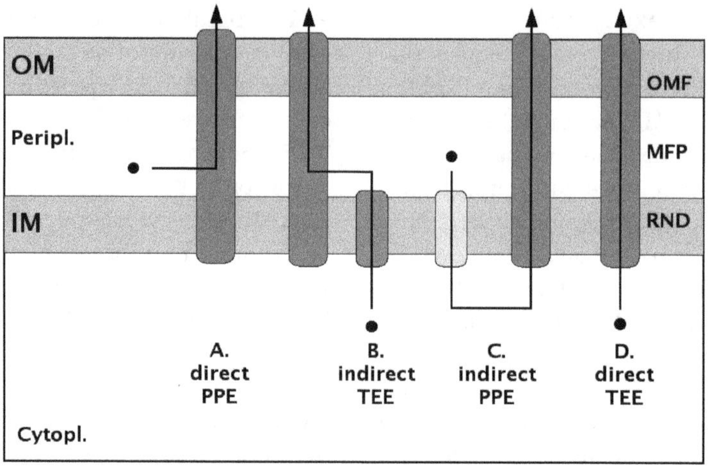

Figure 5.4 Indirect and direct TEE and PEE. The shows that a physiological effect might be the result of different activities of an RND-driven system (long bar spanning IM and OM) alone or in combination of another transporter located solely in the IM. Thus, a decrease in the periplasmic metal ion concentration can be archived by direct export of the ion by the RND system (A. direct PPE) or by a combination of an uptake system (light grey) and an RND-system doing transenvelope efflux (C. indirect PPE). Similarly, cytoplasmic cations may be exported by the RND system doing direct transenvelope efflux (D) or by a combination of inner-membrane efflux and an RND system doing periplasmic efflux (B. indirect TEE).

CusA transports Ag(I) across a membrane *in vitro*

CusA contains three pairs of methionine residues located in the transmembrane domain of the protein, forming a possible export route for copper from the cytoplasm to the central substrate-binding site of CusA in the periplasmic domains PC1/PC2a (Fig. 5.5). From the inside to the outside, these pairs of the cation path are M410/M501, M486/M403, M391/M1009. These residues are conserved among HME4 proteins (Long *et al.*, 2010).

M410/M501 are located in the cytoplasm or at the cytoplasmic face of the inner membrane adjacent to the conserved residues E412-N413, which mark the C-terminal part of the possible proton-translocating TMH IV (Fig. 5.5). M493 of the M403/M486 pair is located in the middle of the inner membrane, adjacent to D405, which is essential for proton transport. However, while D405 points into the direction of K984 of TMH XI, M403 points to the other direction. M391/1009 are sitting on top of the membrane domain of CusA, at the interface between the periplasmic and membrane domain of the protein. The distance between the three pairs is about 15 Å. There is, however, no other pair of methionine residues leading from M391/M1009 to the central substrate-binding site M573/M623/M672. The distance between these methionine residues is about 26 Å. All three methionine residues in this site were essential (Franke *et al.*, 2003; Long *et al.*, 2010). Mutation in the residues M391, M410, M486, however, did not abolish but did decrease copper resistance in the test system (Long *et al.*, 2010).

CusA was clearly able to transport Ag^+ across a proteoliposome membrane in a stopped-flow experiment and any deletion in the methionine residues mentioned above prevented this reaction (Long *et al.*, 2010). Thus, the methionine residues of the cation path and the three of the central substrate-binding site were essential for membrane transport by CusA. However, the methionines of the transport path were less important for copper resistance than those of the central substrate-binding site.

Copper homeostasis in *E. coli*

Cu^{2+} is probably transported across the outer membrane into the periplasm of *E. coli* by facilitated diffusion through outer membrane porins because a change in the porin composition

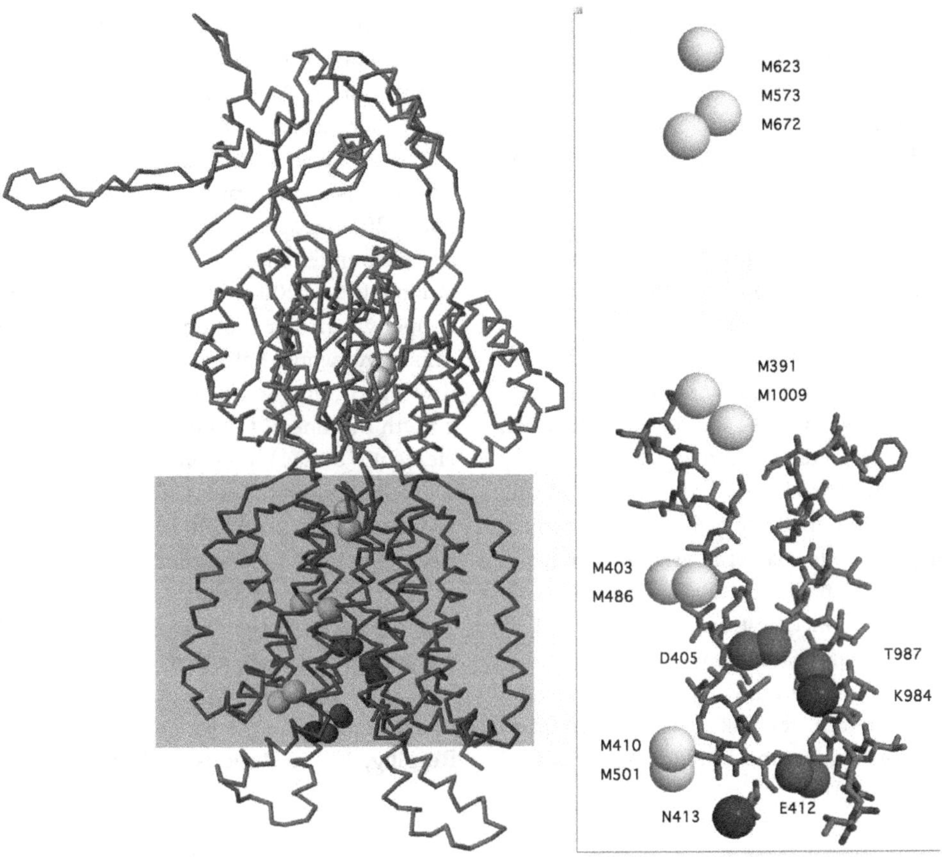

Figure 5.5 The cation path, proton path and central substrate-binding site of CusA. Using protein data base structure entry 3K07, a CusA monomer from *E. coli* is shown as published (Long *et al.*, 2010) as backbone structure. The grey box represents approximately the position of the cytoplasmic membrane. The head of CusA stretches into the periplasm. The large arm on top of the protein contacts the adjacent CusA monomer. The position of important Met and other residues in the CusA monomer is shown and magnified on the right hand. Only the sulphur atoms of three pairs of Met residues are shown that may comprise a transport route for copper cations from the cytoplasm to the periplasm. The distance between these pairs is about 15 Å. A triangle of methionine in the periplasmic part of the protein is probably the central substrate-binding site for copper (Franke *et al.*, 2003, Long *et al.*, 2010). The remaining conserved Asp, Glu, Asn, Thr and Lys residues may form the proton pathway from the periplasm to the cytoplasm.

influences copper resistance (Egler *et al.*, 2005). Since the known copper-containing proteins of *E. coli* reside with the copper-binding sites in this compartment, further transport of copper into the cytoplasm might not be required (Magnani and Solioz, 2007).

The P-type ATPase CopA is the most important factor for copper resistance in *E. coli* (Rensing *et al.*, 2000; Egler *et al.*, 2005; Helbig *et al.*, 2008a). Since CopA exports Cu^{+} from the cytoplasm to the periplasm, this indicates that copper is able to enter the cytoplasm of *E. coli*. *E. coli* does not contain a CTR-type copper uptake systems like eukaryotic cells and the import systems for copper are still unknown in this bacterium (Fig. 5.6). ZupT of the ZIP protein family might be responsible for import of Cu^{2+} (Grass *et al.*, 2005a). However, copper is about 100-fold more toxic under anaerobic compared to aerobic conditions (Outten *et al.*, 2001) so that uptake of Cu^{+} may have occurred with a higher rate than that of Cu^{2+}.

Expression of *copA* and of the gene for the periplasmic multicopper Cu(I) oxidase CueO is controlled by the cytoplasmic copper concentration via the MerR-type regulator CueR (Changela *et al.*, 2003). This protein reacts to copper in the presence of zeptomolar concentrations, which would be far less than one copper per cell. Since cytoplasmic copper is most likely bound to thiol-containing substances such as glutathione with femtomolar affinity constants, the *in vivo* activity of CueR for copper should be in the micromolar range.

Under aerobic conditions, addition of 1 mM Cu^{2+} leads to a linear increase of *copA* – and *cueO*-specific messages up to 334 and 180 copies per cell, respectively, after 2 minutes (Thieme *et al.*, 2008). After this initial period, the copies per cell decline again with a half-life of 6.8 minutes, which is an average value for bacterial mRNAs (Selinger *et al.*, 2003). The maximum level of mRNAs/cell depends on the concentration of copper used as inducer. A second treatment with 1 mM copper after a first one with 0.1 mM leads only to a mild response. Therefore, the number of CopA proteins produced within the initial 2 minutes was obviously sufficient to decrease the cytoplasmic copper concentration below the CueR 'threshold' concentration. At a transcription rate of 40 nucleotides per second at 37°C, it takes about a minute to transcribe and translate the first CopA protein per cell. With six translation events per mRNA copy, 2000 CopAs per cell should be finished after 2 minutes with a turnover number of 3.33/s (Mandal *et al.*, 2002), thus able to export 400,000 copper atoms per minute out off the cytoplasm. This value would calculate to a quota of 664 μM if a cell volume of 1 fl is roughly assumed. After growth in mineral salts medium containing trace element (6 nM) concentrations of copper, *E. coli* cells contain about 170,000 copper per cell (Kirsten *et al.*, 2011). It is therefore likely that freshly produced CopA pumps are very well able to mount sufficient efflux power only 2 minutes after induction to decrease the cytoplasmic copper content again down to the desired quota.

The Cu^{+} exported by CopA into the periplasm is oxidized by CueO to Cu^{2+} again (Grass and Rensing, 2001a; Roberts *et al.*, 2002; Singh *et al.*,

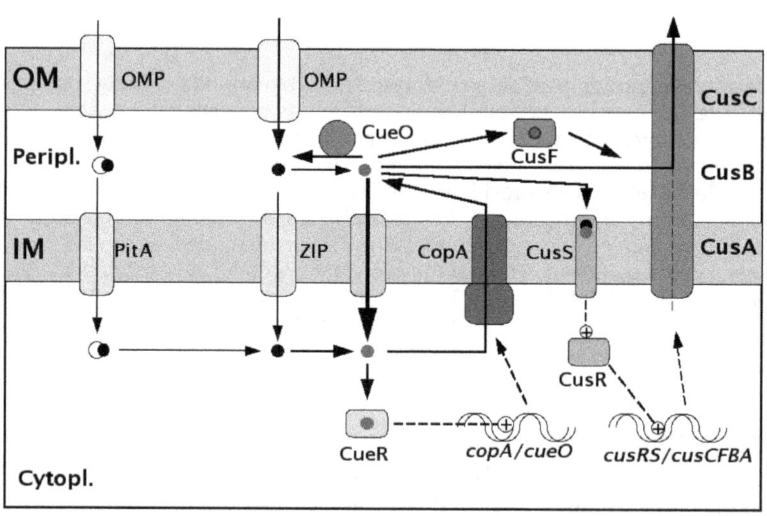

Figure 5.6 Copper homeostasis in *E. coli*. This figure uses the same colour code as Fig. 5.1. Cu(II) and maybe Cu(II):phosphate complexes are transported by OMPs into the periplasm and further on by PitA and ZIP into the cytoplasm but this seems to be a slow uptake process. In either compartment, Cu(II) (black dots) is reduced to Cu(I)(red dots). Periplasmic Cu(I) is rapidly taken up by (an) unknown system(s). Cytoplasmic Cu(I) is sensed by the MerR-type regulator CueR that controls *copA* and *cueR*. CopA is a P-type ATPase that transports Cu(I) back to the periplasm where is gets oxidized by CueO back to Cu(II) in a reaction that depends on molecular oxygen. Periplasmic copper is sensed by the two-component regulatory system CusSR in control of the *cus* determinant. CusF binds periplasmic copper. This copper and that not bound to CusF is exported back to the outside by the RND-driven CusCBA protein complex. A colour version of this figure is located in the plate section at the back of the book.

2004). CueO depends on molecular oxygen for this reaction and thus does not function under anaerobic conditions (Outten *et al.*, 2001). If *E. coli* is treated with 1 mM copper under anaerobic conditions, the *copA*-dependent mRNAs increase steadily in number, reaching a saturation of 1200 copies per cell after 60 minutes (Thieme *et al.*, 2008). If molecular oxygen is provided to the cells, the mRNA number decreases again within 20 minutes to about 200 copies per cell. This indicates that oxidation of Cu(I) to Cu(II) by CueO in the periplasm is essential to prevent re-entry of Cu(I) into the cytoplasm, which would lead to a futile cycle of uptake and CopA-dependent efflux. Again, Cu^{+} is much more rapidly able to enter the cytoplasm than Cu^{2+}, explaining why the monovalent cation is more toxic than the divalent one.

Copper toxicity may be based in three mechanisms. Similar to Cd^{2+}, cytoplasmic Cu^{+} may interfere with thiol residues and compete other metals out off their physiological binding sites in enzymes (Irving and Williams, 1948; Nies, 2007a; Helbig *et al.*, 2008b). Moreover, copper ions may catalyse the Fenton reaction as described by Haber and Weiss (Haber and Weiss, 1932). Target of this reaction seem to be proteins and lipids of the cytoplasmic membrane and the periplasm but not the DNA (Macomber *et al.*, 2007).

It is therefore important to control the periplasmic copper concentration to maintain a sufficient supply for periplasmic copper-dependent protein domains but to prevent copper from doing Fenton reaction and/or becoming reduced to Cu^{+} and taken up into the cytoplasm. Once there, Cu^{+} is sensed by CueR, exported back by CopA and oxidized again to Cu^{2+} by CueO (Fig. 5.6).

CusA does not transport copper across the cytoplasmic membrane *in vivo* under aerobic conditions

The *cus* determinant encodes the components of the CusCBA transenvelope protein complex, the small periplasmic metal chaperone CusF, and the two-component regulatory system CusRS (Franke *et al.*, 2001, 2003; Loftin *et al.*, 2005). CusF is rapidly produced after copper shock in *E. coli* (Egler *et al.*, 2005), sequesters periplasmic copper and transfers it via the MFP CusB to the CusCBA protein complex (Bagai *et al.*, 2008). This process may be either used to funnel periplasmic copper into CusA or to control the transport activity of this RND protein (Kim *et al.*, 2011).

In contrast to the sharp reaction profile of *copA* – and *cueO*-transcription to treatment with 1 mM copper, transcriptional response of *cusA* is much slower. After a delay of 1 minute, the *cusA*-specific mRNA concentration increases from one copy/cell to eight copies/cell reached after 10 minutes. The mRNA level stays at this value for 5 minutes and decreases in the following 15 minutes to two copies/cell, reaching one copy/cell again after 60 minutes. Under anaerobic conditions, the mRNA number increases for 50 minutes reaching 20 copies per cell. Addition of molecular oxygen slows this increase but does not lead to a decrease of the *cusA*-specific mRNA number per cell (Thieme *et al.*, 2008).

This indicates that Cus production follows completely different rules compared to CopA synthesis. A low number of mRNAs per cell is probably compensated by a longer half life, which could be 1 hour as that of the *czcCBAD*'s message in *C. metallidurans* (C. Grosse and D.H. Nies, unpublished). Such a long-living message might be needed to allow translation, export and assembly of the complicated CusCBA or CzcCBA protein complexes (the 5′ end of the *czcD* gene is also present downstream of *czcA* since this region encodes a stem–loop the stabilizes the *czcCBAD*′-mRNA). The lag phase of 1 minute before the transcript number starts to increase can be explained by the time needed to phosphorylate the CusS histidine kinase sensor, followed by phosphate transfer to the CusR response regulator and its migration to the *cus* operator on the chromosome of *E. coli*. The fact that the *cus*-specific mRNA does not decrease when molecular oxygen is provided again to the cells after anaerobic conditions indicates that periplasmic Cu^{+} and Cu^{2+} are sensed by CusS: supply of oxygen allows CueO again to function, periplasmic Cu^{+} is being oxidized to Cu^{2+}, which does not re-enter the cytoplasm as quickly as Cu^{+}, leading to a strong decline in *copA* transcription as a consequence of CueR-mediated repression of *copA*. Since *cus* transcription remains 'on' but periplasmic Cu^{+} is obviously decreased in concentration, CusS should be able to sense Cu^{2+} in addition to Cu^{+}.

Sensing only Cu^{2+} would have the disadvantage that a Cu^{+} detoxification system would not be able to sense its substrate, which should be the predominant form under anaerobic conditions when Cus is needed most. The Cus system is required for copper resistance under anaerobic conditions (Outten *et al.*, 2001) because CueO is not functioning here and a Δ*cueO* mutation has no effect. Under aerobic conditions, deletion of *cus* does not compromise copper resistance, only silver resistance is decreased (Franke *et al.*, 2001, 2003; Grass and Rensing, 2001b; Egler *et al.*, 2005; Helbig *et al.*, 2008a). A Δ*cus* Δ*cueO* double mutant strain, however, is very sensitive to copper. Further deletion of *copA* does not decrease copper resistance any further (Egler *et al.*, 2005; Helbig *et al.*, 2008a). Either CopA is not able to detoxify the cytoplasm sufficiently under aerobic conditions in the absence of CueO or the main target of copper toxicity in the Δ*cus* Δ*cueO* mutant may not be the cytoplasm but the periplasm.

Under anaerobic conditions, deletion of *copA* leads to an about 10-fold decrease in copper resistance but that of *cus* only to a twofold decrease (Outten *et al.*, 2001). Therefore, CopA is also very important for copper resistance under anaerobic conditions and very well able to compensate the high import rate of Cu^{+} under these conditions, which are not different from those under aerobic conditions in the absence of Cus and CueO as far as the enhanced uptake of Cu^{+} into the cytoplasm is concerned. So, CopA incompetence can be ruled out, leaving the explanation that the main target of copper toxicity in a Δ*cus* Δ*cueO* mutant under aerobic conditions is the periplasm, for instance by damage of periplasmic and respiratory chain proteins by Fenton reaction.

Deletion of *copA* under aerobic conditions decreases copper resistance more than single deletions of *cus* or *cueO*. However, additional deletion of *cus* in a Δ*copA* mutant does not decrease copper resistance any further (Grass and Rensing, 2001b; Egler *et al.*, 2005; Helbig *et al.*, 2008a). If CusA is able to perform TEE from the cytoplasm to the outside, its presence in a Δ*copA* strain has to increase copper resistance. As stated above, there is always some *cus* transcript present in the cells and CusS also senses Cu^{2+}, so that the inability of CusA to compensate for a missing CopA cannot be the result of repression of *cus* transcription, e.g. by a high CueO production due to high cytoplasmic copper concentrations. Since uptake of copper under aerobic conditions is slow, indirect PPE (resulting from uptake plus TEE by CusCBA) can be excluded, as well as indirect TEE because there is no longer an inner membrane efflux system present.

Under anaerobic conditions, a Δ*copA* Δ*cus* Δ*cueO* triple deletion mutant of *E. coli* is less resistant to copper that a Δ*copA* single deletion strain (Fig. 5.7). Since CueO is not able to function in the absence of molecular oxygen, the presence of Cus is able to sustain some degree of copper resistance in the absence of CopA. This could indicate PPE or TEE. Either CusCBA remove periplasmic Cu(I) before the monovalent metal cation can be imported into the cytoplasm (PPE) or some removal of cytoplasmic copper by CusA in the absence of CopA (TEE).

Therefore, (i) expression of *cus* is controlled by periplasmic copper cations, (ii) the periplasmic protein CusF is able to deliver periplasmic copper via CusB to the CusCBA efflux complex, (iii) CusCBA performs the same function as CueO, removal of Cu^{+} from the periplasm to prevent Fenton reaction and damage of periplasmic protein domains, and (iv) CusCBA is not able to substitute CopA by performing TEE *in vivo* under aerobic conditions. However, (v) under anaerobic conditions, the data agree with CusCBA performing TEE or PPE.

CzcA transports metal cation across a membrane *in vitro*

CzcA is also able to transport the Czc substrates Co^{2+}, Zn^{2+} and Cd^{2+} into proteoliposomes *in vitro* (Goldberg *et al.*, 1999). The reaction follows a Hill-type sigmoidal substrate-saturation curve with $n=2$ for the two essential cations of zinc and cobalt but a hyperbolic one with $n=1$ for cadmium. The turnover numbers were between 28/s for cadmium and 385/s for zinc and thus much higher than those observed for a copper-transporting P-type ATPase (Mandal *et al.*, 2002), 3.33/s, or even a fast calcium-exporting P-type ATPase with 120/s (Ueoka-Nakanishi and Maeshima, 2000). The substrate affinities, however, were very low, with K_{50} or K_m values between

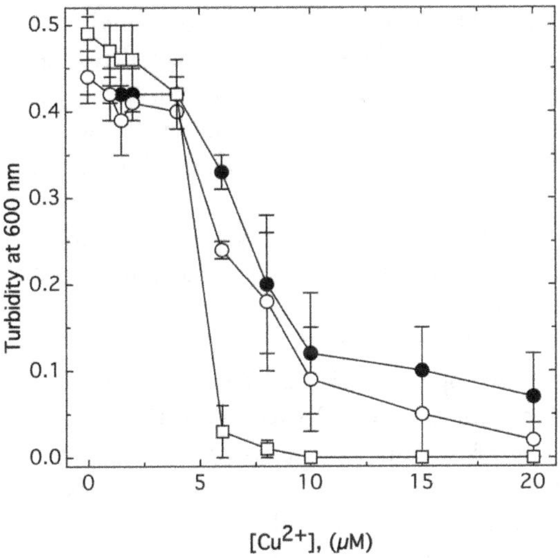

Figure 5.7 Effect of a Δ*cus* deletion on metal resistance of a Δ*copA* deletion strain. *E. coli* strains W3110 wild type (●), KC15 (Δ*copA*, ○) and ECA464 (Δ*copA* Δ*cueO* Δ*cus*, □) were diluted 400-fold into fresh Tris-buffered mineral salts medium containing 2 g/l glucose, 1 mM ascorbic acid and various CuCl2 concentrations. The cultures were incubated for 16 h at 37°C anaerobically in Hungate tubes and growth was determined as turbidity at 600 nm.

6.6 mM (Zn^{2+}) and 18.5 mM (Co^{2+}) (Goldberg *et al.*, 1999).

Deletion of the important amino acid residues in TMH IV, which may represent the proton path, leads to complete loss of metal resistance mediated by the mutant protein in case of residue D402, D408 (equivalent to D405 in CusA) and E418 (Goldberg *et al.*, 1999). The mutant proteins were still able to transport zinc but by facilitated diffusion and not proton-cation antiport. This indicates that cation transport may follow a route different from that used for proton transport. In the presence of glutathione, zinc transport by CzcA was abolished (Nies, D.H., unpublished). Thus and similar to CusA, CzcA is able to transport its substrate cations in a cation-proton-antiport reaction across a proteoliposomic membrane *in vitro*.

Metal homeostasis in *C. metallidurans*

Metal homeostasis in *C. metallidurans* is much more complicated than copper homeostasis in *E. coli* because a variety of metal uptake systems, inner membrane efflux systems and RND-driven efflux systems interact. This complicated metal homeostasis network allows *C. metallidurans* to strive in a variety of mesophilic metal-contaminated environments around the world (Mergeay *et al.*, 1978; Diels and Mergeay, 1990; Brim *et al.*, 1999; Mergeay, 2000; von Rozycki *et al.*, 2005; Monchy *et al.*, 2007; von Rozycki and Nies, 2009; Janssen *et al.*, 2010).

At least six secondary uptake systems are responsible for uptake of divalent transition metal cations such as Zn^{2+} in *C. metallidurans* (von Rozycki *et al.*, 2005; Kirsten *et al.*, 2011). (Kirsten *et al.*, 2011) ZupT of the ZIP protein family (TC#1.A.35) is essential to provide zinc under conditions of low availability (Busch and Saier, 2002; Saier *et al.*, 2006), e.g. complexation of the metal by EDTA or phosphate. Deletion of *zupT* in *C. metallidurans* leads to the accumulation of the beta prime subunit RpoC of the RNA polymerase, indicating severe disturbance of zinc homeostasis in the mutant strain (Herzberg *et al.*, 2012). PitA of the PiT family (TC#2.A.52) imports phosphate-metal cation complexes and seems to be a major provider of metal cations and phosphate to the cell. There are four members of the CorA or MIT (TC#1.A.35) protein family, one of them (ZntB) could be an efflux system rather than an uptake system (Worlock and Smith, 2002) and the structure of the ZntB protein from *Salmonella* is in agreement with this assumption (Wan *et al.*, 2011). $CorA_1$ is the main secondary uptake system for Mg^{2+} while $CorA_2$ and $CorA_3$ are kind of back-up systems (Kirsten *et al.*, 2011).

Only a few signals regulate production of these secondary uptake systems. Zinc in general regulates them all down but cobalt, cadmium and nickel do not interfere. Zinc starvation induced by addition of zinc chelators such as EDTA or phosphate lead to an upregulation of *zupT* expression. Magnesium only regulates the synthesis of $CorA_1$ and phosphate starvation probably that of a phosphate-specific ABC uptake system (Kirsten *et al.*, 2011). Iron, of course, has its own regulatory circuits in *C. metallidurans* (Große *et al.*, 2007).

With the exception of a disturbed folding or synthesis of RpoC, zinc uptake and the cellular zinc content was not changed much in a five-fold Δ*zupT* Δ*pitA* Δ*corA*$_1$ Δ*corA* $_2$ Δ*corA3* mutant or even a six-fold mutant with additional deletion of *zntB* (Herzberg *et al.*, 2012), indicating that other systems could take over and supply zinc to the cell. These could be magnesium-importing P-type ATPases (TC#3.A.3) MgtA (Rmet_5396) or MgtB (Rmet_2211) as in *Salmonella enterica* (Hmiel *et al.*, 1989; Snavely *et al.*, 1989; Snavely *et al.*, 1991), or the HoxN nickel importer of the NiCoT protein family (TC#2.A.52), which supplies Ni^{2+} for hydrogenase synthesis (Eitinger and Friedrich, 1991; Wolfram *et al.*, 1995; Eitinger *et al.*, 2005). Therefore, the final list of all zinc uptake systems in *C. metallidurans* is not available at this time. Nevertheless, disturbance of the synthesis or folding of one of the most important proteins in the cell, RpoC, in the Δ*zupT* (and other) mutants indicates the importance of the five secondary zinc uptake systems for zinc homeostasis in this bacterium.

Concerning the efflux systems of the inner membrane, most of them seem to be known in *C. metallidurans*, their genes have been removed in single or multiple deletion mutants and the effect was studied (Scherer and Nies, 2009). This assigned specific functions to all of them. Three P_{IB2}-type and one P_{IB4}-type ATPases export divalent transition metal cations across the cytoplasmic membrane and four P_{IB1}-type ATPases monovalent ones such as Cu^{+} (von Rozycki *et al.*, 2005; von Rozycki and Nies, 2009). Two of these P_{IB1}-type ATPases could be involved in copper detoxification similar to CopA of *E. coli* while the other two may supply copper to periplasmic copper-depending proteins (Raimunda *et al.*, 2011). Copper homeostasis in *C. metallidurans*, however, is very complicated (Monchy *et al.*, 2006), connected to formation of gold particles (Reith *et al.*, 2009) and will not be covered further here.

The three P_{IB2}-type ATPases ZntA, CadA and PbrA export zinc and cadmium with equal transport rates (Scherer and Nies, 2009) so that their specific roles in zinc, cadmium and lead export (Borremans *et al.*, 2001; Legatzki *et al.*, 2003a), respectively, depends solely on regulation of their synthesis. These three proteins mediate a basic resistance level because they may be able to extract and expel even tightly bound transition metal cations from the cytoplasm. The P_{IB4}-type ATPase CzcP has a 100-fold higher turnover number than the three P_{IB2}-type proteins but does not mediate a high increase in metal resistance alone: this protein might be fast but it is not able to reach all substrate cations in the cell. So, CzcP functions as 'resistance enhancer' (Scherer and Nies, 2009).

Four secondary efflux systems differ in their substrate specificity. CnrT of the drug/metabolite transporter superfamily DMT (TC#2.A.7) is encoded by a gene directly downstream of the *cnrCBA* operon and likely co-expressed with the genes for the CnrCBA RND-driven efflux complex (Liesegang, 1993; Grass *et al.*, 2000, 2005b). CnrT should export nickel and some cobalt. *C. metallidurans* contains three CDF proteins, all belonging to different subgroups of the CDF protein family (Nies, 2003). While FieF may export Fe^{2+} similar to its orthologue from *E. coli* (Grass *et al.*, 2005c), DmeF is mainly involved in cobalt efflux (Munkelt *et al.*, 2004). CzcD, one of the founding members of the CDF protein family, is encoded downstream of *czcCBA* and exports Co^{2+}, Zn^{2+} and Cd^{2+} (Nies *et al.*, 1989; Nies, 1992a; Anton *et al.*, 1999, 2004).

CzcA does not transport cobalt across the cytoplasmic membrane *in vivo* and maybe also not zinc and cadmium

In vivo, CzcA transports no or only very little cobalt from the cytoplasm to the outside: a Δ*dmeF* mutant of *C. metallidurans* strain AE104, which does not contain the megaplasmids pMOL30 and pMOL28 with the *czc* and *cnr* determinants, respectively, is very sensitive to cobalt because the deletion strain does not contain any known cobalt efflux system. Neither ZntA nor CadA exports Co^{2+} *in vitro*, FieF is also found not to be involved in cobalt resistance. Nonetheless, CzcD, CnrT, PbrA and CzcP are encoded by the plasmids missing in strain AE104 and its derivatives (Munkelt *et al.*, 2004; Scherer and Nies, 2009).

Expression of *czcCBAD* or *cnrCBA in trans* from a vector plasmid does not result in any increase in cobalt resistance in the Δ*dmeF* mutant strain. Resistance increases again, however, if CzcD, DmeF, or both are present. Thus, function of CzcCBA (and CnrCBA) essentially depends on the CDF protein DmeF to export cobalt first from the cytoplasm to the periplasm before the metal can be further exported to the outside by one of the two RND-driven systems (Munkelt *et al.*, 2004).

Since *in trans* expression from a vector plasmid is always an artificial condition, the relationship between DmeF and Czc was more closely examined later on in *C. metallidurans* strain AE128, which contains the *czc*-harbouring native plasmid pMOL30 (Scherer and Nies, 2009). In a low-complexing mineral salts medium, growth inhibition by cobalt was determined and expressed as IC_{50} value, the concentration leading to 50% growth inhibition in liquid culture. This value was 2081 ± 132 μM Co^{2+} in all kinds of multiple and single gene deletion strains still containing CzcA and DmeF, and in the parent strain AE128(pMOL30). The IC_{50} values of all Δ*dmeF* mutants was 8.8 ± 1.7 μM in the presence of CzcA and 10 ± 9 μM in the absence of CzcA. Thus, DmeF was again essential for function of the CzcCBA efflux complex in cobalt resistance. In the presence of DmeF, presence of CzcA increased cobalt resistance in various mutant pairs (with or without *czcA*) 9.15 ± 0.98-fold. 10-fold increase in cobalt resistance mediated by the CzcCBA efflux complex essentially and reproducibly depends on DmeF.

The slight increase in cobalt resistance mediated by the *absence* of CzcA resulted from increased expression of *czcD* or of *czcP*, which both were able to export cobalt from the cytoplasm when over expressed. Investigation of regulation of expression of *czc* yielded additional evidence against any export of cobalt from the cytoplasm by CzcA. The two-component regulatory system CzcRS, composed of the histidine kinase sensor CzcS and the response regulator CzcR, is involved in this process. While three promoters upstream of *czcCBAD′* are controlled by different processes and only one of them by CzcR (Große *et al.*, 1999, 2004), *czcP* is under exclusive control of CzcR (Scherer and Nies, 2009). Cobalt is a poor inducer of *czcP* and 500 μM Co^{2+} only double the expression of a *czcP-lacZ* reporter fusion. Deletion of *dmeF* does not change this pattern although the highest inducer concentration were not tolerated by the highly cobalt sensitive Δ*dmeF* strain. Deletion of *czcA*, however, led to a strong upregulation of *czcP-lacZ*. The induction pattern in the Δ*czcA* Δ*dmeF* double mutant was again similar to that of the Δ*czcA* single deletion mutant.

Thus, CzcCBA, but not DmeF, decreases the cobalt concentration in that particular cellular compartment in which CzcS senses the concentration of this cation. Since DmeF is only able to transport cations across the cytoplasmic membrane as far as we know, this is direct evidence for the hypothesis that CzcCBA remove periplasmic cobalt cations by export to the outside (Scherer and Nies, 2009). Incompetence of CzcCBA to do so in the absence of DmeF is direct evidence against *in vivo* transport of cobalt from the cytoplasm to the outside by CzcCBA.

The situation was much more complicated when zinc and cadmium resistance in *C. metallidurans* were concerned. While DmeF and FieF contributed only little to cadmium and nothing to zinc resistance, the CDF protein CzcD and the P_{IB4}-type ATPase CzcP acted as resistance enhancer for both metal cations (Scherer and Nies, 2009). The three P_{IB2}-type ATPases were central for increased zinc and cadmium resistance and were more important for cadmium than for zinc resistance. This agrees with the fact that cadmium toxicity is mainly a cytoplasmic affair (Helbig *et al.*, 2008a,b). Vice versa, CzcA was more important to zinc than to cadmium resistance. Although the P_{IB2}-type ATPases were able to functionally substitute each other, CzcA was not able to do this. Thus, the P-type ATPases and CDF proteins on the one hand differed in function from CzcA on the other hand, indicating that CzcCBA remove cations from the periplasm and the other exporters from the cytoplasm.

The LD_{50} for cadmium in a plasmid-free strain (AE104 Δ*zntA* Δ*cadA* Δ*dmeF* Δ*fieF* = Δe4) containing no known efflux system (ZntB not considered) was 0.082 ± 0.012 μM Cd^{2+} and increased with the addition of a Δ*czcA* Δ*czcD* Δ*czcP* Δ*pbr* mutant plasmid of pMOL30 6-fold to 0.491 ± 0.048 μM (values for zinc 7.1 ± 0.7 μM and 19.5 ± 1.3 μM, respectively). This indicated an important contribution of unknown genes on plasmid pMOL30 to cadmium and zinc resistance, for instance those encoding small periplasmic metal chaperones (Scherer and Nies, 2009). Addition of CzcA increased cadmium resistance 3-fold to an LD_{50} of 1.5 ± 0.61 µM and zinc resistance even 65-fold to 1273 ± 116 µM. Further addition of all the remaining efflux systems led to a 155-fold increase in cadmium but only to a 2.6-fold increase in zinc resistance.

This result may argue also in favour of a TEE mediated by CzcCBA, mainly for zinc, a little bit for cadmium but not for cobalt. If in the bare-naked strain Δe4, which does not contain any known efflux system (again, ZntB not considered), *czcCBAD′* is expressed *in trans* from a vector plasmid, zinc resistance increases enormously to the full wild type resistance level (Scherer and Nies, 2009). This may be a strong argument for TEE but also for PPE (Fig. 5.4): in the case that high expression level leads to a competition of periplasmic zinc cations between Czc and the zinc uptake systems, Zn^{2+} may not be able to reach zinc uptake systems even at high concentrations. So, a possible competition between CzcCBA and zinc uptake systems was investigated.

Deletion of *zupT* for the zinc uptake system required for zinc starvation discriminated between both explanations (Herzberg *et al.*, 2012). As mentioned above, the Δ*zupT* mutant strain of AE104 already exhibited a disturbed zinc homeostasis. Although not different from its parent strain in growth, cellular zinc content and zinc uptake, it accumulated RpoC in large quantities in inclusion bodies. Obviously, the cytoplasmic zinc availability was low and did not allow all RpoC polypeptides to fold correctly and assemble with the other subunits to the mature RNA polymerase core protein. Expression of *czcCBAD′* under these conditions led to a strain severely retarded in growth, which was overgrown rapidly by suppressor mutants. These suppressor mutants were able to remove CzcA by an unknown process, which was not trivial such as plasmid loss or *czcA* promoter mutation. So, either ZupT was required to compete successfully with CzcCBA (doing PPE) for periplasmic zinc ions or to enhance the import rate of zinc into the cytoplasm to overcompensate zinc efflux by a TEE-mediating CzcCBA efflux complex.

Deletion of *pitA* for the metal-phosphate uptake system also led to accumulation of RpoC (Herzberg *et al.*, 2012). The mineral salts medium used for the experiments has a low phosphate content but still enough to complex all divalent metal cations except for a part of the magnesium. So, zinc should be mainly present as zinc phosphate in this growth medium and either PitA is required to take the zinc phosphate complexes up or ZupT to detach the zinc cation from this complex and to import it. Expression of *czcCBAD′* in the Δ*pitA* mutant, however, did not lead to any degradation of CzcA. There was not enough cytoplasmic zinc for RpoC folding but sufficient for the remaining processes. Obviously, the presence of ZupT but not PitA was sufficient to sustain the cytoplasmic zinc content for these remaining processes. If CzcCBA does TEE, this would mean that ZupT but not PitA imports zinc with a sufficient rate to overcompensate TEE by CzcCBA, however, the misfolding of RpoC indicates that both importers are equally important for full zinc supply to *C. metallidurans*. This argues in favour of zinc PPE by CzcCBA and against TEE.

In the strain AE104 Δe4 mutant, further deletion of *zupT* does not lead to RpoC accumulation (Herzberg *et al.*, 2012). PitA and the remaining uptake systems are sufficient to supply enough zinc to the cytoplasm in the absence of zinc efflux systems. In the AE104 Δe4 mutant strain, CadA, ZntA, DmeF and FieF are missing but only CadA and ZntA contribute to zinc resistance. The genes *zntA* and *cadA* are upregulated by zinc but also expressed on a background-level in the absence of added zinc (Legatzki *et al.*, 2003a). ZupT and PitA are required to overcompensate the background-level zinc export by these systems by zinc and zinc phosphate import to allow proper folding of RpoC. If the efflux systems are gone, PitA alone is sufficient, even in the presence of high numbers of CzcCBA efflux complexes. If *czcCBAD′* are expressed in the Δe4 strain, the resulting strain reaches wild type zinc resistance. If, however, *czcCBAD′* is expressed in the Δe4 Δ*zupT* mutant strain, CzcA is degraded again. So, only ZupT, which is able to take up zinc at low availability, but none of the other uptake systems are needed to compete with CzcCBA for zinc ions. Since the cellular zinc uptake rate is similar in all strains carrying one or more deletions in uptake systems, it is not the function of ZupT to provide a sufficiently high zinc influx into the cytoplasm to overcome the zinc efflux by a CzcCBA system performing

TEE. Instead, the role of ZupT is to compete with CzcCBA for periplasmic zinc ions.

Thus, although the high metal resistance of strain AE104 Δe4 that expresses *czcCBAD′ in trans* argues in either direction, that is TEE or PPE by CzcCBA, the phenotypes of various mutant strains with deletions in uptake and efflux systems makes TEE by CzcCBA highly improbable: CzcCBA does PPE!

Energetics and functional modes of HME-RND proteins

Assuming HME-RND proteins functions similar to HAE-RND proteins (Seeger *et al.*, 2006; Murakami, 2008), the three subunits of CusA or CzcA should rotate between L conformation open to the periplasm, the closed or tight conformation T and the O conformation open to the central pore. As stated above, proton transport should be only possible in the T conformation because otherwise the pmf would be uncoupled by uncontrolled proton import. If RND proteins need a protonation step of the central substrate binding site for substrate release similar to other cation/proton antiporters, the reaction cycle of an HME-RND protein doing PPE would be 'LOT', (i) binding of periplasmic cations in L, (ii) protonation of the substrate-binding site by periplasmic protons followed by cation transfer into the central pore in O, and (iii) proton release into the cytoplasm in T (Fig. 5.8). The driving force of further export of the cation from the central pore through the OMF to the outside would be simple entropy and the high local concentration of the cation in the microenvironment composed of the central pore plus the interior of the OMF.

If cytoplasmic cations are being bound, such a process cannot take place in conformation O or L, or periplasmic protons would be able to enter the cytoplasm, leading to uncoupling of the pmf. If cytoplasmic cations access the RND monomer in T, they must wait for the protons to enter the cytoplasm. In this case, conformational change of such a monomer into the L conformation would present the substrate-binding site already occupied to eventual periplasmic cations. Thus, cytoplasmic and periplasmic cations could compete for the substrate-binding site. A TEE cycle would close by rotation of the RND monomer into the O and T conformation.

It is important to realize that such a configuration would mean that an RND protein that performs PPE and TEE simultaneously must also function as a cation uptake system. If a periplasmic cation is bound to the substrate-binding site in conformation L and this conformation rotates back into T, this allow release of the cation into the cytoplasm. Moreover, such a protein could also function as a simple inner membrane cation/proton antiporter: cytoplasmic cations are being bound in conformation T, the cation is being released into the periplasm in conformation L, the empty site is being protonated in O and the cycle is closed by proton import in conformation T.

So, if PPE and TEE are both possible, this should allow also cation import from the periplasmic to the cytoplasm and also cation export from the cytoplasm to the periplasm by cation/proton antiport. Only thermodynamics decide which process is allowed to occur, more specifically the pmf and its $\Delta\Psi$ and $Z{\cdot}\Delta pH$ parts, which may differ depending on the external pH value as outlined above. At an external pH_a 7, $\Delta\Psi = 165$ mV and $Z{\cdot}\Delta pH = 35$ mV if the pmf is 200 mV. In this case, uptake of one or two positive charges driven by the $\Delta\Psi$ portion would be energetically more favourable than an electroneutral proton/cation antiport unless the cytoplasmic metal concentration is 308,000-fold higher than that in the periplasm in case of a divalent metal cation. At pH_a 5, $\Delta\Psi = 47$ mV and $Z\Delta pH = 153$ mV, which would allow an electroneutral cation/proton antiport.

As a consequence, an HME-RND protein performing TEE and PPE at the same time would also be an uptake system for periplasmic metals at pH = 7 and may function as simple inner membrane efflux system at pH = 5. Since *C. metallidurans* and *E. coli* grow aerobically at pH 7, CzcA and CusA should not be allowed to do TEE and PPE at the same time or the cytoplasm would not be protected against incoming periplasmic cations. Both proteins cannot do TEE and PPE at the same time. However, this thermodynamical consideration does not exclude that CusA may perform either function at pH 5, which is reached when *E. coli* grows anaerobically by mixed acid fermentation. Nevertheless, under neutral pH

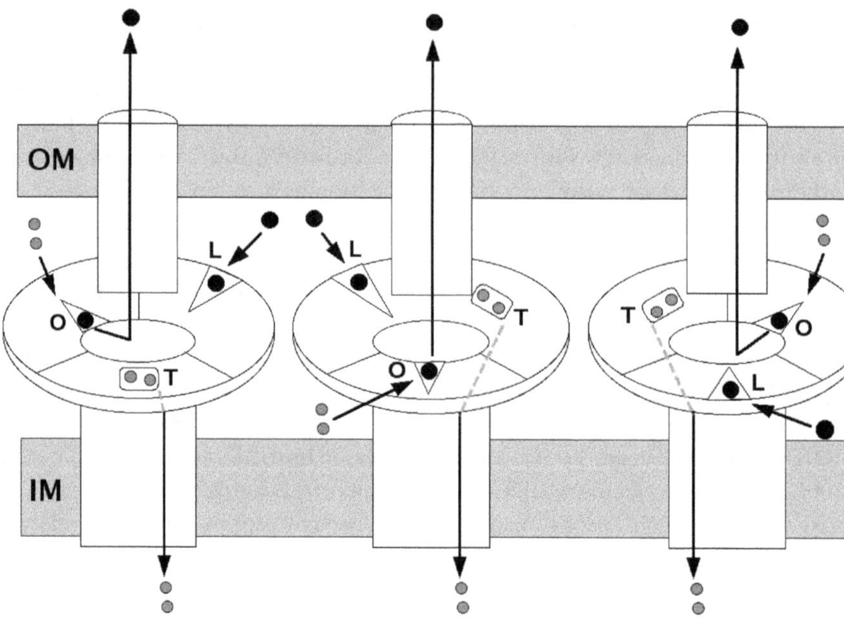

Figure 5.8 Functional model for HME-RND proteins. The potential reaction cycle of a HME-RND protein is shown, which transverses from the left to the right and begins anew from the left. Each of the three monomers rotates successively between three conformations. In 'L' (loose) the substrate-binding site is open to the periplasm and able to bind metal cations (black dots). This changes conformation into 'O' open, the central binding site is protonated (grey dots), and the cation released into the interior of the RND protein for further export through the outer membrane factor (cylinder in the outer membrane on top of the RND protein, membrane fusion protein not indicated). Finally, in 'T' (= tight), the proton(s) are allowed to migrate into the cytoplasm so that the proton motive force drives the transport process after all.

values, TEE and PPE at the same time must be prevented. Since there is clear evidence that PPE is functioning, CzcA and CusA should do PPE exclusively under aerobic conditions at a neutral pH value.

Synthesis by a hypothesis: flux control in RND proteins

So, CusA and CzcA are both able to transport their substrates across a proteoliposomal membrane *in vitro,* which would be the cytoplasmic membrane *in vivo*. However, this process does not contribute very much or not at all to metal resistance mediated by these two proteins at pH 7.

This ability to transport metals across the cytoplasmic membrane does not seem to be any 'historical' artefact of an old RND feature remaining since the time of the last universal common ancestor: the conservation of the methionine residues lining the copper transport route through the transmembrane block of CusA in HME4-RND proteins argues against such an explanation. An 'anabolic' function, providing cytoplasmic copper to periplasmic copper-binding protein as shown for P-type ATPases (Raimunda *et al.*, 2011), is also not likely because CopA in *E. coli* would provide Cu^+ much more easily to the periplasm and not trapped in the interior of the trimeric CusA protein. Moreover, CusCBA are being synthesized under conditions of high periplasmic copper content, so it is unlikely that CusCBA are used to get copper into the periplasm. An explanation for this transport feature for CusA and CzcA, however, could be a regulatory function.

Flux control is a new mechanism that might be involved in fine-tuning of metal homeostasis, functioning in addition to transcriptional control (Nies, 2007b). In two different protein families of magnesium transport systems, metal-binding sites were identified that could downregulate transport rate by the respective system when occupied by the substrate, in this case magnesium (Lunin *et al.*,

2006; Hattori *et al.*, 2007). So, when the cytoplasmic magnesium concentration is high enough, the metal binds to these regulatory sites in the uptake system, preventing further import of yet more magnesium. Similarly, the CDF efflux protein FieF from *E. coli* contains regulatory sites that may activate this exporter when the cytoplasmic substrate concentration becomes too high (Lu and Fu, 2007). Such a procedure makes perfect sense, preventing to waste energy for uptake of unnecessary metal or efflux of essential metal cations.

Zinc, copper and cobalt are essential cations and thus, excessive efflux rates by any export systems for these metals need to be prevented. For efflux systems such as CzcCBA and CusCBA it is especially important to control their activity by flux control because synthesis of this large and complicated protein complexes takes time and energy. Flux control is a much faster process than control of gene expression so that a combination of both processes makes perfect sense for HME-RND-driven efflux systems. The obvious zinc starvation that CzcCBA is generating in a Δe4 Δ*zupT* mutant does not argue against flux control of CzcCBA because in this five-fold deletion mutant the cellular zinc homeostasis is heavily compromised, CzcCBA is constitutively synthesized at a three-fold higher level than even under at the highest zinc concentrations from the native plasmids (Legatzki *et al.*, 2003b), and the three periplasmic metal chaperones that are part of the Czc system and may serve as a periplasmic zinc storage system and buffer are absent.

Flux control would require a sensing mechanism and a possibility to switch off transport, for instance by a conformational change. Zinc and copper are slightly different in this respect because zinc homeostasis is mainly a cytoplasmic affair in *C. metallidurans* but copper a periplasmic one in *E. coli*. So, an ideal sensing mechanism in CzcA would sense the zinc concentration in either compartment. Interestingly, a putative metal-binding site HAEHG is located in a small cytoplasmic loop directly following the important TMH IV. The MFP CzcB contains two repeats of a histidine-rich motif that was important but not essential for full metal resistance mediated by CzcCBA (Nies *et al.*, 1989; Rensing *et al.*, 1997). The CzcB site may serve as sensor switch and funnel (Kim *et al.*, 2011) for periplasmic zinc while the CzcA site may sense the cytoplasmic metal content.

In CusA, CusB may mediate the switching by and funnelling of periplasmic copper ions (Kim *et al.*, 2011). Like AcrB, CusA and CzcA may rotate the conformation of their three monomers between three different states. One monomer should be open to the periplasm, a possible CusB/CzcB-funnel and/or the periplasmic metal chaperones to accept the periplasmic substrate ions. The second monomer should deliver the substrate cation into the central pore of the RND trimer. Similar to P-type ATPases and other transport proteins, this release can be most easily obtained by protonation of carboxylic groups of the central substrate-binding site by periplasmic protons. Monomer three should be open to the cytoplasm to allow these protons into the cytoplasm via TMH IV, driving the transport reaction and freeing the central substrate binding site again for the next round of the reaction cycle (Fig. 5.8).

How can such a peristaltic pump mechanism be blocked to stop the reaction? It can be done most easily by disturbing the proton path of TMH IV. Allowing TMH IV to transport protons would switch the RND protein on. Thus, preventing the flow of these protons would switch the pump off. This leaves the problem how such a switch can be coupled to a sensing mechanism. In CzcA, binding of zinc to the cytoplasmic binding site right after TMH IV could allow TMH IV to function. The Hill coefficient of $n=2$ of zinc transport by CzcA agrees with the assumption that binding of a regulatory zinc allows zinc transport. However, the RpoC folding problem in *C. metallidurans* illustrates that cytoplasmic zinc has to be present in just the right availability. This means that the control of the cytoplasmic zinc concentration should be a redundant process, so the zinc ion has to be cleared from the regulatory site, best by export from this site to the central zinc-binding site in the periplasmic part of CzcA.

Likewise, CusA can only be allowed to function when the periplasmic copper concentration is sufficiently high (Fig. 5.9). In this case, however, the three copper binding methionine pairs just adjacent to TMH IV may be needed to be occupied before CusA is being turned on. So, if there is sufficient copper in the periplasm and cytoplasm, one or all of those sites become occupied. Then, TMH IV comes into its correct position, and CusA is turned on. If the periplasmic copper concentration is too low, CusB closes the funnel, and the copper cations from the regulatory sites jump to the central substrate binding site and from there to the outside. Then, TMH IV is removed from the correct position and the peristaltic pump stops.

So, transport of cations by HME-RND proteins from the cytoplasm to the central binding site of the protein may indeed happen, but this process may be a regulatory event involved in a flux control mechanism that prevents highly developed and far evolved HME-RND efflux pumps from removing too many essential trace element cations from the periplasm. Such an evolutionary step could have been prerequisite for sophisticated HME-RND proteins of the groups HME1, HME2 and HME4 becoming important players in bacterial metal resistance.

The subgroups: structural models, distribution and possible function of HME-RND proteins

The flux-control hypothesis would emphasize the central function of the proton-translocating TMH IV even more. The primary sequence of this helix is conserved among sequences of related HME-RND proteins and was used to sort these proteins into six subgroups designated HME1 to HME5 with HME3 split into HME3a and HME3b (Nies, 2003). After 8 years and currently more than 1000 bacterial genomes are available, the primary sequences of the various HME-RND proteins from *C. metallidurans* were used in an extended BLAST search to evaluate the structure of the 'HME-RND cosmos' in bacterial genomes (Altschul *et al.*, 1997). The relationship of the thousands of HME-RND protein sequences in the current database (May 2011) was very complicated but exhibited a pattern similar to that shown in Fig. 5.3 (data not shown). Very often, sequences from the different branches of the proteobacteria or of CFB-group bacteria, cyanobacteria or others were closely related to each other, indicating frequent occurrence of horizontal gene transfer during evolution of HME-RND proteins. With few exceptions, no HME-RND proteins were found in eukaryotes, *archaea, firmicutes, actinobacteria* and *mollicutes*, that is, all organisms usually without an outer membrane. Exceptions were sequence CRE-06266 from the worm *Caenorhabditis remanei* and ZP_01666833 from the Gram-positive firmicute *Thermosinus carboxidovorans*. The worm sequence starts in the middle of the first periplasmic domain and resembles CzcA. This sequence, however, may be due to a contamination. The gene from the firmicute has genes for a MFP and an OMF upstream and those for a two-component regulatory system downstream. It is an interesting question what a Gram-positive bacterium needs an outer membrane protein, a membrane fusion protein and an RND protein for. Since the name of this bacterium indicates that it is able to oxidize carbon monoxide, the respective gene region might stem via horizontal gene transfer from Gram-negative CO-bacteria in the same ecosystem.

The HME5 cluster of proteins contains mostly cyanobacterial members and the HME2 cluster containing mostly alpha proteobacteria. Sequences from beta – and gamma-proteobacteria are located in groups surrounded by those from alpha-proteobacteria and predicted proteins from *parachlamydiaceae* were at the beginning of the HME2 cluster. The HME2 group was rather small, containing 67 sequences. The HME1 cluster contained 276 sequences from each branch of the proteobacteria plus the worm sequence that was close to a sequence from a gamma proteobacterium. Six clusters containing 177 sequences from various proteobacteria, *verrucomicrobia, spirochaeta, planctomycetales, chlorobiales,* CFB-bacteria and cyanobacteria lead to the HME1/HME2 branching point. So, HME1/HME2-like proteins seem to occur in all sequences of *Cupriavidus pinatubonensis* strain JMP134, which is related to *C. metallidurans*. The second sister contains HmyA and 46 other sequences from proteobacteria, among them many methane – and ammonium oxidizers. HmyA should be removed from group HME3b and is probably member of yet another HME-RND protein cluster.

The group HME3a contains 113 sequences again from all kinds of Gram-negative

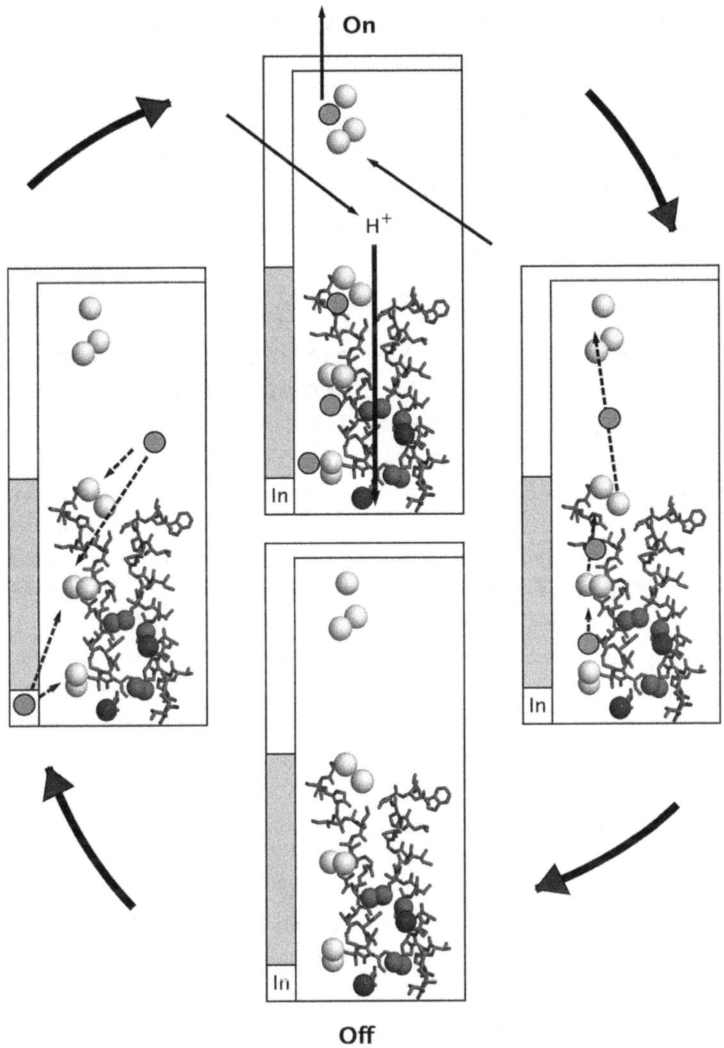

Figure 5.9 Hypothesis for a flux control for CusA. The hypothesis would explain the contradicting data for CusA, which is able to transport copper across a membrane *in vitro* but does not do so *in vivo*. If copper (grey balls outlined in black) is available, it is first loaded to the sites in TMH IV (left), pushing the TMH IV in the correct position to allow proton import into the cytoplasm and copper export from the periplasm to the outside (top, 'on' state). When the copper concentration decreases below a threshold, copper is removed from the sites in TMH IV (right), moving TMH IV out off an ideal position for proton movement into the 'off' state (bottom). Since there are nine copper binding sites in the CusA trimer, fine-tuning of CusA activity might be even so sophisticated that ten different states exists between 'totally off ' (no copper bound) and 'totally on' (all nine sites occupied).

bacteria, mostly proteobacteria and CFB bacteria. The HME4 branch with 140 sequences from proteobacteria starts, however, with the *T. carboxidovorans* strain Nor1 sequence. HME3b proteins and a cluster of HME5 proteins were outside of these HME4/HME3a-HME1/2 groups.

So, the outcome of the searches using the *C. metallidurans* proteins was an extended version of Fig. 5.3 and reproduced the HME-RND 'tree' obtained 8 years ago with just 63 genomic sequences. The HME subgroups HME1, HME2, HME4 and HME5 formed distinct subgroups, which indicated a common ancestral protein. HME1 and HME2 were sister groups with HmyA similar to the common ancestor of both. The HME3a proteins ZneA, ZniA and HmvA were related but deeply branched. The two products from 'pseudogenes' Rmet_3011 and $NimA_1$/A_2 were related, and usually outside of the cluster formed by the other HME-RND proteins. HAE1-RND proteins were always outside of the HME–RND cluster but sometimes leading to it. This allows a speculation about evolution of the HME-RND proteins: HME-RND metal transporters may have originated from HAE1-RND proteins transporting organic substances. The members of the HME3b subgroup but not HmyA represent these early HME-RND proteins. From here, cyanobacterial HME5 proteins branched away, followed by HME4 copper transporters (mostly in proteobacteria), the HME3a proteins (proteobacteria, CFB bacteria), two possible new groups with one of them containing HmyA (Gram-negative bacteria of all kinds), and finally HME1 (proteobacteria) and HME2 (*chlamydia* and proteobacteria, mostly alpha group).

The relationship between the representatives of the HME-RND groups was also visible in the signature sequences of TMH IV. AcrB of the HAE1 groups exhibits the typical double aspartates in a 'AIG-DDA-XEN' motif. The motifs of the HME5 proteins was not different, still showing the double aspartates in a 'AIG-DDX-(M, V)EN'. In the deeply branched HME3b proteins, this double aspartate is gone '(A,g)(I,L)G-D(G,A,s)-VEN'. HME4 proteins show a 'A(I,V)G-DA(A,s)-(V,I)(E,d)N'. The HME1 and HME2 proteins have a similar 'DFG-DGA-VEN' motif, which is more weekly conserved in the more deeply branched HME3a protein with 'GFG-D(G,S,A)(S,A)-(V,M)EN' (Nies, 2003).

HME1 proteins such as CzcA transport Zn^{2+} as their main substrate, HME2 proteins Ni^{2+}. However, when overly expressed, CnrCBA also transport Zn^{2+} (Liesegang, 1993; Grass *et al.*, 2000; Tibazarwa *et al.*, 2000) and a related protein from *Helicobacter pylori* too (Stahler *et al.*, 2006). The fact that the *zni* determinant is induced by zinc and that the MFP ZneB binds zinc indicates that HME3a proteins and early HME1/2 protein may have been zinc transporters. HME2 proteins, which are under control of an ECF sigma factor and not under that of a two-component regulatory system, may have evolved into nickel transporter when this change in regulatory pattern had taken place. Known HME4 proteins transport Cu^{+} and Ag^{+}. The substrates of HME3b and HME5 proteins is unknown with upregulation of the defect *nim* operon by cobalt and a nickel transporter in the vicinity of a HME5-RND-encoding gene giving some direction for speculations (Nies, 2003; Nies *et al.*, 2006).

Speculations and timeline: evolution on earth and changes in metal availability

Two large oxygenation events 2.4 billion (Ga) and 700 million years (Ma) ago, respectively, changed availability of molecular oxygen on earth (Anbar, 2008). Before this event, archaea and bacteria evolved from the last common ancestor LUCA (< 3.85 Ga), bacteria expanded rapidly in the Archean period (3.33–2.85 Ga), eukaryotes diverged from the archaea (> 2.67 Ga), and cyanobacteria emerged (> 2.5 Ga) (David and Alm, 2011). Since the major branches of the bacteria all contain phototrophic representatives at the core, this success of the bacteria may have been the result of the invention of a light-driven electron transport that results in the generation of a proton motive force. Anyhow, ancestors of the cyanobacteria combined the two types of photosystems, the FeS-type and chinon-type, already present in various anaerobic photosynthetic bacteria, added a protein roughly similar to a Mn-containing superoxide dismutase, and were able to use water as electron donor for assimilation of carbon dioxide. This was such a success that cyanobacteria and the chloroplast descendants are still one of the major taxa of life on earth.

Using water as electron donor yields molecular oxygen. Although very toxic, electron transfer from organic carbon components to molecular oxygen allows conservation of more energy than any other respiratory or fermentative process (with the exception of using N_2O and NO as electron acceptors). So, oxygenic photosynthesis and aerobic respiration in combination are able to drive an efficient global carbon cycle that uses the solar radiation energy.

However, first, the toxicity problem of molecular oxygen had to be solved by development of superoxide dismutases, superoxide reductases and catalases. Secondly, availability of the other macro – and microelements changed as a consequence of the rising concentrations of molecular oxygen and declining concentrations of H_2S (Anbar, 2008). Iron was on the anaerobic earth present in μM concentrations as Fe^{2+}, 10,000-fold more available than all the other major transition metals. As a consequence of the oxygenation events, Fe(II) was oxidized to Fe(III), which forms insoluble hydroxides. To solve this problem, bacteria evolved Fe(III)-complexing siderophores that bind Fe^{3+} and makes the metal available for uptake (Braun and Hantke, 2001; Braun and Hantke, 2007).

Alternatively, other transition metal cations may have substituted for iron. During both oxygenation events, nickel remained available, manganese and cobalt declined, and availability of zinc and copper increased several orders of magnitude (Anbar, 2008). Consequently, this led to a change in usage of these transition metal cations over time (David and Alm, 2011): there were time periods in which a rapid evolution of protein families containing a specific metal cation can be deduced. Thus, there seems to be a history of other transition metals replacing Fe^{2+}. First, at 3.3 Ga, many families of zinc-containing protein families originated and usage of cobalt became more common. This was right at the beginning of the Archean period of rapid bacterial evolution but before evolution of molecular oxygen started. Other times of increased evolution rates of zinc-containing proteins was around the first oxygenation event 2.5 Ga ago, that of cobalt-containing protein families 3.0 Ga and 2.0 to 1.5 Ga ago. Nickel had been used before for methanogenesis in archaea but new families of nickel-containing proteins originated around 1.8 Ga ago, and that of copper 1.5 Ga ago. Usage of manganese, however, declined.

Usage of these transition metals in bacteria with a Gram-negative cell wall requires the control of their concentration in the periplasm. Evolution of the HME-RND subgroups fit nicely into the scenario of the changing metal availability and birth of protein families that use these metals. Most HME-RND proteins are somehow connected to zinc transport. So, HME3b proteins as 'oldest' HME-RND protein might have been primitive exporters of Zn^{2+}, and as the induction pattern of the interrupted *nimA* gene indicates, also of Co^{2+}. Both metals are still substrates of 'modern' HME-RND proteins such as CzcA and CnrA. Early on, oxygen-producing cyanobacteria had the need to substitute iron, especially in hydrogen-assimilating hydrogenases that essentially need nickel, so HME5 proteins evolved in this taxon. Improved usage of zinc led to improved management of this metal, leading to evolution of the HME-RND groups HME3a, HME3c and finally the high-performance group HME1 that contains CzcA. Finally, increased usage of nickel and of copper led to the origin of subgroups HME2 and HME4, respectively.

This hypothesis about the function of the proteins of the HME-RND subgroups fit to their role in *C. metallidurans*. This bacterium obtained a variety of HME-RND-encoding genes, but most of them are not expressed. If such a gene would increase metal resistance of this metal-resistant bacterium, turning all this genetic material into silent genes or pseudogenes makes no sense. Only if better determinants are available, switching off the good determinants is sensible. These 'better determinants' generally reside on a plasmid. Such a plasmid should harbour all genes for a perfect metal resistance system to make it useful in all kinds of bacterial hosts. This explains why *C. metallidurans* contains a constitutively expressed *cus* determinant on plasmid pMOL30 although the main copper/silver exporting HME4-RND protein is SilA and encoded by chromosome 2: this makes plasmid pMOL30 also useful in hosts that do not contain a *cus*-like determinant. Everything comes 'ready to use' with this plasmid.

Of the other ten HME-RND proteins with divalent metal cations as predicted substrates, the genes for the two 'old' HME3b proteins NimA and Rmet_3011 were most heavily damaged. Of the three putative zinc exporters, ZinA and HmvA were produced to some extent while *zneA* and the HME3c-encoding gene *hmyA* are silent. The *cnr* and *czc* determinant, which mediate a high metal resistance level in *C. metallidurans*, are complex multicomponent systems with a clear signature of their evolutionary history. The *cnrYXHCBA* determinant on plasmid pMOL28 originated from a gene-duplication event from the silent *ncc* operon in plasmid pMOL30. The core genes of this operon are related to *nccYXH-CBAN* from *Achromobacter* sp. 31A (Schmidt and Schlegel, 1994) so that the *nccYXHCBA* became *cnrYXHCBA* and the regulatory genes *nccYXH* for the ECF sigma factor and the anti-sigma factor complex got deleted (von Rozycki and Nies, 2009). In general, changing the regulatory system of these HME2-RND transporters may have happened to avoid interference with two-component regulatory systems that are often in control of the synthesis of other HME-RND systems.

The *czc* determinant duplicated as a *czcICBA* core operon from *hmuICBA* on chromosome 2. While *hmu* got silenced and interrupted, *czc* acquired more and more genes. First, *czcN* was attached upstream of *czcICBA*. It is still unknown what CzcN is doing but *czcN* is related to *nccN* and might have originated from the *nccYXHCBAN* determinant on the same plasmid. More genes got attached downstream of *czcNICBA* (von Rozycki and Nies, 2009): *czcD* for a CDF protein that exports all three substrates of Czc, Co^{2+}, Zn^{2+} and Cd^{2+}, across the inner membrane (Anton *et al.*, 1999, 2004). The 'old' two component regulatory system for *hmu* was replaced by another one, CzcRS (van der Lelie *et al.*, 1997; Nies and Brown, 1998; von Rozycki and Nies, 2009). Two more genes for periplasmic metal chaperones CzcE and CzcJ indicate the need for better control of the periplasmic zinc content (Große *et al.*, 2004, Zoropogui *et al.*, 2008). Finally, the P_{IB4}-type ATPase CzcP allows rapid export of superfluous cytoplasmic substrate into the periplasm for further handling by CzcCBA (Scherer and Nies, 2009).

Outlook

So, in parallel with the changing usage and availability of zinc and other divalent transition metal cations, HME-RND proteins evolved and the resistance determinants encoding them became more and more sophisticated. However, since HME-RND proteins deal with transition metal cations that became essential trace elements when oxygen decreased the availability of iron, increase in efflux efficiency of HME-RND protein must have been accompanied by increasing control of this activity, or the cells would spend a lot of energy to yield trace element starvation. At this point, the structure of the first HME-RND protein, CusA, the contradictory experimental evidence concerning *in vivo* and *in vitro* function of CzcA and CusA, the genetic environment of HME-RND-encoding genes, the HME-RND subgroups, their primary structure, relationship among each other and distribution, and the bioinorganical consequences of the oxygenation events on earth merge into a scenario useful to explain all of this. However and as always, a lot of experiments are required to verify or falsify all these speculations.

Acknowledgements

Own contribution to this book chapter were supported by several grants from the *Deutsche Forschungsgemeinschaft*. The experiment shown in Fig. 5.7 was performed by Grit Schleuder and Cornelia Grosse.

References

Agranoff, D., Collins, L., Kehres, D., Harrison, T., Maguire, M., and Krishna, S. (2005). The Nramp orthologue of *Cryptococcus neoformans* is a pH-dependent transporter of manganese, iron, cobalt and nickel. Biochem. J. *385*, 225–232.

Allan, C.B., Wu, L.F., Gu, Z.J., Choudhury, S.B., Al-Mjeni, F., Sharma, M.L., Mandrand-Berthelot, M.A., and Maroney, M.J. (1998). An X-ray absorption spectroscopic structural investigation of the nickel site in *Escherichia coli* NikA protein. Inorg. Chem. 37, 5952–5955.

Altschul, S.F., Madden, T.L., Schaffer, A.A., Zhang, J., Zhang, Z., Miller, W., and Lipman, D.J. (1997). Gapped BLAST and PSI-BLAST: a new generation of protein database search programs. Nucl Acid Res. *25*, 3389–3402.

Anbar, A.D. (2008). Oceans. Elements and evolution. Science *322*, 1481–1483.

Anton, A., Große, C., Reißman, J., Pribyl, T., and Nies, D.H. (1999). CzcD is a heavy metal ion transporter involved in regulation of heavy metal resistance in *Ralstonia* sp. strain CH34. J. Bacteriol. *181*, 6876–6881.

Anton, A., Weltrowski, A., Haney, J.H., Franke, S., Grass, G., Rensing, C., and Nies, D.H. (2004). Characteristics of zinc transport by two bacterial cation diffusion facilitators from *Ralstonia metallidurans* and *Escherichia coli*. J. Bacteriol. *186*, 7499–7507.

Argüello, J.M., Eren, E., and Gonzalez-Guerrero, M. (2007). The structure and function of heavy metal transport P-1B-ATPases. Biometals *20*, 233–248.

Bagai, I., Rensing, C., Blackburn, N.J., and McEvoy, M.M. (2008). Direct metal transfer between periplasmic proteins identifies a bacterial copper chaperone. Biochemistry *47*, 11408–11414.

Bleuel, C., Große, C., Taudte, N., Scherer, J., Wesenberg, D., Krauss, G.J., Nies, D.H., and Grass, G. (2005). TolC is involved in enterobactin export across the outer membrane of *Escherichia coli*. J. Bacteriol. *187*, 6701–6707.

Borremans, B., Hobman, J.L., Provoost, A., Brown, N.L., and Van der Lelie, D. (2001). Cloning and functional analysis of the *pbr* lead resistance determinant of *Ralstonia metallidurans* CH34. J. Bacteriol. *183*, 5651–5658.

Braun, V. (1995). Energy-coupled transport and signal transduction through the Gram-negative outer membrane via TonB-ExbB-ExbD-dependent receptor proteins. FEMS Microbiol. Rev. *16*, 295–307.

Braun, V., and Braun, M. (2002). Active transport of iron and siderophore antibiotics. Curr. Opin. Microbiol. *5*, 194–201.

Braun, V., and Hantke, K. (2001). Mechanisms of bacterial iron transport. In Microbial Transport Systems, Winkelmann, G., ed. (Wiley-VCH, Weinheim, Germany), pp. 289–311.

Braun, V., and Hantke, K. (2007). Acquisition of iron by bacteria. In Molecular Microbiology of Heavy Metals, Nies, D.H., and Silver, S., eds. (Springer-Verlag, Berlin and Heidelberg), pp. 189–217.

Braun, V., Hantke, K., and Koster, W. (1998). Bacterial iron transport: mechanisms, genetics, and regulation. Met. Ions Biol. Syst. *35*, 67–145.

Brim, H., Heyndrickx, M., De Vos, P., Wilmotte, A., Springael, D., Schlegel, H.G., and Mergeay, M. (1999). Amplified rDNA restriction analysis and further genotypic characterisation of metal-resistant soil bacteria and related facultative hydrogenotrophs. Syst. Appl. Microbiol. *22*, 258–268.

Busch, W., and Saier, M.H.J. (2002). The transporter classification (TC) system. Crit. Rev. Biochem. Mol. Biol. *37*, 287–337.

Changela, A., Chen, K., Xue, Y., Holschen, J., Outten, C.E., O'Halloran, T.V., and Mondragon, A. (2003). Molecular basis of metal-ion selectivity and zeptomolar sensitivity by CueR. Science *301*, 1383–1387.

Chao, Y., and Fu, D. (2004a). Kinetic study of the antiport mechanism of an *Escherichia coli* zinc transporter, ZitB. J. Biol. Chem. *279*, 12043–12050.

Chao, Y., and Fu, D. (2004b). Thermodynamic studies of the mechanism of metal binding to the *Escherichia coli* zinc transporter YiiP. J. Biol. Chem. *279*, 17173–17180.

Choi, D.W., Do, Y.S., Zea, C.J., McEllistrem, M.T., Lee, S.W., Semrau, J.D., Pohl, N.L., Kisting, C.J., Scardino, L.L., Hartsel, S.C., *et al.* (2006a). Spectral and thermodynamic properties of Ag(I), Au(III), Cd(II), Co(II), Fe(III), Hg(II), Mn(II), Ni(II), Pb(II), U(IV), and Zn(II) binding by methanobactin from *Methylosinus trichosporium* OB3b. J. Inorg. Biochem. *100*, 2150–2161.

Choi, D.W., Zea, C.J., Do, Y.S., Semrau, J.D., Antholine, W.E., Hargrove, M.S., Pohl, N.L., Boyd, E.S., Geesey, G.G., Hartsel, S.C., *et al.* (2006b). Spectral, kinetic, and thermodynamic properties of Cu(I) and Cu(II) binding by methanobactin from *Methylosinus trichosporium* OB3b. Biochemistry *45*, 1442–1453.

Conroy, O., Kim, E.H., McEvoy, M.M., and Rensing, C. (2010). Differing ability to transport nonmetal substrates by two RND-type metal exporters. FEMS Microbiol. Lett. *308*, 115–122.

Dastidar, V., Mao, W., Lomovskaya, O., and Zgurskaya, H.I. (2007). Drug-induced conformational changes in multidrug efflux transporter AcrB from *Haemophilus influenzae*. J. Bacteriol. *189*, 5550–5558.

David, L.A., and Alm, E.J. (2011). Rapid evolutionary innovation during an Archaean genetic expansion. Nature *469*, 93–96.

Davies, B.W., and Walker, G.C. (2007). Disruption of *sitA* compromises *Sinorhizobium meliloti* for manganese uptake required for protection against oxidative stress. J. Bacteriol. *189*, 2101–2109.

Depina, K., Navarro, C., McWalter, L., Boxer, D.H., Price, N.C., Kelly, S.M., Mandrand-Berthelot, M.A., and Wu, L.F. (1995). Purification and characterization of the periplasmic nickel-binding protein NikA of *Escherichia coli* K12. Eur. J. Biochem. *227*, 857–865.

Diels, L., and Mergeay, M. (1990). DNA probe-mediated detection of resistant bacteria from soils highy polluted by heavy metals. Appl. Environ. Microbiol. *56*, 1485–1491.

Echarte, M.M., Rossi, R.C., and Rossi, J.P.F.C. (2007). Phosphorylation of the plasma membrane calcium pump at high ATP concentration. On the mechanism of ATP hydrolysis. Biochemistry *46*, 1034–1041.

Egler, M., Große, C., Grass, G., and Nies, D.H. (2005). Role of ECF sigma factor RpoE in heavy metal resistance of *Escherichia coli*. J. Bacteriol. *187*, 2297–2307.

Eitinger, T., and Friedrich, B. (1991). Cloning, nucleotide sequence, and heterologous expression of a high-affinity nickel transport gene from *Alcaligenes eutrophus*. J. Biol. Chem. *266*, 3222–3227.

Eitinger, T., Suhr, J., Moore, L., and Smith, J.A.C. (2005). Secondary transporters for nickel and cobalt ions: Theme and variations. Biometals *18*, 399–405.

Elkins, C.A., and Nikaido, H. (2002). Substrate specificity of the RND-type multidrug efflux pumps AcrB and AcrD of *Escherichia coli* is determined predominately by two large periplasmic loops. J. Bacteriol. *184*, 6490–6498.

Franke, S., Grass, G., and Nies, D.H. (2001). The product of the *ybdE* gene of the *Escherichia coli* chromosome is involved in detoxification of silver ions. Microbiology *147*, 965–972.

Franke, S., Grass, G., Rensing, C., and Nies, D.H. (2003). Molecular analysis of the copper-transporting CusCFBA efflux system from *Escherichia coli*. J. Bacteriol. *185*, 3804–3812.

Goldberg, M., Pribyl, T., Juhnke, S., and Nies, D.H. (1999). Energetics and topology of CzcA, a cation/proton antiporter of the RND protein family. J. Biol. Chem. *274*, 26065–26070.

Goris, J., De Vos, P., Coenye, T., Hoste, B., Janssens, D., Brim, H., Diels, L., Mergeay, M., Kersters, K., and Vandamme, P. (2001). Classification of metal-resistant bacteria from industrial biotopes as *Ralstonia campinensis* sp. nov., *Ralstonia metallidurans* sp. nov. and *Ralstonia basilensis* Steinle *et al.* 1998 emend. Int. J. Syst. Evol. Microbiol. *51*, 1773–1782.

Grass, G., and Rensing, C. (2001a). CueO is a multi-copper oxidase that confers copper tolerance in *Escherichia coli*. Biochem. Biophys. Res. Commun. *286*, 902–908.

Grass, G., and Rensing, C. (2001b). Genes involved in copper homeostasis in *Escherichia coli*. J. Bacteriol. *183*, 2145–2147.

Grass, G., Große, C., and Nies, D.H. (2000). Regulation of the *cnr* cobalt/nickel resistance determinant from *Ralstonia* sp. CH34. J. Bacteriol. *182*, 1390–1398.

Grass, G., Wong, M.D., Rosen, B.P., Smith, R.L., and Rensing, C. (2002). ZupT Is a Zn(II) uptake system in *Escherichia coli*. J. Bacteriol. *184*, 864–866.

Grass, G., Franke, S., Taudte, N., Nies, D.H., Kucharski, L.M., Maguire, M.E., and Rensing, C. (2005a). The metal permease ZupT from *Escherichia coli* is a transporter with a broad substrate spectrum. J. Bacteriol. *187*, 1604–1611.

Grass, G., Fricke, B., and Nies, D.H. (2005b). Control of expression of a periplasmic nickel efflux pump by periplasmic nickel concentrations. Biometals *18*, 437–448.

Grass, G., Otto, M., Fricke, B., Haney, C.J., Rensing, C., Nies, D.H., and Munkelt, D. (2005c). FieF (YiiP) from *Escherichia coli* mediates decreased cellular accumulation of iron and relieves iron stress. Arch. Microbiol. *183*, 9–18.

Große, C., Grass, G., Anton, A., Franke, S., Navarrete Santos, A., Lawley, B., Brown, N.L., and Nies, D.H. (1999). Transcriptional organization of the *czc* heavy metal homoeostasis determinant from *Alcaligenes eutrophus*. J. Bacteriol. *181*, 2385–2393.

Große, C., Anton, A., Hoffmann, T., Franke, S., Schleuder, G., and Nies, D.H. (2004). Identification of a regulatory pathway that controls the heavy metal resistance system Czc via promoter *czcNp* in *Ralstonia metallidurans*. Arch. Microbiol. *182*, 109–118.

Große, C., Friedrich, S., and Nies, D.H. (2007). Contribution of extracytoplasmic function sigma factors to transition metal homeostasis in *Cupriavidus metallidurans* strain CH34. J. Mol. Microbiol. Biotechnol. *12*, 227–240.

Haber, F., and Weiss, J. (1932). Über die Katalyse des Hydroperoxydes. Naturwissenschaften *20*, 948–950.

Hattori, M., Tanaka, Y., Fukai, S., Ishitani, R., and Nureki, O. (2007). Crystal structure of the MgtE Mg^{2+} transporter. Nature *448*, 1072–1076.

Hausinger, R.P., and Zamble, D.B. (2007). Microbial physiology of nickel and cobalt. In Molecular Microbiology of Heavy Metals, Nies, D.H., and Silver, S., eds. (Springer-Verlag, Berlin), pp. 287–320.

Helbig, K., Bleuel, C., Krauss, G.J., and Nies, D.H. (2008a). Glutathione and transition metal homeostasis in *Escherichia coli*. J. Bacteriol. *190*, 5431–5438.

Helbig, K., Grosse, C., and Nies, D.H. (2008b). Cadmium toxicity in glutathione mutants of *Escherichia coli*. J. Bacteriol. *190*, 5439–5454.

Helmann, J.D., Soonsange, S., and Gabriel, S. (2007). Metallalloregulators: arbiters of metal sufficiency. In Molecular Microbiology of Heavy Metals, Nies, D.H., and Silver, S., eds. (Springer-Verlag, Berlin), pp. 37–71.

Herzberg, M., Kirsten, A., Voigt, A., Scherer, J., and Nies, D.H. (2012). Pleiotropic effects caused by deletion of the central zinc importer in *Cupriavidus metallidurans* CH34. In preparation.

Hmiel, S.P., Snavely, M.D., Florer, J.B., Maguire, M.E., and Miller, C.G. (1989). Magnesium transport in *Salmonella typhimurium*: genetic characterization and cloning of three magnesium transport loci. J. Bacteriol. *171*, 4742–4751.

Hobman, J.L., Yamamoto, K., and Oshima, T. (2007). Transcriptomic responses of bacterial cells to sublethal metal ion stress. In Molecular Microbiology of Heavy Metals, Nies, D.H., and Silver, S., eds. (Springer-Verlag, Berlin), pp. 73–116.

Housecroft, C.E., and Constable, E.C. (2006). Chemistry, 3rd edn (Pearson Education Limited, Essex, England).

Hunte, C., Screpanti, E., Venturi, M., Rimon, A., Padan, E., and Michel, H. (2005). Structure of a Na+/H+ antiporter and insights into mechanism of action and regulation by pH. Nature *435*, 1197–1202.

Ikeda, J.S., Janakiraman, A., Kehres, D.G., Maguire, M.E., and Slauch, J.M. (2005). Transcriptional regulation of *sitABCD* of *Salmonella enterica* serovar *Typhimurium* by MntR and Fur. J. Bacteriol. *187*, 912.

Irving, H., and Williams, R.J.P. (1948). Order of stability of metal complexes. Nature *162*, 746–747.

Iwig, J.S., Rowe, J.L., and Chivers, P.T. (2006). Nickel homeostasis in *Escherichia coli* – the *rcnR-rcnA* efflux pathway and its linkage to NikR function. Mol. Microbiol. *62*, 252–262.

Jackson, R.J., Binet, M.R.B., Lee, L.J., Ma, R., Graham, A.I., McLeod, C.W., and Poole, R.K. (2008). Expression of the PitA phosphate/metal transporter of *Escherichia coli* is responsive to zinc and inorganic phosphate levels. FEMS Microbiol. Lett. *289*, 219–224.

Janssen, P.J., Van Houdt, R., Moors, H., Monsieurs, P., Morin, N., Michaux, A., Benotmane, M.A., Leys, N., Vallaeys, T., Lapidus, A., *et al.* (2010). The complete genome sequence of *Cupriavidus metallidurans* strain CH34, a master survivalist in harsh and anthropogenic environments. PLoS One *5*, e10433.

Junge, W., Sielaff, H., and Engelbrecht, S. (2009). Torque generation and elastic power transmission in the rotary F(O)F(1)-ATPase. Nature *459*, 364–370.

Kehres, D.G., Zaharik, M.L., Finlay, B.B., and Maguire, M.E. (2000). The NRAMP proteins of *Salmonella typhimurium* and *Escherichia coli* are selective manganese transporters involved in the response to reactive oxygen. Mol. Microbiol. *36*, 1085–1100.

Kim, E.-H., Nies, D.H., McEvoy, M., and Rensing, C. (2011). Switch oder funnel: how RND-type transport systems control periplasmic metal homeostasis. J. Bacteriol. *193*, 2381–2387.

Kim, H.J., Graham, D.W., DiSpirito, A.A., Alterman, M.A., Galeva, N., Larive, C.K., Asunskis, D., and Sherwood, P.M. (2004). Methanobactin, a copper-acquisition compound from methane-oxidizing bacteria. Science *305*, 1612–1615.

Kirsten, A., Herzberg, M., Voigt, A., Seravalli, J., Grass, G., Scherer, J., and Nies, D.H. (2011). Contributions of five secondary metal uptake systems to metal homeostasis of *Cupriavidus metallidurans* CH34. J. Bacteriol. *193*, 4652–4663.

Koch, D., Nies, D.H., and Grass, G. (2007). The RcnRA (YohLM) system of *Escherichia coli*: a connection between nickel, cobalt and iron homeostasis. Biometals *20*, 759–771.

Koronakis, V., Sharff, A., Koronakis, E., Luisi, B., and Hughes, C. (2000). Crystal structure of the bacterial membrane protein TolC central to multidrug efflux and protein export. Nature *405*, 914–919.

Legatzki, A., Anton, A., Grass, G., Rensing, C., and Nies, D.H. (2003a). Interplay of the Czc-system and two P-type ATPases in conferring metal resistance to *Ralstonia metallidurans*. J. Bacteriol. *185*, 4354–4361.

Legatzki, A., Franke, S., Lucke, S., Hoffmann, T., Anton, A., Neumann, D., and Nies, D.H. (2003b). First step towards a quantitative model describing Czc-mediated heavy metal resistance in *Ralstonia metallidurans*. Biodegradation *14*, 153–168.

van der Lelie, D., Schwuchow, T., Schwidetzky, U., Wuertz, S., Baeyens, W., Mergeay, M., and Nies, D.H. (1997). Two component regulatory system involved in transcriptional control of heavy metal homoeostasis in *Alcaligenes eutrophus*. Mol. Microbiol. *23*, 493–503.

Lengeler, J.W., Drews, G., and Schlegel, H.G. (1999). Biology of the Prokaryotes (Thieme, Stuttgart).

Li, H., and Jogl, G. (2007). Crystal structure of the zinc-binding transport protein ZnuA from *Escherichia coli* reveals an unexpected variation in metal coordination. J. Mol. Biol. *368*, 1358–1366.

Liesegang, H., Lemke, K., Siddiqui, R.A., and Schlegel, H.-G. (1993). Characterization of the inducible nickel and cobalt resistance determinant *cnr* from pMOL28 of *Alcaligenes eutrophus* CH34. J. Bacteriol. *175*, 767–778.

Loftin, I.R., Franke, S., Roberts, S.A., Weichsel, A., Heroux, A., Montfort, W.R., Rensing, C., and McEvoy, M.M. (2005). A novel copper-binding fold for the periplasmic copper resistance protein CusF. Biochemistry *44*, 10533–10540.

Lonetto, M.A., Brown, K.L., Rudd, K.E., and Buttner, M.J. (1994). Analysis of the *Streptomyces coelicolor sigF* gene reveals the existence of a subfamily of eubacterial RNA polymerase σ factors involved in the regulation of extracytoplasmic functions. Proc. Natl. Acad. Sci. U.S.A. *91*, 7573–7577.

Long, F., Su, C.C., Zimmermann, M.T., Boyken, S.E., Rajashankar, K.R., Jernigan, R.L., and Yu, E.W. (2010). Crystal structures of the CusA efflux pump suggest methionine-mediated metal transport. Nature *467*, 484–488.

Lu, M., and Fu, D. (2007). Structure of the zinc transporter YiiP. Science *317*, 1746–1748.

Lunin, V., Dobrovetsky, E., Khutoreskaya, G., Zhang, R., Joachimiak, A., Doyle, D.A., Bochkarev, A., Maguire, M.E., Edwards, A.M., and Koth, C.M. (2006). Crystal structure of the CorA Mg^{2+} transporter. Nature *440*, 833–837.

Macomber, L., Rensing, C., and Imlay, J.A. (2007). Intracellular copper does not catalyze the formation of oxidative DNA damage in *Escherichia coli*. J. Bacteriol. *189*, 1616–1626.

Macomber, L., Elsey, S.P., and Hausinger, R.P. (2011). Fructose-1,6-bisphosphate aldolase (class II) is the primary site of nickel toxicity in *Escherichia coli*. Mol. Microbiol. *82*, 1291–1300.

Magnani, D., and Solioz, M. (2007). How bacteria handle copper. In Molecular Microbiology of Heavy Metals, Nies, D.H., and Silver, S., eds. (Springer-Verlag, Berlin), pp. 259–285.

Mahren, S., and Braun, V. (2003). The FecI extracytoplasmic-function sigma factor of *Escherichia coli* interacts with the beta' subunit of RNA polymerase. J. Bacteriol. *185*, 1796–1802.

Makui, H., Roig, E., Cole, S.T., Helmann, J.D., Gros, P., and Cellier, M.F. (2000). Identification of the *Escherichia coli* K-12 Nramp orthologue (MntH) as a selective divalent metal ion transporter. Mol. Microbiol. *35*, 1065–1078.

Mandal, A.K., Cheung, W.D., and Argüello, J.M. (2002). Characterization of a thermophilic P-type Ag^+/Cu^+-ATPase from the extremophile *Archaeoglobus fulgidus*. J. Biol. Chem. *277*, 7201–7208.

Marrero, J., Auling, G., Coto, O., and Nies, D.H. (2007). High-level resistance to cobalt and nickel but probably no transenvelope efflux: metal resistance in the Cuban *Serratia marcescens* strain C-1. Microb. Ecol. *53*, 123–133.

Mergeay, M. (2000). Bacteria adapted to industrial biotopes: metal-resistant *Ralstonia*. In Bacterial Stress Responses, Storz, G., and Hengge-Aronis, R., eds. (ASM Press, Washington DC), pp. 403–414.

Mergeay, M., Houba, C., and Gerits, J. (1978). Extrachromosomal inheritance controlling resistance to cadmium, cobalt and zinc ions: evidence for curing a *Pseudomonas*. Arch. Int. Physiol. Biochim. *86*, 440–441.

Mergeay, M., Nies, D., Schlegel, H.G., Gerits, J., Charles, P., and van Gijsegem, F. (1985). *Alcaligenes eutrophus* CH34 is a facultative chemolithotroph with plasmid-bound resistance to heavy metals. J. Bacteriol. *162*, 328–334.

Mergeay, M., Monchy, S., Vallaeys, T., Auquier, V., Benotmane, A., Bertin, P., Taghavi, S., Dunn, J., van der Lelie, D., and Wattiez, R. (2003). *Ralstonia metallidurans*, a bacterium specifically adapted to toxic metals: towards a catalogue of metal-responsive genes. FEMS Microbiol. Rev. *27*, 385–410.

Monchy, S., Benotmane, M.A., Janssen, P., Vallaeys, T., Taghavi, S., van der Lelie, D., and Mergeay, M. (2007). Plasmids pMOL28 and pMOL30 of *Cupriavidus metallidurans* are specialized in the maximal viable response to heavy metals. J. Bacteriol. *189*, 7417–7425.

Monchy, S., Benotmane, M.A., Wattiez, R., van Aelst, S., Auquier, V., Borremans, B., Mergeay, M., Taghavi, S., van der Lelie, D., and Vallaeys, T. (2006). Transcriptomics and proteomic analysis of the pMOL30-encoded copper resistance in *Cupriavidus metallidurans* strain CH34. Microbiology *152*, 1765–1776.

Munkelt, D., Grass, G., and Nies, D.H. (2004). The chromosomally encoded cation diffusion facilitator proteins DmeF and FieF from *Wautersia metallidurans* CH34 are transporters of broad metal specificity. J. Bacteriol. *186*, 8036–8043.

Murakami, S. (2008). Multidrug efflux transporter, AcrB – the pumping mechanism. Curr. Opin. Struct. Biol. *18*, 459–465.

Murakami, S., Nakashima, R., Yamashita, R., and Yamaguchi, A. (2002). Crystal structure of bacterial multidrug efflux transporter AcrB. Nature *419*, 587–593.

Murakami, S., Tamura, N., Saito, A., Hirata, T., and Yamaguchi, A. (2004). Extramembrane central pore of multidrug exporter AcrB in *Escherichia coli* plays an important role in drug transport. J. Biol. Chem. *279*, 3743–3748.

Navarro, C., Wu, L.F., and Mandrand-Berthelot, M.A. (1993). The *nik* operon of *Escherichia* coli encodes a periplasmic binding-protein-dependent transport-system for nickel. Mol. Microbiol. *9*, 1181–1191.

Nies, D.H. (1992a). CzcR and CzcD, gene products affecting regulation of resistance to cobalt, zinc and cadmium (*czc* system) in *Alcaligenes eutrophus*. J. Bacteriol. *174*, 8102–8110.

Nies, D.H. (1992b). Resistance to cadmium, cobalt, zinc and nickel in microbes. Plasmid *27*, 17–28.

Nies, D.H. (1995). The cobalt, zinc, and cadmium efflux system CzcABC from *Alcaligenes eutrophus* functions as a cation-proton-antiporter in *Escherichia coli*. J. Bacteriol. *177*, 2707–2712.

Nies, D.H. (2003). Efflux-mediated heavy metal resistance in prokaryotes. FEMS Microbiol. Rev. *27*, 313–339.

Nies, D.H. (2004a). Essential and toxic effects of elements on microorganisms. In Metals and Their Compounds in the Environment, Anke, K., Ihnat, M., and Stoeppler, M., eds. (Wiley-VCH, Weinheim, Germany), p. Part II.1.

Nies, D.H. (2004b). Incidence and function of sigma factors in *Ralstonia metallidurans* and other bacteria. Arch. Microbiol. *181*, 255–268.

Nies, D.H. (2007a). Bacterial transition metal homeostasis. In Molecular Microbiology of Heavy Metals, Nies, D.H., and Silver, S., eds. (Springer-Verlag, Berlin), pp. 118–142.

Nies, D.H. (2007b). How cells control zinc homeostasis. Science *317*, 1695–1696.

Nies, D.H. (2012). Zinc starvation response in a cyanobacterium revealed. J. Bacteriol. *194*, 2407–2412.

Nies, D.H., and Brown, N. (1998). Two-component systems in the regulation of heavy metal resistance. In Metal Ions in Gene Regulation, Silver, S., and Walden, W., eds. (Chapman Hall, London and New York), pp. 77–103.

Nies, D.H., and Grass, G. (2009). Chapter 5. 4.4.3. Transition metal homeostasis. In EcoSal – *Escherichia coli* and *Salmonella*: Cellular and Molecular Biology, Böck, A., Curtiss III, R., Kaper, J.B., Neidhardt, F.C., Nyström, T., Rudd, K.E., and Squires, C.L., eds. (ASM Press, Washington, DC).

Nies, D.H., and Silver, S. (1989). Plasmid-determined inducible efflux is responsible for resistance to cadmium, zinc, and cobalt in *Alcaligenes eutrophus*. J. Bacteriol. *171*, 896 – 900.

Nies, D.H., and Silver, S. (1995). Ion efflux systems involved in bacterial metal resistances. J. Indust. Microbiol. *14*, 186–199.

Nies, D., Mergeay, M., Friedrich, B., and Schlegel, H.G. (1987). Cloning of plasmid genes encoding resistance to cadmium, zinc, and cobalt in *Alcaligenes eutrophus* CH34. J. Bacteriol. *169*, 4865–4868.

Nies, D.H., Nies, A., Chu, L., and Silver, S. (1989). Expression and nucleotide sequence of a plasmid-determined divalent cation efflux system from *Alcaligenes eutrophus*. Proc. Natl. Acad. Sci. U.S.A. *86*, 7351–7355.

Nies, D.H., Rehbein, G., Hoffmann, T., Baumann, C., and Grosse, C. (2006). Paralogs of genes encoding metal resistance proteins in *Cupriavidus metallidurans* strain CH34. J. Mol. Micobiol. Biotechnol. *11*, 82–93.

Outten, F.W., Huffman, D.L., Hale, J.A., and O'Halloran, T.V. (2001). The independent *cue* and *cus* systems confer copper tolerance during aerobic and anaerobic growth in *Escherichia coli*. J. Biol. Chem. *276*, 30670–30677.

Padan, E., Venturi, M., Gerchman, Y., and Dover, N. (2001). Na^+/H^+ antiporters. Biochim. Biophys. Acta *1505*, 144–157.

Patzer, S.I., and Hantke, K. (1998). The ZnuABC high-affinity zinc uptake system and its regulator Zur in *Escherichia coli*. Mol. Microbiol. *28*, 1199–1210.

Paulsen, I.T., and Saier, M.H., Jr. (1997). A novel family of ubiquitous heavy metal ion transport proteins. J. Membr. Biol. *156*, 99–103.

Paulsen, I.T., Park, J.H., Choi, P.S., and Saier, M.H.J. (1997). A family of Gram-negative bacterial outer membrane factors that function in the export of proteins, carbohydrates, drugs and heavy metals from Gram-negative bacteria. FEMS Microbiol. Lett. *156*, 1–8.

Pontel, L.B., Audero, M.E., Espariz, M., Checa, S.K., and Soncini, F.C. (2007). GolS controls the response to gold by the hierarchical induction of *Salmonella*-specific genes that include a CBA efflux-coding operon. Mol. Microbiol. *66*, 814–825.

Raimunda, D., Gonzalez-Guerrero, M., Leeber, B.W., and Arguello, J.M. (2011). The transport mechanism of bacterial Cu^+-ATPases: distinct efflux rates adapted to different function. Biometals *24*, 467–475.

Ranquet, C., Ollagnier-de-Choudens, S., Loiseau, L., Barras, F., and Fontecave, M. (2007). Cobalt stress in *Escherichia coli*. J. Biol. Chem. *282*, 30442–30451.

Reith, F., Lengke, M.F., Falconer, D., Craw, D., and Southam, G. (2007). The geomicrobiology of gold. ISME J. *1*, 567–584.

Reith, F., Etschmann, B., Grosse, C., Moors, H., Benotmane, M.A., Monsieurs, P., Grass, G., Doonan, C., Vogt, S., Lai, B., *et al.* (2009). Mechanisms of gold biomineralization in the bacterium *Cupriavidus metallidurans* Proc. Natl. Acad. Sci. U.S.A. *106*, 17757–17762.

Rensing, C., Pribyl, T., and Nies, D.H. (1997). New functions for the three subunits of the CzcCBA cation-proton-antiporter. J. Bacteriol. *179*, 6871–6879.

Rensing, C., Fan, B., Sharma, R., Mitra, B., and Rosen, B.P. (2000). CopA: An *Escherichia coli* Cu(I)-translocating P-type ATPase. Proc. Natl. Acad. Sci. U.S.A. *97*, 652–656.

Reyes-Caballero, H., Campanello, G.C., and Giedroc, D.P. (2011). Metalloregulatory proteins: Metal selectivity and allosteric switching. Biophys. Chem. *156*, 103–114.

Roberts, S.A., Weichsel, A., Grass, G., Thakali, K., Hazzard, J.T., Tollin, G., Rensing, C., and Montfort, W.R. (2002). Crystal structure and electron transfer kinetics of CueO, a multicopper oxidase required for copper homeostasis in *Escherichia coli*. Proc. Natl. Acad. Sci. U.S.A. *99*, 2766–2771.

Rodionov, D.A., Hebbeln, P., Gelfand, M.S., and Eitinger, T. (2006). Comparative and functional genomic analysis of prokaryotic nickel and cobalt uptake transporters: evidence for a novel group of ATP-binding cassette transporters. J. Bacteriol. *188*, 317–327.

Rodrigue, A., Effantin, G., and Mandrand-Berthelot, M.A. (2005). Identification of *rcnA* (*yohM*), a nickel and cobalt resistance gene in *Escherichia coli*. J. Bacteriol. *187*, 2912–2916.

Rouviere, P.E., De Las Penas, A., Mecsas, J., Lu, C.Z., Rudd, K.E., and Gross, C.A. (1995). *rpoE*, the gene encoding the second heat-shock sigma factor, sigma E, in *Escherichia coli*. EMBO J. *14*, 1032–1042.

Rowe, J.L., Starnes, G.L., and Chivers, P.T. (2005). Complex transcriptional control links NikABCDE-dependent nickel transport with hydrogenase expression in *Escherichia coli*. J. Bacteriol. *187*, 6317–6323.

von Rozycki, T., and Nies, D.H. (2009). *Cupriavidus metallidurans*: evolution of a metal-resistant bacterium. Antonie Van Leeuwenhoek *96*, 115–139.

von Rozycki, T., Nies, D.H., and Saier, M.H.J. (2005). Genomic analyses of transport proteins in *Ralstonia metallidurans*. Comp. Func. Genom. *6*, 17–56.

Sabri, M., Leveille, S., and Dozois, C.M. (2006). A SitABCD homologue from an avian pathogenic *Escherichia coli* strain mediates transport of iron and manganese and resistance to hydrogen peroxide. Microbiology *152*, 745–758.

Sabri, M., Caza, M., Proulx, J., Lymberopoulos, M.H., Bree, A., Moulin-Schouleur, M., Curtiss, R., 3rd, and Dozois, C.M. (2008). Contribution of the SitABCD, MntH, and FeoB metal transporters to the virulence of avian pathogenic *Escherichia coli* O78 strain chi7122. Infect. Immun. *76*, 601–611.

Saier, M.H., Jr., Tam, R., Reizer, A., and Reizer, J. (1994). Two novel families of bacterial membrane proteins concerned with nodulation, cell division and transport. Mol. Microbiol. *11*, 841–847.

Saier, M.H., Jr., Tran, C.V., and Barabote, R.D. (2006). TCDB: the Transporter Classification Database for membrane transport protein analyses and information. Nucl Acid Res. *34*, D181–D186.

Scherer, J., and Nies, D.H. (2009). CzcP is a novel efflux system contributing to transition metal resistance in *Cupriavidus metallidurans* CH34. Mol. Microbiol. *73*, 601–621.

Schmidt, T., and Schlegel, H.G. (1994). Combined nickel-cobalt-cadmium resistance encoded by the *ncc* locus of *Alcaligenes xylosoxidans* 31A. J. Bacteriol. *176*, 7045–7054.

Schwarz, G., Hagedoorn, P.-L., and Fischer, K. (2007). Molybdate and tungstate: uptake, homeostasis, cofactors, and enzymes. In Molecular Microbiology of Heavy Metals, Nies, D.H., and Silver, S., eds. (Springer, Berlin) pp. 421–452.

Seeger, M.A., Schiefner, A., Eicher, T., Verrey, F., Diederichs, K., and Pos, K.M. (2006). Structural asymmetry of AcrB trimer suggests a peristaltic pump mechanism. Science *313*, 1295–1298.

Seeger, M.A., von Ballmoos, C., Eicher, T., Brandstatter, L., Verrey, F., Diederichs, K., and Pos, K.M. (2008). Engineered disulfide bonds support the functional rotation mechanism of multidrug efflux pump AcrB. Nat. Struct. Mol. Biol. *15*, 199–205.

Selinger, D.W., Saxena, R.M., Cheung, K.J., Church, G.M., and Rosenow, C. (2003). Global RNA half-life analysis in *Escherichia coli* reveals positional patterns of transcript degradation. Genome Res. *13*, 216–223.

Sennhauser, G., Bukowska, M.A., Briand, C., and Grutter, M.G. (2009). Crystal structure of the multidrug exporter MexB from *Pseudomonas aeruginosa*. J. Mol. Biol. *389*, 134–145.

Shepherd, M., Heath, M.D., and Poole, R.K. (2007). NikA binds heme: A new role for an *Escherichia coli* periplasmic nickel-binding protein. Biochemistry *46*, 5030–5037.

Silver, S. (2003). Bacterial silver resistance: molecular biology and uses and misuses of silver compounds. FEMS Microbiol. Rev. *27*, 341–353.

Silver, S. (2011). BioMetals: a historical and personal perspective. Biometals *24*, 379–390.

Silver, S., and Phung, L.T. (1996). Bacterial heavy metal resistance: New surprises. Ann. Rev. Microbiol. *50*, 753–789.

Singh, S.K., Grass, G., Rensing, C., and Montfort, W.R. (2004). Cuprous oxidase activity of CueO from *Escherichia coli*. J. Bacteriol. *186*, 7815–7817.

Smith, R.L., Banks, J.L., Snavely, M.D., and Maguire, M.E. (1993). Sequence and topology of the CorA magnesium transport system of *Salmonella typhimurium* and *Escherichia coli*. Identification of a new class of transport protein. J. Biol. Chem. *268*, 14071–14080.

Snavely, M.D., Florer, J.B., Miller, C.G., and Maguire, M.E. (1989). Magnesium transport in *Salmonella typhimurium*: $^{28}Mg^{2+}$ transport by CorA, MgtA, and MgtB systems. J. Bacteriol. *171*, 4761–4766.

Snavely, M.D., Miller, C.G., and Maguire, M.E. (1991). The *mgtB* Mg^{2+} transport locus of *Salmonella typhimurium* encodes a P-type ATPase. J. Biol. Chem. *266*, 815–823.

Southam, G., Lengke, M.F., Fairbrother, L., and Reith, F. (2009). The biogeochemistry of gold. Elements *5*, 303–307.

Stahler, F.N., Odenbreit, S., Haas, R., Wilrich, J., Van Vliet, A.H., Kusters, J.G., Kist, M., and Bereswill, S. (2006). The novel *Helicobacter pylori* CznABC metal efflux pump is required for cadmium, zinc, and nickel resistance, urease modulation, and gastric colonization. Infect. Immun. *74*, 3845–3852.

Su, C.C., Li, M., Gu, R.Y., Takatsuka, Y., McDermott, G., Nikaido, H., and Yu, E.W. (2006). Conformation of the AcrB multidrug efflux pump in mutants of the putative proton relay pathway. J. Bacteriol. *188*, 7290–7296.

Su, C.C., Long, F., Zimmermann, M.T., Rajashankar, K.R., Jernigan, R.L., and Yu, E.W. (2011). Crystal structure of the CusBA heavy metal efflux complex of *Escherichia coli*. Nature *470*, 558–563.

Taghavi, S., Mergeay, M., Nies, D., and van der Lelie, D. (1997). *Alcaligenes eutrophus* as a model system for bacterial interactions with heavy metals in the environment. Res. Microbiol. *6*, 536–551.

Takatsuka, Y., and Nikaido, H. (2006). Threonine-978 in the transmembrane segment of the multidrug efflux pump AcrB of *Escherichia coli* is crucial for drug transport as a probable component of the proton relay network. J. Bacteriol. *188*, 7284–7289.

Takatsuka, Y., and Nikaido, H. (2007). Site-directed disulfide cross-linking shows that cleft flexibility in the periplasmic domain is needed for the multidrug efflux pump AcrB of *Eschetichia coli*. J. Bacteriol. *189*, 8677–8684.

Takatsuka, Y., and Nikaido, H. (2009). Covalently linked trimer of the AcrB multidrug efflux pump provides support for the functional rotating mechanism. J. Bacteriol. *191*, 1729–1737.

Taudte, N., and Grass, G. (2010). Point mutations change specificity and kinetics of metal uptake by ZupT from *Escherichia coli*. Biometals *23*, 643–656.

Thieme, D., Neubauer, P., Nies, D.H., and Grass, G. (2008). Sandwich hybridization assay for sensitive detection of dynamic changes in mRNA transcript levels in crude *Escherichia coli* cell extracts in response to copper ions. Appl. Environ. Microbiol. *74*, 7463–7470.

Thorgersen, M.P., and Downs, D.M. (2007). Cobalt targets multiple metabolic processes in *Salmonella enterica*. J. Bacteriol. *189*, 7774–7781.

Tibazarwa, C., Wuertz, S., Mergeay, M., Wyns, L., and van der Lelie, D. (2000). Regulation of the *cnr* cobalt and nickel resistance determinant of *Ralstonia eutropha* (*Alcaligenes eutrophus*) CH34. J. Bacteriol. *182*, 1399–1409.

Tottey, S., Harvie, D.R., and Robinson, N.J. (2007). Understanding how cells allocate metals. In Molecular Microbiology of Heavy Metals, Nies, D.H., and Silver, S., eds. (Springer-Verlag, Berlin), pp. 3–36.

Townsend, D.E., Esenwine, A.J., George, J.I., Bross, D., Maguire, M.E., and Smith, R.L. (1995). Cloning of the *mgtE* Mg^{2+} transporter from *Providencia stuartii* and the distribution of *mgtE* in Gram-negative and Gram-positive bacteria. J. Bacteriol. *177*, 5350–5354.

Toyoshima, C., Nomura, N., and Tsuda, T. (2004). Lumenal gating mechanism revealed in calcium pump crystal structures with phosphate analogues. Nature *432*, 361–368.

Trepreau, J., Girard, E., Maillard, A.P., de Rosny, E., Petit-Haertlein, I., Kahn, R., and Coves, J. (2011). Structural basis for metal sensing by CnrX. J. Mol. Biol. *408*, 766–779.

Tseng, T.-T., Gratwick, K.S., Kollman, J., Park, D., Nies, D.H., Goffeau, A., and Saier, M.H.J. (1999). The RND superfamily: an ancient, ubiquitous and diverse family that includes human disease and development proteins. J. Mol. Microbiol. Biotechnol. *1*, 107–125.

Tsukazaki, T., Mori, H., Echizen, Y., Ishitani, R., Fukai, S., Tanaka, T., Perederina, A., Vassylyev, D.G., Kohno, T., Maturana, A.D., *et al.* (2011). Structure and function of a membrane component SecDF that enhances protein export. Nature *474*, 235–238.

Ueoka-Nakanishi, H., and Maeshima, M. (2000). Quantification of Ca^{2+}/H^{+} antiporter VCAX1p in vacuolar membranes and its absence in roots of Mung Bean. Plant Cell Physiol. *41*, 1067–1071.

Umena, Y., Kawakami, K., Shen, J.R., and Kamiya, N. (2011). Crystal structure of oxygen-evolving photosystem II at a resolution of 1.9 A. Nature *473*, 55–60.

Waldron, K.J., Rutherford, J.C., Ford, D., and Robinson, N.J. (2009). Metalloproteins and metal sensing. Nature *460*, 823–830.

Wan, Q., Ahmad, M.F., Fairman, J., Gorzelle, B., de la Fuente, M., Dealwis, C., and Maguire, M.E. (2011). X-ray crystallography and isothermal titration calorimetry studies of the *Salmonella* zinc transporter ZntB. Structure *19*, 700–710.

Watt, I.N., Montgomery, M.G., Runswick, M.J., Leslie, A.G.W., and Walker, J.E. (2010). Bioenergetic cost of making an adenosine triphosphate molecule in animal mitochondria. Proc. Natl. Acad. Sci. U.S.A. *107*, 16823–16827.

Weast, R.C. (1984). CRC handbook of chemistry and physics, 64 edn (CRC Press, Inc., Boca Raton, FL).

Wei, Y.N., and Fu, D. (2005). Selective metal binding to a membrane-embedded aspartate in the *Escherichia coli* metal transporter YiiP (FieF). J. Biol. Chem. *280*, 33716–33724.

Williams-Jones, A.E., Bowell, R.J., and Migdisov, A.A. (2009). Gold in solution. Elements *5*, 281–287.

Wolfram, L., Friedrich, B., and Eitinger, T. (1995). The *Alcaligenes eutrophus* protein HoxN mediates nickel transport in *Escherichia coli*. J. Bacteriol. *177*, 1840–1843.

Worlock, A.J., and Smith, R.L. (2002). ZntB is a novel Zn^{2+} transporter in *Salmonella enterica serovar typhimurium*. J. Bacteriol. *184*, 4369–4373.

Wu, L.F., Navarro, C., and Mandrand Berthelot, M.A. (1991). The *hydC* region contains a multi-cistronic operon (*nik*) involved in nickel transport in *Escherichia coli*. Gene *107*, 37–42.

Wu, L.F., Navarro, C., Depina, K., Quenard, M., and Mandrand, M.A. (1994). Antagonistic effect of nickel on the fermentative growth of *Escherichia coli* K12 and comparison of nickel and cobalt toxicity on the aerobic and anaerobic growth. Environ. Health Perspect. *102*, 297–300.

Xu, Y., Feng, L., Jeffrey, P.D., Shi, Y.G., and Morel, F.M.M. (2008). Structure and metal exchange in the cadmium carbonic anhydrase of marine diatoms. Nature *452*, 56–61.

Yu, E.W., Aires, J.R., and Nikaido, H. (2003). AcrB multidrug efflux pump of *Escherichia coli:* composite substrate-binding cavity of exceptional flexibility generates its extremely wide substrate specificity. J. Bacteriol. *185*, 5657–5664.

Yu, E.W., Aires, J.R., McDermott, G., and Nikaido, H. (2005). A periplasmic drug-binding site of the AcrB multidrug efflux pump: a crystallographic and site-directed mutagenesis study. J. Bacteriol. *187*, 6804–6815.

Yum, S.W., Xu, Y.B., Piao, S.F., Sim, S.H., Kim, H.M., Jo, W.S., Kim, K.J., Kweon, H.S., Jeong, M.H., Jeon, H.S., *et al.* (2009). Crystal structure of the periplasmic component of a tripartite macrolide-specific efflux pump. J. Mol. Biol. *387*, 1286–1297.

Zgurskaya, H.I., and Nikaido, H. (1999). Bypassing the periplasm: reconstitution of the AcrAB multidrug efflux pump of *Escherichia coli*. Proc. Natl. Acad. Sci. U.S.A. *96*, 7190–7195.

Zoropogui, A., Gambarelli, S., and Coves, J. (2008). CzcE from *Cupriavidus metallidurans* CH34 is a copper-binding protein. Biochem. Biophys. Res. Commun. *365*, 735–739.

The Role of Efflux Pumps in the Nosocomial Pathogens *Staphylococcus aureus* and *Acinetobacter baumannii*

6

Bart A. Eijkelkamp, Karl A. Hassan, Ian T. Paulsen and Melissa H. Brown

Abstract

Infections caused by opportunistic nosocomial pathogens can complicate the treatment of patients admitted to intensive care units. Most nosocomial bacterial pathogens possess an extended collection of resistance strategies to circumvent the effects of continuous exposure to the antimicrobial stresses present in these environments. Of these, active efflux has proven to be a successful detoxification mechanism employed by both Gram-positive and Gram-negative nosocomial pathogens, such as *Staphylococcus aureus* and *Acinetobacter baumannii*, respectively. Bioinformatic analyses of the genomes of these organisms determined that they each encode a large number of putative drug efflux systems including representatives from each of the five main families of efflux systems. Nonetheless, the repertoires of putative efflux systems encoded by these bacteria differ, primarily with respect to the dominant families of protective efflux systems; a trend that is likely to extend to most Gram-negative and Gram-positive bacteria. Various *S. aureus* transporters were among the first drug efflux systems to be described. Consequently, these proteins have been extensively characterized in work that has significantly advanced our fundamental knowledge of the structure and function of these complex proteins. In contrast, studies of *A. baumannii* efflux systems have been largely genetic, predominantly focused on explaining the high level of multidrug resistance observed in this bacterium without examining the biochemical or structural nature of the transporters themselves. Regulatory control of the genes encoding drug transporters is of major importance for multidrug resistance in both *S. aureus* and *A. baumannii*, since the overproduction of these proteins is typically detrimental to cell growth under non-selective conditions. Accordingly, it is not unusual for these systems to be controlled at a number of levels, both globally and locally by a range of regulatory factors. Here, we compare and contrast the efflux capabilities of these two bacterial pathogens.

Introduction

Despite the best efforts of clinicians, bacterial infections continue to plague intensive care units worldwide. Bacterial pathogens pose not only serious health risks but also a significantly increased financial burden, due to extended hospitalization. Bacterial nosocomial infections are caused by a range of both Gram-positive and Gram-negative pathogens, such as *Staphylococcus aureus* and *Acinetobacter baumannii*, respectively. Although not obligate human pathogens, strains of *S. aureus* and *A. baumannii* have become specialists in the hospital environment, a phenomenon at least partly related to their innate or acquired drug resistance capabilities. These pathogens are exposed to high levels of antimicrobial stress, including antibiotic treatment in human host niches and exposure to biocides as part of decontamination regimes. Paradoxically, in light of their remarkable capacity to develop or acquire drug resistance, the extensive use of antibiotics and disinfectants may give these strains a competitive advantage in the clinic, preventing the growth of competing organisms, and freeing up available resources for those that are resistant.

Alarmingly, strains of both *S. aureus* and *A. baumannii* that are resistant to the majority of clinically useful antimicrobials have been isolated (Clark *et al.*, 2003; Michalopoulos and Falagas, 2010). Drug resistance in these, and other bacterial pathogens, is mediated by a number of general mechanisms, including target site bypass, antimicrobial inactivation or degradation and reduction of the intracellular antimicrobial concentration, either by reducing uptake through the cell wall, membrane modification, or active efflux of the antimicrobial. The latter mechanism, drug efflux that is mediated by integral membrane transport proteins, is a resistance strategy of particular importance in both *S. aureus* and *A. baumannii* strains isolated from hospitals. Although not always associated with high levels of resistance, active efflux gives cells an advantage to subsequently allow development of synergistic means of resistance, e.g. target site mutations (Markham and Neyfakh, 1996).

Currently, most bacterial drug efflux systems are recognized as being members of one of five families of membrane transport proteins, including the major facilitator superfamily (MFS), the resistance-nodulation-division (RND) superfamily, the ATP-binding cassette (ABC) superfamily, the multidrug and toxin extrusion (MATE) family and the small multidrug resistance (SMR) family (Paulsen *et al.*, 1996b; Brown *et al.*, 1999). Representative efflux systems from each of these families have been identified in both Gram-positive and Gram-negative bacterial strains, including *S. aureus* and *A. baumannii*. However, the complement of transporters from each of the five families of efflux systems is distinct to each organism and the systems considered to be primarily responsible for drug and toxin efflux are unique.

An important consideration in the study of efflux systems in bacteria is the architecture of the cell wall. Both Gram-positive and Gram-negative bacterial cells are encapsulated by a cytoplasmic (inner) membrane, which accommodates the energy-transducing components of drug efflux systems. However, Gram-negative bacteria possess a second protective (outer) membrane that serves as an additional permeability barrier to prevent antimicrobials and other toxins from reaching intracellular targets. To exploit this outer membrane, Gram-negative organisms have developed efflux systems that include not only an inner-membrane bound energy-transducing efflux protein, but also periplasmic (membrane fusion protein; MFP) and outer membrane-bound (outer membrane protein; OMP) components. When assembled, these components form tripartite complexes that span both inner and outer membranes to potentially facilitate removal of substrates from the cytoplasm, inner membrane and/or periplasm to the extracellular environment (Blair and Piddock, 2009; Pos, 2009). Efflux proteins from at least three of the five families of drug exporters, the RND, MFS and ABC superfamilies, have been found to form tripartite complexes in Gram-negative bacteria (Magnet *et al.*, 2001; Tanabe *et al.*, 2009; Tikhonova *et al.*, 2009). However, these complexes are most commonly associated with RND efflux systems. Indeed, drug-exporting RND transporters are almost universally associated with cognate MFPs and OMPs in Gram-negative bacteria (Poole, 2002), although all three components are not always encoded in the same genomic locus.

Bioinformatic analyses of bacterial efflux systems

The explosion of genome sequence information over the past decade has allowed the number and type of transport systems in a broad range of organisms to be bioinformatically determined (Ren and Paulsen, 2007). The transporter automated annotation pipeline (TransAAP) was developed for this purpose (Ren *et al.*, 2007). This pipeline uses a range of bioinformatic tools including blast, HMM, Pfam, TIGRfam HMM and COG searches, as well as other analyses such as hydropathy analysis to predict the complete complement of transport proteins encoded in an organism from its genome sequence (the output of these analyses are stored in a web-based database; www.membranetransport.org).

We have used TransAAP to predict the number and type of drug efflux systems encoded in representative *Staphylococcus* and *Acinetobacter* strains. Overall, *Acinetobacter* genomes contain more putative drug efflux systems than their staphylococcal counterparts. Indeed, clinical *A. baumannii* isolates were found to encode at least 50 putative drug efflux systems. Interestingly, these efflux systems are highly conserved between clinical strains with approximately 80% of systems conserved within the core genomes of available sequenced strains (Paulsen *et al.*, unpublished data), highlighting their potential as widespread drug resistance mechanisms. Furthermore, a significant number of putative efflux systems are conserved between clinical *Acinetobacter* isolates and non-clinical or environmental strains such as *A. baumannii* SDF and *A. baylyi* ADP1 (Fig. 6.1). Recent studies of orthologous exporters from these strains have demonstrated close functional overlap (Hassan *et al.*, 2011), indicating that environmental isolates, such as *A. baylyi* ADP1, which are more amenable to genetic manipulation than clinical *A. baumannii* isolates (de Berardinis *et al.*, 2008), represent excellent model systems to examine the function of these transport systems.

In comparison to *Acinetobacter*, staphylococcal strains were found to encode fewer putative drug efflux systems (Fig. 6.1). This lower abundance of efflux proteins is partly explained by the smaller average size of staphylococcal genomes. Nonetheless, in addition to chromosomally encoded drug efflux systems, staphylococcal multiresistance plasmids frequently encode drug efflux systems, providing a vehicle for their dissemination and spread between strains (Hassan *et al.*, 2007a). Highlighting the importance of these plasmids in the failure of current therapies, a number of these plasmid-encoded drug efflux systems are routinely identified in contemporary clinical strains worldwide (Mayer *et al.*, 2001; Noguchi *et al.*, 2005; Longtin *et al.*, 2011). Similar to *A. baumannii* strains, the putative drug efflux systems identified in *S. aureus* are highly conserved and a significant number are conserved in related species, e.g. *S. epidermidis* (Fig. 6.1).

Staphylococcus aureus

S. aureus is the archetypal opportunistic Gram-positive nosocomial pathogen, responsible for a range of infections of varying severities. Owing to the propensity of *S. aureus* strains to rapidly develop or acquire resistance to effectively any antimicrobial compound they encounter, these bacteria are among the oldest adversaries of clinicians in the post-antibiotic age (Lowy, 2003). As previously discussed, unlike Gram-negative bacteria, staphylococcal strains do not produce a protective outer membrane barrier. Therefore, active efflux systems are of significant importance in keeping cytosolic concentrations of antimicrobials below toxic levels.

S. aureus encodes several of the best characterized bacterial drug efflux systems. Indeed, the first bacterial multidrug efflux system described was found in this organism; the plasmid-encoded multidrug efflux pump QacA (Tennent *et al.*, 1985), and the first multidrug efflux ABC system

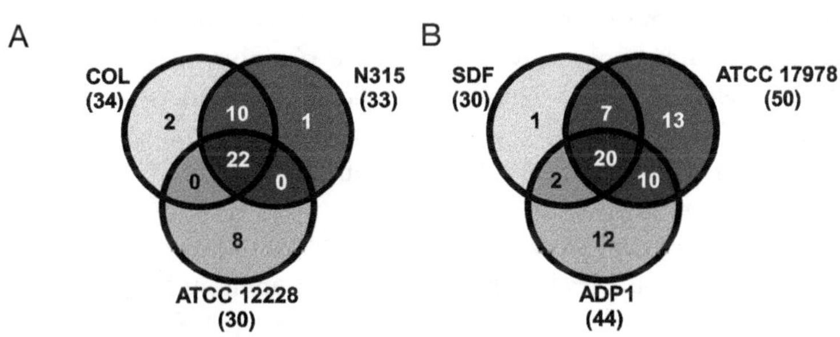

Figure 6.1 Venn diagrams depicting conservation of putative drug efflux systems in the staphylococcal strains, *S. aureus* COL, *S. aureus* N315 and *S. epidermidis* ATCC 12228 (A) and the *Acinetobacter* strains, *A. baumannii* ATCC 17978, *A. baumannii* SDF and *A. baylyi* ADP1 (B). Putative efflux systems were identified using TransAAP (Ren and Paulsen, 2007), with manual association of efflux proteins likely to assemble into a single export system.

to be characterized by crystallographic structure analyses, Sav1866 (Dawson and Locher, 2006). In addition to these proteins, our bioinformatic predictions suggest that the chromosomes of sequenced *S. aureus* strains contain more than 30 drug efflux systems (Fig. 6.1). The majority of these systems are classified as members of the MFS or ABC superfamily of transport systems, with very few transporters classified within the other three major families of bacterial drug efflux proteins (Hassan *et al.*, 2007a). This is in contrast to Gram-negative organisms, where RND efflux systems typically constitute the major protective efflux systems. Nonetheless, a number of potentially important SMR family transporters are encoded on staphylococcal plasmids and at least one MATE family protein, MepA, has been shown to confer protective levels of drug resistance in *S. aureus*.

Chromosomally encoded efflux systems in *S. aureus*

Several characterized drug efflux systems in *S. aureus*, such as LmrS, NorA, NorB, NorC, MdeA, SdrM, LmrS, MepA and Sav1866 are chromosomally encoded. The genes encoding the majority of these proteins are well conserved within the core genomic regions of clinical isolates, and are therefore potential sources of drug resistance across a broad collection of strains (Fig. 6.1). In addition to the biophysical properties of these proteins, a significant level of effort has been devoted to understanding their regulation, which is of interest for several reasons. Firstly, due to their potential toxicity, these efflux systems are generally under tight regulatory control and high-level drug resistance within a strain may only be achieved after appropriate environmental signals or regulatory mutations (Truong-Bolduc *et al.*, 2003; Kaatz *et al.*, 2006; Kosmidis *et al.*, 2010). Additionally, a detailed comprehension of the regulatory elements and physical cues controlling expression of these proteins could lead to the tailoring of clinical practices that either directly disrupt or limit their expression. Finally, since the physiological roles of many drug efflux systems, particularly multidrug efflux systems remain poorly understood, studies of the regulatory signals promoting expression of these systems may provide insight into their natural functions (Neyfakh, 1997).

NorA

Transporters classified within the MFS are among the best characterized chromosomal efflux systems in *S. aureus* and are the most numerous. Consequently, these systems are likely to be the primary efflux systems in *S. aureus*, in terms of protection against drugs and toxic compounds. The NorA efflux pump, encoded by the *norA* multidrug resistance gene, was the first chromosomally encoded drug exporter to be identified within *S. aureus* and is the best characterized of these efflux pumps in terms of both function and regulation. The NorA protein is a member of the drug:H^+ antiporter (DHA) 1 family of the MFS and comprises 12 transmembrane segments (TMS) (Yoshida *et al.*, 1990). NorA was initially identified by screening a library of cloned DNA fragments from a norfloxacin resistant *S. aureus* strain for their capacity to confer resistance to norfloxacin in susceptible *S. aureus* and *Escherichia coli* strains (Yoshida *et al.*, 1990). In addition to norfloxacin, NorA was initially shown to confer resistance to a range of quinolone antibiotics (Yoshida *et al.*, 1990; Ng *et al.*, 1994). Subsequent studies determined that this protein is able to recognize not only quinolones, but several cationic antimicrobials and the antibiotic chloramphenicol (Table 6.1) (Neyfakh *et al.*, 1993). The NorA protein has been reconstituted into liposomes and shown to mediate transport of Hoeschst 33342 when supplied with an artificial proton gradient, demonstrating that it can function independently of other cellular components (Yu *et al.*, 2002). Transport mediated by NorA is sensitive to nigericin, but not valinomycin, suggesting that it is coupled specifically to the ΔpH component of the proton motive force (pmf) (Yoshida *et al.*, 1990; Ng *et al.*, 1994). Studies examining the level of *norA* expression in multidrug resistant *S. aureus* strains have determined that this exporter is a probable cause of multidrug resistance in a number of clinical strains (Noguchi *et al.*, 1999; Patel *et al.*, 2010). Consequently, the NorA protein is a common target for studies aimed at developing efflux system inhibitors (Brincat *et al.*, 2011).

Table 6.1 Characterized multidrug transporters from *S. aureus*

Family	Transporter	Genomic location	Demonstrated substrates[a]	Reference(s)
ABC	Sav1866	Chromosome	Et, Hs (Dx, Vb)[b]	Dawson and Locher (2006), Velamakanni *et al.* (2008)
MFS	LmrS	Chromosome	Cm, Ct, Er, Et, Fa, Ff, Gtf, Km, Ln, Lz, Oxy, Rf, SDS, Sm, Tp, TPP	Floyd *et al.* (2010)
	MdeA	Chromosome	Af, Bk, Cf[c], Dq, Dr, Du, Et, Fa, Hs, Mp, Nb, Nf[d], R6G, TPP, Vm	Huang *et al.* (2004), Yamada *et al.* (2006b)
	NorA	Chromosome	AcO, AM-1155, Cf, Cm, CS-940, Cx, En, Et, Fl, Lfx, NA, Nf, OA, Of, PA, Pfx, PpA, Pu, R6G, Sf, TPP	Yoshida *et al.* (1990), Kaatz *et al.* (1993), Neyfakh *et al.* (1993)
	NorB	Chromosome	Cf, Ct, Et, Gmf, Gr, Mf, Nf, Prf, Sf, TPM	Truong-Bolduc *et al.* (2005), Truong-Bolduc *et al.* (2006)
	NorC	Chromosome	Cf, Gr, Mf, Nf, Prf, Sf	Truong-Bolduc *et al.* (2006)
	QacA	Plasmid	1i-39/JC-1-134, 1a-62/JC-1-127, Ac, Af, AcY, Bk, Ch, CTA, CV, DAEP, DAP, DAPI, DAZ, DBP, $DiOC_3$, Et, He, Pe, Pf, Phe, Prp, PyY, QR, R6G, SO, St, TMA-DPH, TPA, TPM, TPMP, TPP	Littlejohn *et al.* (1992), Mitchell *et al.* (1998), Brown and Skurray (2001)
	SdrM	Chromosome	Af, Et, Nf	Yamada *et al.* (2006a)
	TetA(K)	Plasmid	Dc, Tc	Yamaguchi *et al.* (1995)
MATE	MepA	Chromosome	Af, Bk, Cf, Ct, CV, Ch, DAPI, dMc, Dq, dTc, Et, Hs, Nf, Pe, PyY, Rh, Tgc, TMA-DPH, TPP	Kaatz *et al.* (2005a), McAleese *et al.* (2005)
SMR	QacC (QacD, Smr, Ebr)	Plasmid	Bk, Cph, CTA, CTP, CV, Dq, Et, Ox, Pf, R6G	Paulsen *et al.* (1995), Fuentes *et al.* (2005)

Ac: amicarbalide, AcO: acridine orange, AcY: acridine yellow, Af: acriflavine, Bk: benzalkonium, Cf: ciprofloxacin, Ch: chlorhexidine, Cm: chloramphenicol, Cph: cephaloridine, Ct: cetrimide, CTA: cetyltrimethylammonium, CTP: cetylpyridium, CV: crystal violet, Cx: cinoxacin, DAEP: dimethylaminostyrl-1-ethylpyridinium, DAP: diamidinodiphenylamine, DAPI: 4',6-diamidino-2-phenylindole, DAZ: diminazene, DBP: dibromopropamidine, Dc: doxycycline, $DiOC_3$: 3', 3'-dipropyloxacarbocyanine, dMc: 9-*N*-*N*-dimethylglycylamido-minocycline, Dq: dequalinium, Dr: doxorubicin, dTc: 9 – *N*-*N*-dimethylglycylamido-6-demethyl-6-deoxytetracycline N, Du: daunorubicin, Dx: doxorubicin, En: enoxacin, Et: ethidium, Fa: Fusidic acid, Fl: fleroxacin, Gmf: gemifloxacin, Gr garenoxacin, He: hexamidine, Hs: Hoechst 33342, Lfx: lomefloxacin, Mf: moxifloxacin, Mp: mupirocin, NA: nalidixic acid, Nb: novobiocin, Nf: norfloxacin, OA: oxolinic acid, Of, ofloxacin, Ox: oxacillin, Oxy: oxytetracycline, PA: piromidic acid, Pe: pentamidine, Pf: proflavine, Pfx: perfloxacin, Phe: phenamidine, PpA: pipemidic acid, Prf: premafloxacin, Prp: propamidine, Pu: puromycin, PyY: pyronin Y, QR: quinaldine red, R6G: rhodamine 6G, Rh: rhodamine, Rf: rifampin, Sf: sparfloxacin, SO: safranin O, St: stilbamidine, Su: sulphadimethoxine, Tc: tetracycline, Tgc: tigecycline, TMA-DPH: 1-(4-trimethylammoniumphenyl)-6-phenyl-1,3,5-hexatriene, TPA: tetraphenylarsonium, TPM: tetraphenylammonium, TPMP: triphenylmethylphosphonium, TPP: tetraphenylphosphonium, Vb: vinblastine, Vm: virginiamycin.

[a]Confers at least twofold increase in resistance from strains devoid of resistance determinant or is able to mediate active transport of the substrate.

[b]Determined as an increase in ATPase activity of purified protein (Dawson and Locher, 2006).

[c]Not seen by Huang *et al.* (2004).

[d]In *E. coli* only (Yamada *et al.*, 2006b).

NorB and NorC

In addition to NorA, the NorB and NorC export systems have also been shown to confer resistance to a number of quinolone antibiotics in *S. aureus* (Table 6.1) (Truong-Bolduc *et al.*, 2006). The NorB protein has also been shown to confer resistance to several biocides (Truong-Bolduc *et al.*, 2005). NorB is a 14 TMS transporter classified within the DHA2 family of the MFS (Truong-Bolduc *et al.*, 2005), whereas, NorC is classified within the DHA1 family (Truong-Bolduc *et al.*, 2006). The NorB and NorC resistance proteins were identified in the *S. aureus* strain QT1, which contains a mutation to the *mgrA* global regulatory gene. This strain displayed increased resistance levels to a range of drugs, however, the increased quinolone resistance levels were not accounted for by NorA overexpression. The *norB* and *norC* genes were identified as being overexpressed in a microarray analysis comparing gene expression levels in the *S. aureus* QT1 strain and its parental strain (Truong-Bolduc *et al.*, 2003, 2006).

MdeA

The MdeA exporter was identified by screening an open reading frame (ORF) expression library including more than 2300 ORFs from S. *aureus* strain Buttle. Plasmid clones carrying the *mdeA* gene confer increased levels of resistance to a broad range of antimicrobials in *S. aureus* strains (Table 6.1) (Huang *et al.*, 2004; Yamada *et al.*, 2006b). The MdeA protein is predicted to comprise 14 TMS and is a member of the DHA2 family of the MFS. MdeA has been shown to facilitate the transport of both ethidium and Hoeschst 33342 in a manner dependent on the pmf (Yamada *et al.*, 2006b).

SdrM and LmrS

The genome sequence information generated for a range of *S. aureus* strains can be exploited for the identification of new drug resistance determinants including drug efflux systems. This strategy has been applied to at least two of the most recently described chromosomal MFS drug efflux systems, SdrM and LmrS (Yamada *et al.*, 2006a; Floyd *et al.*, 2010). The SdrM efflux system was identified in the *S. aureus* N315 genome sequence (annotated with the gene tag SA1972). This protein is a member of the DHA2 family of the MFS and confers resistance to norfloxacin, acriflavine and ethidium when overexpressed in *S. aureus* (Yamada *et al.*, 2006a) (Table 6.1). The SdrM protein mediated efflux of acriflavine has been shown to be sensitive to CCCP, demonstrating that this efflux is powered by the pmf (Yamada *et al.*, 2006a). Similar to SdrM, the LmrS transporter is composed of 14 TMS and is classified within the DHA2 family of the MFS. The gene encoding this protein was identified in the genome sequence of *S. aureus* COL, annotated with the locus tag SACOL2157. When overexpressed in *E. coli* KAM32 cells, LmrS conferred up to 16-fold increases in resistance to a broad range of antimicrobial compounds and antibiotics from a large number of functional classes (Floyd *et al.*, 2010) (Table 6.1). Transport of ethidium mediated by LmrS in *E. coli* was sensitive to CCCP, indicating that this transport reaction is also powered by the pmf.

MepA

To date, the MepA transport protein is the only characterized MATE family efflux system encoded on the *S. aureus* chromosome. This efflux system was identified by microarray analyses of strains adapted to tolerate higher concentrations of ethidium and moxifloxacin or norfloxacin (Kaatz *et al.*, 2005a). These strains were found to overexpress three genes that formed an operon like structure, including a *marR* family regulator gene *mepR*, the *mepA* MATE family transporter and *mepB*, a gene of unknown function. These genes have also been found to be overexpressed in strains adapted to increased concentrations of tigecycline (McAleese *et al.*, 2005). The MepA protein was shown to confer resistance to cationic antimicrobials, as well as fluoroquinolone and glycycline antibiotics (Kaatz *et al.*, 2005a; McAleese *et al.*, 2005). MepA facilitates the active transport of ethidium and is sensitive to CCCP, indicating that it is coupled to the pmf (Kaatz *et al.*, 2005a).

Sav1866

The Sav1866 efflux system was the first ABC superfamily multidrug efflux system to be characterized by crystallographic analyses (Dawson and Locher, 2006). Sav1866 polypeptides comprise a single transmembrane domain containing six TMS and a nucleotide-binding domain. These polypeptides function as homodimers. The crystal structure of Sav1866 demonstrated that the two protomers of the homodimer form extensive and intimate associations between both the transmembrane domains and nucleotide binding domains. In the available Sav1866 structure, the putative substrate translocation or binding region is exposed to the extracellular space, indicating that this structure likely represents the outward-facing conformation of the protein. Upon crystallization, only limited functional data were available for the Sav1866 transporter. However, recent studies have confirmed this protein to be a multidrug efflux system (Velamakanni *et al.*, 2008). Sav1866 was shown to mediate ethidium transport from whole cells and transport of Hoechst 33342 when reconstituted into membrane vesicles, demonstrating that it is a self-sufficient transport protein (Velamakanni *et al.*, 2008).

Plasmid-encoded efflux pumps carried by *S. aureus* strains

Multiresistance plasmids are very common in clinical *S. aureus* strains. These plasmids can be quite large and carry genes encoding for the maintenance and spread of the plasmid. Additionally, these plasmids may harbour large numbers of drug resistance genes, including those that encode drug efflux systems, providing a vehicle for their dissemination to new strains. A number of well characterized drug efflux systems, including the QacA, QacB and QacC multidrug transporters are plasmid encoded in *S. aureus*. The carriage of one or more genes encoding these efflux systems has been linked to increased biocide resistance levels in clinical strains worldwide (Noguchi *et al.*, 2005) and these genes have been directly associated with the failure of antimicrobial treatment regimes (Batra *et al.*, 2010). In addition to these transporters, *S. aureus* plasmids are known to encode a range of other efflux systems, such as the TetA(K) tetracycline efflux system.

QacA and QacB

QacA and QacB are classified within the DHA2 family of the MFS. These genes have been found in clinical isolates of *S. aureus* across the world and it is not unusual for approx. 50% of *S. aureus* isolates within a clinical setting to carry a *qacA/qacB* resistance determinant (Mayer *et al.*, 2001; Noguchi *et al.*, 2005; Teixeira *et al.*, 2010). The QacA/QacB exporters have been long known to confer resistance to a range of cationic biocides (Paulsen *et al.*, 1996a; Mitchell *et al.*, 1998; Mitchell *et al.*, 1999) and recently some QacB proteins have been suggested to confer low level resistance to fluoroquinolones (Nakaminami *et al.*, 2010). Additionally, the presence of the QacA protein in the staphylococcal membrane has been shown to influence sensitivity to a thrombin-induced platelet microbicidal protein (Kupferwasser *et al.*, 1999; Bayer *et al.*, 2006).

The QacA/QacB proteins provide an excellent platform for studying the evolution of multidrug transport proteins and the nature of multidrug interactions in these transporters. Nature has provided an array of QacA/QacB variants that differ in their relative capacities to transport various subsets of substrates and a significant level of effort has been devoted to understanding the differential capacities of these variants to recognize subsets of antimicrobial compounds. This phenomenon was first investigated in studies of the prototypical QacA protein, encoded by the large multiresistance plasmid pSK1 (Tennent *et al.*, 1985), and the prototypical QacB protein, encoded on pSK23 (Paulsen *et al.*, 1998). These two proteins differ by only seven amino acid residues. Nonetheless, the QacA protein was known to confer resistance to a range of both mono- and bivalent cationic antimicrobials, whereas, resistance mediated by the QacB protein was restricted primarily to monovalent cationic antimicrobials (Paulsen *et al.*, 1998). Mutagenic analyses determined that the expanded drug resistance potential of QacA was a result of a single amino acid change, an aspartic acid incorporation at position 323, central to TMS 10 (Paulsen *et al.*, 1996a; Mitchell *et al.*, 1999). This position is predicted to line the bivalent drug binding or translocation region in QacA and facilitate interactions with bivalent drugs, possibly via neutralization of the additional positive charge carried by these proteins (Mitchell *et al.*, 1999; Xu *et al.*, 2006). Alternatively, this residue may also participate in proton coupling.

Interestingly, natural QacA/QacB variants containing acidic residues within alternative transmembrane regions have also been recognized as playing a role in conferring resistance to mono- and bivalent cationic substrates. One of these residues is also located within TMS 10, approximately one helical turn from position 323, at position 320, suggesting that this side of TMS 10 may face the bivalent cationic drug binding region. However, it has been shown that an acidic residue incorporated within TMS 12 of QacA/QacB at position 377 could also facilitate the recognition of bivalent cationic substrates after neutralization of the residue at position 323 (Hassan *et al.*, 2007b). Recent studies have suggested that the nature of the amino acid residue at position 320 also influences the ability of QacA/QacB proteins to recognize fluoroquinolones (Nakaminami *et al.*, 2010). These findings highlight the potential flexibility of the QacA/QacB drug binding region.

The location of the genes encoding QacA/QacB on plasmids may provide for higher rates of mutation and potentially faster rates of evolution of these genes. The multicopy nature of plasmids allows for the exploration of favourable mutations, where if one resistance gene becomes inactivated through mutation, additional copies of the plasmid will exist to cover this loss of function. The identification of a range of natural QacA/QacB variants may reflect various selective pressures operating on *S. aureus* strains within clinical settings.

TetA(K)

The *tetA(K)* gene, which encodes TetA(K), was identified in the early 1980s on the *S. aureus* plasmid pT181 (Khan and Novick, 1983). Studies have shown this plasmid to be capable of integration into host chromosomes, facilitating the prolonged and stable inheritance of the gene (Khan and Novick, 1983). The TetA(K) transporter confers resistance to tetracycline and the related antibiotic doxycycline. These antibiotics are recognized by TetA(K) as complexes with divalent metal ions (Me^{2+}; e.g. Mg^{2+}), that form within the cell, and are exchanged for protons in an electrogenic transport reaction (Yamaguchi *et al.*, 1995; Jin *et al.*, 2002). Interestingly, TetA(K) can also exchange protons for Na^+ or K^+ and can use a K^+ gradient to mediate transport of Tc–Me^{2+} complexes, or Na^+ or K^+ ions (Jin *et al.*, 2002; Krulwich *et al.*, 2005). A number of mutagenesis studies have probed the structure and function of the TetA(K) protein and used this protein to examine the function of well conserved sequence motifs in the DHA2 family of transporters (Ginn *et al.*, 2000; De Jesus *et al.*, 2005; Hassan *et al.*, 2006).

QacC

The QacC transporter was one of the first SMR family drug efflux systems to be functionally characterized. Members within the SMR family have been subdivided into three subclasses; the small multidrug pumps (SMPs), paired small multidrug pumps (PSMRs) and suppressors of *groEL* mutations (SUGs). Transporters belonging to the SMP and the PSMRs families are characterized based on their ability to confer resistance to a broad range of lipophilic drugs and quaternary ammonium compounds; QacC belongs to the SMP subclass. The *qacC* gene has been observed on a number of staphylococcal plasmids, both conjugative and non-conjugative (Littlejohn *et al.*, 1992; Leelaporn *et al.*, 1995) and is frequently associated with drug resistant clinical isolates of *S. aureus* (Noguchi *et al.*, 1999; Mayer *et al.*, 2001; Noguchi *et al.*, 2005; Longtin *et al.*, 2011). Like all members of the SMR family, QacC is a small protein consisting of only four TMS (Paulsen *et al.*, 1995; Poget *et al.*, 2010). Consequently, QacC is likely to function as an oligomer. QacC mediates the efflux of a number of cationic antimicrobials via a reaction stimulated by both the ΔpH and $\Delta\psi$ components of the pmf (Grinius and Goldberg, 1994). Although limited, some mutagenic analyses have been conducted on QacC and determined the importance of several highly conserved amino acid residues (Grinius and Goldberg, 1994; Paulsen *et al.*, 1995).

Regulation of efflux pumps in *S. aureus*

Regulation of efflux systems in *S. aureus*, particularly those encoded on the chromosome, is typically multifaceted, involving a number of regulatory systems, both local and global. Several chromosomally encoded efflux systems in *S. aureus* have proven to be excellent case studies for understanding exporter gene expression in this organism. This is particularly true for NorA, the regulation of which has been the subject of an array of studies since its discovery. Promoter region mutations at the *flqB* locus have been associated with increased levels of *norA* transcript; these mutations may increase transcript stability or act as a binding site for a transcriptional regulatory protein (Fournier *et al.*, 2001; Kaatz *et al.*, 2005b). The *norA* gene is also known to be regulated by the ArlS-ArlR two component regulatory system and the MgrA transcriptional regulator, which along with other regulatory systems form a complex network controlling the expression of not only drug efflux systems, but a range of virulence factors (Fournier *et al.*, 2000; Truong-Bolduc *et al.*, 2003; Kaatz *et al.*, 2005b; Truong-Bolduc and Hooper, 2010). Interestingly, MgrA is subject to phosphorylation and dephosphorylation by a number of enzymes, and displays differential binding capacities for the promoter regions of drug efflux system genes, such as *norA* and *norB*, depending on its phosphorylation state (Truong-Bolduc and Hooper, 2010). The NorG protein represents another important global regulator of efflux system expression in *S. aureus*. This system significantly affects expression of *norB* and *norC* and is known to influence expression of other global regulators such as *mgrA* and *arlS*, highlighting the complexity of the regulatory network controlling efflux system expression (Truong-Bolduc *et al.*, 2011).

In addition to global regulatory systems, drug efflux system genes can also be tightly controlled by local transcriptional regulators (Grkovic *et al.*, 2002). These regulators commonly bind a similar array of compounds as their cognate efflux system. Binding of these ligands typically induces conformational changes in the regulatory proteins that lead to their association or dissociation from the transporter promoter region (Schumacher *et al.*, 2001). Well characterized local regulatory proteins include MepR, a member of the MarR family of regulatory proteins that controls transcription of the *mepA* transporter gene (Kaatz *et al.*, 2006; Kumaraswami *et al.*, 2009) and QacR, a member of the TetR family of regulators, which controls expression of *qacA* (Grkovic *et al.*, 1998; Schumacher *et al.*, 2001).

Acinetobacter baumannii

A. baumannii is a Gram-negative non-fermenter that belongs to the group of gamma-proteobacteria. An increasing prevalence of *A. baumannii* has been observed in the clinical setting throughout the last two decades (Dijkshoorn *et al.*, 2007). This has changed its significance from being considered a contaminant in diagnostic specimens to being recognized as a significant pathogen. Consequently, *A. baumannii* has become the focal point of many studies of clinical, epidemiological and molecular nature. *Acinetobacter* species can be isolated from both environmental and clinical samples, with the natural reservoir of pathogenic *A. baumannii* remaining unclear. Multiple *Acinetobacter* species can be of pathogenic significance; nevertheless, *A. baumannii* is most frequently isolated as a human pathogen (Bergogne-Berezin and Towner, 1996; Joly-Guillou, 2005; Diancourt *et al.*, 2010; Turton *et al.*, 2010; Chuang *et al.*, 2011). Although uncommon, this Gram-negative bacillus can also be part of the normal flora of the skin and the gastrointestinal and upper respiratory tracts (Seifert *et al.*, 1997; Berlau *et al.*, 1999). The diversity of these habitats, in which different detoxification and survival strategies are required, may have resulted in the presence of a wide variety of resistance and persistence mechanisms. This has given *A. baumannii* the ability to adapt to the hospital environment and it is now recognized as one of the most troublesome bacteria in intensive care units worldwide (Cunha *et al.*, 1980; Patterson *et al.*, 1991; Neely, 2000; Dijkshoorn *et al.*, 2007).

A. baumannii is notorious for having an exceptionally high level of intrinsic resistance to antimicrobial compounds. A number of studies have indicated that the spread of a limited number of successful clonal lineages has resulted in the globally increased prevalence of *A. baumannii*, this specifically concerns the multidrug resistant isolates (Dijkshoorn *et al.*, 1996; van Dessel *et al.*, 2004; Nemec *et al.*, 2004; Dijkshoorn *et al.*, 2007; Garnacho-Montero and Amaya-Villar, 2010; Michalopoulos and Falagas, 2010; Goel *et al.*, 2011; Neonakis *et al.*, 2011). The distribution of three common multidrug resistant *A. baumannii* lineages, named international clone I, II and III, has been well documented (Turton *et al.*, 2007).

Several different resistance mechanisms have been studied in *A. baumannii* of which active efflux is a known contributor to the development of multidrug resistance (Gordon and Wareham, 2010). Efflux mediated resistance has been reported for almost all antibiotics, including fluoroquinolones, aminoglycosides and beta-lactams.

Even the very promising antibiotic tigecycline appears to be transported by various *A. baumannii* efflux systems (Ruzin *et al.*, 2007; Damier-Piolle *et al.*, 2008; Coyne *et al.*, 2010b; Sun *et al.*, 2010).

The prevalence of drug transporters in clinical *A. baumannii* isolates

The total number of efflux systems identified in strain *A. baumannii* ATCC 17978 using TransAAP equates to close to 2% of all predicted ORFs of the genome, which is significantly higher than the percentage of predicted transporters than that seen in other sequenced bacteria, e.g. the *E. coli* K12 genome (approx. 1%). This may be due to the diversity of habitats of different *Acinetobacter* species, e.g. soil, human skin and wastewater, and therefore the broad selection of foreign toxic compounds that it can encounter. We found the *A. baumannii* ATCC 17978 genome encodes 50 putative drug transporter systems (Fig. 6.1b). These can be broken down to 11 ABC transporters, four SMR proteins, four MATE proteins, 22 MFS members and nine RND proteins. These efflux systems are highly conserved across fully sequenced clinical isolates of *A. baumannii*. Additionally, we recently determined the prevalence of these 50 putative transporters across 54 diverse clinical *A. baumannii* isolates using a PCR method with primers targeting the highly conserved regions of each gene (Eijkelkamp *et al.*, unpublished data). Out of the 50 putative transporters, 23 were found in all strains and an additional 23 were found in more than 50% of the clinical isolates. Since the majority of genes encoding drug transporters are well conserved between a diverse range of clinical *A. baumannii* isolates, many efflux systems appear to be part of the *A. baumannii* core genome, corroborating previous data (Adams *et al.*, 2008).

In Gram-negative bacteria, MFPs are often involved in stabilizing the interaction between inner membrane drug transporters, such as members of the RND superfamily or ABC superfamily, and OMPs. The resulting tripartite protein complex facilitates transport of substrates to the extracellular space, instead of only to the periplasm and may also facilitate the capture of substrates from within the periplasm of Gram-negative cells (Tal and Schuldiner, 2009). Therefore, overexpression of these complete protein complexes may potentiate higher levels of drug resistance than inner membrane drug transporters only. The genes encoding MFPs are often transcriptionally linked to drug transporters. Therefore, the 50 genes of the *A. baumannii* ATCC 17978 genome encoding putative drug transporter proteins were examined for the presence of adjacent genes encoding putative MFPs. As seen in other Gram-negative bacteria, only genes encoding members of the MFS, ABC and RND families possessed adjacent genes encoding putative MFPs in *A. baumannii* strain ATCC 17978. A total of four of the 22 putative MFS transporters are located adjacent to putative MFP genes, which includes the previously studied efflux proteins AedB, AedC and AedE (Table 6.2; Hassan *et al.*, 2011). Four predicted ABC transport systems also contain a putatively co-transcribed MFP. Two of these ABC transport systems, encoded by A1S_1242 and A1S_2621/2622, are located in operons that also contain putative OMPs. Consequently, these systems may facilitate high levels of drug resistance. As mentioned above, RND efflux systems are often transcribed from operons that also contain genes encoding MFPs and OMPSs; this includes the previously characterized RND members AdeB, AdeG, AdeJ and AdeM (Table 6.2).

Regulation of genes encoding transporters

Due to their typically low level of basal expression, efflux systems generally require high increases in expression level before a significant increase in resistance to antimicrobials can be observed (Coyne *et al.*, 2010b; Sun *et al.*, 2010). Therefore, measuring mRNA expression levels of transporters or examining mutations in regulatory elements associated with high expression levels is of great importance when investigating the significance of efflux systems in drug resistance. In clinical *A. baumannii* isolates, AdeB has arisen as the most significant drug transporter, as high expression levels of the *adeB* gene have been correlated with resistance to multiple antibiotics, including ciprofloxacin, cefepime and tigecycline (Higgins *et al.*, 2004; Pannek *et al.*, 2006; Ruzin *et al.*, 2007; Bratu *et al.*, 2008; Sun *et al.*, 2010).

Table 6.2 Characterized multidrug transporters from *A. baumannii*

Family	Transporter	Putative regulator	Demonstrated substrates	References
SMR	AbeS	No	Af, Bk, Do, Er, Et, Nb, SDS	Srinivasan *et al.* (2009)
MATE	AbeM	No	Af, Cf, Ch, DAPI, Du, Dx, Et, Hs, Nf, Tp	Su *et al.* (2005), Eijkelkamp *et al.* (2011a)
	AbeM2	No	Cf	Eijkelkamp *et al.* (2011a)
	AbeM4	No	Cf	Eijkelkamp *et al.* (2011a)
MFS	AedC	No	Cm, Tc	Hassan *et al.* (2011)
	AedF/AmvA	Yes	Af, Ch, DAPI, Et, Mv	Rajamohan *et al.* (2010a), Hassan *et al.* (2011)
	CraA	No	Cm, Tc	Roca *et al.* (2009)
RND	AdeB	Yes	Cf, Cm, Co, Er, Gm, Nf, Tc, Tg	Magnet *et al.* (2001)
	AdeG	Yes	Cm, Nf, Su, Tc	Coyne *et al.* (2010b)
	AdeJ	No	Cl, Co, Tc, Tr	Damier-Piolle *et al.* (2008)
	AdeM	Yes	Ch, Nf	Eijkelkamp *et al.* (2011a)

Af: acriflavine, Bk: benzalkonium, Cf: ciprofloxacin, Ch: chlorhexidine, Cl: clindamycin, Cm: chloramphenicol, Co: cefotaxime, Do: deoxycholate, DAPI: 4′,6-diamidino-2-phenylindole, Du: daunorubicin, Dx: doxorubicin, Er: erythromycin, Et: ethidium, Gm: gentamicin, Hs: Hoechst 33342, Mv: methyl viologen, Nb: novobiocin, Nf: norfloxacin, SDS: sodium dodecyl sulphate, Su: sulphamethoxazole, Tc: tetracycline, Tg: tigecycline, Tr: ticarcillin, Tp: trimethoprim

The *adeB* determinant is part of an operon, which includes genes encoding a MFP (*adeA*) and OMP (*adeC*). Of note, this cluster is not fully conserved between all *A. baumannii* strains, e.g. *adeC* is not present in *A. baumannii* strains ATCC 17978 and ATCC 19606. Constitutive upregulation of the *adeABC* operon has been observed in various multidrug resistant clinical *A. baumannii* isolates. It has been shown that this can result from mutations in the two-component regulatory system encoded by the divergently transcribed gene cluster *adeRS* (Ruzin *et al.*, 2007). Interestingly, clinical *A. baumannii* strains overexpressing *adeB* without mutations in *adeRS* have also been identified, indicating that expression of *adeB* is likely to be controlled by multiple mechanisms (Sun *et al.*, 2010). Up-regulation of two other RND transport systems in *A. baumannii*, encoded by *adeIJK* and *adeFGH*, has been observed in both laboratory-generated mutants and clinical isolates (Coyne *et al.*, 2010b; Rajamohan *et al.*, 2010b). The *adeIJK* operon appears to be constitutively expressed at higher levels in certain clinical strains. To date, the regulator or differences in regulatory elements responsible for this have not been identified. A LysR-like regulator, *adeL*, is divergently expressed from the *adeFGH* operon. Mutations in the regulatory elements of the intergenic region between *adeL* and *adeF* have been related to overexpression of this RND system (Coyne *et al.*, 2010b). We have examined the genes encoding putative transporters in the *A. baumannii* ATCC 17978 genome for the presence of adjacent regulators (Table 6.2); 17 putative regulators were identified adjacent to putative drug transporters, including regulators adjacent to operonic clusters containing a gene encoding a transporter. A TetR-like regulator was found to be divergently transcribed from a well conserved gene cluster containing three putative ABC transporters (A1S_1671–1673). Various MFS and RND transporter genes were also associated with putative regulators, including *aedB*, *aedF* and *adeM*. None of the MATE or SMR family members were encoded with putative regulators in close proximity. However, regulators found elsewhere in the genome may play an important role in directing transcription of these drug transport proteins. Indeed, transcription of several of these transporters has been shown to be responsive to various environmental stimuli.

For example, *abeM4* was found to be upregulated under highly saline conditions (Hood *et al.*, 2010). Nonetheless, a link between drug resistance in clinical *A. baumannii* strains and expression levels of a different member of the MATE family, *abeM*, could not established (Bratu *et al.*, 2008).

Functionally characterized *A. baumannii* drug efflux systems

The major facilitator superfamily

The genome of *A. baumannii* strain ATCC 17978 contains 22 MFS drug transporters that can be further subdivided into the DHA1–3 groups based on phylogenetic analyses. Members of this family are not only involved in drug resistance. Some DHA proteins are encoded within various siderophore biosynthesis gene clusters in other bacteria and in *E. coli*, a DHA2 transporter has been demonstrated to play a role in the efflux of the siderophore enterobactin (Furrer *et al.*, 2002). To date, all six DHA2 proteins, identified in the currently available *A. baumannii* genome sequences, have been functionally examined (Rajamohan *et al.*, 2010a; Hassan *et al.*, 2011) (Table 6.2). Two of these, AedC and AedF (and its orthologue AmvA), appear to be multidrug transporters, as both proteins confer resistance to multiple distinct compounds (Table 6.2). The DHA1 member CraA showed specificity for chloramphenicol when analysed in a recombinant *E. coli* host (Roca *et al.*, 2009). The *tet(A)* gene encoding a tetracycline transporter of the MFS and its putative regulator encoded by *tetR(A)* were identified in a genomic segment from *A. baumannii* analysed in a recombinant *E. coli* host (Ribera *et al.*, 2003). Interestingly, *tet(A)* and *tetR(A)* are located on a transposon within the analysed DNA fragment, this possibly facilitates enhanced mobilization of this genomic element between different strains.

The small multidrug resistance family

Currently, four putative *A. baumannii* SMR proteins have been identified. AbeS and A1S_2502 belong to the SMP subclass, and A1S_0710 belongs to the SUG subclass. The product of A1S_2844 is related to the PSMRs (Bay and Turner, 2009), a subgroup potentiating high levels of drug resistance that require heterodimerization. In *A. baumannii*, A1S_2844 appears to have lost the second part of the heterodimer, therefore, functionality of the A1S_2844 requires experimental confirmation. Only one of the SMR proteins, AbeS, has been functionally characterized (Table 6.2). Various antimicrobial agents, detergents and dyes were found to be substrates of this protein by examination of an insertionally inactivated *abeS A. baumannii* mutant strain and a recombinant *E. coli* strain expressing *abeS* (Srinivasan *et al.*, 2009).

The multidrug and toxic compound extrusion family

Three of the four members of the MATE family from *A. baumannii* ATCC 17978 have been confirmed as drug transporters (Table 6.2). AbeM from *A. baumannii* was heterologously expressed in *E. coli* (Su *et al.*, 2005; Eijkelkamp *et al.*, 2011a), where it exhibited increased resistance to a wide range of compounds, including fluoroquinolones, several dyes such as 4,6-diamidino-2-phenylindole, Hoechst 33342 and ethidium, and other toxic compounds, such as doxorubicin, daunorubicin and acriflavine. AbeM2 and AbeM4 also conferred resistance to the fluoroquinolone antibiotic ciprofloxacin (Eijkelkamp *et al.*, 2011a). Previously, *abeM2* has only been identified in *A. baumannii* strain ATCC 17978 and *A. baylyi* strain ADP1 (Eijkelkamp *et al.*, 2011b). Recently, with more *A. baumannii* genome sequences being released, we could identify orthologues of AbeM2 in two more *A. baumannii* strains, 6013150 (NZ_ACYQ02000002) and 6013113 (NZ_ACYR02000218).

The resistance–nodulation–cell division superfamily

RND efflux systems are considered the most significant antimicrobial transporters in Gram-negative bacteria. To date, five *Acinetobacter* RND proteins have been characterized, AdeB, AdeE, AdeJ, AdeG and AdeM (Magnet *et al.*, 2001; Chau *et al.*, 2004; Damier-Piolle *et al.*, 2008; Coyne *et al.*, 2010a,b; Eijkelkamp *et al.*, 2011a) (Table 6.2). Although most studies describe the importance of AdeB in multidrug resistance, expression studies of AdeB, AdeG and AdeJ using *A. baumannii* mutants lacking the genes encoding these proteins indicated that AdeJ has a resistance potential similar or greater than AdeB (Coyne *et al.*, 2011).

An *A. baumannii* gene cluster containing eight ORFs (A1S_0112–0119) located adjacent to the genes encoding acyl-homoserine lactone synthase (*abaI*) and its regulator *abaR*, plays a role in surface motility (Clemmer *et al.*, 2011). The genes within this gene cluster appear to be regulated by quorum-sensing as a putative AbaR binding site was identified upstream of A1S_0112. This cluster has been predicted to function in biosynthesis of secondary metabolites, such as lipopeptides and/or polyketides (Clemmer *et al.*, 2011). Two genes within this cluster encode proteins with a predicted function in transport of the synthesized molecules to the extracellular space (A1S_0116–0117). A1S_0116 encodes a predicted member of the RND superfamily and A1S_0117 encodes a putative OMP. The 54 clinical *A. baumannii* isolates in our collection were examined for the presence of A1S_0116 using PCR (Eijkelkamp *et al.*, 2011c). Whereas all international clone I isolates were positive for A1S_0116, less than 25% of international clone II isolates harboured A1S_0116. No direct correlation could be established between presence or absence of A1S_0116 and the level of biofilm formation or motility characteristics (Eijkelkamp *et al.*, 2011c).

Efflux pumps functioning in siderophore transport

Iron plays a crucial role in many cellular processes, such as metabolism, cell-to-cell signalling and biofilm formation (Braun and Braun, 2002; Hazan *et al.*, 2010). Soluble iron (Fe^{2+}) is available in only limited quantities in many environments. Consequently, ongoing competition for this micronutrient takes place, both in human host niches and on abiotic surfaces, such as medical equipment. Therefore, iron acquisition from the environment is crucial for persistence of pathogens such as *A. baumannii*. Most clinical *A. baumannii* strains have the ability to grow under iron-limiting conditions (Dorsey *et al.*, 2003). Immunoblot analysis using serum from patients with an *Acinetobacter* infection showed that the iron acquisition mechanisms are being expressed during a state of infection, suggesting that these mechanisms also play a role in virulence (Smith and Alpar, 1991).

A. baumannii cells express outer membrane proteins that directly bind iron or haem from the extracellular space (Clarke *et al.*, 2001; Koster, 2005). However, most *A. baumannii* strains also possess mechanisms for iron acquisition that involve siderophores; the secretion of apo-siderophores is facilitated by active transporters. Genes encoding putative members of the MFS, ABC and RND superfamily can be found in the genetic clusters harbouring the genes required for biosynthesis of siderophores in many different bacterial genera, e.g. the P43 and YhcA exporters encoded in *E. coli* and *Erwinia chrysanthemi*, respectively (Furrer *et al.*, 2002; Franza *et al.*, 2005). Various genes encoding active efflux systems have been identified in the *A. baumannii* siderophore biosynthesis gene clusters (Eijkelkamp *et al.*, 2011b). These clusters were found to be heavily overexpressed in strain ATCC 17978 when grown under iron-limiting conditions (Eijkelkamp *et al.*, 2011b). The expression of the genes within these clusters appears to be regulated by the ferric uptake regulator (FUR), as putative FUR binding sites were identified upstream from the operons that make up the *A. baumannii* siderophore biosynthesis cluster. Two transporters located within the *A. baumannii* siderophore biosynthesis clusters, AedD and AbeM2, from the MFS and MATE families, respectively, were also examined for their ability to transport antimicrobial compounds (Eijkelkamp *et al.*, 2011a; Hassan *et al.*, 2011). Phylogenetic analysis carried out with the DHA2-family protein AedD showed that it was related to the *E. chrysanthemi* siderophore efflux protein, YhcA (Franza *et al.*, 2005; Hassan *et al.*, 2011). None of the extensive range of antimicrobial compounds tested were found to be substrates of AedD, as determined by examination of recombinant *E. coli* expressing AedD and an *A. baylyi aedD* null mutant. AbeM2 is the first reported putative siderophore transporter of the bacterial MATE family. The complete siderophore biosynthesis gene cluster (A1S_2562–2581), containing *abeM2* comprises 15 genes involved in siderophore biosynthesis (A1S_2567–2581), three genes involved in recognition and uptake of ferric siderophores (A1S_2563, A1S_2564 and

A1S_2566) and two genes encoding the aforementioned efflux pumps *abeM2* and A1S_2565, showed high levels of overexpression under iron-limiting conditions. FUR appears to be the main transcriptional regulator for *abeM2*, since FUR binding sites could be identified upstream of the operonic cluster containing *abeM2* (Eijkelkamp *et al.*, 2011b). This emphasizes the notion that most efflux systems have adapted to exposure to novel toxic compounds, like antibiotics, and therefore function opportunistically in conferring multidrug resistance (Poole, 2008).

Conclusions

This review on the *S. aureus* and *A. baumannii* efflux systems illustrated some of the key differences that exist between efflux systems in Gram-negative and Gram-positive bacteria, respectively. The bioinformatic analyses performed for this chapter showed that, in general, *Acinetobacter* isolates possess a large number of putative efflux systems as compared to staphylococci. Members of the MFS are most abundant in *S. aureus* and functional characterization of representatives, such as QacA, has confirmed their importance in multidrug resistant *S. aureus*. Both bioinformatic analyses and knockout studies have shown the importance of RND transporters in antibiotic resistance in *A. baumannii*. RND pumps in Gram-negative bacteria, including *A. baumannii*, are often associated with periplasmic and outer-membrane proteins to form a continuous transport system that spans both membranes. The genes encoding these components are often positioned in an operon of which high levels of upregulation can be associated with multidrug resistance. Up-regulation of these operonic clusters has been linked to mutations in their regulatory elements and to mutations in the adjacent regulators themselves. Overall, efflux mediated resistance continues to threaten patients in the intensive care units. Therefore, functional and structural characterization of these proteins is essential for development of efflux pump inhibitors suitable for treatment of multidrug resistant nosocomial pathogens.

References

Adams, M.D., Goglin, K., Molyneaux, N., Hujer, K.M., Lavender, H., Jamison, J.J., MacDonald, I.J., Martin, K.M., Russo, T., Campagnari, A.A., *et al.* (2008). Comparative genome sequence analysis of multidrug-resistant *Acinetobacter baumannii*. J. Bacteriol. *190*, 8053–8064.

Batra, R., Cooper, B.S., Whiteley, C., Patel, A.K., Wyncoll, D., and Edgeworth, J.D. (2010). Efficacy and limitation of a chlorhexidine-based decolonization strategy in preventing transmission of methicillin-resistant *Staphylococcus aureus* in an intensive care unit. Clin. Infect. Dis. *50*, 210–217.

Bay, D.C., and Turner, R.J. (2009). Diversity and evolution of the small multidrug resistance protein family. BMC Evol. Biol. *9*, 140.

Bayer, A.S., Kupferwasser, L.I., Brown, M.H., Skurray, R.A., Grkovic, S., Jones, T., Mukhopadhay, K., and Yeaman, M.R. (2006). Low-level *Staphylococcus aureus* resistance to thrombin-induced platelet microbicidal protein-1 (tPMP-1) *in vitro* associated with *qacA* gene carriage is independent of multidrug efflux pump activity. Antimicrob. Agents Chemother. *50*, 2448–2454.

Bergogne-Berezin, E., and Towner, K.J. (1996). *Acinetobacter* spp. as nosocomial pathogens: microbiological, clinical, and epidemiological features. Clin. Microbiol. Rev. *9*, 148–165.

Berlau, J., Aucken, H., Malnick, H., and Pitt, T. (1999). Distribution of *Acinetobacter* species on skin of healthy humans. Eur. J. Clin. Microbiol. Infect. Dis. *18*, 179–183.

de Berardinis, V., Vallenet, D., Castelli, V., Besnard, M., Pinet, A., Cruaud, C., Samair, S., Lechaplais, C., Gyapay, G., Richez, C., *et al.* (2008). A complete collection of single-gene deletion mutants of *Acinetobacter baylyi* ADP1. Mol. Syst. Biol. *4*, 174.

Blair, J.M., and Piddock, L.J. (2009). Structure, function and inhibition of RND efflux pumps in Gram-negative bacteria: an update. Curr. Opin. Microbiol. *12*, 512–519.

Bratu, S., Landman, D., Martin, D.A., Georgescu, C., and Quale, J. (2008). Correlation of antimicrobial resistance with beta-lactamases, the OmpA-like porin, and efflux pumps in clinical isolates of *Acinetobacter baumannii* endemic to New York City. Antimicrob. Agents. Chemother. *52*, 2999–3005.

Braun, V., and Braun, M. (2002). Iron transport and signaling in *Escherichia coli*. FEBS Lett. *529*, 78–85.

Brincat, J.P., Carosati, E., Sabatini, S., Manfroni, G., Fravolini, A., Raygada, J.L., Patel, D., Kaatz, G.W., and Cruciani, G. (2011). Discovery of novel inhibitors of the NorA multidrug transporter of *Staphylococcus aureus*. J. Med. Chem. *54*, 354–365.

Brown, M.H., and Skurray, R.A. (2001). Staphylococcal multidrug efflux protein QacA. J. Mol. Microbiol. Biotechnol. *3*, 163–170.

Brown, M.H., Paulsen, I.T., and Skurray, R.A. (1999). The multidrug efflux protein NorM is a prototype of a new family of transporters. Molec. Microbiol. *31*, 394–396.

Chau, S.L., Chu, Y.W., and Houang, E.T. (2004). Novel resistance-nodulation-cell division efflux system AdeDE in *Acinetobacter* genomic DNA group 3. Antimicrob. Agents Chemother. *48*, 4054–4055.

Chuang, Y.C., Sheng, W.H., Li, S.Y., Lin, Y.C., Wang, J.T., Chen, Y.C., and Chang, S.C. (2011). Influence of genospecies of *Acinetobacter baumannii* complex on clinical outcomes of patients with *Acinetobacter bacteremia*. Clin. Infect. Dis. *52*, 352–360.

Clark, N.M., Hershberger, E., Zervosc, M.J., and Lynch, J.P., 3rd. (2003). Antimicrobial resistance among Gram-positive organisms in the intensive care unit. Curr. Opin. Crit. Care. *9*, 403–412.

Clarke, T.E., Tari, L.W., and Vogel, H.J. (2001). Structural biology of bacterial iron uptake systems. Curr. Top. Med. Chem. *1*, 7–30.

Clemmer, K.M., Bonomo, R.A., and Rather, P.N. (2011). Analysis of surface motility in *Acinetobacter baumannii*. Microbiology *157*, 2534–2544.

Coyne, S., Guigon, G., Courvalin, P., and Perichon, B. (2010a). Screening and quantification of the expression of antibiotic resistance genes in *Acinetobacter baumannii* with a microarray. Antimicrob. Agents. Chemother. *54*, 333–340.

Coyne, S., Rosenfeld, N., Lambert, T., Courvalin, P., and Perichon, B. (2010b). Overexpression of resistance-nodulation-cell division pump AdeFGH confers multidrug resistance in *Acinetobacter baumannii*. Antimicrob. Agents. Chemother. *54*, 4389–4393.

Coyne, S., Courvalin, P., and Perichon, B. (2011). Efflux-mediated antibiotic resistance in *Acinetobacter* spp. Antimicrob. Agents. Chemother. *55*, 947–953.

Cunha, B.A., Klimek, J.J., Gracewski, J., McLaughlin, J.C., and Quintiliani, R. (1980). A common source outbreak of *Acinetobacter* pulmonary infections traced to Wright respirometers. Postgrad. Med. J. *56*, 169–172.

Damier-Piolle, L., Magnet, S., Bremont, S., Lambert, T., and Courvalin, P. (2008). AdeIJK, a resistance-nodulation-cell division pump effluxing multiple antibiotics in *Acinetobacter baumannii*. Antimicrob. Agents. Chemother. *52*, 557–562.

Dawson, R.J., and Locher, K.P. (2006). Structure of a bacterial multidrug ABC transporter. Nature *443*, 180–185.

De Jesus, M., Jin, J., Guffanti, A.A., and Krulwich, T.A. (2005). Importance of the GP dipeptide of the antiporter motif and other membrane-embedded proline and glycine residues in tetracycline efflux protein Tet(L). Biochemistry *44*, 12896–12904.

van Dessel, H., Dijkshoorn, L., van der Reijden, T., Bakker, N., Paauw, A., van den Broek, P., Verhoef, J., and Brisse, S. (2004). Identification of a new geographically widespread multiresistant *Acinetobacter baumannii* clone from European hospitals. Res. Microbiol. *155*, 105–112.

Diancourt, L., Passet, V., Nemec, A., Dijkshoorn, L., and Brisse, S. (2010). The population structure of *Acinetobacter baumannii*: expanding multiresistant clones from an ancestral susceptible genetic pool. PLoS One 5, e10034.

Dijkshoorn, L., Aucken, H., Gerner-Smidt, P., Janssen, P., Kaufmann, M.E., Garaizar, J., Ursing, J., and Pitt, T.L. (1996). Comparison of outbreak and nonoutbreak *Acinetobacter baumannii* strains by genotypic and phenotypic methods. J. Clin. Microbiol. *34*, 1519–1525.

Dijkshoorn, L., Nemec, A., and Seifert, H. (2007). An increasing threat in hospitals: multidrug-resistant *Acinetobacter baumannii*. Nat. Rev. Microbiol. *5*, 939–951.

Dorsey, C.W., Beglin, M.S., and Actis, L.A. (2003). Detection and analysis of iron uptake components expressed by *Acinetobacter baumannii* clinical isolates. J. Clin. Microbiol. *41*, 4188–4193.

Eijkelkamp, B.A., Hassan, K.A., Paulsen, I.T., and Brown, M.H. (2011a). Development of a high-throughput cloning strategy for characterization of *Acinetobacter baumannii* drug transporter proteins. J. Mol. Microbiol. Biotechnol. *20*, 211–219.

Eijkelkamp, B.A., Hassan, K.A., Paulsen, I.T., and Brown, M.H. (2011b). Investigation of the human pathogen *Acinetobacter baumannii* under iron limiting conditions. BMC Genomics *12*, 126.

Eijkelkamp, B.A., Stroeher, U.H., Hassan, K.A., Papadimitrious, M.S., Paulsen, I.T., and Brown, M.H. (2011c). Adherence and motility characteristics of clinical *Acinetobacter baumannii* isolates. FEMS Microbiol. Lett. *323*, 44–51.

Floyd, J.L., Smith, K.P., Kumar, S.H., Floyd, J.T., and Varela, M.F. (2010). LmrS is a multidrug efflux pump of the major facilitator superfamily from *Staphylococcus aureus*. Antimicrob. Agents Chemother. *54*, 5406–5412.

Fournier, B., Aras, R., and Hooper, D.C. (2000). Expression of the multidrug resistance transporter NorA from *Staphylococcus aureus* is modified by a two-component regulatory system. J. Bacteriol. *182*, 664–671.

Fournier, B., Klier, A., and Rapoport, G. (2001). The two component system ArlS-ArlR is a regulator of virulence gene expression in *Staphylococcus aureus*. Mol. Microbiol. *41*, 247–261.

Franza, T., Mahe, B., and Expert, D. (2005). *Erwinia chrysanthemi* requires a second iron transport route dependent of the siderophore achromobactin for extracellular growth and plant infection. Mol. Microbiol. *55*, 261–275.

Fuentes, D.E., Navarro, C.A., Tantalean, J.C., Araya, M.A., Saavedra, C.P., Perez, J.M., Calderon, I.L., Youderian, P.A., Mora, G.C., and Vasquez, C.C. (2005). The product of the *qacC* gene of *Staphylococcus epidermidis* CH mediates resistance to beta-lactam antibiotics in Gram-positive and Gram-negative bacteria. Res. Microbiol. *156*, 472–477.

Furrer, J.L., Sanders, D.N., Hook-Barnard, I.G., and McIntosh, M.A. (2002). Export of the siderophore enterobactin in *Escherichia coli*: involvement of a 43 kDa membrane exporter. Mol. Microbiol. *44*, 1225–1234.

Garnacho-Montero, J., and Amaya-Villar, R. (2010). Multiresistant *Acinetobacter baumannii* infections: epidemiology and management. Curr. Opin. Infect. Dis. *23*, 332–339.

Ginn, S.L., Brown, M.H., and Skurray, R.A. (2000). The TetA(K) tetracycline/H(+) antiporter from

Staphylococcus aureus: mutagenesis and functional analysis of motif C. J. Bacteriol. *182*, 1492–1498.

Goel, N., Wattal, C., Oberoi, J.K., Raveendran, R., Datta, S., and Prasad, K.J. (2011). Trend analysis of antimicrobial consumption and development of resistance in non-fermenters in a tertiary care hospital in Delhi, India. J. Antimicrob. Chemother. *66*, 1625–1630.

Gordon, N.C., and Wareham, D.W. (2010). Multidrug-resistant *Acinetobacter baumannii*: mechanisms of virulence and resistance. Int. J. Antimicrob. Agents. *35*, 219–226.

Grinius, L.L., and Goldberg, E.B. (1994). Bacterial multidrug resistance is due to a single membrane protein which functions as a drug pump. J. Biol. Chem. *269*, 29998–30004.

Grkovic, S., Brown, M.H., Roberts, N.J., Paulsen, I.T., and Skurray, R.A. (1998). QacR is a repressor protein that regulates expression of the *Staphylococcus aureus* multidrug efflux pump QacA. J. Biol. Chem. *273*, 18665–18673.

Grkovic, S., Brown, M.H., and Skurray, R.A. (2002). Regulation of bacterial drug export systems. Microbiol. Molec. Biol. Rev. *66*, 671–701.

Hassan, K.A., Robinson, K.L., Smith, A.N., Gibson, J.H., Skurray, R.A., and Brown, M.H. (2006). Glycine-rich transmembrane helix 10 in the staphylococcal tetracycline transporter TetA(K) lines a solvent-accessible channel. Biochemistry *45*, 15661–15669.

Hassan, K.A., Skurray, R.A., and Brown, M.H. (2007a). Active export proteins mediating drug resistance in staphylococci. J. Mol. Microbiol. Biotechnol. *12*, 180–196.

Hassan, K.A., Skurray, R.A., and Brown, M.H. (2007b). Transmembrane helix 12 of the *Staphylococcus aureus* multidrug transporter QacA lines the bivalent drug-binding pocket. J. Bacteriol. *189*, 9131–9134.

Hassan, K.A., Brzoska, A.J., Wilson, N.L., Eijkelkamp, B.A., Brown, M.H., and Paulsen, I.T. (2011). Roles of DHA2 family transporters in drug resistance and iron homeostasis in *Acinetobacter* spp. J. Mol. Microbiol. Biotechnol. *20*, 116–124.

Hazan, R., He, J., Xiao, G., Dekimpe, V., Apidianakis, Y., Lesic, B., Astrakas, C., Deziel, E., Lepine, F., and Rahme, L.G. (2010). Homeostatic interplay between bacterial cell–cell signaling and iron in virulence. PLoS Pathog. *6*, e1000810.

Higgins, P.G., Wisplinghoff, H., Stefanik, D., and Seifert, H. (2004). Selection of topoisomerase mutations and overexpression of *adeB* mRNA transcripts during an outbreak of *Acinetobacter baumannii*. J. Antimicrob. Chemother. *54*, 821–823.

Hood, M.I., Jacobs, A.C., Sayood, K., Dunman, P.M., and Skaar, E.P. (2010). *Acinetobacter baumannii* increases tolerance to antibiotics in response to monovalent cations. Antimicrob. Agents. Chemother. *54*, 1029–1041.

Huang, J., O'Toole, P.W., Shen, W., Amrine-Madsen, H., Jiang, X., Lobo, N., Palmer, L.M., Voelker, L., Fan, F., Gwynn, M.N., *et al.* (2004). Novel chromosomally encoded multidrug efflux transporter MdeA in *Staphylococcus aureus*. Antimicrob. Agents. Chemother. *48*, 909–917.

Jin, J., Guffanti, A.A., Bechhofer, D.H., and Krulwich, T.A. (2002). Tet(L) and tet(K) tetracycline-divalent metal/H+ antiporters: characterization of multiple catalytic modes and a mutagenesis approach to differences in their efflux substrate and coupling ion preferences. J. Bacteriol. *184*, 4722–4732.

Joly-Guillou, M.L. (2005). Clinical impact and pathogenicity of *Acinetobacter*. Clin. Microbiol. Infect. *11*, 868–873.

Kaatz, G.W., Seo, S.M., and Ruble, C.A. (1993). Efflux-mediated fluoroquinolone resistance in *Staphylococcus aureus*. Antimicrob. Agents. Chemother. *37*, 1086–1094.

Kaatz, G.W., McAleese, F., and Seo, S.M. (2005a). Multidrug resistance in *Staphylococcus aureus* due to overexpression of a novel multidrug and toxin extrusion (MATE) transport protein. Antimicrob. Agents. Chemother. *49*, 1857–1864.

Kaatz, G.W., Thyagarajan, R.V., and Seo, S.M. (2005b). Effect of promoter region mutations and *mgrA* overexpression on transcription of *norA*, which encodes a *Staphylococcus aureus* multidrug efflux transporter. Antimicrob. Agents. Chemother. *49*, 161–169.

Kaatz, G.W., DeMarco, C.E., and Seo, S.M. (2006). MepR, a repressor of the *Staphylococcus aureus* MATE family multidrug efflux pump MepA, is a substrate-responsive regulatory protein. Antimicrob. Agents. Chemother. *50*, 1276–1281.

Khan, S.A., and Novick, R.P. (1983). Complete nucleotide sequence of pT181, a tetracycline-resistance plasmid from *Staphylococcus aureus*. Plasmid. *10*, 251–259.

Kosmidis, C., DeMarco, C.E., Frempong-Manso, E., Seo, S.M., and Kaatz, G.W. (2010). *In silico* genetic correlations of multidrug efflux pump gene expression in *Staphylococcus aureus*. Int. J. Antimicrob. Agents. *36*, 222–229.

Koster, W. (2005). Cytoplasmic membrane iron permease systems in the bacterial cell envelope. Front. Biosci. *10*, 462–477.

Krulwich, T.A., Lewinson, O., Padan, E., and Bibi, E. (2005). Do physiological roles foster persistence of drug/multidrug-efflux transporters? A case study. Nat. Rev. Microbiol. *3*, 566–572.

Kumaraswami, M., Schuman, J.T., Seo, S.M., Kaatz, G.W., and Brennan, R.G. (2009). Structural and biochemical characterization of MepR, a multidrug binding transcription regulator of the *Staphylococcus aureus* multidrug efflux pump MepA. Nucleic Acids Res. *37*, 1211–1224.

Kupferwasser, L.I., Skurray, R.A., Brown, M.H., Firth, N., Yeaman, M.R., and Bayer, A.S. (1999). Plasmid-mediated resistance to thrombin-induced platelet microbicidal protein in staphylococci: role of the *qacA* locus. Antimicrob. Agents Chemother. *43*, 2395–2399.

Leelaporn, A., Firth, N., Paulsen, I.T., Hettiaratchi, A., and Skurray, R.A. (1995). Multidrug resistance plasmid pSK108 from coagulase-negative staphylococci; relationships to *Staphylococcus aureus qacC* plasmids. Plasmid *34*, 62–67.

Littlejohn, T.G., Paulsen, I.T., Gillespie, M.T., Tennent, J.M., Midgley, M., Jones, I.G., Purewal, A.S., and

Skurray, R.A. (1992). Substrate specificity and energetics of antiseptic and disinfectant resistance in *Staphylococcus aureus*. FEMS Microbiol. Lett. *95*, 259–266.

Longtin, J., Seah, C., Siebert, K., McGeer, A., Simor, A., Longtin, Y., Low, D.E., and Melano, R.G. (2011). Distribution of antiseptic resistance genes *qacA*, *qacB*, and *smr* in methicillin-resistant *Staphylococcus aureus* isolated in Toronto, Canada, from 2005 to 2009. Antimicrob. Agents Chemother. *55*, 2999–3001.

Lowy, F.D. (2003). Antimicrobial resistance: the example of *Staphylococcus aureus*. J. Clin. Invest. *111*, 1265–1273.

McAleese, F., Petersen, P., Ruzin, A., Dunman, P.M., Murphy, E., Projan, S.J., and Bradford, P.A. (2005). A novel MATE family efflux pump contributes to the reduced susceptibility of laboratory-derived *Staphylococcus aureus* mutants to tigecycline. Antimicrob. Agents Chemother. *49*, 1865–1871.

Magnet, S., Courvalin, P., and Lambert, T. (2001). Resistance-nodulation-cell division-type efflux pump involved in aminoglycoside resistance in *Acinetobacter baumannii* strain BM4454. Antimicrob. Agents Chemother. *45*, 3375–3380.

Markham, P.N., and Neyfakh, A.A. (1996). Inhibition of the multidrug transporter NorA prevents emergence of norfloxacin resistance in *Staphylococcus aureus*. Antimicrob. Agents Chemother. *40*, 2673–2674.

Mayer, S., Boos, M., Beyer, A., Fluit, A.C., and Schmitz, F.J. (2001). Distribution of the antiseptic resistance genes *qacA*, *qacB* and *qacC* in 497 methicillin-resistant and – susceptible European isolates of *Staphylococcus aureus*. J. Antimicrob. Chemother.*47*, 896–897.

Michalopoulos, A., and Falagas, M.E. (2010). Treatment of *Acinetobacter* infections. Expert Opin. Pharmacother. *11*, 779–788.

Mitchell, B.A., Brown, M.H., and Skurray, R.A. (1998). QacA multidrug efflux pump from *Staphylococcus aureus*: comparative analysis of resistance to diamidines, biguanidines and guanylhydrazones. Antimicrob. Agents Chemother. *42*, 475–477.

Mitchell, B.A., Paulsen, I.T., Brown, M.H., and Skurray, R.A. (1999). Bioenergetics of the staphylococcal multidrug export protein QacA: identification of distinct binding sites for monovalent and divalent cations. J. Biol. Chem. *274*, 3541–3548.

Nakaminami, H., Noguchi, N., and Sasatsu, M. (2010). Fluoroquinolone efflux by the plasmid-mediated multidrug efflux pump QacB variant QacBIII in *Staphylococcus aureus*. Antimicrob. Agents Chemother. *54*, 4107–4111.

Neely, A.N. (2000). A survey of Gram-negative bacteria survival on hospital fabrics and plastics. J. Burn Care Rehabil. *21*, 523–527.

Nemec, A., Dijkshoorn, L., and van der Reijden, T.J. (2004). Long-term predominance of two pan-European clones among multi-resistant *Acinetobacter baumannii* strains in the Czech Republic. J. Med. Microbiol. *53*, 147–153.

Neonakis, I.K., Spandidos, D.A., and Petinaki, E. (2011). Confronting multidrug-resistant *Acinetobacter baumannii*: a review. Int. J. Antimicrob. Agents *37*, 102–109.

Neyfakh, A.A. (1997). Natural functions of bacterial multidrug transporters. Trends Microbiol. *5*, 309–313.

Neyfakh, A.A., Borsch, C.M., and Kaatz, G.W. (1993). Fluoroquinolone resistance protein NorA of *Staphylococcus aureus* is a multidrug efflux transporter. Antimicrob. Agents Chemother. *37*, 128–129.

Ng, E.Y., Trucksis, M., and Hooper, D.C. (1994). Quinolone resistance mediated by *norA*: physiologic characterization and relationship to *flqB*, a quinolone resistance locus on the *Staphylococcus aureus* chromosome. Antimicrob. Agents Chemother. *38*, 1345–1355.

Noguchi, N., Hase, M., Kitta, M., Sasatsu, M., Deguchi, K., and Kono, M. (1999). Antiseptic susceptibility and distribution of antiseptic-resistance genes in methicillin-resistant *Staphylococcus aureus*. FEMS Microbiol. Lett. *172*, 247–253.

Noguchi, N., Suwa, J., Narui, K., Sasatsu, M., Ito, T., Hiramatsu, K., and Song, J.H. (2005). Susceptibilities to antiseptic agents and distribution of antiseptic-resistance genes *qacA/B* and *smr* of methicillin-resistant *Staphylococcus aureus* isolated in Asia during 1998 and 1999. J. Med. Microbiol. *54*, 557–565.

Pannek, S., Higgins, P.G., Steinke, P., Jonas, D., Akova, M., Bohnert, J.A., Seifert, H., and Kern, W.V. (2006). Multidrug efflux inhibition in *Acinetobacter baumannii*: comparison between 1-(1-naphthylmethyl)-piperazine and phenyl-arginine-beta-naphthylamide. J. Antimicrob. Chemother. *57*, 970–974.

Patel, D., Kosmidis, C., Seo, S.M., and Kaatz, G.W. (2010). Ethidium bromide MIC screening for enhanced efflux pump gene expression or efflux activity in *Staphylococcus aureus*. Antimicrob. Agents Chemother. *54*, 5070–5073.

Patterson, J.E., Vecchio, J., Pantelick, E.L., Farrel, P., Mazon, D., Zervos, M.J., and Hierholzer, W.J., Jr. (1991). Association of contaminated gloves with transmission of *Acinetobacter calcoaceticus* var. *anitratus* in an intensive care unit. Am. J. Med. *91*, 479–483.

Paulsen, I.T., Brown, M.H., Dunstan, S.J., and Skurray, R.A. (1995). Molecular characterization of the staphylococcal multidrug resistance export protein QacC. J. Bacteriol. *177*, 2827–2833.

Paulsen, I.T., Brown, M.H., Littlejohn, T.G., Mitchell, B.A., and Skurray, R.A. (1996a). Multidrug resistance proteins QacA and QacB from *Staphylococcus aureus*: membrane topology and identification of residues involved in substrate specificity. Proc. Natl. Acad. Sci. U.S.A. *93*, 3630–3635.

Paulsen, I.T., Brown, M.H., and Skurray, R.A. (1996b) Proton-dependent multidrug efflux systems. Microbiol. Rev. *60*, 575–608.

Paulsen, I.T., Brown, M.H., and Skurray, R.A. (1998). Characterization of the earliest known *Staphylococcus aureus* plasmids encoding multidrug efflux systems. J. Bacteriol. *180*, 3477–3479.

Plessner, O., Klapatch, T., and Guerinot, M.L. (1993). Siderophore utilization by *Bradyrhizobium japonicum*. Appl. Environ. Microbiol. *59*, 1688–1690.

Poget, S.F., Harris, R., Cahill, S.M., and Girvin, M.E. (2010). 1H, 13C, 15N backbone NMR assignments of the *Staphylococcus aureus* small multidrug-resistance pump (Smr) in a functionally active conformation. Biomol. NMR Assign. *4*, 139–142.

Poole, K. (2002). Outer membranes and efflux: the path to multidrug resistance in Gram-negative bacteria. Curr. Pharm. Biotechnol. 3, 77–98.

Poole, K. (2008). Bacterial multidrug efflux pumps serve other functions. Microbe 3, 179–185.

Pos, K.M. (2009). Drug transport mechanism of the AcrB efflux pump. Biochim. Biophys. Acta *1794*, 782–793.

Rajamohan, G., Srinivasan, V.B., and Gebreyes, W.A. (2010a). Molecular and functional characterization of a novel efflux pump, AmvA, mediating antimicrobial and disinfectant resistance in *Acinetobacter baumannii*. J. Antimicrob. Chemother. *65*, 1919–1925.

Rajamohan, G., Srinivasan, V.B., and Gebreyes, W.A. (2010b). Novel role of *Acinetobacter baumannii* RND efflux transporters in mediating decreased susceptibility to biocides. J. Antimicrob. Chemother. *65*, 228–232.

Ren, Q., and Paulsen, I.T. (2007). Large-scale comparative genomic analyses of cytoplasmic membrane transport systems in prokaryotes. J. Mol. Microbiol. Biotechnol. *12*, 165–179.

Ren, Q., Chen, K., and Paulsen, I.T. (2007). TransportDB: a comprehensive database resource for cytoplasmic membrane transport systems and outer membrane channels. Nuc. Acids. Res. 35, D274–279.

Ribera, A., Roca, I., Ruiz, J., Gibert, I., and Vila, J. (2003). Partial characterization of a transposon containing the *tet(A)* determinant in a clinical isolate of *Acinetobacter baumannii*. J. Antimicrob. Chemother. *52*, 477–480.

Roca, I., Marti, S., Espinal, P., Martinez, P., Gibert, I., and Vila, J. (2009). CraA, a major facilitator superfamily efflux pump associated with chloramphenicol resistance in *Acinetobacter baumannii*. Antimicrob. Agents Chemother. *53*, 4013–4014.

Ruzin, A., Keeney, D., and Bradford, P.A. (2007). AdeABC multidrug efflux pump is associated with decreased susceptibility to tigecycline in *Acinetobacter calcoaceticus–Acinetobacter baumannii* complex. J. Antimicrob. Chemother. *59*, 1001–1004.

Schumacher, M.A., Miller, M.C., Grkovic, S., Brown, M.H., Skurray, R.A., and Brennan, R.G. (2001). Structural mechanisms of QacR induction and multidrug recognition. Science *294*, 2158–2163.

Seifert, H., Dijkshoorn, L., Gerner-Smidt, P., Pelzer, N., Tjernberg, I., and Vaneechoutte, M. (1997). Distribution of *Acinetobacter* species on human skin: comparison of phenotypic and genotypic identification methods. J. Clin. Microbiol. *35*, 2819–2825.

Smith, A.W., and Alpar, K.E. (1991). Immune response to *Acinetobacter calcoaceticus* infection in man. J. Med. Microbiol. *34*, 83–88.

Srinivasan, V.B., Rajamohan, G., and Gebreyes, W.A. (2009). Role of AbeS, a novel efflux pump of the SMR family of transporters, in resistance to antimicrobial agents in *Acinetobacter baumannii*. Antimicrob. Agents Chemother. *53*, 5312–5316.

Su, X.Z., Chen, J., Mizushima, T., Kuroda, T., and Tsuchiya, T. (2005). AbeM, an H+-coupled *Acinetobacter baumannii* multidrug efflux pump belonging to the MATE family of transporters. Antimicrob. Agents Chemother. *49*, 4362–4364.

Sun, J.R., Chan, M.C., Chang, T.Y., Wang, W.Y., and Chiueh, T.S. (2010). Overexpression of the *adeB* gene in clinical isolates of tigecycline-nonsusceptible *Acinetobacter baumannii* without insertion mutations in *adeRS*. Antimicrob. Agents Chemother. *54*, 4934–4938.

Tal, N., and Schuldiner, S. (2009). A coordinated network of transporters with overlapping specificities provides a robust survival strategy. Proc. Natl. Acad. Sci. U.S.A. *106*, 9051–9056.

Tanabe, M., Szakonyi, G., Brown, K.A., Henderson, P.J., Nield, J., and Byrne, B. (2009). The multidrug resistance efflux complex, EmrAB from *Escherichia coli* forms a dimer *in vitro*. Biochem. Biophys. Res. Commun. *380*, 338–342.

Teixeira, C.F., Pereira, T.B., Miyazaki, N.H., and Villas Boas, M.H. (2010). Widespread distribution of *qacA/B* gene among coagulase-negative *Staphylococcus* spp. in Rio de Janeiro, Brazil. J. Hosp. Infect. *75*, 333–334.

Tennent, J.M., Lyon, B.R., Gillespie, M.T., May, J.W., and Skurray, R.A. (1985). Cloning and expression of *Staphylococcus aureus* plasmid-mediated quaternary ammonium resistance in *Escherichia coli*. Antimicrob. Agents Chemother. *27*, 79–83.

Tikhonova, E.B., Dastidar, V., Rybenkov, V.V., and Zgurskaya, H.I. (2009). Kinetic control of TolC recruitment by multidrug efflux complexes. Proc. Natl. Acad. Sci. U.S.A. *106*, 16416–16421.

Truong-Bolduc, Q.C., and Hooper, D.C. (2010). Phosphorylation of MgrA and its effect on expression of the NorA and NorB efflux pumps of *Staphylococcus aureus*. J. Bacteriol. *192*, 2525–2534.

Truong-Bolduc, Q.C., Zhang, X., and Hooper, D.C. (2003). Characterization of NorR protein, a multifunctional regulator of *norA* expression in *Staphylococcus aureus*. J. Bacteriol. *185*, 3127–3138.

Truong-Bolduc, Q.C., Dunman, P.M., Strahilevitz, J., Projan, S.J., and Hooper, D.C. (2005). MgrA is a multiple regulator of two new efflux pumps in *Staphylococcus aureus*. J. Bacteriol. *187*, 2395–2405.

Truong-Bolduc, Q.C., Strahilevitz, J., and Hooper, D.C. (2006). NorC, a new efflux pump regulated by MgrA of *Staphylococcus aureus*. Antimicrob. Agents Chemother. *50*, 1104–1107.

Truong-Bolduc, Q.C., Dunman, P.M., Eidem, T., and Hooper, D.C. (2011). Transcriptional profiling analysis of the global regulator NorG, a GntR-like protein of *Staphylococcus aureus*. J. Bacteriol. *193*, 6207–6214.

Turton, J.F., Gabriel, S.N., Valderrey, C., Kaufmann, M.E., and Pitt, T.L. (2007). Use of sequence-based typing and multiplex PCR to identify clonal lineages of outbreak strains of *Acinetobacter baumannii*. Clin. Microbiol. Infect. *13*, 807–815.

Turton, J.F., Shah, J., Ozongwu, C., and Pike, R. (2010). Incidence of *Acinetobacter* species other than *A.*

baumannii among clinical isolates of *Acinetobacter*: evidence for emerging species. J. Clin. Microbiol. *48*, 1445–1449.

Velamakanni, S., Yao, Y., Gutmann, D.A., and van Veen, H.W. (2008). Multidrug transport by the ABC transporter Sav1866 from *Staphylococcus aureus*. Biochemistry 47, 9300–9308.

Yamada, Y., Hideka, K., Shiota, S., Kuroda, T., and Tsuchiya, T. (2006a). Gene cloning and characterization of SdrM, a chromosomally-encoded multidrug efflux pump, from *Staphylococcus aureus*. Biol. Pharm. Bull. *29*, 554–556.

Yamada, Y., Shiota, S., Mizushima, T., Kuroda, T., and Tsuchiya, T. (2006b). Functional gene cloning and characterization of MdeA, a multidrug efflux pump from *Staphylococcus aureus*. Biol. Pharm. Bull. *29*, 801–804.

Yamaguchi, A., Shiina, Y., Fujihira, E., Sawai, T., Noguchi, N., and Sasatsu, M. (1995). The tetracycline efflux protein encoded by the *tet(K)* gene from *Staphylococcus aureus* is a metal-tetracycline/H+ antiporter. FEBS Lett. *365*, 193–197.

Yoshida, H., Bogaki, M., Nakamura, S., Ubukata, K., and Konno, M. (1990). Nucleotide sequence and characterization of the *Staphylococcus aureus norA* gene, which confers resistance to quinolones. J. Bacteriol. *172*, 6942–6949.

Yu, J.-L., Grinius, L., and Hooper, D.C. (2002). NorA functions as a multidrug efflux protein in both cytoplasmic membrane vesicles and reconstituted proteoliposomes. J. Bacteriol. *184*, 1370–1377.

Xu, Z., O'Rourke, B., Skurray, R.A., and Brown, M.H. (2006). Role of transmembrane segment 10 in efflux mediated by the staphylococcal multidrug transport protein QacA. J. Biol. Chem. *281*, 792–799.

Mycobacterium tuberculosis Drug Efflux Pumps: An Update

7

Maria Rosalia Pasca, Silvia Buroni and Giovanna Riccardi

Abstract

It is well known that drug efflux systems contribute to the development of multiresistance patterns in several bacterial pathogens. The selection and diffusion of *Mycobacterium tuberculosis* multidrug-resistant (MDR-TB), extensively drug-resistant (XDR-TB) and, more recently, totally drug-resistant (TDR) strains constitute a serious threat for tuberculosis global control. Mycobacteria, such as *M. tuberculosis* and *Mycobacterium smegmatis*, possess several putative drug efflux transporters, but their role in resistance is still a hard topic and needs to be further investigated as resistance to several drugs is usually the result of the combination of independent mutations in genes encoding either the drug target or the enzymes involved in drug activation. However, as the genetic basis of resistance to some antitubercular agents is not fully known for some clinical isolates, we cannot rule out an efflux mechanism in these strains. Several drug efflux transporters have been described in mycobacteria as responsible for resistance to aminoglycosides, chloramphenicol, fluoroquinolones, isoniazid, linezolid, rifampicin, tetracycline and other compounds but most of them were isolated in laboratory rather than in hospitals. This chapter highlights recent advances in our understanding of efflux-mediated drug resistance in mycobacteria, including the distribution of efflux systems in these organisms, their substrate profiles and their contribution to drug resistance.

Tuberculosis and the problem of the drug resistance

Tuberculosis is a bacterial infectious disease that mainly involves the lungs, but may spread to other organs. Its etiologic agent in humans is *Mycobacterium tuberculosis*, also known as Koch's bacillus as it was discovered by the German physician Robert Koch in 1882. Only 1–5% of primary tuberculosis cases develop the active disease soon after infection. In the majority of cases, the latent infection occurs without evident symptoms. Dormant bacilli can produce active tuberculosis in 2–23% of these cases, usually many years after infection. The risk of reactivation increases if the immune system is compromised; while in immune competent people the reactivation risk is between 5–10% in a lifetime; in patients with *M. tuberculosis*–HIV co-infection, the percentage rises to 10% per year (El-Sadr and Tsiouris, 2008).

Tuberculosis has increased at a significant rate, with 8 million new cases and 2 million deaths per year (Riccardi *et al.*, 2009). Further, it is noteworthy that hot spots of the disease are found scattered in different continents, while the selection and spread of drug-resistant *M. tuberculosis* strains menaces to make the treatment of this disease extremely difficult and a threat to public health worldwide (Caminero *et al.*, 2010). The development of drug-resistant *M. tuberculosis* strains is essentially favoured by an inappropriate drug prescription of physicians and irregular intake of drugs by patients (Goldman *et al.*, 2007).

Drugs utilized to treat tuberculosis are classified into first-line and second-line agents. Treatment of drug-susceptible tuberculosis involves an initial phase with isoniazid, rifampicin, pyrazinamide and ethambutol for the first 2 months followed by a continuation phase of isoniazid and rifampicin for the last four months. Up to 95% of people infected by sensitive *M. tuberculosis* strains can be cured in 6 months with this four-drug regimen (Koul *et al.*, 2011). *M. tuberculosis* strains which are resistant to both isoniazid and rifampicin, with or without resistance to other drugs, have been termed multidrug-resistant strains (MDR-TB). Frequently, MDR tuberculosis results from either primary infection with drug-resistant bacteria or may develop in the course of a patient's treatment when non-optimal treatment durations or regimens are used. MDR tuberculosis requires a further two years of treatment with second-line drugs such as: quinolones, aminoglycosides, ethionamide, D-cycloserine and basic peptides. Cure rates for MDR-TB are lower than drug-susceptible tuberculosis, typically ranging from 50% to 70% (Koul *et al.*, 2011).

The emergence of extensively drug-resistant tuberculosis (XDR-TB), caused by MDR strains also resistant to two major second-line agents (aminoglycosides and fluoroquinolones), greatly alarmed the WHO. These resistant strains have been identified in all regions of the World, most often in Asia and countries of the former Soviet Union. According to the WHO, 425,000 new MDR-TB cases occur every year, where up to 14% of all new cases are not responding to standard drug treatment (Jassal and Bishai, 2009). MDR tuberculosis has been associated with a high mortality in individuals with HIV, while patients with XDR tuberculosis were 64% more likely to die during treatment than patients with MDR tuberculosis (Jassal and Bishai, 2009; LoBue, 2009).

The spreading of very dangerous *M. tuberculosis* strains named totally drug-resistant (TDR) constitutes a deadly threat to the patients. These strains are fully capable of being transmitted and causing active diseases in individuals with secondary cases. Velayati and collaborators (2009) reported that the 95% of XDR and TDR strains were isolated from patients with a previous history of tuberculosis. TDR strains are defined as *M. tuberculosis* isolates which are resistant to all first-line (isoniazid, rifampicin, streptomycin, ethambutol and pyrazinamide) and second-line drugs tested (ofloxacin, cycloserine, prothionamide, amikacin, kanamycin, ethionamide, and para-aminosalicylic acid). Generally, the spectrum of resistance reflects the drugs that the patients have used and the way in which therapy was controlled (Velayati *et al.*, 2009).

The treatment of tuberculosis becomes more complicated as the antibiotic resistance profile of *M. tuberculosis* broadens. TDR and XDR tuberculosis are generally thought to have high mortality rates. With the exception of the fluoroquinolones, no new antitubercular drug has been introduced in clinical practice in the past 45 years. The probability of successful treatment further decreases with the emergence of new drug-resistant strains.

Bacterial resistance to antibiotics typically involves drug inactivation or modification, target alteration, or a decrease in drug accumulation associated with a decrease in permeability and/or an increase in efflux. Other resistance mechanisms include inhibition of the activation of pro-drugs into active drugs. *M. tuberculosis* shows a high degree of intrinsic resistance to several antibiotics and chemotherapeutic agents attributed to the low permeability of its cell wall, because of its specific lipid-rich composition and structure (Jarlier and Nikaido, 1994). However, recent reports suggested that efflux pumps may also be involved in this phenomenon; the balance between drug transport into the cell and drug efflux is not clearly understood yet, and further studies are required in the case of mycobacteria (De Rossi *et al.*, 2006). In this chapter, the contribution of efflux pumps to *M. tuberculosis* drug resistance will be analysed.

Antitubercular drug resistance by efflux mechanisms

Efflux pumps in *M. tuberculosis*

The intrinsic resistance of *M. tuberculosis* to most antibiotics is generally attributed to the low permeability of the mycobacterial cell wall, because of its specific lipid-rich composition and structure (De Rossi *et al.*, 2006). This low permeability, which limits drug uptake, seems to be one of the main factors involved in resistance (Brennan, 2003). Along with cell wall permeability, active efflux systems also provide resistance by extruding the drug molecules that enter the cell. The intracellular concentration of a given drug depends on the balance between its influx and efflux. It is therefore of primary importance to increase our understanding of the processes of drug influx through porins and drug efflux *via* drug transporters (Niederweis, 2003).

Several mycobacterial drug efflux pumps,

belonging to different transporter families, have been identified and characterized experimentally. There are four main families of drug transporters in *M. tuberculosis*: the ATP binding cassette (ABC) superfamily, the major facilitator superfamily (MFS), the small multidrug resistance (SMR) family, and the resistance–nodulation–cell division (RND) family. The MFS, SMR, and RND families are secondary transporters, typically energized by the proton motive force. In contrast, ATP is utilized as the energy donor for members of the ABC family of multidrug efflux pumps that are often considered primary transporters. Several drug efflux pumps have been identified or characterized in mycobacteria, belonging to the four families just described (Table 7.1).

MFS drug transporters

The major facilitator superfamily (MFS) constitutes one of the largest families of membrane transporters. MFS permeases possess 12 or 14 putative or established transmembrane segments and transport many different compounds, including simple sugars, oligosaccharides, inositols, drugs, amino acids, nucleosides, organophosphate esters, Krebs cycle metabolites, and a large variety of inorganic anions and cations (Pao *et al.*, 1998; Saier *et al.*, 1999). Some MFS transporters may confer drug resistance to bacteria (De Rossi *et al.*, 2006).

The first MFS efflux pump characterized was QacA of the human pathogen *Staphylococcus aureus* (Littlejohn *et al.*, 1992). QacA confers high-level resistance to various toxic organic cations, including intercalating dyes (e.g. ethidium bromide) and a number of commonly used antiseptics and disinfectants such as cetrimide, benzalkonium chloride and chlorhexidine. Thus, *S. aureus* seems to have found a way of circumventing the antiseptics and disinfectants prevalent in the hospital environment by the export of such compounds (Brown and Skurray, 2001). A survey of 98 clinical isolates of methicillin-resistant *S. aureus* revealed that 70% of the strains were antiseptic-resistant and one-third carried a transmissible plasmid containing the *qacA* gene (Noguchi *et al.*, 1999), which could

Table 7.1 *M. tuberculosis* drug efflux pumps

Efflux pumps	Family	Resistance	References
Rv1634	MFS	Fluoroquinolones	De Rossi *et al.* (2002)
Tap	MFS	Aminoglycosides and tetracycline	Aínsa *et al.* (1998), De Rossi *et al.* (2002)
P55	MFS	Aminoglycosides and tetracycline	Silva *et al.* (2001)
Rv2333c (Stp)	MFS	Spectinomycin and tetracycline	De Rossi *et al.* (2002), Ramón-García *et al.* (2007)
Rv2459 (JefA)	MFS	Isoniazid and ethambutol	De Rossi *et al.* (2002), Gupta *et al.* (2010b)
Rv2846c (EfpA)	MFS	Ethidium bromide, acriflavine, ciprofloxacin, norfloxacin, and gentamicin	De Rossi *et al.* (2002), Li *et al.* (2004)
MmR	SMR	TPP, ethidium bromide, erythromycin, acriflavine, safranin O, and pyronin Y	De Rossi *et al.* (1998b)
MmpL7	RND	Isoniazid	Pasca *et al.* (2005)
MmpL5	RND	Azoles	Milano *et al.* (2009)
DrrAB	ABC	Tetracycline, erythromycin, ethambutol, norfloxacin, streptomycin, chloramphenicol, and anthracyclines	Choudhuri *et al.* (2002)
Rv2686c-2687c-2688c	ABC	Fluoroquinolones	Pasca *et al.* (2004)
Rv0194	ABC	Ampicillin, streptomycin, vancomycin, novobiocin, tetracycline, erythromycin, and chloramphenicol	Danilchanka *et al.* (2008)

be transferred into plasmid-free *S. aureus* strains, thereby also playing an important role in acquired drug resistance.

Many members of the MFS have been described in several mycobacterial species. In *M. smegmatis*, the LfrA efflux pump, conferring resistance to several fluoroquinolones (Takiff *et al.*, 1996) and the Tet(V) protein, conferring resistance to tetracycline (De Rossi *et al.*, 1998a), have 14 and 11 hydrophobic membrane-spanning regions, respectively. The *M. fortuitum* Tap conferred resistance to aminoglycosides and tetracycline (Aínsa *et al.*, 1998) and, by uptake/efflux experiments, it was shown that Tap uses the electrochemical gradient across the cytoplasmic membrane to extrude tetracycline out of the cell. This efflux activity was inhibited by carbonyl cyanide m-chlorophenylhydrazone (CCCP) and reserpine (Ramón-García *et al.*, 2006).

In the *M. tuberculosis* genome 16 open reading frames encoding putative drug efflux pumps belonging to the MFS have been identified (Table 7.2) (Cole *et al.*, 1998; De Rossi *et al.*, 2002).

De Rossi and collaborators (2002) demonstrated that one of these *M. tuberculosis* MFS pumps, Rv1634, decreased susceptibility to various fluoroquinolones when overexpressed in *M. smegmatis*, and accumulation data showed that this pump is involved in norfloxacin and ciprofloxacin efflux. In particular, the overexpression of *Rv1634* in *M. smegmatis* and in *Mycobacterium bovis* BCG conferred resistance to ciprofloxacin (De Rossi *et al.*, 2002).

The overexpression of *M. tuberculosis Rv1258c* (a Tap orthologue) in *M. smegmatis* conferred detectable levels of resistance to tetracycline and aminoglycosides (De Rossi *et al.*, 2002). This efflux pump transports the same antibiotics as a *M. fortuitum* orthologue, although the levels of resistance conferred by the *M. tuberculosis* protein are notably lower (De Rossi *et al.*, 2002). Accumulation data demonstrated that this transporter is involved in tetracycline efflux.

Table 7.2 Putative MFS transporters in *M. tuberculosis*

M. tuberculosis MFS transporters	Function	References
Rv0037c	Integral membrane protein	De Rossi *et al.* (2002)
Rv0191	Probable chloramphenicol resistance protein	De Rossi *et al.* (2002)
Rv0783c (EmrB)	Multidrug resistance efflux protein	De Rossi *et al.* (2002)
Rv0849	Integral membrane transport protein	De Rossi *et al.* (2002)
Rv1250	Probable drug efflux protein	De Rossi *et al.* (2002)
Rv1258c (Tap)	Multidrug resistance pump	Ainsa *et al.* (1998), De Rossi *et al.* (2002)
Rv1410c (P55)	Drug efflux protein	Silva *et al.* (2001), De Rossi *et al.* (2002)
Rv1634	Fluoroquinolone drug efflux protein	De Rossi *et al.* (2002)
Rv1877	Probable drug efflux protein	De Rossi *et al.* (2002), Li *et al.* (2004)
Rv2333c (Stp)	Spectinomycin-tetracycline efflux pump	De Rossi *et al.* (2002), Ramón-García *et al.* (2007)
Rv2456c	Probable transmembrane transport protein	De Rossi *et al.* (2002)
Rv2459 (JefA)	Probable drug efflux protein	De Rossi *et al.* (2002), Gupta *et al.* (2010b)
Rv2846c (EfpA)	EfpA putative efflux protein	De Rossi *et al.* (2002), Li *et al.* (2004)
Rv2994	Probable fluoroquinolone efflux protein	De Rossi *et al.* (2002)
Rv3239c	Probable drug efflux protein	De Rossi *et al.* (2002)
Rv3728	Possible sugar transporter	De Rossi *et al.* (2002)

M. tuberculosis P55 is a MFS transporter that extrudes several drugs out of cells (Silva *et al.*, 2001). Genes homologous to P55 (*Rv1410c*) are present in all species of the *M. tuberculosis* complex and other mycobacteria such as *M. leprae* and *M. avium*. The overexpression of *M. bovis* P55 in *M. smegmatis* conferred aminoglycoside and tetracycline resistance. The levels of resistance to streptomycin and tetracycline decreased in the presence of the pump inhibitors CCCP, verapamil, and reserpine. *M. smegmatis* cells overexpressing P55 accumulated less tetracycline than the control cells (Silva *et al.*, 2001). A *M. bovis* BCG mutant deleted in the *Rv1410c* gene was more susceptible to a range of toxic compounds, including rifampicin and clofazimine. In addition, the P55 efflux pump mutant was more susceptible to redox compounds and displayed increased intracellular redox potential, suggesting an essential role of the efflux pump in detoxification processes coupled to oxidative balance within the cell (Ramón-García *et al.*, 2009). Finally, cells that lacked the *P55* gene displayed smaller colony sizes and had a growth defect in liquid culture. This, together with an increased susceptibility to the cell wall-targeting compounds bacitracin and vancomycin, suggested that P55 is needed for proper cell wall assembly and normal growth *in vitro* (Ramón-García *et al.*, 2009).

Two *M. tuberculosis* MFS transporters, Rv1877 and EfpA, have been studied in *M. smegmatis* (Li *et al.*, 2004). The orthologues of Rv1877 and EfpA were expressed at detectable levels in *M. smegmatis* mc^{2}155. The deletion of the *M. smegmatis efpA* orthologue produced increased susceptibility to ethidium bromide, acriflavine, ciprofloxacin, norfloxacin, and gentamicin, but unexpectedly also resulted in decreased susceptibility to rifamycins, isoniazid, and chloramphenicol [2 – to 4-fold increase in minimal inhibitory concentration (MICs)]. Deletion of the *Rv1877* orthologue produced detectable increased susceptibility to ethidium bromide, acriflavine and erythromycin (Li *et al.*, 2004). These results gave some evidence that Rv1877 and EfpA could be responsible for drug efflux. The transcriptional analysis by microarray of an *M. tuberculosis* culture grown in presence of isoniazid revealed the induction of the expression of *efpA* and of other genes (Wilson *et al.*, 1999).

The *M. tuberculosis* Rv2333c (Stp) MFS pump was characterized in *M. bovis* BCG (Ramón-García *et al.*, 2007) where its overexpression conferred resistance to spectinomycin and tetracycline and decreased the [^{3}H]tetracycline accumulation. The *M. bovis* BCG *Rv2333c* inactivated strain also showed a decrease in the MIC values of spectinomycin and tetracycline, and an increase in [^{3}H]tetracycline accumulation. These results indicated that spectinomycin and tetracycline were substrates of the Rv2333c efflux pump (Ramón-García *et al.*, 2007).

Recently, another MFS transporter has been characterized, JefA (Rv2459). The overexpression of *Rv2459* in *M. tuberculosis* conferred resistance to isoniazid and ethambutol. In the presence of CCCP and verapamil the MIC for these drugs decreased, demonstrating that JefA could extrude isoniazid and ethambutol out of the cell (Gupta *et al.*, 2010b).

SMR drug transporters

The efflux pumps belonging to the SMR family are the smallest multidrug transporters, about 110 amino acids, which extrude various drugs in exchange with protons, thereby rendering bacteria resistant to these compounds. These transporters are prokaryotic homo-oligomeric or hetero-oligomeric efflux systems. The subunits of these systems are composed of 100–120 amino acids and contain four membrane-spanning helices. The functionally characterized members of the SMR family catalyse drug efflux by means of H^+ (De Rossi *et al.*, 2006).

Only one example of a SMR family protein has been reported in *M. tuberculosis*, the Mmr efflux transporter (De Rossi *et al.*, 1998b). The overexpression of *M. tuberculosis mmr* in *M. smegmatis* conferred resistance to tetraphenyl phosphonium (TPP), ethidium bromide, erythromycin, acriflavine, safranin O, and pyronin Y. TPP accumulation experiments showed that Mmr actively extrudes TPP, in a process driven by the proton motive force (De Rossi *et al.*, 1998b). The *M. tuberculosis* Mmr protein was produced in *E. coli*, in which it conferred resistance to ethidium bromide, acriflavine and methyl viologen. The purified Mmr protein has also been demonstrated to function as a proton/drug antiporter *in vitro* (Ninio *et al.*, 2001).

RND drug transporters

The members of the RND family are transport proteins with 12 transmembrane spans and include a number of multidrug efflux proteins of particularly broad substrate specificity. RND transporters have been found in all major kingdoms of living organisms, but seem to be involved in drug resistance only in Gram-negative bacteria (De Rossi *et al.*, 2006). In these bacteria, some RND multidrug efflux systems require two auxiliary constituents: a membrane fusion protein and an outer membrane protein (Tseng *et al.*, 1999). These two proteins may act together, enabling the bacterium to transport drugs across both membranes of the cell envelope directly into the external medium.

The AcrAB/TolC drug efflux pump of *E. coli* provides a prototype for such export systems, with AcrB constituting the membrane pump itself, AcrA the membrane fusion protein and TolC the outer membrane component (Nikaido and Zgurskaya, 2001). This structural organization allows the extrusion of substrates directly into the external medium, bypassing the periplasm. The analysis of the crystal structure of AcrB suggested that RND transporters may also pump substrates out of the periplasm (Murakami *et al.*, 2002).

The complexity of RND multidrug efflux pumps in Gram-negative bacteria contrasts with most of the efflux pumps of Gram-positive bacteria, which are simpler in organization and have only one component located in the cytoplasmic membrane, like the MFS pumps. Although mycobacteria cluster phylogenetically with Gram-positive bacteria, they are structurally more similar to Gram-negative bacteria, as they are protected by an outer lipid bilayer made of mycolic acids and a cell envelope composed of non-covalently bound lipids and glycolipids (De Rossi *et al.*, 2006).

The genome sequence of *M. tuberculosis* H37Rv contains 15 genes encoding putative transmembrane proteins predicted to be transport proteins of the RND superfamily (Cole *et al.*, 1998). These proteins have been designated MmpL (mycobacterial membrane proteins, large). The MmpL proteins comprise about 950 amino acid residues and are predicted to contain 12 membrane-spanning α-helices. These MmpL proteins have two large extracytoplasmic domains, which probably face the periplasmic space: the first, comprising about 140 residues, is located between transmembrane segments 1 and 2; the second, located between transmembrane segments 7 and 8, is composed of about 300–400 residues. The C-terminal residues are located in the cytoplasm. In four cases, the *mmpL* genes are preceded in the operon by sequences encoding MmpS proteins, which are small mycobacterial membrane proteins with a hydrophobic α-helix close to the NH_2-terminus, and COOH-terminal domains of about 120 residues predicted to be exposed to the exterior. The hydrophobic nature of the Mmp proteins and the close association of their genes with genes encoding proteins involved in lipid metabolism suggest that these proteins may be involved in the transport of fatty acids (Tekaia *et al.*, 1999). The MmpS proteins of *M. tuberculosis* may be analogous to *E. coli* AcrA (the membrane fusion protein) and the MmpL proteins may be analogous to the *E. coli* AcrB efflux pump (De Rossi *et al.*, 2006).

The involvement of several MmpL proteins in *M. tuberculosis* pathogenesis has been demonstrated. In fact, in order to decipher the role of the MmpL proteins in *M. tuberculosis*, Domenech and collaborators generated mutant strains inactivated in *mmpL* genes (Domenech *et al.*, 2005). Insertional inactivation has revealed that only one of them (MmpL3) is apparently essential for viability. A role in virulence in mice was demonstrated for four of these proteins (MmpL4, MmpL7, MmpL8, and MmpL11). MmpL4 and MmpL7, which transport phthiocerol dimycocerosate, were found to have both impaired growth kinetics and impaired lethality. Mutants with *MmpL8* inactivated, which transports a precursor of the sulphatides, or *MmpL11* were found to establish infection normally but to be significantly attenuated for lethality in time-to-death studies (Domenech *et al.*, 2005).

M. tuberculosis SigF is strongly induced within cultured human macrophages and upon nutrient starvation, and SigF has been implicated in *M. tuberculosis* entry into stationary phase; consequently, SigF appears to contribute to the immune pathology of tuberculosis (Williams *et al.*, 2007). Induction of *sigF* resulted in significant upregulation of several genes encoding MmpL proteins, including *mmpL2*, *mmpL5*, and *mmpL11*. Although these *mmpL* genes are not essential for *in vitro* growth, *in vivo* growth of an *mmpL2* mutant was observed to be compromised when signature-tagged transposon mutants of *M. tuberculosis* were screened (Camacho *et al.*, 1999). Similarly, *M. tuberculosis* transposon mutants with disruptions in *mmpL5* and *mmpL11* were shown to be significantly attenuated for growth in mouse lungs (Lamichhane *et al.*, 2005). Moreover Lamichhane *et al.* (2005) demonstrated that other four transposon mutants with *mmpL4*, *mmpL7*, *mmpL8*, and *mmpL10* genes inactivated, had a compromised ability to multiply in mouse lungs.

In *M. tuberculosis*, MmpL7 catalyses the export of phthiocerol dimycocerosate (PDIM), a lipid constituent of the outer membrane (Camacho *et al.*, 2001). Inactivation of the *mmpL7* gene attenuates *M. tuberculosis* virulence in the mouse model (Camacho *et al.*, 1999). Converse *et al.* (2003) and Domenech *et al.* (2004) demonstrated that MmpL8 is required for the transport of a precursor of the sulphoglycolipids and for sustained bacterial growth and persistence in mice, as an MmpL8 mutant was only weakly pathogenic in mice. The sulphoglycolipids play a role in the virulence or pathogenesis of the tubercle bacilli. MmpL8 is also required for the complete assembly of the tetra-acylated forms of sulphoglycolipids (Layre *et al.*, 2011). Until now the involvement in drug efflux has been demonstrated only for two MmpL protein: MmpL5 (Milano *et al.*, 2009) and MmpL7 (Pasca *et al.*, 2005).

To investigate the mechanisms of resistance to azoles in mycobacteria, several spontaneous azoles resistant mutants from *M. tuberculosis* and *M. bovis* BCG were isolated and characterized (Milano *et al.*, 2009). All of the analysed resistant mutants exhibited an increased transcription of the *mmpS5–mmpL5* genes. By microarray analysis and real-time PCR it was shown that the overexpression of *mmpS5–mmpL5* genes, encoding a hypothetical RND drug transporter, is responsible for resistance to azoles *via* an efflux mechanism. The overexpression of the *mmpS5* and *mmpL5* genes was also detected in *M. bovis* BCG mutants which exhibited resistance to different azoles such as ketoconazole and econazole. Interestingly, these mutants accumulated a reduced amount of econazole within the cell, suggesting that the mechanism of resistance to azoles is due to an active efflux of the drug because of an overexpression of MmpS5–MmpL5 caused by the lack of an active form of Rv0678 (Milano *et al.*, 2009).

The *M. tuberculosis mmpL7* overexpression in *M. smegmatis* conferred a high resistance level to isoniazid. The resistance level decreased in the presence of the efflux pump inhibitors reserpine and CCCP. Energy-dependent efflux of isoniazid from *M. smegmatis* cells overexpressing the *mmpL7* gene was also observed. In this last experiment the reserpine addition inhibited the efflux pump (Pasca *et al.*, 2005).

ABC drug transporters

The ABC transporters constitute a large superfamily of multisubunit permeases which transport various molecules (ions, amino acids, peptides, drugs, antibiotics, lipids, polysaccharides, proteins, etc.), using ATP as energy source (Schmitt and Tampè, 2002). ABC transporters appear to consist of at least four domains: two membrane-spanning domains (MSDs) and two nucleotide-binding domains (NBDs). Each of the highly hydrophobic MSDs consists of six putative transmembrane segments that form the pathway through which the substrate crosses the membrane. The two NBDs lie at the periphery of the cytoplasmic face of the membrane. They bind ATP and couple ATP hydrolysis to substrate translocation (Kerr, 2002).

Genes encoding ABC transporters occupy about 2.5% of the *M. tuberculosis* genome (Cole *et al.*, 1998; Braibant *et al.*, 2000). Based on structural similarities to the typical subunits of ABC transporters present in all living organisms, at least 37 complete and incomplete ABC transporters have been identified in *M. tuberculosis* (Braibant *et al.*, 2000). As reported in Table 7.3, only a few of these transporters have been characterized and shown to be involved in drug resistance in *M. tuberculosis*.

Table 7.3 ABC transporters in *M. tuberculosis*

M. tuberculosis ABC transporters	Function	References
DrrABC	Probable daunorubicin-DIM ABC transporter	Choudhuri *et al.* (2002)
Rv2686c-Rv2687c-Rv2688c	Probable drug efflux ABC transporter	Pasca *et al.* (2004)
Rv0194	Probable drug-efflux ABC transporter	Danilchanka *et al.* (2008)

The first ABC transporter identified in *M. tuberculosis* is encoded by a doxorubicin-resistance operon, *drrABC*, that, when overexpressed in *M. smegmatis*, conferred resistance to a broad range of unrelated compounds, including tetracycline, erythromycin, ethambutol, norfloxacin, streptomycin, chloramphenicol, ethidium bromide, daunorubicin, and doxorubicin (Choudhuri *et al.*, 2002). The resistant phenotype was reversed by verapamil and reserpine, two potent inhibitors of transporters. Doxorubicin uptake in *M. smegmatis* overexpressing *drrAB* was inhibited by reserpine (Choudhuri *et al.*, 2002).

The main physiological role of the Drr proteins of *M. tuberculosis* may be the export of complex lipids to the exterior of the cell. In fact, a *M. tuberculosis* strain with an insertion in the *drrC* gene appeared to be involved in the transport of phthiocerol dimycocerosates (Camacho *et al.*, 2001). A library of signature-tagged transposon mutants of *M. tuberculosis* was constructed and screened for low levels of multiplication in mouse lungs. One of the 16 mutants with attenuated virulence had an insertion in the *drrC* gene, indicating that the DrrC protein is involved in virulence (Camacho *et al.*, 1999).

The *M. tuberculosis Rv2686c-Rv2687c-Rv2688c* operon, encoding an ABC transporter, conferred resistance to ciprofloxacin and, to a lesser extent, to norfloxacin, moxifloxacin, and sparfloxacin, when overexpressed in *M. smegmatis*. The resistance level decreased in the presence of the efflux pump inhibitors reserpine, CCCP, and verapamil (Pasca *et al.*, 2004). Energy-dependent efflux of ciprofloxacin from *M. smegmatis* cells containing the *Rv2686c-Rv2687c-Rv2688c* operon was observed. The addition of reserpine decreased the ciprofloxacin accumulation, showing that the Rv2686c-Rv2687c-Rv2688c proteins actively pump out ciprofloxacin, probably by using ATP hydrolysis as an energy source (Pasca *et al.*, 2004). The overexpression of this operon in *M. bovis* BCG also conferred resistance to fluoroquinolones (M.R. Pasca, unpublished).

Danilchanka and collaborators (2008) generated a library of 7500 transposon mutants in *M. bovis* BCG in order to identify the molecular mechanisms of *M. tuberculosis* resistance to β-lactams. Thirty-three unique insertion sites were determined that conferred medium – or high-level resistance to ampicillin. Insertion of the transposon in front of *bcg0231* (*Rv0194* in *M. tuberculosis*) increased transcription of the gene and concomitantly the resistance of *M. bovis* BCG to ampicillin, streptomycin, vancomycin, novobiocin, tetracycline, erythromycin, and chloramphenicol. *M. smegmatis* cells overexpressing *Rv0194* significantly reduced accumulation of ethidium bromide and conferred multidrug resistance. When reserpine was added to the *Rv0194*-expressing strain, accumulation of ethidium bromide quickly reached levels observed for wild-type *M. smegmatis* (Danilchanka *et al.*, 2008).

Another ABC transporter, the *M. tuberculosis iniA* gene (*Rv0342*) participates in the development of tolerance to both isoniazid and ethambutol through a MDR-pump like mechanism (Colangeli *et al.*, 2005). This gene is strongly induced along with *iniB* and *iniC* by treatment of *M. bovis* BCG or *M. tuberculosis* with isoniazid or ethambutol. *M. tuberculosis iniA* overexpressing cells survived longer than control strains upon exposure to inhibitory concentrations of either isoniazid or ethambutol. A *M. tuberculosis* strain containing an *iniA* deletion showed increased susceptibility to isoniazid. The overexpression of *M. tuberculosis iniA* in *M. bovis* BCG conferred resistance to ethidium bromide and tolerance to isoniazid. The pump inhibitor reserpine reversed both tolerance to isoniazid and resistance to ethidium bromide in *M. bovis* BCG.M. *tuberculosis iniA* knockouts were killed more rapidly by isoniazid and accumulated higher intracellular concentrations of ethidium bromide compared to wild-type controls (Colangeli *et al.*, 2005).

Recently, an ABC transporter, Rv1218c, was demonstrated to be an efflux pump (Balganesh *et al.*, 2010). A Δ*Rv1218c* mutant of *M. tuberculosis* displayed a 4 – to 8-fold increase in the inhibitory and bactericidal potency of different classes of compounds. The MICs were reversed to wild-type values when the full-length *Rv1218c* gene was reintroduced into the Δ*Rv1218c* mutant on a multicopy plasmid. Most of the compounds had significantly better bactericidal activity in the Δ*Rv1218c* mutant than in the wild-type H37Rv, suggesting the involvement of *Rv1218c* gene product in effluxing these compounds out of *M. tuberculosis* cells (Balganesh *et al.*, 2010).

Transcriptional regulation of efflux system

The expression of efflux systems can be subjected to multiple levels of regulation, as reviewed in Chapter 1 of this book. Indeed, a variety of local and global transcriptional regulators and other modulators are involved in this process, which is a very complex one, as described, for example, with the regulation of AcrAB-TolC efflux system of *E. coli* (Li and Nikaido, 2004). In particular, most regulators of the efflux pumps fall into the TetR family of transcriptional repressors (Ramos *et al.*, 2005), but two component systems (Eguchi *et al.*, 2003; Li and Nikaido, 2004), as well as activator proteins belonging to the AraC-XylS family have been described in *E. coli* and *Salmonella typhimurium* (Nishino *et al.*, 2007, 2009).

Although many studies were conducted on the mechanisms of regulation of efflux pumps in laboratory-derived mutants, less is known on the increased expression of these transporters in clinical isolates. However, mutations were reported in local as well as in global regulatory genes or in the promoter region of the transporter gene. Also the presence of insertion elements upstream of the transporter gene were found in overexpressed efflux systems (Piddock, 2006).

The data available on different microorganisms show that multidrug transporters are often expressed under precise and elaborate control at the level of transcription (Grkovic *et al.*, 2002). Examples of both repressors (Lomovskaya *et al.*, 1995; Lucas *et al.*, 1997; Grkovic *et al.*, 1998) and activators (Ahmed *et al.*, 1994; Kohler *et al.*, 1999) of transcription whose genes are adjacent to that for the transporter have been described (Ma *et al.*, 1996).

Overexpression of multidrug resistance pumps, resulting in increased bacterial resistance, is usually due to mutations in these regulatory genes (Hagman *et al.*, 1995; Poole *et al.*, 1996; Grkovic *et al.*, 1998). For these reasons, the study of the regulation of MDR efflux gene expression is an important issue in the field of antibiotic resistance. Furthermore, an increasing number of efflux pump genes has also been found to be controlled by global transcriptional activator proteins (Grkovic *et al.*, 2002) or by two-component regulatory systems (Li and Nikaido, 2004). As only a few examples of efflux pumps regulators have been described in *M. tuberculosis* (see below), here we report the well-known regulation of LfrA efflux pump of *M. smegmatis*.

LfrA is an MFS transporter that confers resistance to ethidium bromide, acriflavine, and some fluoroquinolones when overexpressed from a multicopy plasmid (Takiff *et al.*, 1996). The upstream region of *lfrA* contains a gene coding for a putative TetR family transcriptional repressor, named LfrR (Sander *et al.*, 2000; Li and Nikaido, 2004). In a paper published in 2006, we demonstrated that LfrR is the transcriptional regulator of LfrA (Buroni *et al.*, 2006).

When the sequence upstream of the *lfrRA* transcriptional start site was compared with sequences of other mycobacterial promoter structures, a putative –10 box (TATATT) could be identified. No homologous –35 regions with a canonical distance of 16 to 18 bp have been found; a –35 box (GGGACA) similar to the consensus sequence of *E. coli* was identified, but it was at a 28 bp distance from the –10 box (Buroni *et al.*, 2006). Promoters that lack a canonical –35 sequence but are still functional have been reported in mycobacteria (Bashyam *et al.*, 1996). For mycobacterial promoters, where apparent conservation in the –35 region is absent, many of them possess TGN nucleotides immediately upstream of the −10 region, and thus they are termed 'extended −10 promoters' (Smith *et al.*, 2005). This 'extended −10 promoter' is also present as TGC upstream of the −10 region of the *lfrR* gene.

Since many of the known bacterial mechanisms of drug resistance are inducible by the corresponding drugs (Kieboom *et al.*, 1998), the expression of the *lfrA* efflux pump gene was shown to be induced by ethidium bromide, acriflavine, ciprofloxacin, doxorubicin, and rhodamine 123 (Buroni *et al.*, 2006), previously reported to be substrates of the LfrA efflux pump (Takiff *et al.*, 1996; Sander *et al.*, 2000).

It was also possible to conclude that LfrR acts as a moderator to maintain balanced production of LfrA to meet the physiological needs and facilitate the adaptation of *M. smegmatis* to environmental changes, including antibiotic treatments. It was previously reported that the expression of MDR efflux pumps can be conditionally induced by structurally diverse substrates of these pumps (Ahmed *et al.*, 1994; Ma *et al.*, 1996; Grkovic *et al.*, 1998; Brown *et al.*, 1999; Rosenberg *et al.*, 2003). This induction is due to the direct interaction of the substrates with repressor molecules, which interfere with the binding of repressors to operator DNA and which results in increased levels of expression of MDR pump genes.

Moreover, the crystal structure of the apoprotein form of this repressor was solved by our group (Bellinzoni *et al.*, 2009) and it revealed a structurally asymmetrical homodimer exhibiting local unfolding and a blocked drug-binding site, emphasizing the significant conformational plasticity of the protein necessary for DNA and multidrug recognition. The intrinsic flexibility of this homodimer provides a dynamic mechanism to broaden multidrug binding specificity and may be a general property of transcriptional repressors regulating microbial efflux pump expression (Bellinzoni *et al.*, 2009). In the following paragraphs we report the few examples of efflux pump regulators described in *M. tuberculosis*.

In 2005, Morris and collaborators described the dose-dependent induction of *M. tuberculosis whiB7* by subinhibitory concentrations of three antibiotics, namely erythromycin, tetracycline and streptomycin, and by fatty acids (Morris *et al.*, 2005). *whiB7* encodes a putative transcriptional activator. As antibiotic resistance was observed depending on *whiB7* expression, upregulation of this gene was thought to be required for the induction of other genes that could be responsible for *M. tuberculosis* resistance. Together these observations suggested that *whiB7* encoded a regulator and, in order to determine whether the induction of *whiB7* was correlated with the expression of genes associated with antibiotic resistance, microarray expression profiles of a *whiB7* deletion mutant and a strain overexpressing *whiB7* were compared to *M. tuberculosis* H37Rv wild-type strain. Among the genes significantly regulated by *whiB7*, three could provide intrinsic antibiotic resistance: *tap* (*Rv1258c*), encoding an efflux pump that confers low-level resistance to aminoglycosides and tetracycline (De Rossi *et al.*, 2002), an unstudied ORF encoding a putative macrolide transporter (*Rv1473*) with an ATP-binding cassette, and *erm* (*Rv1988*), homologous to ribosomal methyltransferases and conferring MLS (macrolide, lincosamide, and streptogramin) resistance by modification of 23S rRNA (Doucet-Populaire *et al.*, 2002; Buriankova *et al.*, 2004). The *whiB7* regulon, including antibiotic resistance genes, was shown to be activated also by palmitic acid which is considered to be a major source of carbon for *M. tuberculosis* in the mammalian macrophage (Wheeler, 1994). Therefore, the authors speculated that *whiB7* regulon could be induced when *M. tuberculosis* enters macrophages or other lipid rich cells thereby allowing mycobacteria to resist some chemotherapeutic strategies (Morris *et al.*, 2005).

A second example is given by the regulator of a MmpL efflux transporters (see above). Our group showed that the overexpression of *mmpS5–mmpL5* genes is responsible for resistance to azoles *via* an efflux mechanism (Milano *et al.*, 2009). Furthermore, the upregulation of *mmpS5–mmpL5* (*Rv0677c* and *Rv0676c*) genes was demonstrated to be linked to mutations in *Rv0678* gene, the transcriptional regulator of this efflux system (Milano *et al.*, 2009). *mmpL5* and *mmpS5* belong to the same transcriptional unit, while *Rv0678* is located 84 bp upstream *mmpS5* and is transcribed in the opposite orientation (http://genolist.pasteur.fr/TubercuList/).

Rv0678 shows homologies with transcriptional repressors belonging to the MarR (Multiple Antibiotic Resistance) family (Poole *et al.*, 1996). This family of regulatory proteins is found in bacteria and archaea, and plays important roles in the development of antibiotic resistance as shown in *Pseudomonas aeruginosa*, where the inactivation of MexR, a MarR family member, causes the overexpression of a multidrug efflux system, which is the major determinant for the broad resistance phenotype observed in this opportunistic pathogen (Boshoff *et al.*, 2004). Azole resistant mutants from both *M. tuberculosis* and *M. bovis* BCG revealed the presence of mutations either in the Rv0678 gene or in its putative promoter/operator region. These mutants accumulated a reduced amount of econazole within the cell, suggesting that the mechanism of resistance to azoles is most probably due to an active efflux of the drug because of an overexpression of MmpS5–MmpL5 caused by the lack of an active form of Rv0678. As hypothesized, *Rv0678* gene encodes a transcriptional repressor and its overexpression, as well as that of *mmpL5* and *mmpS5* in the *M. tuberculosis* mutants, was attributed to an inefficient negative auto-regulation, due to a defective copy either of *Rv0678* or of its promoter/operator (Milano *et al.*, 2009).

M. tuberculosis Rv1986 is homologous to *lysE* from *Corynebacterium glutamicum* (Vrljic *et al.*, 1996), and encodes a putative efflux pump. A LysR-type regulator, encoded by *Rv1985c* gene, is adjacent and divergently transcribed from *Rv1986*. The LysR-type transcriptional regulator family is a well-characterized group of transcriptional regulators, highly conserved and ubiquitous among bacteria (Maddocks and Oyston, 2008). They are global transcriptional regulators, acting either as activators or repressors of single or operonic genes (Maddocks and Oyston, 2008). *lysE* expression during growth was monitored by using a fragment containing the promoter region of *lysE* of *M. tuberculosis* plus the upstream *lysR*-type regulator gene fused to the *xylE*. The *lysRE–xylE* fusion demonstrated that *lysE* expression was maximal in early stationary phase, after which, it rapidly declined (Blokpoel *et al.*, 2003).

The *M. bovis P55* gene, located downstream of the gene that encodes the immunogenic lipoprotein P27, has been characterized in 2001 by Silva and collaborators (see above). In the genomes of *M. tuberculosis* and *M. bovis*, the *P55* gene forms an operon with *P27* and both genes are transcribed from the operon promoter (Bigi *et al.*, 2000). As some bacteria have drug-sensor proteins that induce the expression of the corresponding efflux pump, and genes encoding the drug sensor and the efflux pump are located adjacent to each other (Lewis, 1999), the P27 protein could be a sensor of specific signals that would activate the expression of the *P55* gene (Silva *et al.*, 2001).

The regulation of the *M. tuberculosis iniBAC* operon, which encodes an important example of antibiotic-regulated genes, has also been described. This operon has been demonstrated to be specifically induced by isoniazid and ethambutol (Alland *et al.*, 1998). The gene required for transcriptional repression of the *iniBAC* promoter (P_{iniBAC}) was identified by Colangeli and co-workers (2007). They demonstrated that the *lsr2* gene product downregulates transcription of the *M. tuberculosis iniBAC* genes as well as another isoniazid-induced pump, *efpA*. Lsr2 exhibited properties similar to bacterial histone-like proteins, suggesting that it regulates gene expression by controlling chromosomal topology and showing that it has important regulatory functions in mycobacteria. In fact, the presence of *lsr2* on a multicopy plasmid repressed ethambutol-mediated P_{iniBAC} induction below that of the wild-type strain (Colangeli *et al.*, 2007), thus indicating that the *lsr2* gene controls both basal and antibiotic-induced levels of *iniBAC* expression. A deletion *lsr2* mutant was also shown to be more resistant to ethambutol due to de-repression of *iniA*, while complementation restored ethambutol susceptibility. Microarray studies were performed and genes involved in cell wall processes, metabolism, and transport resulted to be upregulated. The results achieved indicated that many of the genes controlled by *lsr2* were not related to ethambutol treatment, suggesting that *lsr2* is involved in the control of a broad range of cellular processes (Colangeli *et al.*, 2007). Moreover, microarray results were consistent with the hypothesis that *lsr2* encodes a protein with relatively non-specific rather than sequence-specific DNA-binding properties that preferentially binds to AT-rich sequences in a manner similar to that of some other histone-like proteins (Lucchini *et al.*, 2006). However, the Lsr2 sequence is unique, exhibiting no significant similarities to any histone-like protein. Thus, Lsr2 represents a novel class of histone-like proteins (Colangeli *et al.*, 2007). As antibiotic tolerance can occur through other mechanisms such as overproduction of various inhibitors, enzymes, and regulatory proteins in non-mycobacterial bacteria (Balaban *et al.*, 2004), Colangeli and co-workers suggested it is possible that a multifunctional *lsr2* also regulates these and other pathways in mycobacteria.

In conclusion, an improved understanding of the role of efflux regulators and of the stress responses associated with these genes may provide important insights into the mechanisms of action of antibiotics and the way that mycobacteria adapt to certain types of stresses such as antibiotic treatment. This knowledge could in turn be used to design more effective antibiotic treatments for both drug-susceptible and drug-resistant *M. tuberculosis*.

Efflux inhibitors

As described in the previous paragraphs, many genes encoding drug efflux pumps have been described in *M. tuberculosis* genome. However, their role in drug resistance has not been fully demonstrated, but many researchers agree that in order to fight the increasing drug resistance problem successfully, a deeper knowledge on these transporters is required. A first step towards this aim is the characterization of molecules able to inhibit efflux pumps activity.

The modulation assay is a method to identify potential efflux pump inhibitors (EPIs). A concentration, normally 4-fold lower than the MIC, is chosen. Serial doubling dilutions of a drug known to be a substrate for an efflux transporter are added and the results are interpreted in the same manner as MIC determinations. Also accumulation studies have been used to identify potential EPIs. An increase in drug accumulation only in the presence of an inhibitor indicates that the inhibitor is a blocker of an efflux mechanism (Kaatz *et al.*, 2003). An ideal efflux pump inhibitor must enhance the activity of substrates of the pump and not of antibiotics that are not efflux substrates. It must not change MICs in strains lacking efflux pumps but increase accumulation and decrease extrusion of efflux pump substrates (Lomovskaya and Watkins, 2001).

Several compounds showing efflux inhibitory activity have been developed and patented: WO200.406.2674 (substituted polyamines as inhibitors of bacterial efflux pumps) (Nelson and Alekshun, 2004); WO200.814.1012 (quaternary alkyl ammonium bacterial efflux pump inhibitors) (Glinka *et al.*, 2008); WO200.911.0002 (novel efflux pump inhibitors) (Koul *et al.*, 2009); and WO201.005.4102 (polybasic bacterial efflux pump inhibitors) (Glinka *et al.*, 2010).

CCCP is used to dissipate the proton motive force and inhibit the efflux of several drugs but it causes cell death and is described as highly cytotoxic and as a substrate of efflux pumps (Krulwich *et al.*, 1990). Reserpine is a plant alkaloid known to inhibit P-glycoprotein in eukaryotic cells (Stavri *et al.*, 2007). Its inhibitory activity was originally demonstrated against the Bmr efflux pump, which mediates tetracycline efflux in *Bacillus subtilis* (Neyfakh *et al.*, 1991). In mycobacteria, reserpine has been used to reduce resistance to isoniazid in *M. tuberculosis* strains (Viveiros *et al.*, 2002). Verapamil has been shown to inhibit P-glycoprotein and also several bacterial ABC efflux pumps, including MmpL7 of *M. tuberculosis* (Pasca *et al.*, 2004). Verapamil also showed activity in *M. smegmatis* and *M. avium* complex (Rodrigues *et al.*, 2009).

Among the compounds active against *M. smegmatis* and *M. tuberculosis* there are also some plant derivatives, such as luteolin, biochanin A, and farnesol (Lechner *et al.*, 2008; Jin *et al.*, 2010). Until now, no natural products have been taken up for further development as much of the data acquired are only preliminary. Biochemical studies to provide evidence of an EPI/protein interaction are required as well as toxicity and small *in vivo* studies to ascertain the pharmacokinetic and toxicity data for potential EPIs.

Phenothiazines are heterotricyclic compounds derived from the dye methylene blue (Ehrlic, 1956). Among them, chlorpromazine has been shown to directly inhibit the *in vitro* growth of *M. tuberculosis* (Viveiros and Amaral, 2001).

However, it has never been considered as an antitubercular compound due to side effects (Amaral *et al.*, 2004). The mechanism of action of phenothiazines has not been completely clarified. It is known that phenothiazines intercalate between nucleic bases of the DNA helix (Rohs and Sklenar, 2004), thus inhibiting all DNA-based processes as well as the degree of coiling and uncoiling of DNA promoted by gyrases (Boshoff *et al.*, 2004). The concentrations of the phenothiazine required for the inhibition of bacterial replication vary greatly depending on the type of bacterium. However, phenothiazines inhibit bacterial Ca^{2+}-dependent enzyme systems, which are needed for generating the proton motive force, which is the energy source used by many efflux pumps (Bhatnagar and Singh, 2003). These compounds also enhance the killing of intracellular bacteria, such as *M. tuberculosis*. This can be ascribed to a concentration effect, possibly due to the inhibition of K^+ and Ca^{2+} transport processes (Wittekindt *et al.*, 2006). In fact, chlorpromazine, the phenothiazine most studied with respect to its antimicrobial properties, inhibits the binding of calcium to calcium-binding proteins (Trimble and Grinstein, 2007), inhibits bacterial kinases (Stowe *et al.*, 1989) and phosphatases (Bullough *et al.*, 1985), and calcium (Molnàr and Pràgai, 1973) and potassium transport (Kristiansen *et al.*, 1982). Mycobacteria are among the less resistant bacterial species to phenothiazines. However, the MICs are well beyond the level that can be achieved in the patient, namely, 0.5 mg/l of plasma (Rivera-Calimlim *et al.*, 1976). It has been shown that phenothiazine induces filamentation and cluster formation both in Gram-negative and Gram-positive bacteria, but the precise mechanism is not known. However, as at high concentrations these agents intercalate between nucleic bases and inhibit gyrase (Csiszar and Molnar, 1992) it may be assumed that filamentation promoted by chlorpromazine involves gyrase and relaxation of supercoiled DNA, as in the case of fluoroquinolones.

A phenothiazine, thioridazine, showed activity against intracellular MDR-TB and XDR-TB isolates and it enhanced the killing of intracellular MDR-TB, as well as showing the ability to cure TB infection in *in vivo* model (Ordway *et al.*, 2003; Martins *et al.*, 2007). This drug is a neuroleptic which has been used for the therapy of psychosis. It has a low price, good availability and a generally good tolerability in humans. Thioridazine inhibits efflux pumps of both prokaryotes and eukaryotes (Amaral *et al.*, 2007), and among the pumps affected are those responsible for the transport of K^+ and Ca^{2+} ions which are essential for the acidification of the phagolysosome that results in the activation of lysosome hydrolases (Reeves *et al.*, 2002). This is an alternative approach which targets the macrophage that contains the intracellular mycobacterium, thus by-passing the mutational response expected with the targeting of the bacterium itself. Until now, thioridazine has been used only for the therapy of patients who did not respond to other antibiotic therapy and whose conditions were critical (Amaral *et al.*, 2010). Thioridazine can be used safely and its use does not produce any side effects that differ from those produced by other neuroleptics (Salih *et al.*, 2007). However, because approximately 6% of Eastern Europeans have a mutation in the P450 cytochrome (Berecz *et al.*, 2004), the metabolism of thioridazine will be slower, with the possibility of the heart slowing down to a level that cannot sustain life (Sawicke and Sturla, 2008). Recently, thioridazine was administered to 12 XDR-TB patients who complied with the therapy, with a successful treatment in 10 of them (Abbate *et al.*, 2007). However, further investigation is required to develop other and improved efflux inhibitors, more effective than thioridazine, in order to prevent the efflux-mediated resistance in multidrug resistant strains.

In 2010, Sharma and colleagues described the use of piperine in combination with rifampicin to treat *M. tuberculosis* infections (Sharma *et al.*, 2010). In a previous work they demonstrated piperine as a *S. aureus* NorA efflux pump inhibitor (Khan *et al.*, 2006; Kumar *et al.*, 2008). Piperine is the trans-trans isomer of 1-piperoyl-piperidine and it is isolated from black pepper (Atal *et al.*, 1985; Singh *et al.*, 1986). Rifampicin has been shown to upregulate the expression of the efflux pump Rv1258c in clinical isolates of *M. tuberculosis* (Jiang *et al.*, 2008). For the first time, Sharma and collaborators described the potentiating effect of piperine on the bioefficacy of rifampicin and its role in putative efflux pump Rv1258c inhibition. First of all, the MIC of rifampicin was evaluated and it was reduced by 4 – to 8-fold in the presence of piperine. This reduction in the MIC was more prominent for rifampicin resistant *M. tuberculosis* as compared with *M. tuberculosis* H37Rv wild-type strain. However, piperine on its own did not show any antibacterial activity when tested up to 100 mg/l. Moreover, rifampicin alone was bactericidal at 1 mg/l, while 0.25 and 0.5 mg/l concentrations of rifampicin in combination with 25 mg/l piperine were sufficient for bactericidal activity (Sharma *et al.*, 2010). Accumulation and efflux of ethidium bromide are good indicators of the involvement of efflux pumps in the resistance mechanism. In this way, another important piece of evidence was that the increases in MICs of rifampicin and ethidium bromide of resistant strains was reversed by piperine and reserpine. Since efflux is the only known mechanism for ethidium bromide resistance, the reversal of its MIC by piperine indicated its role as an efflux pump inhibitor (Sharma *et al.*, 2010). Moreover, *Rv1258c* gene in *M. tuberculosis* resistant strain was significantly induced in the presence of rifampicin, whereas the expression was not significant in *M. tuberculosis* H37Rv: this is a typical efflux pump feature. Lastly, owing to non-availability of the crystal structure of Rv1258c, the 3D structure of this protein was predicted on the basis of homology modelling. The predicted structure of Rv1258c revealed five probable binding pockets. The docking studies of reserpine, a known inhibitor, and piperine on these binding pockets revealed one of the binding sites showed better binding affinity for these inhibitors. In particular, better binding affinity was observed for piperine than for reserpine (Sharma *et al.*, 2010). In conclusion, inhibition of the rifampicin efflux pump Rv1258c by an efflux pump inhibitor can be useful in increasing the efficacy of rifampicin against *M. tuberculosis* isolates. However, the effective concentration of piperine described by Sharma and co-authors may not be clinically achievable, but this study paves the way for the exploration of more potent inhibitors of Rv1258c, which may result in combinations (drug and an efflux pump inhibitor) with high clinical utility.

In summary, various compounds have been used to inhibit efflux activity *in vitro*, but none of them can be used in clinical practice for this purpose. There is a concern regarding the adverse effects and selectivity of these compounds, since some of them have been shown to inhibit both eukaryotic and bacterial efflux systems. The search for safe and effective efflux inhibitors that could be used as 'helper compounds' administered in combination with conventional antibiotics to which the organism was initially resistant has proven to be a challenge.

Efflux pumps overexpressed in *M. tuberculosis* clinical isolates

Bacterial resistance may be an intrinsic (natural) feature of an organism, or may result from spontaneous mutations or the acquisition of exogenous resistance genes. As *M. tuberculosis* does not possess plasmids and horizontal gene transfer is thought to be rare, all resistances appear to emerge through mutations in chromosomal genes (Gillespie, 2007). The *M. tuberculosis* clinical isolates resistant to antibiotics generally present mutations in the genes coding for the drug target or enzymes responsible for drug inactivation. For example, the resistance to fluoroquinolones in 42–85% of clinical isolates of *M. tuberculosis* has been shown to be associated with mutations in a 120 bp region in the *gyrA* gene encoding DNA gyrase, the fluoroquinolone target (Wade and Zhang, 2004). Active drug efflux could account for the fluoroquinolone resistance in the remaining isolates.

The first *M. tuberculosis* drug-resistant clinical isolate with an overexpressed efflux pump was identified by Siddiqi and collaborators in 2004. This isolate was resistant to the following drugs: rifampicin, isoniazid, ofloxacin, and minomycin. In the presence of rifampicin and ofloxacin, this isolate showed increased transcription of the gene *Rv1258c* coding for Tap pump (Aìnsa *et al.*, 1998). In particular, the RT-PCR analysis revealed a tenfold increase in the *Rv1258c* transcript level in the presence of rifampicin, and a six-fold increase in the presence of ofloxacin (Siddiqi *et al.*, 2004).

In another paper the researchers found overexpression of some efflux pumps in a MDR clinical isolate (Jiang *et al.*, 2008). The clinical isolate No.

1499 showed a complete deletion of *katG*, which can cause isoniazid resistance, and a mutation at codon 531 (TCG→TTG) in the *rpo* gene responsible for rifampicin resistance. By Real-time PCR the authors demonstrated that the expression of *Rv1258c* (Tap) (Aìnsa *et al.*, 1998) and *Rv1410c* (P55) (Silva *et al.*, 2001) in MDR isolate No. 1499 was significantly increased in the presence of isoniazid or rifampicin. Overexpression of both Rv1258c and Rv1410c was observed as shown by a 3-fold increase in the presence of rifampicin and a five – to six-fold increase in the presence of isoniazid, respectively (Jiang *et al.*, 2008).

In order to understand the contribution of efflux pumps to drug resistance in *M. tuberculosis*, microarray analysis validated by Real-time PCR was performed in five multidrug-resistant isolates of *M. tuberculosis* (resistant to rifampicin, isoniazid, ethambutol, streptomycin, and ofloxacin) (Gupta *et al.*, 2010a). The authors constructed a custom DNA microarray containing 25 drug efflux pump genes of *M. tuberculosis* and monitored changes in the expression of these genes on exposure of common antitubercular drugs. Ten efflux pump genes were found to be overexpressed in the presence of common antitubercular drugs, but none of the studied efflux pump genes were found to be suppressed. Whereas isoniazid stress induced a total of five genes, *Rv2459* (Gupta *et al.*, 2010b), *Rv1819c*, *Rv3728* (De Rossi *et al.*, 2002), *Rv3065*, and *Rv2846* (De Rossi *et al.*, 2002; Li *et al.*, 2004), rifampicin stress resulted in a low-level induction of two of these five genes, *Rv1819c* and *Rv3728*. Streptomycin stress induced a set of three genes, *Rv2688* (Pasca *et al.*, 2004), *Rv2938*, and *Rv2994* (De Rossi *et al.*, 2002), and ethambutol stress induced *Rv2938*, *Rv2459*, *Rv3728*, and *Rv3065*. Ofloxacin stress altered the expression of two efflux pump genes, *Rv2477* and *Rv2209* (Gupta *et al.*, 2010a).

Until now, there are few evidences of the contribution of efflux pumps to drug resistance in *M. tuberculosis* clinical isolates. It could be very interesting to test the overexpression of efflux pumps in the *M. tuberculosis* multidrug resistant clinical isolates without mutations in genes coding for drug targets.

Concluding remarks

Recent research into efflux mechanisms in mycobacteria, using laboratory strains, has provided promising insights, but the relevance of the efflux mechanism to the resistance of *M. tuberculosis* clinical strains is ongoing. New experimental approaches are needed to assess the involvement of efflux pumps in intrinsic/acquired drug resistance in mycobacteria (De Rossi *et al.*, 2006). For example, in *P. aeruginosa* the overproduction of efflux pumps is responsible for drug resistance in 14–75% of clinical isolates (Mesaros *et al.*, 2007). These findings have stimulated research on efflux pump inhibitors (Coban *et al.*, 2004). The inhibition of efflux pumps significantly decreases the level of intrinsic resistance to a similar extent to gene inactivation. However, no inhibitors that are not toxic to eukaryotic cells have yet been identified. Recent studies regarding mycobacteria, such as that by Siddiqi *et al.* (2004), are therefore paving the way for studies of efflux pumps in clinical isolates of *M. tuberculosis*. However, although several mycobacterial pumps have been characterized, the clinical consequences of efflux-mediated resistance are mostly unknown because of variable levels of expression and the lack of specific markers for use in laboratory practice. The question of the importance of the role played by efflux pumps in intrinsic and acquired antibiotic resistance in mycobacteria remains unresolved and consequently other research in this direction have to be performed.

Acknowledgements

The authors acknowledge EC-VII Framework, Contract no. 260872, for funding.

References

Abbate, E., Vescovo, M., Natiello, M., Cufre, M., Garcia, A., Ambroggi, M., Poggi, S., Sìmboli, N., and Ritacco, V. (2007). Tuberculosis extensamente resistente (XDR TB) en Argentina: aspectos destacables epidemiologicos, bacteriologicos, terapeuticos, y evolutivos. Revista Argentina de Medicina Respiratoria *1*, 19–25.

Ahmed, M., Borsch, M., Taylor, S.S., Vazquez-Laslop, N., and Neyfakh, A.A. (1994). A protein that activates expression of a multidrug efflux transporter upon binding the transporter substrates. J. Biol. Chem. *269*, 28506–28513.

Aínsa, J.A., Blokpoel, M.C., Otal, I., Young, D.B., De Smet, K.A., and Martín, C. (1998). Molecular cloning and characterization of Tap, a putative multidrug efflux pump present in *Mycobacterium fortuitum* and *Mycobacterium tuberculosis*. J. Bacteriol. *180*, 5836–5843.

Alland, D., Kramnik, I., Weisbrod, T.R., Otsubo, L., Cerny, R., Miller, L.P., Jacobs, Jr, W.R. , and Bloom, B.R. (1998). Identification of differentially expressed mRNA in prokaryotic organisms by customized amplification libraries (DECAL): the effect of isoniazid on gene expression in *Mycobacterium tuberculosis*. Proc. Natl. Acad. Sci. U.S.A. *95*, 13227–13232.

Amaral, L., Viveiros, M., and Molnar, J. (2004). Antimicrobial activity of phenothiazines. In Vivo *18*, 725–732.

Amaral, L., Engi, H., Viveiros, M., and Molnar, J. (2007). Comparison of multidrug resistant efflux pumps of cancer and bacterial cells with respect to the same inhibitory agents. In Vivo *21*, 237–244.

Amaral, L., Boeree, M.J., Gillespie, S.H., Udwadiad, Z.F., and van Soolingen, D. (2010). Thioridazine cures extensively drug-resistant tuberculosis (XDR-TB) and the need for global trials is now! Int. J. Antimicrob. Agents. *35*, 524–526.

Atal, C.K., Dubey, R.K., and Singh, J. (1985). Biochemical basis of enhanced drug bioavailability by piperine: evidence that piperine is a potent inhibitor of drug metabolism. J. Pharmacol. Exp. Ther. *232*, 258–262.

Balaban, N.Q., Merrin, J., Chait, R., Kowalik, L., and Leibler, S. (2004). Bacterial persistence as a phenotypic switch. Science *305*, 1622–1625.

Balganesh, M., Kuruppath, S., Marcel, N., Sharma, S., Nair, A., and Sharma, U. (2010). Rv1218c, an ABC transporter of *Mycobacterium tuberculosis* with implications in drug discovery. Antimicrob. Agents Chemother. *54*, 5167–5172.

Bashyam, M.D., Kaushal, D., Das Gupta, S.K., and Tyagi, A.K. (1996). A study of mycobacterial transcriptional apparatus: identification of novel features in promoter elements. J. Bacteriol. *178*, 4847–4853.

Bellinzoni, M., Buroni, S., Schaeffer, F., Riccardi, G., De Rossi, E., and Alzari, P.M. (2009). Structural plasticity and distinct drug-binding modes of LfrR, a mycobacterial efflux pump regulator. J. Bacteriol. *191*, 7531–7537.

Berecz, R., Dorado, P., De La Rubia, A., Càceres, M.C., Degrell, I., and Lerena, L., A. (2004). The role of cytochrome P450 enzymes in the metabolism of risperidone and its clinical relevance for drug interactions. Curr. Drug Targets. *5*, 573–579.

Bhatnagar, K., and Singh, V.P. (2003). Ca^{2+}-dependence and inhibition of transformation by trifluoperazine and chlorpromazine in *Thermoactinomyces vulgaris*. Curr. Microbiol. *46*, 265–269.

Bigi, F., Alito, A., Romano, M.I., Zumarraga, M., Caimi, K., and Cataldi, A. (2000). The gene encoding P27 lipoprotein and a putative antibiotic-resistance gene form an operon in *Mycobacterium tuberculosis* and *Mycobacterium bovis*. Microbiology *146*, 1011–1018.

Blokpoel, M.C., O'Toole, R., Smeulders, M.J., and Williams, H.D. (2003). Development and application of unstable GFP variants to kinetic studies of mycobacterial gene expression. J. Microbiol. Methods. *54*, 203–211.

Boshoff, H.,I., Myers, T.G., Copp, B.R., McNeil, M.R., Wilson, M.A., and Barry, 3rd, C.E. (2004). The transcriptional responses of *Mycobacterium tuberculosis* to inhibitors of metabolism: novel insights into drug mechanisms of action. J. Biol. Chem. *279*, 40174–40184.

Braibant, M., Gilot, P., and Content, J. (2000). The ATP binding cassette (ABC) transport systems of *Mycobacterium tuberculosis*. FEMS Microbiol. Rev. *24*, 449–467.

Brennan, P.J. (2003). Structure, function, and biogenesis of the cell wall of *Mycobacterium tuberculosis*. Tuberculosis. *83*, 91–97.

Brown, M.H., and Skurray, R.A. (2001). Staphylococcal multidrug efflux protein QacA. J. Mol. Microbiol. Biotechnol. *3*, 163–170.

Brown, M.H., Paulsen, T., and Skurray, R.A. (1999). The multidrug efflux protein NorM is a prototype of a new family of transporters. Mol. Microbiol. *31*, 394–395.

Bullough, D.A., Kwan, M., Laikind, P.K., Yoshida, M., and Allison, W.S. (1985). The varied responses of different F1-ATPases to chlorpromazine. Arch. Biochem. Biophys. *236*, 567–575.

Buriankova, K., Doucet-Populaire, F., Dorson, O., Gondran, A., Ghnassia, J.C., Weiser, J., and Pernodet, J.L. (2004). Molecular basis of intrinsic macrolide resistance in the *Mycobacterium tuberculosis* complex. Antimicrob. Agents Chemother. *48*, 143–150.

Buroni, S., Manina, G., Guglierame, P., Pasca, M.R., Riccardi, G., and De Rossi, E. (2006). LfrR is a repressor that regulates expression of the efflux pump LfrA in *Mycobacterium smegmatis*. Antimicrob. Agents Chemother. *50*, 4044–4052.

Camacho, L.R., Ensergueix, D., Perez, E., Gicquel, B., and Guilhot, C. (1999). Identification of a virulence gene cluster of *Mycobacterium tuberculosis* by signature-tagged transposon mutagenesis. Mol. Microbiol. *34*, 257–267.

Camacho, L.R., Constant, P., Raynaud, C., Laneelle, M.A., Triccas, J.A., Gicquel, B., Daffé, M., and Guilhot, C. (2001). Analysis of the phthiocerol dimycocerosate locus of *Mycobacterium tuberculosis*. Evidence that this lipid is involved in the cell wall permeability barrier. J. Biol. Chem. *276*, 19845–19854.

Caminero, J.A., Sotgiu, G., Zumla, A., and Migliori, G.B. (2010). Best drug treatment for multidrug-resistant and extensively drug-resistant tuberculosis. Lancet Infect. Dis. *10*, 621–629.

Choudhuri, B.S., Bhakta, S., Barik, R., Basu, J., Kundu, M., and Chakrabarti, P. (2002). Overexpression and functional characterization of an ABC (ATP-binding cassette) transporter encoded by the genes *drrA* and *drrB* of *Mycobacterium tuberculosis*. Biochem. J. *367*, 279–285.

Coban, A.Y., Ekinci, B., and Durupinar, B. (2004). A multidrug efflux pump inhibitor reduces fluoroquinolone resistance in *Pseudomonas aeruginosa* isolates. Chemotherapy *50*, 22–26.

Colangeli, R., Helb, D., Sridharan, S., Sun, J., Varma-Basil, M., Hazbón, M.H., Harbacheuski, R., Megjugorac, N.J., Jacobs, Jr, W.R., Holzenburg, A., *et al.* (2005). The *Mycobacterium tuberculosis iniA* gene is essential for activity of an efflux pump that confers drug tolerance to both isoniazid and ethambutol. Mol. Microbiol. *55*, 1829–1840.

Colangeli, R., Helb, D., Vilchèze, C., Hazbón, M.H., Lee, C.G., Safi, H., Sayers, B., Sardone, I., Jones, M.B., Fleischmann, R.D., *et al.* (2007). Transcriptional regulation of multi-drug tolerance and antibiotic-induced responses by the histone-like protein Lsr2 in *M. tuberculosis*. PLoS Pathog. *3*, e87.

Cole, S.T., Brosch, R., Parkhill, J., Garnier, T., Churcher, C., Harris, D., Gordon, S.V., Eiglmeier, K., Gas, S., Barry, C.E. 3rd, *et al.* (1998). Deciphering the biology of *Mycobacterium tuberculosis* from the complete genome sequence. Nature *393*, 537–544.

Converse, S.E., Mougous, J.D., Leavell, M.D., Leary, J.A., Bertozzi, C.R., and Cox, J.S. (2003). MmpL8 is required for sulfolipid-1 biosynthesis and *Mycobacterium tuberculosis* virulence. Proc. Natl. Acad. Sci. U.S.A. *100*, 6121–6126.

Csiszar, K., and Molnar, J. (1992). Mechanism of action of tricyclic drugs on *Escherichia coli* and *Yersinia enterocolitica* plasmid maintenance and replication. Anticancer. Res. *12*, 2267–2272.

Danilchanka, O., Mailaender, C., and Niederweis, M. (2008). Identification of a novel multidrug efflux pump of *Mycobacterium tuberculosis*. Antimicrob. Agents Chemother. *52*, 2503–2511.

De Rossi, E., Blokpoel, M.C., Cantoni, R., Branzoni, M., Riccardi, G., Young, D.B., De Smet, K.A., and Ciferri, O. (1998a). Molecular cloning and functional analysis of a novel tetracycline resistance determinant, *tet(V)*, from *Mycobacterium smegmatis*. Antimicrob. Agents Chemother. *42*, 1931–1937.

De Rossi, E., Branzoni, M., Cantoni, R., Milano, A., Riccardi, G., and Ciferri, O. (1998b). *mmr*, a *Mycobacterium tuberculosis* gene conferring resistance to small cationic dyes and inhibitors. J. Bacteriol. *180*, 6068–6071.

De Rossi, E., Arrigo, P., Bellinzoni, M., Silva, P.A., Martin, C., Ainsa, J.A., Guglierame, P., and Riccardi, G. (2002). The multidrug transporters belonging to major facilitator superfamily in *Mycobacterium tuberculosis*. Mol. Med. *8*, 714–724.

De Rossi, E., Aínsa, J.A., and Riccardi, G. (2006). Role of mycobacterial efflux transporters in drug resistance: an unresolved question. FEMS Microbiol. Rev. *30*, 36–52.

Domenech, P., Reed, M.B., Dowd, C.S., Manca, C., Kaplan, G., Barry, C.E. 3rd. (2004). The role of MmpL8 in sulfatide biogenesis and virulence of *Mycobacterium tuberculosis*. J. Biol. Chem. *279*, 21257–21265.

Domenech, P., Reed, M.B., and Barry, C.E. 3rd. (2005). Contribution of the *Mycobacterium tuberculosis* MmpL protein family to virulence and drug resistance. Infect. Immun. *73*, 3492–3501.

Doucet-Populaire, F., Buriánková, K., Weiser, J., and Pernodet, J.L. (2002). Natural and acquired macrolide resistance in mycobacteria. Curr. Drug Targets Infect. Disord. *2*, 355–370.

Eguchi, Y., Oshima, T., Mori, H., Aono, R., Yamamoto, K., Ishihama, A., and Utsumi, R. (2003). Transcriptional regulation of drug efflux genes by EvgAS, a two-component system in *Escherichia coli*. Microbiology *149*, 2819–2828.

El-Sadr, W.M., and Tsiouris, S.J. (2008). HIV-associated tuberculosis: diagnostic and treatment challenges. Semin. Respir. Crit. Care Med. *29*, 525–531.

Ehrlich, P. (1956). The Collected Papers of Paul Ehrlich, Himmelweit, F., Marquardi, M., and Dale, H., eds. (Pergammon Press, London).

Gillespie, S.H. (2007). Tuberculosis: evolution in millennia and minutes. Biochem. Soc. Trans. *35*, 1317–1320.

Glinka, T., Rodny, O., Bostian, K., and Wallace, D.M. (2008). Quaternary alkyl ammonium bacterial efflux pump inhibitors and therapeutic uses thereof. WO200.814.1012.

Glinka, T., Rodny, O., Bostian, K.A., Wallace, D.M., Higuchi, R., Chow, C., Mak, C.C., Hirst, G., and Eastman, B. (2010). Polybasic bacterial efflux pump inhibitors and therapeutic uses thereof. WO201.005.4102.

Goldman, R.C., Plumley, K.V., and Laughon, B.E. (2007). The evolution of extensively drug resistant tuberculosis (XDR-TB): history, status and issues for global control. Infect. Disord. Drug Targets *7*, 73–91.

Grkovic, S., Brown, M.H., Roberts, N.J., Paulsen, I.T., and Skurray, R.A. (1998). QacR is a repressor protein that regulates expression of the *Staphylococcus aureus* multidrug efflux pump QacA. J. Biol. Chem. *273*, 18665–18673.

Grkovic, S., Brown, M.H., and Skurray, R.A. (2002). Regulation of bacterial drug export systems. Microbiol. Mol. Biol. Rev. *66*, 671–701.

Gupta, A.K., Katoch, V.M., Chauhan, D.S., Sharma, R., Singh, M., Venkatesan, K., and Sharma, V.D. (2010a). Microarray analysis of efflux pump genes in multidrug-resistant *Mycobacterium tuberculosis* during stress induced by common anti-tuberculous drugs. Microb. Drug Resist. *16*, 21–28.

Gupta, A.K., Reddy, V.P., Lavania, M., Chauhan, D.S., Venkatesan, K., Sharma, V.D., Tyagi, A.K., and Katoch, V.M. (2010b). *jefA* (*Rv2459*), a drug efflux gene in *Mycobacterium tuberculosis* confers resistance to isoniazid and ethambutol. Indian J. Med. Res. *132*, 176–188.

Hagman, K.E., Pan, W., Spratt, B.G., Balthazar, J.T., Judd, R.C., and Shafer, W.M. (1995). Resistance of *Neisseria gonorrhoeae* to antimicrobial hydrophobic agents is modulated by the *mtrRCDE* efflux system. Microbiology *141*, 611–622.

Jarlier, V., and Nikaido, H. (1994). Mycobacterial cell wall: structure and role in natural resistance to antibiotics. FEMS Microbiol. Lett. *123*, 11–18.

Jassal, M., and Bishai, W.R. (2009). Extensively drug-resistant tuberculosis. Lancet Infect. Dis. *9*, 19–30.

Jiang, X., Zhang, W., Zhang, Y., Gao, F., Lu, C., Zhang, X., and Wang, H. (2008). Assessment of efflux pump gene expression in a clinical isolate *Mycobacterium tuberculosis* by real-time reverse transcription PCR. Microb. Drug Resist. *14*, 7–11.

Jin, J., Zhang, J.Y., Guo, N., Sheng, H., Li, L., Liang, J.C., Wang, X.L., Li, Y., Liu, M.Y., Wu, X.P., *et al.* (2010). Farnesol, a potential efflux pump inhibitor in *Mycobacterium smegmatis*. Molecules *15*, 7750–7762.

Kaatz, G.W., Moudgal, V.V., Seo, Hansen, S.M. , J.B., and Kristiansen, J.E. (2003). Phenylpiperidine selective serotonin reuptake inhibitors interfere with multidrug efflux pump activity in *Staphylococcus aureus*. Int. J. Antimicrob. Agents. *22*, 254–261.

Kerr, I.D. (2002). Structure and association of ATP-binding cassette transporter nucleotide-binding domains. Biochim. Biophys. Acta *156*, 47–64.

Khan, I.A., Mirza, Z.M., Kumar, A., Verma, V., and Qazi, G.N. (2006). Piperine, a phytochemical potentiator of ciprofloxacin against *Staphylococcus aureus*. Antimicrob. Agents Chemother. *50*, 810–812.

Kieboom, J., Dennis, J.J., Zylstra, G.J., and De Bont, J.A.M. (1998). Active efflux of organic solvents by *Pseudomonas putida* S12 is induced by solvents. J. Bacteriol. *180*, 6769–6772.

Köhler, T., Epp, S.F., Curty, L.K., and Pechère, J.-C. (1999). Characterization of MexT, the regulator of the MexE-MexF-OprN multidrug efflux system of *Pseudomonas aeruginosa*. J. Bacteriol. *181*, 6300–6305.

Koul, A., Arnoult, E., Lounis, N., Guillemont, J., and Andries, K. (2011). The challenge of new drug discovery for tuberculosis. Nature *469*, 483–490.

Koul, S., Thota, N., Mallepally, V.R., Sangwan, P.L., Taneja, S.C., Khan, I.A., Kumar, A., Raja, A.F., Saxena, A.K., Agrawal, S.K., *et al.* (2009). Novel efflux pump inhibitors. WO200.911.0002.

Kristiansen, J.E., Mortensen, I., and Nissen, B. (1982). Membrane stabilizers inhibit potassium efflux from *Staphylococcus aureus* strain No. U2275. Biochim. Biophys. Acta *685*, 379–382.

Krulwich, T.A., Quirk, P.G., and Guffanti, A.A. (1990). Uncoupler-resistant mutants of bacteria. Microbiol. Rev. *54*, 52–65.

Kumar, A., Khan, I.A., Koul, S., Koul, J.L., Taneja, S.C., Ali, I., Ali, F., Sharma, S., Mirza, Z.M., Kumar, M., *et al.* (2008). Novel structural analogues of piperine as inhibitors of the NorA efflux pump of *Staphylococcus aureus*. J. Antimicrob. Chemother. *16*, 1270–1276.

Lamichhane, G., Tyagi, S., and Bishai, W.R. (2005). Designer arrays for defined mutant analysis to detect genes essential for survival of *Mycobacterium tuberculosis* in mouse lungs. Infect. Immun. 73, 2533–2540.

Layre, E., Cala-De Paepe, D., Larrouy-Maumus, G., Vaubourgeix, J., Mundayoor, S., Lindner, B., Puzo, G., and Gilleron, M. (2011). Deciphering sulfoglycolipids of *Mycobacterium tuberculosis*. J. Lipid. Res. *52*, 1098–1110.

Lechner, D., Gibbons, S., and Bucar, F. (2008). Plant phenolic compounds as ethidium bromide efflux inhibitors in *Mycobacterium smegmatis*. J. Antimicrob. Chemother. *62*, 345–348.

Lewis, K. (1999). Multidrug resistance: versatile drug sensors of bacterial cells. Curr. Biol. *9*, R403–R407.

Li, X.Z., and Nikaido, H. (2004). Efflux-mediated drug resistance in bacteria. Drugs *64*, 159–204.

Li, X.Z., Zhang, L., and Nikaido, H. (2004). Efflux pump-mediated intrinsic drug resistance in *Mycobacterium smegmatis*. Antimicrob. Agents Chemother. *48*, 2415–2423.

Littlejohn, T.G., Paulsen, I.T., Gillespie, M.T., Tennet, J.M., Midgley, M., Jones, I.G., Purewal, A.S., and Skurray, R.A. (1992). Substrate specificity and energetics of antiseptic and disinfectant resistance in *Staphylococcus aureus*. FEMS Microbiol. Lett. *95*, 259–266.

LoBue, P. (2009). Extensively drug-resistant tuberculosis. Curr. Opin. Infect. Dis. *22*, 167–173.

Lomovskaya, O., and Watkins, W. (2001). Inhibition of efflux pumps as a novel approach to combat drug resistance in bacteria. J. Mol. Microbiol. Biotechnol. *3*, 225–236.

Lomovskaya, O., Lewis, K., and Matin, A. (1995). EmrR is a negative regulator of the *Escherichia coli* multidrug resistance pump EmrAB. J. Bacteriol. *177*, 2328–2334.

Lucas, C.E., Balthazar, J.T., Hagman, K.E., and Shafer, W.M. (1997). The MtrR repressor binds the DNA sequence between the *mtrR* and *mtrC* genes of *Neisseria gonorrhoeae*. J. Bacteriol. *179*, 4123–4128.

Lucchini, S., Rowley, G., Goldberg, M.D., Hurd, D., Harrison, M., and Hinton, J.C. (2006). HNS mediates the silencing of laterally acquired genes in bacteria. PLoS Pathog. *2*, e81.

Ma, D., Alberti, M., Lynch, C., Nikaido, H., and Hearst, J.E. (1996). The local repressor AcrR plays a modulating role in the regulation of acrAB genes of *Escherichia coli* by global stress signals. Mol. Microbiol. *19*, 101–112.

Maddocks, S.E., and Oyston, P.C. (2008). Structure and function of the LysR-type transcriptional regulator (LTTR) family proteins. Microbiology *154*, 3609–3623.

Martins, M., Viveiros, M., Kristiansen, J.E., Molnar, J., and Amaral, L. (2007). The curative activity of thioridazine on mice infected with *Mycobacterium tuberculosis*. *In vivo* *21*, 771–775.

Mesaros, N., Nordmann, P., Plésiat, P., Roussel-Delvallez, M., Van Eldere, J., Glupczynski, Y., Van Laethem, Y., Jacobs, F., Lebecque, P., Malfroot, A., *et al.* (2007). *Pseudomonas aeruginosa*: resistance and therapeutic options at the turn of the new millennium. Clin. Microbiol. Infect. *13*, 560–578.

Milano, A., Pasca, M.R., Provvedi, R., Lucarelli, A.P., Manina, G., de Jesus Lopes Ribeiro, A.L., Manganelli, R., and Riccardi, G. (2009). Azole resistance in *Mycobacterium tuberculosis* is mediated by the MmpS5–MmpL5 efflux system. Tuberculosis (Edinb.) *89*, 84–90.

Molnàr, J., and Pràgai, B. (1973). Repression of alkaline phosphatise synthesis by adenine, adenosine and adenine nucleotides in an adenine auxotroph of *Bacillus anthracis*. Acta Microbiol. Acad. Sci. Hung. *20*, 171–181.

Morris, R.P., Nguyen, L., Gatfield, J., Visconti, K., Nguyen, K., Schnappinger, D., Ehrt, S., Liu, Y., Heifets, L., Pieters, J., Schoolnik, G., and Thompson, C.J. (2005). Ancestral antibiotic reistance in *Mycobacterium tuberculosis*. Proc. Natl. Acad. Sci. U.S.A. *102*, 12200–12205.

Murakami, S., Nakashima, R., Yamashita, E., and Yamaguchi, A. (2002). Crystal structure of bacterial multidrug efflux transporter AcrB. Nature *419*, 587–593.

Nelson, M.L., and Alekshun, M.N. (2004). Substituted polyamines as inhibitors of bacterial efflux pumps. WO200.406.2674.

Neyfakh, A.A., Bidnenko, V.E., and Chen, L.B. (1991). Efflux-mediated multi-drug resistance in *Bacillus subtilis*: similarities and dissimilarities with the mammalian system. Proc. Natl. Acad. Sci. U.S.A. *88*, 4781–4785.

Niederweis, M. (2003). Mycobacterial porins – new channel proteins in unique outer membranes. Mol. Microbiol. *49*, 1167–1177.

Nikaido, H., and Zgurskaya, H.I. (2001). AcrAB and related multidrug efflux pumps of *Escherichia coli*. J. Mol. Microbiol. Biotechnol. 3, 215–218.

Ninio, S., Rotem, D., and Schuldiner, S. (2001). Functional analysis of novel multidrug transporters from human pathogens. J. Biol. Chem. *276*, 48250–48256.

Nishino, K., Nikaido, E., and Yamaguchi, A. (2007). Regulation of multidrug efflux systems involved in multidrug and metal resistance of *Salmonella enterica* serovar Typhimurium. J. Bacteriol. *189*, 9066–9075.

Nishino, K., Senda, Y., Hayashi-Nishino, M., and Yamaguchi, A. (2009). Role of the AraC-XylS family regulator YdeO in multi-drug resistance of *Escherichia coli*. Antibiot, J. (Tokyo) *62*, 251–257.

Noguchi, N., Hase, M., Kitta, M., Sasatsu, M., Deguchi, K., and Kono, M. (1999). Antiseptic susceptibility and distribution of antiseptic resistance genes in methicillin-resistant *Staphylococcus aureus*. FEMS Microbiol. Lett. *172*, 247–253.

Ordway, D., Viveiros, M., Leandro, C., and Amaral, L. (2003). Clinical concentrations of thioridazine kill intracellular multi-drug resistant *Mycobacterium tuberculosis*. Antimicrob. Agents Chemother. *47*, 917–922.

Pao, S.S., Paulsen, I.T., and Saier, Jr, M.H. (1998). Major facilitator superfamily. Microbiol. Mol. Biol. Rev. *62*, 1–34.

Pasca, M.R., Guglierame, P., Arcesi, F., Bellinzoni, M., De Rossi, E., and Riccardi, G. (2004). Rv2686c-Rv2687c-Rv2688c, an ABC fluoroquinolone efflux pump in *Mycobacterium tuberculosis*. Antimicrob. Agents Chemother. *48*, 3175–3178.

Pasca, M.R., Guglierame, P., De Rossi, E., Zara, F., and Riccardi, G. (2005). *mmpL7* gene of *Mycobacterium tuberculosis* is responsible for isoniazid efflux in *Mycobacterium smegmatis*. Antimicrob. Agents Chemother. *49*, 4775–4777.

Piddock, L.J. (2006). Clinically relevant chromosomally encoded multidrug resistance efflux pumps in bacteria. Clin. Microbiol. Rev. *19*, 382–402.

Poole, K., Tetro, K., Zhao, Q., Neshat, S., Heinrichs, D.E., and Bianco, N. (1996). Expression of the multidrug resistance operon *mexA–mexB–oprM* in *Pseudomonas aeruginosa*: *mexR* encodes a regulator of operon expression. Antimicrob. Agents Chemother. *40*, 2021–2028.

Ramón-García, S., Martín, C., Aínsa, J.A., and De Rossi, E. (2006). Characterization of tetracycline resistance mediated by the efflux pump Tap from *Mycobacterium fortuitum*. J. Antimicrob. Chemother. *57*, 252–259.

Ramón-García, S., Martín, C., De Rossi, E., and Aínsa, J.A. (2007). Contribution of the Rv2333c efflux pump (the Stp protein) from *Mycobacterium tuberculosis* to intrinsic antibiotic resistance in *Mycobacterium bovis* BCG. J. Antimicrob. Chemother. *59*, 544–547.

Ramón-García, S., Martín, C., Thompson, C.J., and Aínsa, J.A. (2009). Role of the *Mycobacterium tuberculosis* P55 efflux pump in intrinsic drug resistance, oxidative stress responses, and growth. Antimicrob. Agents Chemother. *53*, 3675–3682.

Ramos, J.L., Martinez-Bueno, M., Molina-Henares, A.J., Terán, W., Watanabe, K., Zhang, X., Gallegos. M.T., Brennan, R., and Tobes, R. (2005). The TetR family of transcriptional repressors. Microbiol. Mol. Biol. Rev. *69*, 326–356.

Reeves, E.P., Lu, H., Jacobs, H.L., Messina, C.G., Bolsover, S., Gabella, G., Potma, E.O., Warley, A., Roes, J., and Segal, A.W. (2002). Killing activity of neutrophils is mediated through activation of proteases by K^+ flux. Nature *416*, 291–297.

Riccardi, G., Pasca, M.R., and Buroni, S. (2009). *Mycobacterium tuberculosis*: drug resistance and future perspectives. Future Microbiol. *4*, 597–614.

Rivera-Calimlim, L., Nasrallah, H., Strauss, J., and Lasagna, L. (1976). Clinical response and plasma levels: effect of dose, dosage schedules, and drug interactions on plasma chlorpromazine levels. Am. J. Psychiatry *133*, 646–652.

Rodrigues, L., Sampaio, D., Couto, I., Machado, D., Kern, W.V., Amaral, L., and Viveiros, M. (2009). The role of efflux pumps in macrolide resistance in *Mycobacterium avium* complex. Int. J. Antimicrob. Agents. *34*, 529–533.

Rohs, R., and Sklenar, H. (2004). Methylene blue binding to DNA with alternating AT base sequence: minor groove binding is favored over intercalation. J. Biomol. Struct. Dyn. *21*, 699–711.

Rosenberg, E.Y., Bertenthal, D., Nilles, M.L., Bertrand, K.P., and Nikaido, H. (2003). Bile salts and fatty acids induce the expression of *Escherichia coli* AcrAB multidrug efflux pump through their interaction with Rob regulatory protein. Mol. Microbiol. *48*, 1609–1619.

Saier, M.H., Jr, Beatty, J.T., Goffeau, A., Harley, K.T., Heijne, W.H., Huang, S.C., Jack, D.L., Jähn, P.S., Lew, K., Liu, J., *et al.* (1999). The major facilitator superfamily. J. Mol. Microbiol. Biotechnol. *1*, 257–279.

Salih, I.S., Thanacoody, R.H., McKay, G.A., and Thomas, S.H. (2007). Comparison of the effects of thioridazine and mesoridazine on the QT interval in healthy adults after single oral doses. Clin. Pharmacol. Ther. *82*, 548–554.

Sander, P., De Rossi, E., Boddinghaus, B., Cantoni, R., Branzoni, M., Bottger, E.C., Takiff, H., Rodriquez, R., Lopez, G., and Riccardi, G. (2000). Contribution of the multidrug efflux pump LfrA to innate mycobacterial drug resistance. FEMS Microbiol. Lett. *193*, 19–23.

Sawicke, L., and Sturla, S. (2008). Potentially lethal cardiac side-effects caused by psychiatric drugs. Vertex *19*, 387–393.

Schmitt, L., and Tampé, R. (2002). Structure and mechanism of ABC transporters. Curr. Opin. Struct. Biol. *12*, 754–760.

Sharma, S., Kumar, M., Sharma, S., Nargotra, A., Koul, S., and Khan, I.A. (2010). Piperine as an inhibitor of Rv1258c, a putative multidrug efflux pump of *Mycobacterium tuberculosis*. J. Antimicrob. Chemother. *65*, 1694–1701.

Siddiqi, N., Das, R., Pathak, N., Banerjee, S., Ahmed, N., Katoch, V.M., and Hasnain, S.E. (2004). *Mycobacterium tuberculosis* isolate with a distinct genomic identity overexpresses a tap-like efflux pump. Infection *32*, 109–111.

Silva, P.E., Bigi, F., Santangelo, M.P., Romano, M.I., Martín, C., Cataldi, A., and Aínsa, J.A. (2001). Characterization of P55, a multidrug efflux pump in *Mycobacterium bovis* and *Mycobacterium tuberculosis*. Antimicrob. Agents Chemother. *45*, 800–804.

Singh, J., Dubey, R.K., and Atal, C.K. (1986). Piperine-mediated inhibition of glucuronidation activity in isolated epithelial cells of the guinea-pig small intestine: evidence that piperine lowers the endogenous UDP-glucuronic acid content. J. Pharmacol. Exp. Ther. *236*, 488–493.

Smith, I., Bishai, W.R., and Nagaraja, V. (2005). Control of mycobacterial transcription, In Tuberculosis and the Tubercle Bacillus, Cole, S.T., Eisenach, K.D., McMurray, D.N., and Jacobs, W.R., Jr., eds. (ASM Press, Washington, DC), pp. 219–231.

Stavri, M., Piddock, L.J., and Gibbons, S. (2007). Bacterial efflux pump inhibitors from natural sources. J. Antimicrob. Chemother. *59*, 1247–1260.

Stowe, D.J., Atkinson, T., and Mann, N.H. (1989). Protein kinase activities in cell-free extracts of *Streptomyces coelicolor* A3[2]. Biochimie *71*, 1101–1105.

Takiff, H.E., Cimino, M., Musso, M.C., Weisbrod, T., Martinez, R., Delgado, M.B., Salazar, L., Bloom, B.R., and Jacobs, W.R. (1996). Efflux pump of the proton antiporter family confers low-level fluoroquinolone resistance in *Mycobacterium smegmatis*. Proc. Natl. Acad. Sci. U.S.A. *93*, 362–366.

Tekaia, F., Gordon, S.V., Garnier, T., Brosch, R., Barrell, B.G., and Cole, S.T. (1999). Analysis of the proteome of *Mycobacterium tuberculosis* in silico. Tuber. Lung Dis. *79*, 329–342.

Trimble, W.S., and Grinstein, S. (2007). TB or not TB: calcium regulation in mycobacterial survival. Cell *130*, 12–14.

Tseng, T.T., Gratwick, K.S., Kollman, J., Park, D., Nies, D.H., Goffeau, A., and Saier, Jr, M.H. . (1999). The RND permease superfamily: an ancient, ubiquitous and diverse family that includes human disease and development proteins. J. Mol. Microbiol. Biotechnol. *1*, 107–125.

Velayati, A.A., Masjedi, M.R., Farnia, P., Tabarsi, P., Ghanavi, J., Ziazarifi, A.H., and Hoffner, S.E. (2009). Emergence of new forms of totally drug-resistant tuberculosis bacilli: super extensively drug-resistant tuberculosis or totally drug-resistant strains in Iran. Chest *136*, 420–425.

Viveiros, M., and Amaral, L. (2001). Enhancement of antibiotic activity against poly-drug resistant *Mycobacterium tuberculosis* by phenothiazines. Int. J. Antimicrob. Agents. *17*, 225–228.

Viveiros, M., Portugal, I., Bettencourt, R., Victor, T.C., Jordaan, A.M., Leandro, C., Ordway, D., and Amaral, L. (2002). Isoniazid-induced transient high-level resistance in *Mycobacterium tuberculosis*. Antimicrob. Agents Chemother. *46*, 2804–2810.

Vrljic, M., Sahm, H., and Eggeling, L. (1996). A new type of transporter with a new type of cellular function: L-lysine export from *Corynebacterium glutamicum*. Mol. Microbiol. *22*, 815– 826.

Wade, M.M., and Zhang, Y. (2004). Mechanisms of drug resistance in *Mycobacterium tuberculosis*. Front. Biosci. *9*, 975–994.

Williams, E.P., Lee, J.H., Bishai, W.R., Colantuoni, C., and Karakousis, P.C. (2007). *Mycobacterium tuberculosis* SigF regulates genes encoding cell wall-associated proteins and directly regulates the transcriptional regulatory gene *phoY1*. J. Bacteriol. *189*, 4234–4242.

Wilson, M., DeRisi, J., Kristensen, H.H., Imboden, P., Rane, S., Brown, P.O., and Schoolnik, G.K. (1999). Exploring drug-induced alterations in gene expression in *Mycobacterium tuberculosis* by microarray hybridization. Proc. Natl. Acad. Sci. U.S.A. *96*, 12833–12838.

Wheeler, P.R. (1994). Metabolism of *Mycobacterium tuberculosis*. In Tuberculosis: Pathogenesis, Protection and Control, Bloom, B.R., ed. (ASM Press, Washington, DC), pp. 353–385.

Wittekindt, O.H., Schmitz, A., Lehmann-Horn, F., Hansel, W., and Grissmer, S. (2006). The human Ca^{2+}-activated K^{+} channel, IK, can be blocked by the tricyclic antihistamine promethazine. Neuropharmacology *50*, 458–467.

Salmonella Efflux Pumps

8

Stephanie Baugh and Laura J.V. Piddock

Abstract

Salmonella has around 350 efflux pumps which are important in the physiology of the bacterial cell. Nine of these pumps are known to confer low-level antibiotic resistance to three or more unrelated classes of drugs and so are known as multidrug resistant (MDR) efflux pumps. *Salmonella* possesses at least one MDR efflux pump from four of the five MDR transporter families; RND family, MATE family, MF superfamily and ABC superfamily. The MDR efflux system of most clinical importance, AcrAB-TolC, can export a wide variety of substrates. Antibiotic susceptibility testing of mutants in which specific genes have been inactivated reveals some overlapping substrate specificity of MDR efflux pumps. The efflux systems are under tight and well-ordered control from a variety of local and global regulatory genes, typically transcriptional activators such as RamA, although the precise nature of control is still to be elucidated. As well as being important in resistance of *Salmonella* to many antibiotics, dyes and detergents, MDR efflux pumps are also involved in virulence and biofilm formation revealing that they are fundamental to the biology of this important pathogen. Furthermore, these data suggest that AcrAB-TolC, or regulation thereof, could be targets for new inhibitors.

Salmonella

Theobald Smith first isolated *Salmonella* from pigs in 1806. He named the genus after his supervisor, Daniel Elmer Salmon who was the head of the Bureau of Animal Industry where Smith worked (Salmon and Smith, 1886). *Salmonella* is a Gram negative rod shaped bacterium and a member of the Enterobacteriaceae family. *Salmonella* can be divided into two distinct species, *Salmonella enterica,* which contains six subspecies, and *Salmonella bongori* (Hardy, 2004). Within these two species there are many different serovars; these differ from each other based on the surface antigens they possess (Tindall *et al.*, 2005). Serovars are distinguished serologically by combinations of somatic, flagella and capsular antigens present on their surface (Tindall *et al.*, 2005). To date there have been over 2500 different *Salmonella* serovars described (http://www.who.int). *Salmonella* has many different natural reservoirs including the gastrointestinal tracts of animals including chickens and other poultry (Gast, 2007), pigs (Boyen *et al.*, 2008) and rodents (Valdezate *et al.*, 2007) and on the skin of reptiles including lizards, snakes and turtles (Lamm *et al.*, 1972; Pedersen *et al.*, 2009) and amphibians including frogs (Ozek *et al.*, 1969; Woodward *et al.*, 1997; CDC, 2009).

Salmonella infections

In humans *Salmonella* serovars cause disease. Such infections have been amongst the major causes of food poisoning in the developing world for almost all of the 20th century, although their importance has only been recognized in the last few decades (Hardy, 2004). Infections caused by *Salmonella* are known collectively as salmonellosis, of which there are two main disease types, non-typhoidal and typhoidal (Valdezate *et al.*, 2007). Salmonellosis is a major problem worldwide, with thousands of bacteriologically confirmed cases per year even in developed countries (Fig. 8.1).

Non-typhoidal salmonellosis

Non-typhoidal salmonellosis (NTS) in humans is most commonly caused by the serovars *Salmonella enterica* serovar Enteritidis and *Salmonella enterica* serovar Typhimurium (Oundo *et al.*, 2000; Gordon *et al.*, 2008). NTS is the second largest cause of bacterial gastroenteritis in many countries (Lucey *et al.*, 2000; Nagakubo *et al.*, 2002). Laboratory based surveillance of NTS, although useful, hugely underestimates the disease burden

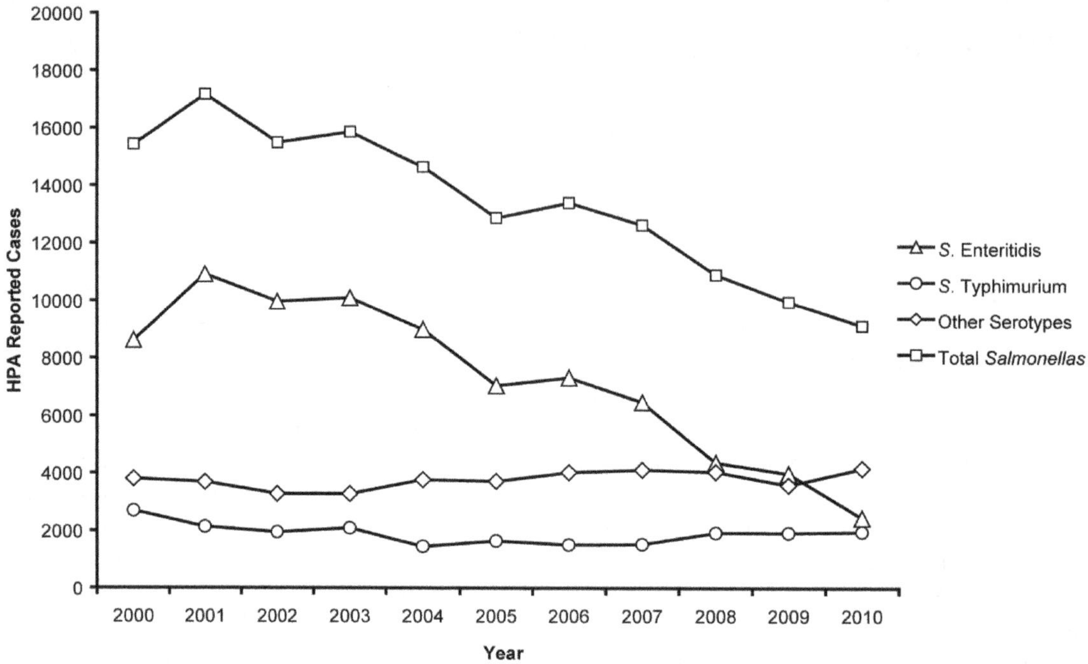

Figure 8.1 Non-typhoidal salmonellosis cases in England and Wales from 2000 to 2010 (data from hpa.org.uk).

so a report published in 2010 by Majowicz *et al* extrapolated existing data from literature, special studies and laboratory-based surveillance to estimate the global burden of gastroenteritis caused by *Salmonella*. The study predicts that 93.8 million cases of NTS occur globally each year with around 155,000 deaths (Majowicz *et al.*, 2010).

Salmonella infections are frequently transmitted via contaminated poultry and eggs (Valdezate *et al.*, 2007), however *Salmonella* is a commensal in the intestines of many animals and so can be transmitted via a variety of different meats and meat containing foods (Hardy, 2004). Salmonellosis can also be contracted via consumption of vegetables or other products contaminated with animal manure and human waste (Barton Behravesh *et al.*, 2011). Due to the food-borne nature of *Salmonella* infections, outbreaks as a result of contaminated food are common. Symptoms of NTS are mainly sickness, diarrhoea, (Valdezate *et al.*, 2007) abdominal pain and fever; in healthy individuals the infection is usually self limiting (Mead *et al.*, 1999). However, the infection can be invasive and spread to the blood stream where intervention with antibiotics is needed. Interestingly, infections with *S.* Typhimurium strains in sub-Saharan Africa has recently been found to cause a systemic infection (mimicking enteric fever, see below) instead of a gastrointestinal infection (Feasey *et al.*, 2010). Antibiotics commonly used for treatment are ciprofloxacin or ceftriaxone (Levy *et al.*, 2004).

Enteric fever

Enteric fever is caused by *Salmonella enterica* serovar Typhi (Typhoid fever) and *Salmonella enterica* serovar Paratyphi (Paratyphoid fever) (Crump and Mintz, 2010). Both infections are transmitted via the faecal–oral route and have no animal or environmental reservoirs; they are entirely adapted to humans (Mandell, G.L., *et al.*, 2005). Symptoms include profuse sweating, sickness and a rash of flat rose coloured spots (Mathura *et al.*, 2003; Darby and Sheorey, 2008). These species of *Salmonella* can move quickly into the bloodstream so rapid antibiotic treatment is necessary as this increases survival rates considerably. Intervention with antibiotic therapy is used in conjunction with oral rehydration therapy to treat the enteric fever. Antibiotic use depends upon local antibiotic resistance rates; when resistance is uncommon, fluoroquinolones such as ciprofloxacin are used (Thaver *et al.*, 2008; Parry and Beeching, 2009), where resistance is seen frequently third-generation cephalosporins such as ceftriaxone are used (Soe and Overturf, 1987). Where both ciprofloxacin and cephalosporin resistance is seen, azithromycin has been used with mixed outcomes (Effa and Bukirwa, 2008).

Antimicrobial resistance in *Salmonella*

Resistance to antibiotics is an emerging and increasing problem in *Salmonella*. This has been seen in both typhoidal and non-typhoidal serovars (Stevens *et al.*, 2009; Harish and Menezes, 2011). Resistance has been seen for β-lactams, tetracyclines, aminoglycosides and fluoroquinolones and can be mediated by the acquisition of multiple antibiotic resistance genes on genomic islands (as seen in *S.* Typhimurium DT104) and transmissible elements such as plasmids and integrons. Antibiotic resistance has profound clinical implications and drug resistant infections have a poor prognosis. Although more infections are caused by *S.* Enteritidis in Europe, *S.* Typhimurium is of particular clinical relevance as it is not only the second largest cause of NTS but of the cases reported in the European Union, almost half are multidrug resistant. This presents challenges if the infection is extra-intestinal and needs to be treated with antibiotics (http://www.enter.net). Infections caused by fluoroquinolone resistant strains of *S.* Typhimurium are three times more likely to become systemic or even fatal when compared to strains without such resistance (Helms *et al.*, 2004). As well as specific antibiotic resistance genes, efflux is also an important mediator of clinically relevant antibiotic resistance in *Salmonella* (Piddock *et al.*, 1987).

Salmonella efflux pumps

As described in Chapter 1, efflux pumps play an essential role in the physiology of bacteria. They do this by mediating the entry and extrusion of essential nutrients, metabolic waste and xenobiotics. As for other Gram negative bacteria there are four classes of solute transporters defined in the Transport Classification (TC) system; channels, secondary transporters, primary active transporters and group translocators. Channels do not depend upon energy but use the gradient of the substrate to drive translocation. Primary active transporters use a source of energy such as the hydrolysis of ATP. Secondary transporters exploit an ion or electrochemical gradient such as the sodium or proton motive force. Group translocators alter their substrates during the transport process (e.g. substrate phosphorylation).

A genomic analysis of all membrane transport systems within prokaryotes predicts that *S.* Typhimurium has 350 membrane transporters (73 transporters per Mb of genome) (Ren and Paulsen, 2007). Of these, 83 (23.7%) are primary active transporters (all using ATP as their energy source), 15 (4.3%) are channels, 29 (8.3%) are group translocators (all using phosphorylation to modify their substrates), 218 (62.3%) are secondary transporters (using either sodium or proton motive force) and two (0.6%) are as yet unclassified (Ren and Paulsen, 2007).

Multidrug resistant efflux pumps in *Salmonella*

Of the 350 predicted transport proteins within *Salmonella* there are nine known and experimentally characterized multidrug resistance (MDR) efflux pumps which individually confer low level MDR to three or more different classes of antibiotics, dyes or detergents (Baucheron *et al.*, 2004; Eaves *et al.*, 2004; Buckley *et al.*, 2006; Nishino *et al.*, 2006). These nine systems belong to four of the five distinct families of efflux pumps known to confer multidrug resistance (Fig. 8.2); the ATP binding cassette (ABC) family, the multidrug and toxic compound extrusion (MATE) family, the major facilitator superfamily (MFS), the resistance nodulation division (RND) family and the small multidrug resistance (SMR) family (Piddock, 2006b) (see Chapter 1). Except for MdsABC (**mds**-**m**ulti**d**rug transporter for **S**almonella) which is unique (Nishino *et al.*, 2006), all of the identified MDR efflux systems in *Salmonella* also exist in *E. coli* (Nishino *et al.*, 2006) (see Chapter 4). *Salmonella* has at least one MDR pump from each family with the exception of the SMR family of efflux pumps. Their ability to export a very

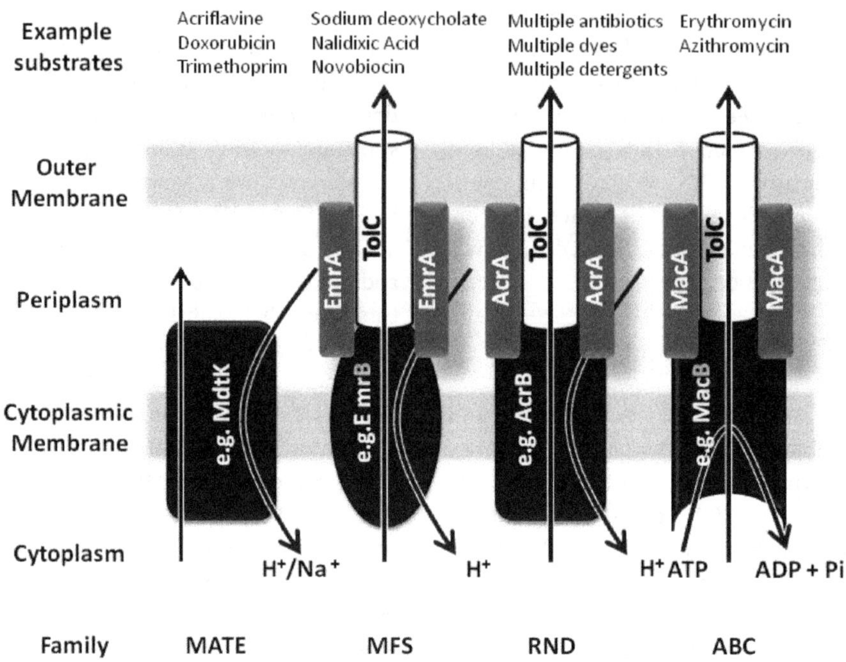

Figure 8.2 Four families of efflux pumps found in *Salmonella* with example substrates.

broad range of other toxic compounds including antibiotics and biocides allows the bacterium to survive in hostile environments (Blair and Piddock, 2009; Piddock, 2006a).

Resistance nodulation division (RND) family

The RND family of efflux pumps are a group of secondary transporters (Ren and Paulsen, 2007) (see Chapter 1). Work with *E. coli* showed that RND pumps are organized as a tripartite system with three components; an inner membrane pump protein, a periplasmic accessory protein and an outer membrane porin protein (Koronakis *et al.*, 2004). The inner membrane pump protein captures substrates from the periplasmic leaflet of the inner membrane phospholipid bilayer and exports them via the outer membrane porin protein into the extracellular milieu (Aires and Nikaido, 2005). The RND systems are proton antiporters as they use the proton gradient across the inner membrane to power efflux, substituting one hydrogen ion for one molecule of substrate in an antiport reaction (Paulsen, 2003). RND pumps assemble as a trimer, linked to an outer membrane channel which is also trimeric (Murakami *et al.*, 2002, 2006). Both of these proteins are stabilized by an adapter protein, the stoichiometry of this adapter protein is as yet uncertain.

Salmonella possess five different MDR RND efflux proteins, AcrB, AcrD, AcrF, MdsB and MdtBC. AcrB and AcrD associate with AcrA and TolC to form their tripartite system. AcrF, MdsB and MdtBC have their own accessory proteins, AcrE, MdsA and MdtA, respectively, but all complex with TolC as their outer membrane porin (OMP). MdsB, as well as associating with TolC in the outer membrane, can also associate with its own OMP, MdsC (Nishino *et al.*, 2006) (Table 8.1). There is a high level of homogeneity between the RND efflux pumps of *S.* Typhimurium and the corresponding homologues in *E. coli,* for example AcrA has 94% identity and AcrB has 97% identity (Eaves *et al.*, 2004; Piddock, 2006a,b). Although this similarity between *E. coli* and *S.* Typhimurium efflux pumps exists, the substrate profiles can differ between them (Nishino *et al.*, 2007).

Table 8.1 Associations and substrate specificity of all characterized MDR pumps of *Salmonella*

Name	Family	Accessory protein	Outer membrane protein	Substrates	References
AcrB	RND	AcrA	TolC	Acriflavin, benzalkonium chloride chloramphenicol, crystal violet, doxorubicin, erythromycin, ethidium bromide, methylene blue, minocycline, novobiocin, nalidixic acid, norfloxacin rhodamine 6G, tigecycline, sodium dodecyl sulphate, sodium deoxycholate, tetracycline, tetraphenylphosphonium bromide	Eaves *et al.* (2004), Nishino *et al.* (2006, 2009), Bailey *et al.* (2008), Horiyama *et al.* (2010a)
AcrD	RND	AcrA	TolC	Novobiocin, sodium dodecyl sulphate, sodium deoxycholate	Eaves *et al.* (2004), Nishino *et al.* (2006, 2009), Bailey *et al.* (2008)
AcrF	RND	AcrE	TolC	Acriflavin, benzalkonium chloride, chloramphenicol, crystal violet, doxorubicin, ethidium bromide, erythromycin, methylene blue, nalidixic acid, norfloxacin, novobiocin, rhodamine 6G, tetracycline, tetraphenylphosphonium bromide, tigecycline	Eaves *et al.* (2004), Nishino *et al.* (2006, 2009), Bailey *et al.* (2008), Horiyama *et al.* (2010a)
MdsB	RND	MdsA	MdsC/TolC	Acriflavin, benzalkonium chloride, crystal violet novobiocin, methylene blue, rhodamine 6G, sodium dodecyl sulphate	Nishino *et al.* (2009)
MdtBC	RND	MdtA	TolC	Novobiocin, sodium dodecyl sulphate, sodium deoxycholate	Nishino *et al.* (2009)
EmrB	MFS	EmrA	TolC	naladixic acid, novobiocin, rhodamine 6G, sodium dodecyl sulphate	Nishino *et al.* (2009)
MdfA	MFS	N/A	N/A	chloramphenicol, doxorubicin, norfloxacin, tetracycline	Nishino *et al.* (2009)
SmvA	MFS	N/A	N/A	acriflavin, ethidium bromide, malachite green, pyronin B	Villagra *et al.* (2008)
MdtK	MATE	N/A	N/A	acriflavin, doxorubicin, norfloxacin, sodium deoxycholate trimethoprim	Nishino *et al.* (2009)
MacB	ABC	MacA	TolC	azithromycin, erythromycin	Nishino *et al.* (2009)

AcrB, AcrD and AcrF

The most well characterized of the RND pumps in *Salmonella* is AcrB and its tripartite complex AcrAB-TolC (Fig. 8.2). It comprises the three proteins; AcrB (the inner membrane pump protein), AcrA (the periplasmic protein) and TolC (the OMP). Typically, the genes that encode the three proteins comprising the RND pump systems are co-localized within an operon with the local regulator followed by the periplasmic adapter protein gene, the pump gene and finally the OMP gene. However, in Enterobacteriaceae such as *E. coli* and *Salmonella* this is not the case for AcrAB-TolC; *acrR* (the local repressor gene), *acrA* and *acrB* are co-located within an operon, but *tolC* is elsewhere on the chromosome (Nishino *et al.*, 2006).

AcrAB-TolC has many different substrates making this efflux pump (and other RND homologues) a key mediator of multidrug resistance in Gram negative bacteria including many Enterobacteriaceae (Paulsen *et al.*, 1996). AcrB can export a huge range of substrates (Table 8.1) including cationic dyes (including crystal violet, ethidium bromide, rhodamine 6G), antibiotics (including penicillins, cephalosporins, fluoroquinolones, macrolides, chloramphenicol, tetracyclines, novobiocin, fusidic acid, oxazolidinones and rifampicin) and detergents (including

Triton X-100, sodium dodecyl sulphate and bile acids). Clearly there is little correlation between the structures of this diverse array of substrates, the only feature that they have in common is the inclusion of lipophilic domains within the molecule (Nikaido, 1996; Eaves *et al.*, 2004; Nishino *et al.*, 2006; Bailey *et al.*, 2008).

The presence of these lipophilic domains allows them to associate with the hydrophobic interior of the substrate-binding site within AcrB (Nikaido and Takatsuka, 2009). All three proteins in the AcrAB-TolC system are essential for efficient efflux of antimicrobial compounds across both membranes, and the lack of any component compromises the activity of the entire complex although not to the same degree (Ma *et al.*, 1995, 1996; Blair *et al.*, 2009).

Genomic analysis of AcrB, AcrD and AcrF indicates that there is a high level of homology between these three efflux proteins. AcrB is 80% and 64% homologous to AcrF and AcrD, respectively and AcrD is 65% homologous to AcrF (Eaves *et al.*, 2004).

MdtABC

MdtBC associates with TolC in the outer membrane and confers resistance to novobiocin, SDS and sodium deoxylcholate. The *mdtABC* operon encodes all three proteins of the efflux system and unusually codes for two efflux pump proteins that associate into one complex along with its specific periplasmic adapter protein MdtA (Horiyama *et al.*, 2010b).

MdsABC

MdsB was first described in 2006 and appears specific to *Salmonella* (Nishino *et al.*, 2006). The tripartite complex which it forms contains is own periplasmic adapter protein, MdsA, which can associate with either its specific outer membrane protein, MdsC, or TolC (Nishino *et al.*, 2003). MdsB has a variety of substrates including novobiocin, acriflavin, crystal violet, methylene blue, rhodamine 6G, benzalkonium chloride and SDS (Nishino *et al.*, 2006).

Major facilitator superfamily (MFS)

MFS pumps also fall under the category of secondary transporters (Ren and Paulsen, 2007). Around 25% of all known prokaryotic membrane transporters belong to the MFS, which is the largest and most varied family of transporters containing 74 subfamilies (Reddy *et al.*, 2012). There are around 15,000 MFS pumps identified to date (Saier *et al.*, 1999). MFS pumps are similar to RND pumps in that they are driven by the proton motive force exchanging one hydrogen ion for one substrate molecule and they span the periplasmic membrane with 12 transmembrane regions (with some exceptions containing 14 domains) (Piddock, 2006a, Reddy *et al.*, 2012). However, they do not always form a tripartite complex with an adapter protein and outer membrane porin and can exist as a single protein within the cytoplasmic membrane. MdfA and EmrAB are the two MFS MDR pumps described to date in *Salmonella*. MdfA exists as a single cytoplasmic efflux protein and confers resistance to chloramphenicol, doxorubicin, norfloxacin and tetracyclin in *S.* Typhimurium. Unlike MdfA, EmrB can associate with its own periplasmic adapter protein, EmrA, and TolC in the outer membrane (Horiyama *et al.*, 2010b). This EmrAB–TolC complex in *S.* Typhimurium can transport substrates such as novobiocin, nalidixic acid, SDS and sodium deoxycholate across both membranes into the extracellular milieu (Nishino *et al.*, 2006).

Multidrug and toxic compound extrusion (MATE) family

MATE pumps, similar to the RND and MFS pumps, are also secondary transporters with 12 transmembrane domains and use the proton motive force as an energy source. However, some MATE pumps use a sodium ion gradient to power efflux. MATE pumps in contrast to RND and MFS exist as a single efflux protein. *Salmonella* has one known MATE MDR efflux pump, MdtK. MdtK is a NorE MDR MATE transporter that uses the proton motive force to translocate substrates across the periplasmic membrane. MdtK can confer low level resistance norfloxacin, acriflavin and doxorubicin (Nishino *et al.*, 2006).

ATP-binding cassette (ABC) family

ABC transporters are primary active transporters and typically have six transmembrane domains spanning the periplasmic membrane which differs from the 12 domains that RND and MFS pumps possess and assemble as dimers. The characteristic of the ABC family of efflux pumps is the motor domain of the protein that binds and hydrolyses ATP, as it is the only family of transporters that uses ATP as its energy source to drive the export of substrates (Bouige *et al.*, 2002). MacB is the only known ABC MDR efflux pump in *Salmonella* and associates with MacA as its accessory protein and TolC as its OMP. MacB forms the tripartite system, MacAB-TolC, and has been shown to export 14 – and 15-membered lactone macrolides, for example erythromycin and azithromycin. Work in *E. coli* has shown that MacB is the inner membrane pump protein localized in the cytoplasm and functions as a dimer. MacA is thought to form a hexamer and interact with the α barrel tip region of TolC and stimulates the activity of MacB ATPase (Tikhonova *et al.*, 2007).

Efflux pump components are promiscuous

As stated previously, TolC is not encoded in the same genetic locus as AcrAB and acts as an OMP for several other efflux systems in addition to AcrAB. In *S.* Typhimurium, TolC is known to interact with eight multidrug efflux systems including AcrAB (Nishino *et al.*, 2006; Horiyama *et al.*, 2010b). MdsAB has also been shown to have an association with TolC in *S.* Typhimurium (Nishino *et al.*, 2006). AcrD (Hirakawa *et al.*, 2003; Nishino *et al.*, 2003), AcrEF (Nishino *et al.*, 2003), MdsAB, MdtABC (Nagakubo *et al.*, 2002), EmrAB, (Nishino and Yamaguchi, 2002) and MacAB (Kobayashi *et al.*, 2001b) have been found to associate with TolC in *E. coli* and it is assumed do likewise in *Salmonella*. Clearly TolC is not restricted by the family of efflux pump it associates with as it can form tripartite systems with proteins from RND, MFS and ABC families. In *E. coli*, AcrA has also been shown to be promiscuous and can combine with three other transporter proteins; AcrD (Elkins and Nikaido, 2002), AcrF (Kobayashi *et al.*, 2001a) and MdtF (Elkins and Nikaido, 2002). Again it is assumed that the *Salmonella* AcrA is similarly promiscuous.

Regulation of efflux in *Salmonella*

Multidrug transporters of *Salmonella* and other bacteria are usually under precise and complex transcriptional control. In essence, regulation is via local repressors, transcriptional activators or two-component regulatory systems. The latter two comprise global regulators with numerous genes within their regulons. In *E. coli* and *S.* Typhimurium, *acrAB* has a local repressor AcrR, Olliver *et al.* showed that a mutation in *acrR* along with a mutation in *gyrA* (topoisomerase gene) can confer resistance to fluoroquinolones as well as other unrelated classes of antibiotics (Olliver *et al.*, 2004).

In *E. coli*, *acrAB* and *tolC* are part of the regulons controlled by MarA, SoxS and Rob. The global regulatory regulation systems can respond to many environmental stresses to both increase expression of efflux pumps and downregulate porin expression to stop accumulation of potentially toxic substrates within the cytoplasm. They ordinarily exist with a repressor and activator gene (and sometimes an operator sequence) in an operon. Although homologues of these transcriptional activators occur in *Salmonella*, mutations conferring overproduction of these genes and hence of *acrAB* and *tolC* are rarely detected (Giraud *et al.*, 2000; Piddock *et al.*, 2000). In many Enterobacteriaceae, including *S.* Typhimurium, another transcriptional activator, RamA, influences expression of *acrAB* (van der Straaten *et al.*, 2004; Bailey *et al.*, 2008;). RamA is itself regulated by RamR which is a TetR-family repressor. Expression of *ramA* is repressed and so RamA is produced at a relatively low level (Bailey *et al.*, 2010); upon exposure to an inducer (e.g. indole or chlorpromazine) transcription of *ramA* is derepressed (Nikaido and Nishino 2008; Bailey 2008). RamA then binds upstream of its target genes to a recognized DNA sequence known as the 'ram box' and activates transcription of genes in its regulon including *acrAB* and *tolC*. At high concentrations of RamA, expression of additional genes are activated, including *acrEF* and ompF (Bailey *et al.*, 2010).

acrB, *acrD* and *acrF* are co-ordinately regulated and the expression of these three pumps effects the expression of the global transcriptional activators *marA* and *soxS* (Eaves *et al.*, 2004) as does RamA

(Piddock, L.J.V., unpublished observations). The two component regulatory system BaeSR influences the expression of both *acrD* and *mdtBC*, another RND efflux pump of *S.* Typhimurium. Up-regulation of this two component system also confers resistance to the metals copper and zinc via *acrD* and *mdtBC* (Nishino *et al.*, 2007). Expression of *macB* is also controlled by a two component regulatory system, PhoPQ, which is an essential regulator of virulence in *Salmonella* (Nishino *et al.*, 2006).

Inducers of efflux pumps in *Salmonella*

The complex control of efflux pump genes allows rapid response to inducers so as to increase expression and hence efflux of noxious molecules. Indole and chlorpromazine both induce *ramA* (Bailey *et al.*, 2008; Nikaido *et al.*, 2008). However, on induction of *ramA*, *acrAB* transcription is repressed and so the effect of chlorpromazine is itself complex. Expression of *acrD* and *mdtABC* is induced by high intracellular levels of copper and zinc (2 mM and 1 mM, respectively) via the regulator BaeSR, indicating that AcrD and MdtABC, respectively, have a physiological role in metal homeostasis in *Salmonella* as well as multidrug resistance (Nishino *et al.*, 2007).

Other roles for MDR efflux pumps in Salmonella

Virulence and pathogenicity

It is thought that the physiological role of efflux pumps in *Salmonella* is to mediate the extrusion of certain molecules produced by the host organism (including bile, hormones, fatty acids and host defence molecules), in order to persist and invade host cells and also to export toxins, metabolites and other environmental compounds (Piddock, 2006a; Blair and Piddock, 2009). Both *Salmonella* Typhi and Typhimurium are able to export bile salts and other host defence molecules and are able to survive in concentrations of bile much higher than those they would encounter in the animal gastrointestinal tract (van Velkinburgh and Gunn, 1999).

Loss of TolC in *Salmonella* confers a virulence defect; *tolC* mutants poorly adhered *in vitro* to embryonic intestinal cells and mouse monocyte macrophages and were also unable to invade (Buckley *et al.*, 2006; Nishino *et al.*, 2006). *In vivo*, *tolC* mutants colonized the avian gut poorly and also did not persist (Buckley *et al.*, 2006) and such mutants were attenuated in the mouse model of infection (Nishino, 2006). *acrB* mutants were able to adhere to tissue culture cells *in vitro*, but could not invade (Buckley *et al.*, 2006); they only colonized the avian gut transiently and were unable to persist. Other RND family efflux systems were also required for virulence in a mouse model; *mdtABC* and *mdsABC* were required for full virulence, whereas *acrAB* and *acrEF* mutants conferred a delay in the mortality of the mice (Nishino *et al.*, 2006). MacAB, an ABC transporter, was also essential for full virulence of *Salmonella* in the BALB/c mouse model (Nishino *et al.*, 2006).

Loss of TolC and AcrB conferred altered expression of several known virulence genes in *Salmonella*, probably explaining the observed attenuation in infection models(Webber *et al.*, 2009). There was reduced expression of *Salmonella* pathogenicity island (SPI) 1 genes (*sipABCD*, *prgHJ*, *spaO* and *invB*) and SPI-1 regulator genes (*hilC* and *invF*) as well as chemotaxis and motility genes, including *cheWY* and *flgLMK*. Loss of AcrA only had reduced expression of *Salmonella* pathogenicity island 2 genes (Webber *et al.*, 2009).

Biofilm formation

Inactivation of efflux pumps by efflux pump inhibitors abolishes biofilm formation in *E. coli*, *Klebsiella*, *Staphylococcus aureus* and *Pseudomonas putida* (Kvist *et al.*, 2008). Recent work in our laboratory with *S.* Typhimurium has shown this same biofilm defect results from genetic inactivation of each of the nine known MDR pumps (Baugh, S. *et al.*, unpublished data). Using AcrAB-TolC as a model system we showed that *tolC* and *acrB* mutants of *S.* Typhimurium were unable to produce a competent biofilm in different laboratory models. However, *acrA* mutants are able to produce a biofilm. Lack of biofilms was associated with overexpression of *ramA* and the repression of genes encoding curli fimbriae (an essential proteinaceous extracellular matrix component of *Salmonella* biofilms). These data show that *S.*

Typhimurium multidrug resistance via active efflux and biofilm formation are co-regulated.

Concluding remarks

Multidrug efflux pumps have been found to be fundamental to a variety of processes in many bacteria including drug resistance, pathogenicity and biofilm formation. *Salmonella* is an excellent example of this diversity of function and the study of *Salmonella* efflux pumps has allowed us to understand more about the role they play in the basic biology of this pathogen. As *S.* Typhimurium is commonly used as a model pathogen, data on MDR efflux pumps in this species provide a paradigm for other bacterial species.

References

Aires, J.R., and Nikaido, H. (2005). Aminoglycosides are captured from both periplasm and cytoplasm by the AcrD multidrug efflux transporter of *Escherichia coli*. J. Bacteriol. *187*, 1923–1929.

Bailey, A.M., Paulsen, I.T., and Piddock, L.J. (2008). RamA confers multidrug resistance in *Salmonella enterica* via increased expression of *acrB*, which is inhibited by chlorpromazine. Antimicrob. Agents Chemother. *52*, 3604–3611.

Bailey, A.M., Ivens, A., Kingsley, R., Cottell, J.L., Wain, J., and Piddock, L.J. (2010). RamA, a member of the AraC/XylS family, influences both virulence and efflux in *Salmonella enterica* serovar Typhimurium. J. Bacteriol. *192*, 1607–1616.

Barton Behravesh, C., Mody, R.K., Jungk, J., Gaul, L., Redd, J.T., Chen, S., Cosgrove, S., Hedican, E., Sweat, D., Chavez-Hauser, L., *et al.* (2011). 2008 outbreak of *Salmonella* Saintpaul infections associated with raw produce. N. Eng. J. Med. *364*, 918–927.

Baucheron, S., Tyler, S., Boyd, D., Mulvey, M.R., Chaslus-Dancla, E., and Cloeckaert, A. (2004). AcrAB-TolC directs efflux-mediated multidrug resistance in *Salmonella enterica* Serovar Typhimurium DT104. Antimicrob. Agents Chemother. *48*, 3729–3735.

Blair, J.M., and Piddock, L.J. (2009). Structure, function and inhibition of RND efflux pumps in Gram-negative bacteria: an update. Curr. Opin. Microbiol. *12*, 512–519.

Blair, J.M., La Ragione, R.M., Woodward, M.J., and Piddock, L.J. (2009). Periplasmic adaptor protein AcrA has a distinct role in the antibiotic resistance and virulence of *Salmonella enterica* serovar Typhimurium. J. Antimicrob. Chemother. *64*, 965–972.

Bouige, P., Laurent, D., Piloyan, L., and Dassa, E. (2002). Phylogenetic and functional classification of ATP-binding cassette (ABC) systems. Curr. Protein Peptide Sci. *3*, 541–559.

Boyen, F., Haesebrouck, F., Maes, D., Van Immerseel, F., Ducatelle, R., and Pasmans, F. (2008). Non-typhoidal *Salmonella* infections in pigs: A closer look at epidemiology, pathogenesis and control. Vet. Microbiol. *130*, 1–19.

Buckley, A.M., Webber, M.A., Cooles, S., Randall, L.P., La Ragione, R.M., Woodward, M.J., and Piddock, L.J. (2006). The AcrAB-TolC efflux system of *Salmonella enterica* serovar Typhimurium plays a role in pathogenesis. Cell. Microbiol. *8*, 847–856.

C.D.C. (2009). Multistate outbreak of human *Salmonella* Typhimurium infections associated with aquatic frogs – United States, 2009. MMWR Morb. Mortal. Wkly. Rep. *58*, 1433–1436.

Crump, J., and Mintz, E. (2010). Emerging infections: global trends in typhoid and paratyphoid fever. Clin. Infect. Dis. *50*, 241–246.

Darby, J., and Sheorey, H. (2008). Searching for *Salmonella*. Aust. Fam. Physician *10*, 806–810.

Eaves, D.J., Ricci, V., and Piddock, L.J. (2004). Expression of *acrB*, *acrF*, *acrD*, *marA*, and *soxS* in *Salmonella enterica* serovar Typhimurium: role in multiple antibiotic resistance. Antimicrob. Agents Chemother. *48*, 1145–1150.

Effa, E.E., and Bukirwa, H. (2008). Azithromycin for treating uncomplicated typhoid and paratyphoid fever (enteric fever). Cochrane Database Syst. Rev. CD006083.

Elkins, C.A., and Nikaido, H. (2002). Substrate specificity of the RND-type multidrug efflux pumps AcrB and AcrD of *Escherichia coli* is determined predominantly by two large periplasmic loops. J. Bacteriol. *184*, 6490–6498.

Feasey, N.A., Archer, B.N., Heyderman, R.S., Sooka, A., Dennis, B., Gordon, M.A., and Keddy, K.H. Typhoid fever and invasive nontyphoid salmonellosis, Malawi and South Africa. Emerg. Infect. Dis. *16*, 1448–1451.

Gast, R. (2007). Serotype-specific and serotype-independent strategies for preharvest control of food-bourne *Salmonella* in poultry. Avian Dis. *51*, 817–828.

Giraud, E., Cloeckaert, A., Kerboeuf, D., and Chaslus-Dancla, E. (2000). Evidence for active efflux as the primary mechanism of resistance to ciprofloxacin in Salmonella enterica serovar typhimurium. Antimicrob. Agents Chemother. *44*, 1223–1228.

Gordon M.A., Graham, S.M., Walsh, A.L., Wilson, L., Phiri, A., Molyneux, E., Zijlstra, E.E., Heyderman, R.S., Hart, C.A., and Molyneux, M.E. (2008). Epidemics of invasive *Salmonella enterica* serovar enteritidis and *S. enterica* Serovar typhimurium infection associated with multidrug resistance among adults and children in Malawi. Clin. Infect. Dis. *46*, 963–969.

Hardy, A. (2004). *Salmonella*: a continuing problem. Clin. Infect. Dis. *80*, 541–545.

Harish, B.N., and Menezes, G.A. (2011). Antimicrobial resistance in typhoidal salmonellae. Indian J. Med. Microbiol. *29*, 223–229.

Helms, M., Simonsen, J., and Molbak, K. (2004). Quinolone resistance is associated with increased risk of invasive illness or death during infection with *Salmonella* serotype Typhimurium. J. Infect. Dis. *190*, 1652–1654.

Hirakawa, H., Nishino, K., Hirata, T., and Yamaguchi, A. (2003). Comprehensive studies of drug resistance mediated by overexpression of response regulators of two-component signal transduction systems in *Escherichia coli*. J. Bacteriol. *185*, 1851–1856.

Horiyama, T., Nikaido, E., Yamaguchi, A., and Nishino, K. (2010a). Roles of *Salmonella* multidrug efflux pumps in tigecycline resistance. J. Antimicrob. Chemother. *66*, 105–110.

Horiyama, T., Yamaguchi, A., and Nishino, K. (2010b). TolC dependency of multidrug efflux systems in *Salmonella enterica* serovar Typhimurium. J. Antimicrob. Chemother. *65*, 1372–1376.

Kobayashi, K., Tsukagoshi, N., and Aono, R. (2001a). Suppression of hypersensitivity of *Escherichia coli acrB* mutant to organic solvents by integrational activation of the *acrEF* operon with the IS1 or IS2 element. J. Bacteriol. *183*, 2646–2653.

Kobayashi, N., Nishino, K., and Yamaguchi, A. (2001b). Novel macrolide-specific ABC-type efflux transporter in *Escherichia coli*. J. Bacteriol. *183*, 5639–5644.

Koronakis, V., Eswaran, J., and Hughes, C. (2004). Structure and function of TolC: the bacterial exit duct for proteins and drugs. Ann. Rev. Biochem. *73*, 467–489.

Kvist, M., Hancock, V., and Klemm, P. (2008). Inactivation of efflux pumps abolishes bacterial biofilm formation. Appl. Environ. Microbiol. *74*, 7376–7382.

Lamm, S.H., Taylor, A., Jr., Gangarosa, E.J., Anderson, H.W., Young, W., Clark, M.H., and Bruce, A.R. (1972). Turtle-associated salmonellosis. I. An estimation of the magnitude of the problem in the United States, 1970–1971. Am. J. Epidemiol. *95*, 511–517.

Levy, D.D., Sharma, B., and Cebula, T.A. (2004). Single-nucleotide polymorphism mutation spectra and resistance to quinolones in Salmonella enterica serovar Enteritidis with a mutator phenotype. Antimicrob. Agents Chemother. *48*, 2355–2363.

Lucey, B., Feurer, C., Greer, P., Moloney, P., Cryan, B., and Fanning, S. (2000). Antimicrobial resistance profiling and DNA amplification fingerprinting (DAF) of thermophilic *Campylobacter* spp. in human, poultry and porcine samples from the Cork region of Ireland. J. Appl. Microbiol. *89*, 727–734.

Ma, D., Alberti, M., Lynch, C., Nikaido, H., and Hearst, J.E. (1996). The local repressor AcrR plays a modulating role in the regulation of *acrAB* genes of *Escherichia coli* by global stress signals. Mol. Microbiol. *19*, 101–112.

Ma, D., Cook, D.N., Alberti, M., Pon, N.G., Nikaido, H., and Hearst, J.E. (1995). Genes *acrA* and *acrB* encode a stress-induced efflux system of *Escherichia coli*. Mol. Microbiol. *16*, 45–55.

Majowicz, S.E., Musto, J., Scallan, E., Angulo, F.J., Kirk, M., O'Brien, S.J., Jones, T.F., Fazil, A., and Hoekstra, R.M. (2010). The global burden of nontyphoidal *Salmonella* gastroenteritis. Clin. Infect. Dis. *50*, 882–889.

Mandell G.L., Bennett, J.E., and Dolin, R. (2005). Douglas and Bennett's principles and practice of infectious diseases, 6th edn (Churchill Livingstone, Pennsylvania, Elsevier).

Mathura, K., Gurubacharya, D., Shrestha, A., Pant, S., Basnet, P., and Karki, D. (2003). Clinical profile of typhoid patients. Kathmandu Univ. Med. J. *2*, 135–137.

Mead, P.S., Slutsker, L., Dietz, V., McCaig, L.F., Bresee, J.S., Shapiro, C., Griffin, P.M., and Tauxe, R.V. (1999). Food-related illness and death in the United States. Emerg. Infect. Dis. *5*, 607–625.

Murakami, S., Nakashima, R., Yamashita, E., and Yamaguchi, A. (2002). Crystal structure of bacterial multidrug efflux transporter AcrB. Nature *419*, 587–593.

Murakami, S., Nakashima, R., Yamashita, E., Matsumoto, T., and Yamaguchi, A. (2006). Crystal structures of a multidrug transporter reveal a functionally rotating mechanism. Nature *443*, 173–179.

Nagakubo, S., Nishino, K., Hirata, T., and Yamaguchi, A. (2002). The putative response regulator BaeR stimulates multidrug resistance of *Escherichia coli* via a novel multidrug exporter system, MdtABC. J. Bacteriol. *184*, 4161–4167.

Nikaido, H. (1996). Multidrug efflux pumps of Gram-negative bacteria. J. Bacteriol. *178*, 5853–5859.

Nikaido, H., and Takatsuka, Y. (2009). Mechanisms of RND multidrug efflux pumps. Biochimica Biophysica Acta *1794*, 769–781.

Nikaido, E., Yamaguchi, A., and Nishino, K. (2008). AcrAB multidrug efflux pump regulation in *Salmonella enterica* serovar Typhimurium by RamA in response to environmental signals. J. Biol. Chem. *283*, 24245–24253.

Nishino, K., and Yamaguchi, A. (2002). EvgA of the two-component signal transduction system modulates production of the yhiUV multidrug transporter in *Escherichia coli*. J. Bacteriol. *184*, 2319–2323.

Nishino, K., Yamada, J., Hirakawa, H., Hirata, T., and Yamaguchi, A. (2003). Roles of TolC-dependent multidrug transporters of *Escherichia coli* in resistance to beta-lactams. Antimicrob. Agents Chemother. *47*, 3030–3033.

Nishino, K., Latifi, T., and Groisman, E.A. (2006). Virulence and drug resistance roles of multidrug efflux systems of *Salmonella enterica* serovar Typhimurium. Mol. Microbiol. *59*, 126–141.

Nishino, K., Nikaido, E., and Yamaguchi, A. (2007). Regulation of multidrug efflux systems involved in multidrug and metal resistance of *Salmonella enterica* serovar Typhimurium. J. Bacteriol. *189*, 9066–9075.

Nishino, K., Nikaido, E., and Yamaguchi, A. (2009). Regulation and physiological function of multidrug efflux pumps in *Escherichia coli* and *Salmonella*. Biochimica Biophysica Acta *1794*, 834–843.

Olliver, A., Valle, M., Chaslus-Dancla, E., and Cloeckaert, A. (2004). Role of an acrR mutation in multidrug resistance of *in vitro*-selected fluoroquinolone-resistant mutants of *Salmonella enterica* serovar Typhimurium. FEMS Microbiol. Lett. *238*, 267–272.

Oundo, J.O., Kariuki, S., Maghenda, J.K., and Lowe, B.S. (2000). Antibiotic susceptibility and genotypes of non-typhi *Salmonella* isolates from children in Kilifi on the Kenya coast. Trans. R. Soc. Trop. Med. Hyg. *94*, 212–215.

Ozek, O., Cetin, E.T., Ang, O., Toreci, K., and Sanli, Z. (1969). *Salmonella* serotypes isolated from frogs (*Rana ridibunda*). Zentralbl. Bakteriol. Orig. *210*, 557–559.

Parry, C.M., and Beeching, N.J. (2009). Treatment of enteric fever. BMJ *338*, b1159.

Paulsen, I.T. (2003). Multidrug efflux pumps and resistance: regulation and evolution. Curr. Opin. Microbiol. *6*, 446–451.

Paulsen, I.T., Brown, M.H., and Skurray, R.A. (1996). Proton-dependent multidrug efflux systems. Microbiol. Rev. *60*, 575–608.

Pedersen, K., Lassen-Nielsen, A.M., Nordentoft, S., and Hammer, A.S. (2009). Serovars of *Salmonella* from captive reptiles. Zoonoses Public Health *56*, 238–242.

Piddock, L.J. (2006a). Clinically relevant chromosomally encoded multidrug resistance efflux pumps in bacteria. Clin. Microbiol. Rev. *19*, 382–402.

Piddock, L.J. (2006b). Multidrug-resistance efflux pumps – not just for resistance. Nat. Rev. *4*, 629–636.

Piddock, L.J., Wijnands, W.J., and Wise, R. (1987). Quinolone/ureidopenicillin cross-resistance. Lancet *2*, 907.

Piddock, L.J., White, D.G., Gensberg, K., Pumbwe, L., and Griggs, D.J. (2000). Evidence for an efflux pump mediating multiple antibiotic resistance in *Salmonella enterica* serovar Typhimurium. Antimicrob. Agents Chemother. *44*, 3118–3121.

Reddy, V.S., Shlykov, M.A., Castillo, R., Sun, E.I., and Saier, M.H., Jr. (2012). The major facilitator superfamily (MFS) revisited. FEBS J. *279*, 2022–2035.

Ren, Q., and Paulsen, I.T. (2007). Large-scale comparative genomic analyses of cytoplasmic membrane transport systems in prokaryotes. J. Mol. Microbiol. Biotechnol. *12*, 165–179.

Saier, M.H., Jr., Beatty, J.T., Goffeau, A., Harley, K.T., Heijne, W.H., Huang, S.C., Jack, D.L., Jahn, P.S., Lew, K., Liu, J., *et al.* (1999). The major facilitator superfamily. J. Mol. Microbiol. Biotechnol. *1*, 257–279.

Salmon, D., and Smith, T. (1886). The bacterium of swine-plague. Am. Monthly Microscopical J. 7, 204–205.

Soe, G.B., and Overturf, G.D. (1987). Treatment of typhoid fever and other systemic salmonelloses with cefotaxime, ceftriaxone, cefoperazone, and other newer cephalosporins. Rev. Infect. Dis. *9*, 719–736.

Stevens, M.P., Humphrey, T.J., and Maskell, D.J. (2009). Molecular insights into farm animal and zoonotic Salmonella infections. Philos. Trans. R. Soc. Lond. B Biol. Sci. *364*, 2709–2723.

van der Straaten, T., Zulianello, L., van Diepen, A., Granger, D.L., Janssen, R., and van Dissel, J.T. (2004). *Salmonella enterica* serovar Typhimurium RamA, intracellular oxidative stress response, and bacterial virulence. Infect. Immun. *72*, 996–1003.

Thaver, D., Zaidi, A.K., Critchley, J.A., Azmatullah, A., Madni, S.A., and Bhutta, Z.A. (2008). Fluoroquinolones for treating typhoid and paratyphoid fever (enteric fever). Cochrane Database Syst. Rev. CD004530.

Tikhonova, E.B., Devroy, V.K., Lau, S.Y., and Zgurskaya, H.I. (2007). Reconstitution of the *Escherichia coli* macrolide transporter: the periplasmic membrane fusion protein MacA stimulates the ATPase activity of MacB. Mol. Microbiol. *63*, 895–910.

Tindall, B.J., Grimont, P.A.D., Garrity, G.M., and Euzeby, J.P. (2005). Nomenclature and taxonomy of the genus *Salmonella*. Int. J. Syst. Evol. Microbiol. *55*, 521–524.

Valdezate, S., Arroyo, M., Gonzalez-Sanz, R., Ramiro, R., Herrera-Leon, S., Usera, M., Fuente, D.l., and Echeita, A. (2007). Antimicrobial resistance and phage and molecular typing of *Salmonella* strains isolated from food for human consumption in Spain. J. Food Prot. *70*, 2741–2748.

van Velkinburgh, J.C., and Gunn, J.S. (1999). PhoP-PhoQ-regulated loci are required for enhanced bile resistance in *Salmonella* spp. Infect. Immun. *67*, 1614–1622.

Villagra, N.A., Hidalgo, A.A., Santiviago, C.A., Saavedra, C.P., and Mora, G.C. (2008). SmvA, and not AcrB, is the major efflux pump for acriflavine and related compounds in *Salmonella enterica* serovar Typhimurium. J. Antimicrob. Chemother. *62*, 1273–1276.

Webber, M.A., Bailey, A.M., Blair, J.M., Morgan, E., Stevens, M.P., Hinton, J.C., Ivens, A., Wain, J., and Piddock, L.J. (2009). The global consequence of disruption of the AcrAB-TolC efflux pump in *Salmonella enterica* includes reduced expression of SPI-1 and other attributes required to infect the host. J. Bacteriol. *191*, 4276–4285.

Woodward, D.L., Khakhria, R., and Johnson, W.M. (1997). Human salmonellosis associated with exotic pets. J. Clin. Microbiol. 35, 2786–2790.

Pseudomonas aeruginosa Efflux Pumps

9

Keith Poole

Abstract

Antibiotic efflux systems are common in *Pseudomonas aeruginosa*, with chromosomally encoded multidrug efflux systems of the Resistance Nodulation Division (RND) family, specifically MexAB-OprM, MexCD-OprJ, MexEF-OprN and MexXY-OprM, of particular importance in clinical settings. Despite the broad substrate specificity of many of these, their clinical importance is limited to fluoroquinolone resistance (MexAB-OprM, MexCD-OprJ and MexEF-OprN), β-lactam resistance (MexAB-OprM, MexXY-OprM) and aminoglycoside resistance (MexXY-OprM). Expression of these systems is governed by the products of regulatory genes (*mexAB-oprM*: *mexR, nalC, nalD*; *mexCD-oprJ*: *nfxB*; *mexEF-oprN*: *mexT*; *mexXY*: *mexZ*) whose mutation is typically responsible for acquired multidrug resistance in lab and clinical isolates. With few exceptions these efflux systems are not inducible by substrate antimicrobials, consistent with antimicrobial efflux not being their intended function. Indeed, recent data highlight their induction by environmental stresses (oxidative stress, nitrosative stress, envelope stress) suggestive of a role in stress response systems in this organism. Significantly, such stresses may provide a selective pressure for antibiotic-resistant efflux mutants *in vivo* independent of antibiotic exposure. Given the importance of these efflux systems in intrinsic and acquired multidrug resistance in *P. aeruginosa*, strategies aimed at interfering with efflux-mediated resistance are being investigated.

Introduction

Pseudomonas aeruginosa is a common nosocomial pathogen (Hidron *et al.*, 2008; Zhanel *et al.*, 2008; Jones *et al.*, 2009; Zhanel *et al.*, 2010) that causes infections with a high mortality rate (Mutlu and Wunderink, 2006; Kerr and Snelling, 2009; Mahar *et al.*, 2010; Lambert *et al.*, 2011). This latter is, in part, attributable to the organism's intrinsically high resistance to many antimicrobials (Giamarellos-Bourboulis *et al.*, 2006) and the development of increased, particularly multidrug resistance in health care settings (Shorr, 2009; Kerr and Snelling, 2009; Keen, III *et al.*, 2010; Kallen *et al.*, 2010; Hirsch and Tam, 2010), both of which complicate antipseudomonal chemotherapy. Indeed, numerous studies point to a link between antimicrobial, particularly multidrug resistance and increased morbidity/mortality, as well as increased length of hospital stay and increased hospital costs (Evans *et al.*, 2007; Slama, 2008; Kerr and Snelling, 2009; Shorr, 2009; Hirsch and Tam, 2010; Lautenbach *et al.*, 2010; Mauldin *et al.*, 2010; Tumbarello *et al.*, 2011). Antimicrobial resistance in clinical *P. aeruginosa* isolates is multifactorial with numerous mechanisms contributing, including efflux (Poole, 2011). While efflux mechanisms of resistance come in two 'flavours', drug/class-specific and multidrug (Poole, 2005b), the latter are generally more important as regards intrinsic and acquired resistance to clinically relevant antimicrobials in *P. aeruginosa* (Poole, 2007, 2011). These are typically chromosomally encoded, and while representatives of the ATP-binding cassette (ABC), small multidrug resistance (SMR), multidrug and toxic efflux (MATE) and major facilitator (MF) families are all represented in this organism, resistance–nodulation–cell division (RND) family systems predominate as efflux determinants of resistance in clinical isolates (Poole, 2005b, 2007, 2011). Efflux contributions to resistance may be intrinsic as a result of constitutive efflux gene expression or acquired following mutational upregulation of otherwise quiescent genes (Poole, 2004) although recent studies also point to induction of multidrug efflux systems in *P. aeruginosa* by various environmental signals (Poole, 2008) (Table 9.1), which may also impact the resistance phenotype of this organism. Indeed, efflux systems of the RND family have increasingly been shown to be stress inducible and, so, may play a role in stress responses in this organism.

Non-RND family efflux systems

Efflux determinants of antimicrobial resistance of the MF, SMR, ABC and MATE families are uncommon in *P. aeruginosa* and with two exceptions are encoded by plasmid-/transposon-borne genes (Table 9.2). These are not generally clinically significant determinants of antimicrobial resistance. MF family exporters reported in *P. aeruginosa* include CmlA and Tet, determinants of resistance to chloramphenicol and tetracycline, respectively. CmlA was first described as a non-enzymatic determinant of resistance to chloramphenicol that was found on transposon Tn*1696* (Rubens *et al.*, 1979), an element only later shown to encode the CmlA chloramphenicol exporter of the MF superfamily (Bissonnette *et al.*, 1991). TetA was the first efflux determinant of tetracycline resistance and, indeed, antimicrobial resistance to be described in bacteria (in *E. coli*) (Ball *et al.*, 1980; McMurry *et al.*, 1980) and this, as well as other *tet* determinants of efflux-mediated tetracycline resistance have been described in *P. aeruginosa* where they are invariably plasmid-encoded (Mendez *et al.*, 1980; Jones *et al.*, 1992; Roberts, 1996). A single, chromosomally encoded MATE family drug exporter, PmpM, has been described in *P. aeruginosa* (He *et al.*, 2004). Capable of accommodating fluoroquinolones (FQs) and biocides (benzalkonium chloride; BC), PmpM is a homologue of the original MATE family drug exporter, NorM, from *Vibrio parahaemolyticus* (Morita *et al.*, 1998), although unlike NorM, which is a Na^+-drug antiporter (Morita *et al.*, 2000), PmpM is a H^+-drug antiporter (He *et al.*, 2004). Several SMR family drug exporters

Table 9.1 Inducers of RND type multidrug pump gene expression in *P. aeruginosa*

Efflux system	Inducers	References
MexAB-OprM	Chlorinated phenols, including PCP	Muller *et al.* (2007), Ghosh *et al.* (2011)
	Growth in tobacco plant	Weir *et al.* (2008)
	Cumene hydroperoxide (redox)	Chen *et al.* (2008)
MexCD-OprJ	MDAs, including biocides (chlorhexidine, benzalkonium chloride, alexidine, cetrimide, PHMB), detergents (SDS), solvents (ethanol, hexane, xylene), cationic antimicrobials (polymyxin B, antimicrobial peptides [melittin, V8 and V681])	Morita *et al.* (2003), Fraud *et al.* (2008)
	Azithromycin	Kai *et al.* (2009)
	Dyes (ethidium bromide, rhodamine)	Morita *et al.* (2003)
MexEF-OprN	GSNO, DETA (nitrosative stress)	Fetar *et al.* (2011)
	Chloramphenicol	Fetar *et al.* (2011)
	Airway epithelial cells	Frisk *et al.* (2004)
	Furanone antagonist of quorum-sensing	Hentzer *et al.* (2003)
MexXY(-OprM)	Ribosome-targeting antimicrobials, including chloramphenicol, erythromycin, azithromycin, tetracycline, tigecycline, gentamicin, streptomycin, kanamycin	Dean *et al.* (2003), Jeannot *et al.* (2005), Morita *et al.* (2006)
	Peroxide (oxidative stress)	Fraud and Poole (2010)
	Airway epithelial cells	Frisk *et al.* (2004)
	CF sputum	Son *et al.* (2007)

Table 9.2 Efflux mechanisms of antimicrobial resistance found in *P. aeruginosa*

Pump[a]	Antimicrobial substrates	Family	Reference(s)
Cml (CmlA)[b]	CM	MF	Schwarz *et al.* (2004), Wu *et al.* (2008), Libisch *et al.* (2008)
Tet[b]	TC	MF	Roberts (1996)
EmrE	AG	SMR	Li *et al.* (2003a)
QacE[b]	QAC	SMR	Kazama *et al.* (1998)
QacF[b]	QAC	SMR	Schluter *et al.* (2005), Jeong *et al.* (2009)
QacG[b]	QAC	SMR	Laraki *et al.* (1999)
PmpM[c]	FQ, BC	MATE	He *et al.* (2004)
PA1875–1876–1877	AG[d]	ABC	Zhang and Mah (2008)
MexAB-OprM	FQ, ML, BL, TC, CM, TS, NV, TP, TG	RND	Poole *et al.* (1993), Gotoh *et al.* (1995), Li *et al.* (1995), Schweizer (1998), Masuda *et al.* (2000), Dean *et al.* (2003), Chuanchuen *et al.* (2001)
MexCD-OprJ	FQ, ML, BL, TC, CM, TS, NV, CH, TG, DP[e]	RND	Poole *et al.* (1996), Chuanchuen *et al.* (2001), Dean *et al.* (2003), Fraud *et al.* (2008)
MexEF-OprN	FQ, CM, TM, TS	RND	Köhler *et al.* (1997a), Chuanchuen *et al.* (2001)
MexXY-OprM	FQ, ML, BL, AG, TC, TG	RND	Mine *et al.* (1999), Aires *et al.* (1999), Westbrock-Wadman *et al.* (1999), Masuda *et al.* (2000), Dean *et al.* (2003)
MexJK-OprM[f]	TC, ER	RND	Chuanchuen *et al.* (2002)
MexJK-OpmH[f]	TS	RND	Chuanchuen *et al.* (2005)
MexHI-OpmD[g]	NF	RND	Sekiya *et al.* (2003), Aendekerk *et al.* (2005)
MexMN-OprM[g]	CM	RND	Mima *et al.* (2005)
MexPQ-OpmE[g]	ML, FQ, Cu[h]	RND	Mima *et al.* (2005)
MexVW-OprM[g]	FQ, TC, CM, ER, CP	RND	Li *et al.* (2003b)
MuxABC-OpmB[g]	AZ, ML, NV, TC	RND	Mima *et al.* (2009)
TriABC-OpmH	TS	RND	Mima *et al.* (2007)

AG, aminoglycosides, AZ, aztreonam, BL, β-lactams, BC, benzalkonium chloride, CH, chlorhexidine, CM, chloramphenicol, CP, cefpirome, ER, erythromycin, FQ, fluoroquinolones, ML, macrolides, NF, norfloxacin, NV, novobiocin, QAC, quaternary ammonium compounds, TC, tetracycline, TP, trimethoprim, TS, triclosan.
[a]Unless otherwise indicated efflux systems are encoded by endogenous chromosomal genes.
[b]Efflux genes carried by mobile genetic elements.
[c]Contribution to resistance was confirmed following deletion and reintroduction of cloned *pmpM* from/into a mutant lacking *mexAB*, *mexCD-oprJ*, *mexEF-oprN* and *mexXY*.
[d]Implicated in biofilm-specific resistance to aminoglycosides.
[e]2,2′-dipyridyl. An *nfxB* mutant hyperexpressing *mexCD-oprJ* was selected on minimal medium containing the iron chelator 2,2′-dipyrdyl (Poole *et al.*, 1993).
[f]Contribution to resistance was observed following mutational overexpression or cloning of the efflux genes in a mutant lacking *mexAB-oprM* and *mexCD-oprJ*.
[g]Contribution to resistance was observed following mutational overexpression or cloning of the efflux genes in a mutant lacking *mexAB*, *mexCD-oprJ*, *mexEF-oprN* and *mexXY*.
[h]*mexPQ-opmE* is inducible by Cu and pump mutants show elevated sensitivity to Cu (Thaden *et al.*, 2010).

have been identified in *P. aeruginosa*, most being plasmid-encoded Qac exporters of biocides (i.e. quaternary ammonium compounds such as BC) (Table 9.2). The lone chromosomal SMR family exporter linked to drug resistance in *P. aeruginosa* is EmrE (also know as EmrE_{Pae}), a determinant of modest intrinsic resistance to aminoglycosides (Li *et al.*, 2003a) and a homologue of the EmrE drug exporter of *E. coli* (Yerushalmi *et al.*, 1995). A tripartite ABC-family efflux system that is more highly expressed in biofilm versus planktonic cells, PA1875-PA1876-PA1877, has also been linked to biofilm-specific aminoglycoside resistance (Zhang and Mah, 2008).

RND family efflux systems

As in other organisms, RND family drug exporters in *P. aeruginosa* operate as three-component pumps, with a cytoplasmic membrane (CM) drug-proton antiporter (the RND component) linked to an outer membrane channel-forming protein (the outer membrane factor; OMF) by a CM-tethered periplasmic protein (the membrane fusion protein; MFP) (Blair and Piddock, 2009; Misra and Bavro, 2009). Eleven RND family drug exporters have been described in *P. aeruginosa* although only four (MexAB-OprM, MexXY-OprM, MexCD-OprJ and MexEF-OprN) contribute to meaningful antimicrobial resistance (i.e. resistance to clinically relevant antimicrobials at relevant levels) and have been linked to resistance in clinical isolates (Poole, 2005b, 2007, 2011).

MexAB-OprM

The first of the multidrug exporters to be identified in *P. aeruginosa* (Poole *et al.*, 1993), the MexAB-OprM efflux system is the best studied of the *P. aeruginosa* RND family exporters. Its constituent proteins have been individually crystallized (Akama *et al.*, 2004a,b; Higgins *et al.*, 2004; Sennhauser *et al.*, 2009; Phan *et al.*, 2010), revealing structures remarkably similar to those of other RND family multidrug efflux systems, notably the AcAB-TolC system of *E. coli* (Koronakis *et al.*, 2000; Mikolosko *et al.*, 2006; Murakami *et al.*, 2006; Seeger *et al.*, 2006). A three-component structure has been proposed for this latter efflux system (Symmons *et al.*, 2009) and structure-derived models of pump operation have been developed, describing the AcrB component as a peristaltic pump comprising three asymmetrical monomers each of which represents sequential states in a directional transport cycle that begins with substrate capture at the outer leaflet of the CM and ends with substrate delivery to the outer membrane porter, TolC (Nikaido and Takatsuka, 2009; Pos, 2009; Schulz *et al.*, 2010). Purified MexB catalyses proton-gradient-dependent drug transport in proteoliposomes, and Gram-positive bacteria harbouring *mexB* are capable of exporting ethidium bromide indicating that MexB is active in the absence of MexA and OprM (Welch *et al.*, 2010). Still, its contribution to net drug efflux and resistance *in vivo* is dependent on these other components (Li and Poole, 2001; Nehme *et al.*, 2004).

MexAB-OprM exhibits one of the broadest substrate profiles of the RND pumps in *P. aeruginosa*, accommodating a variety of non-clinical agents [e.g. organic solvents (Li *et al.*, 1998; Li and Poole, 1999), dyes (Srikumar *et al.*, 1997; Srikumar *et al.*, 1998; Srikumar and Poole, 1999; Li *et al.*, 2003a; Vidal-Aroca *et al.*, 2009), detergents (Srikumar *et al.*, 1998), metabolic inhibitors (Schweizer, 1998), acylhomoserine lactones (AHLs) associated with quorum-sensing (QS) (Evans *et al.*, 1998; Pearson *et al.*, 1999), environmental pollutants (Muller *et al.*, 2007)] as well as a variety of clinically relevant antimicrobials and biocides (Table 9.2). The *mexAB-oprM* genes occur as an operon, which is expressed constitutively at moderate levels in wild type *P. aeruginosa* grown under typical laboratory conditions (i.e. in rich media) and, so, MexAB-OprM contributes to intrinsic resistance to various antimicrobials. This has been confirmed in mutant studies where deletion of the efflux genes has been shown to produce a multidrug-susceptible phenotype (Srikumar *et al.*, 1997). Mutational hyperexpression of the efflux operon has been noted in lab and clinical mutant strains [*nalB* (Jalal and Wretlind, 1998; Ziha-Zarifi *et al.*, 1999; Saito *et al.*, 1999; Jalal *et al.*, 2000; Srikumar *et al.*, 2000), *nalC* (Srikumar *et al.*, 2000; Cao *et al.*, 2004; Llanes *et al.*, 2004) and *nalD* (Sobel *et al.*, 2005a)], with elevated expression of the pump leading to increased (i.e. acquired) antimicrobial resistance. Though often selected by FQs *in vitro* (Kohler *et al.*, 1997) and *in vivo* (Join-Lambert *et al.*, 2001), strains hyperexpressing MexAB-OprM have also been identified amongst *in vitro*-isolated tetracycline- (Hamzehpour *et al.*, 1995; Jalal *et al.*, 1999; Alonso *et al.*, 1999) and chloramphenicol-resistant strains (Jalal *et al.*, 1999).

Clinical significance

Despite the broad range of antimicrobials that are accommodated by MexAB-OprM, the pump has been linked to resistance to a limited number of agents in clinical isolates, primarily FQs (Poole, 2000; Dupont *et al.*, 2005; Henrichfreise *et al.*, 2007), often in conjunction with other resistance mechanisms (i.e. target site mutations such as *gyrA/B* and *parC/E*) (Poole, 2000; Dupont *et al.*, 2005). The pump has also been linked to β-lactam resistance in clinical strains, notably as a contributor to resistance to antipseudomonal penicillins (Cavallo *et al.*, 2002; Boutoille *et al.*, 2004; Cavallo *et al.*, 2007; Hocquet *et al.*, 2007b) and to the carbapenem meropenem (Pournaras *et al.*, 2005), the latter often in conjunction with other resistance mechanisms, typically loss of the carbapenem channel OprD (Gutierrez *et al.*, 2007; Maniati *et al.*, 2007; Hammami *et al.*, 2009; Rodriguez-Martinez *et al.*, 2009; Zhao *et al.*, 2009; Tomas *et al.*, 2010; Wang *et al.*, 2010b). Mutations in *mexR*, *nalC* and *nalD* have all been noted in *mexAB-oprM*-expressing clinical isolates resistant to β-lactams (Quale *et al.*, 2006; Tomas *et al.*, 2010; Campo Esquisabel *et al.*, 2011). A recent paper also highlights the involvement of MexAB-OprM efflux system in the development of tolerance to colistin in the cap subpopulation of biofilm cells, although the nature of the tolerance and MexAB-OprM's contribution to it is unclear (colistin is not a MexAB-OprM substrate) (Pamp *et al.*, 2008).

Regulation

Expression of *mexAB-oprM* is controlled directly or indirectly by repressors encoded by the *mexR*, *nalC* and *nalD* genes (Fig. 9.1A), with inactivating mutations in any of these resulting in increased *mexAB-oprM* expression and multidrug resistance (Srikumar *et al.*, 2000; Cao *et al.*, 2004; Sobel *et al.*, 2005a). Efflux gene expression and the attendant antimicrobial resistance are higher, however, in *mexR* versus *nalC* or *nalD* mutants.

mexR, the target of mutation in MexAB-OprM-expressing multidrug-resistant *nalB* mutants (Jalal and Wretlind, 1998; Ziha-Zarifi *et al.*, 1999; Saito *et al.*, 1999; Srikumar *et al.*, 2000), occurs upstream of and is divergently transcribed from *mexAB-oprM* (Poole *et al.*, 1996b) (Fig. 9.1B). MexR is a member of the MarR family of regulators (Miller and Sulavik, 1996) and binds as a dimer (Lim *et al.*, 2002) to two sites in the *mexR-mexA* intergenic region, near *mexR* and overlapping promoters for both *mexR* and *mexAB-oprM* (Evans *et al.*, 2001; Saito *et al.*, 2001; Sanchez *et al.*, 2002b) (Fig. 9.1). MexR binding to its target DNA is influenced both by its oxidation state (Chen *et al.*, 2008, 2010) and by a 54-amino-acid polypeptide modulator, ArmR, that functions as an antirepressor (Daigle *et al.*, 2007).

MexR apparently senses and responds to oxidative stress as a result of interprotomer disulphide bond formation between two redox-active cysteines, with oxidation promoting MexR dissociation from promoter DNA and, thus, *mexAB-oprM* expression (Chen *et al.*, 2008) (Fig. 9.2A). A crystallographic study shows that oxidative stress-promoted disulphide bond formation in MexR alters MexR structure such that the repressor is unable to bind to target DNA (Chen *et al.*, 2010). Still, while hydroperoxide or hydrogen peroxide (both capable of oxidizing MexR) treatment of MexR compromises its binding to promoter DNA *in vitro*, their induction of *mexAB-oprM* expression *in vivo* is minimal (Chen *et al.*, 2008). Indeed, array studies of *P. aeruginosa* treated with peroxide (Chang *et al.*, 2005a) and other oxidative stress agents (Chang *et al.*, 2005b) fail to demonstrate any *mexAB-oprM* induction, making it uncertain as to what the redox-sensitive MexR is responsive to *in vivo* in its promotion of *mexAB-oprM* expression.

ArmR (aka PA3719) is encoded by the second gene of the two-gene operon PA3720–*armR* (Daigle *et al.*, 2007) whose expression is negatively regulated by the product of the *nalC* repressor gene (Cao *et al.*, 2004). Mutations in *nalC*, which occurs upstream of and is divergently transcribed from PA3720-*armR*, lead to *mexAB-oprM* expression and multidrug resistance as a result of the derepression of PA3720-*armR* (Cao *et al.*, 2004), with ArmR complexing with MexR and obviating MexR binding to its target efflux promoter PI (Daigle *et al.*, 2007) (Fig. 9.1C). The ArmR antirepressor binding to MexR promotes conformational changes in the repressor that alter the spacing of its DNA-binding helices such that they are no

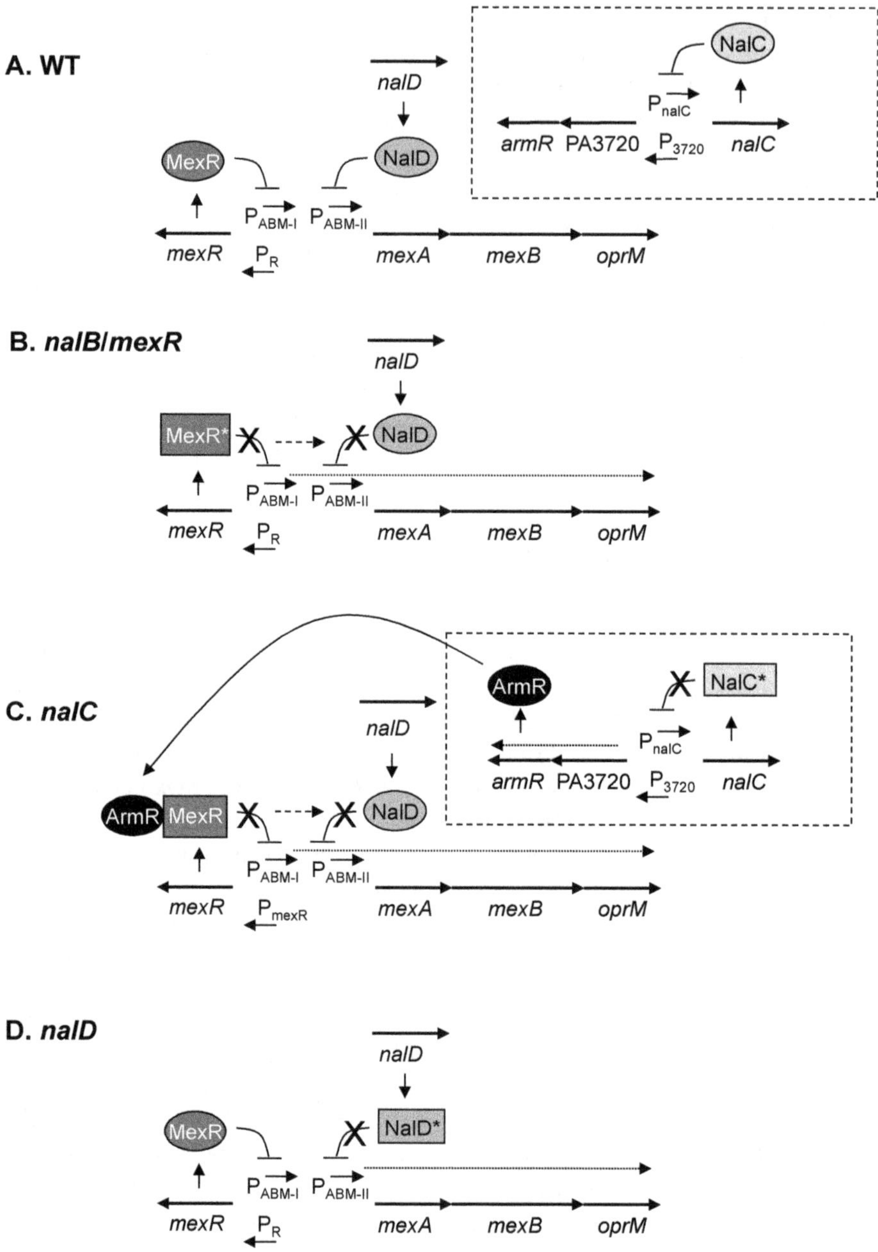

Figure 9.1 Repressor control of *mexAB-oprM* expression. (A) Expression of *mexAB-oprM* is controlled directly by the MexR and NalD repressors that act at promoters PI (ABM-I) and PII (ABM-II), respectively, upstream of the efflux genes and indirectly by the NalC repressor that controls expression of the PA3720-*armR* operon and, therefore, limits production of the ArmR anti-repressor that modulates MexR repressor activity. (B) In *mexR* (also known as *nalB*) mutants, MexR binding to promoter P_{ABM-I} (and its own promoter, P_R) is lost with a resulting interference with NalD binding to P_{ABM-II} and, so, full derepression of *mexAB-oprM*. (C) In *nalC* mutants, NalC binding to the PA3720-*armR* promoter (P_{3720}) and, so, repression of PA3720-*armR* is lost, leading to production of ArmR, which binds to MexR and obviates its binding to P_{ABM-I}. This interferes with NalD binding to P_{ABM-II} leading to derepression of *mexAB-oprM*. NalC also acts on the *nalC* promoter (P_{nalC}) to effect negative autoregulation. In *nalC* mutants the levels of ArmR are insufficient to modulate all cellular MexR and, so, full derepression of *mexAB-oprM* is not realized. (D) In *nalD* mutants, NalD binding to P_{ABM-II} is lost leading to derepression of *mexAB-oprM*. Since expression is limited to that initiated at P_{ABM-II} only, full derepression does not occur.

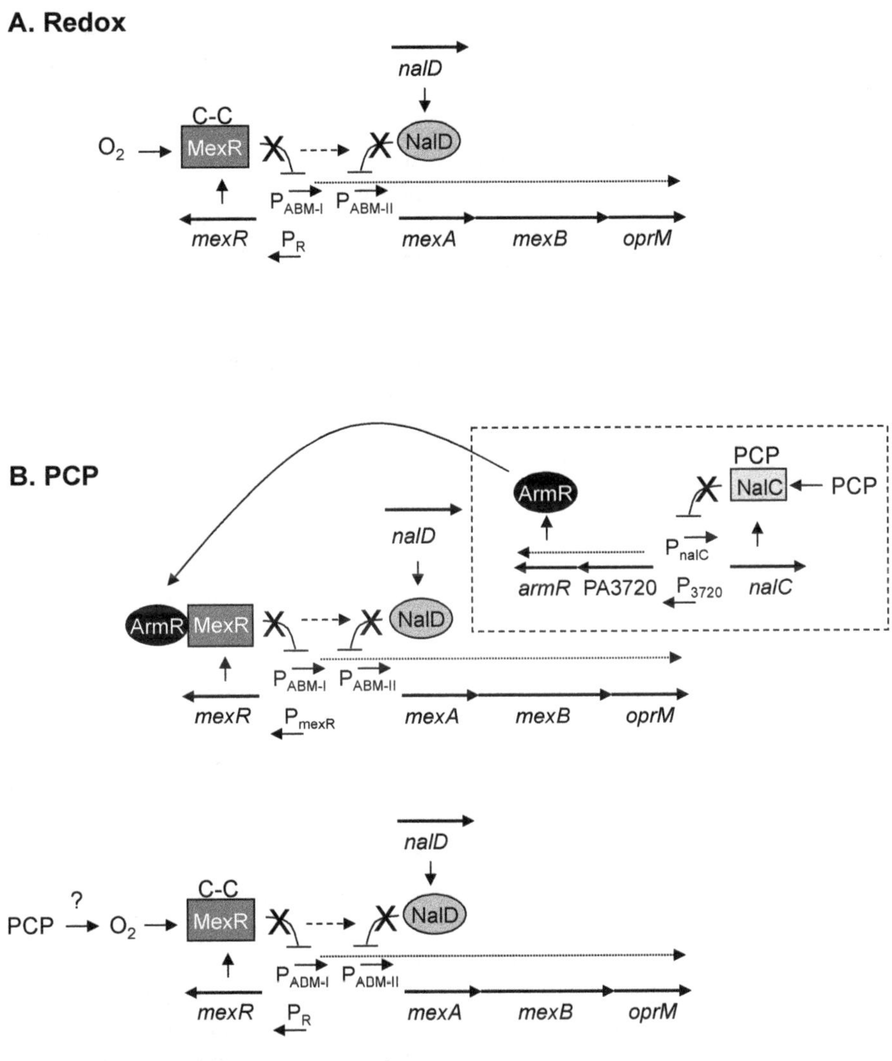

Figure 9.2 Environmental control of *mexAB-oprM* expression. (A) MexR is a redox-responsive repressor that upon oxidation (O_2) of a critical cysteine pair (C-C) is rendered unable to bind promoter P_{ABM-I}. This in turn interferes with NalD binding to P_{ABM-II} and derepression of *mexAB-oprM*. (B, top) NalC is a PCP-responsive repressor whose binding to the PA3720-*armR* promoter (P_{3720}) is obviated upon its complexation with PCP. The resultant derepression of PA3720-*armR* increases ArmR levels and modulation of MexR binding to P_{ABM-I}, with an attendant interference with NalD binding to P_{ABM-II} and, so, derepression of *mexAB-oprM*. (B, bottom) Because *mexAB-oprM* expression is still PCP inducible in an *armR* knockout but not an *mexR* knockout it is suggested that MexR (which does not bind PCP) also responds to PCP-dependent oxidation of its redox-sensitive cysteines, with this interfering with MexR binding to P_{ABM-I} and, ultimately, NalD binding to P_{ABM-II}.

longer compatible with DNA binding (Wilke *et al.*, 2008). This contrasts with the impact of oxidation on MexR structure, which involves changes to the DNA-binding helix–turn–helix motif itself (Chen *et al.*, 2010).

NalD, like MexR, is a direct repressor of *mexAB-oprM* expression, capable of binding to a second promoter (PII) upstream of the efflux genes but more proximal than PI (Morita *et al.*, 2006a) (Fig. 9.1D). The *nalD* gene (also known as PA3574) is not linked to *mexR/mexAB-oprM* or *nalC*/PA3720-armR and there is no information currently as to its regulation or the signals to which it responds. Interestingly, NalD binding to PII appears to be somewhat dependent upon MexR binding to PI such that in the absence of functional MexR there is almost full derepression of *mexAB-oprM* (Morita *et al.*, 2006a).

Recently, two homologous loci, *roc1* and *roc2*, which control expression of the *cup* genes involved in fimbriae synthesis, have been implicated in regulation of *mexAB-oprM* (Sivaneson *et al.*, 2010). Roc1 encodes a RocS1 sensor and a RocA1 response regulator that work together to controls *cup* gene expression, with the *roc2*-encoded sensor, RocS2, also regulating cup gene expression via RocA1 (Sivaneson *et al.*, 2010). Interestingly, RocS1 and RocS2 also act via the *roc2*-encoded sensor, RocA2, to repress *mexAB-oprM* expression – expression of this efflux locus increases in a *rocA2* mutant, with a concomitant increase in antimicrobial resistance (Sivaneson *et al.*, 2010).

Stress inducibility

Accommodation of AHLs explains reduced virulence gene expression (Evans *et al.*, 1998) and virulence (Sanchez *et al.*, 2002a) of *nalB* mutants hyperexpressing *mexAB-oprM* and speak to this pump having other than drug efflux as an intended natural function. Consistent with this, mutants hyperexpressing MexAB-OprM were readily selected *in vivo* in a rat model of acute *P. aeruginosa* pneumonia in the absence of any antibiotic selection (Join-Lambert *et al.*, 2001). The specific nature of the selective *in vivo* growth advantage provided by this efflux system was, however, unclear. Similarly, a MexAB-OprM-deficient mutant shows reduced invasiveness in a kidney cell monolayer invasion model and reduced virulence in a mouse model (Hirakata *et al.*, 2002), and efflux pump inhibitors (EPIs), including a MexAB-OprM-specific EPI, reduced invasiveness of wild type and *nalB P. aeruginosa* (Hirakata *et al.*, 2009). *mexAB-oprM* expression increases in host (tobacco) relative to non-host in a plant model of pathogenesis (Weir *et al.*, 2008) although, again, the plant signals/constituents responsible for efflux gene induction are unknown. Indeed, only very recently have specific compounds that induce *mexAB-oprM* expression been identified. The demonstration that pentachlorophenol (PCP) (Muller *et al.*, 2007) and other chlorinated phenols (Ghosh *et al.*, 2011) could induce expression of this efflux operon and that PCP is a MexAB-OprM substrate (Muller *et al.*, 2007) provided the first clues as to a possible natural substrate and, so, function for this efflux system. PCP induction of *mexAB-oprM* is complicated and, possibly, indirect. This uncoupler of oxidative phosphorylation induces expression of PA3720-*armR* as a result of its interaction with NalC and interference with its binding to the PA3720-*armR* promoter (Ghosh *et al.*, 2011; Starr *et al.*, 2012). This result both ruled out *mexAB-oprM* (and PA3720-*armR*) responding to the adverse downstream effects of uncoupling oxidative phosphorylation and suggested that PCP induction of *mexAB-oprM* occurred as a result of increased production of ArmR and its subsequent anti-MexR repressor activity (Ghosh *et al.*, 2011) (Fig. 9.2B). PCP is, however, still capable of inducing *mexAB-oprM* expression in an *armR* deletion mutant (Starr *et al.*, 2012) suggesting that its upregulation of this efflux system occurs independently of ArmR or that there is an additional mechanism for PCP induction of *mexAB-oprM*. One possibility is that PCP promotes oxidative stress that impacts the redox sensitive MexR directly (Fig. 9.2B).

MexXY-OprM

The components of the MexXY-OprM multidrug efflux system are encoded by the *mexXY* operon and the *oprM* gene of the *mexAB-oprM* efflux operon (Mine *et al.*, 1999; Masuda *et al.*, 2000a), the latter functioning as the OMF component of several three-component RND-type multidrug efflux systems in *P. aeruginosa* (Table 9.2). MexXY-OprM accommodates a broad range of antimicrobials, including β-lactams, macrolides, FQs, chloramphenicol, tetracylines (Masuda *et al.*, 2000b), tigecycline (Dean *et al.*, 2003), and somewhat unique for multidrug efflux systems, aminoglycosides (Aires *et al.*, 1999; Mine *et al.*, 1999; Westbrock-Wadman *et al.*, 1999). Unlike most of the other RND-type multidrug efflux systems, *mexXY* is inducible by many of the antimicrobials that it exports and, indeed, was the first RND-type efflux system shown to be inducible by its antimicrobials substrates (Masuda *et al.*, 2000a; Jeannot *et al.*, 2005; Morita *et al.*, 2006b) (Table 9.1). Its inducibility by many of its substrate antimicrobials means that it plays a role in intrinsic resistance to these agents, and indeed lab-constructed mutants lacking this pump show enhanced susceptibly to these agents (Masuda *et al.*, 2000a; Morita *et al.*, 2006b). MexXY-OprM and MexAB-OprM are thus the major efflux determinants of intrinsic antimicrobial resistance in *P. aeruginosa*.

Clinical significance

Despite its ability to accommodate multiple classes of antimicrobials, MexXY-OprM has been predominantly linked to pan-aminoglycoside resistance in clinical isolates, especially pulmonary isolates recovered from cystic fibrosis (CF) patients (Sobel *et al.*, 2003; Poole, 2005a; Hocquet *et al.*, 2006; Henrichfreise *et al.*, 2007; Islam *et al.*, 2009). Indeed, while drug modification by aminoglycoside-modifying enzymes (AMEs) is a common mechanism of aminoglycoside resistance in *P. aeruginosa* (Poole, 2005a) this is almost unheard of in CF isolates (Shawar *et al.*, 1999; Poole, 2005a; Henrichfreise *et al.*, 2007; Islam *et al.*, 2009). Moreover, it is now accepted that earlier reports of so-called impermeability-type aminoglycoside resistance in CF isolates, which was characterized by reduced aminoglycoside uptake (Tseng *et al.*, 1973; Bryan *et al.*, 1976; Maloney *et al.*, 1989), was MexXY-OprM-mediated (Poole, 2005a). The MexXY-OprM efflux system has also been linked to β-lactam resistance in clinical isolates of *P. aeruginosa* (as one of several contributors) (Maniati *et al.*, 2007; Vettoretti *et al.*, 2009) and while this efflux system has been shown to accommodate carbapenems (except imipenem) (Okamoto *et al.*, 2002), it has not been linked to carbapenem resistance in clinical strains. MexXY production has been noted in ticarcillin-resistant *P. aeruginosa* (Hocquet *et al.*, 2007b) although a contribution to resistance was not proven and this efflux system is more commonly associated with resistance to the fourth generation cephalosporin cefepime in clinical isolates (Hocquet *et al.*, 2006; Pena *et al.*, 2009). Indeed, cefepime commonly selects for MexXY-derepressed mutants *in vitro* (Queenan *et al.*, 2010). In one study, this efflux system was apparently responsible for an unusual cefepime-resistant ceftazidime-susceptible phenotype seen in a number of clinical isolates (Pena *et al.*, 2009). MexXY-OprM was also responsible for reduced susceptibility to ceftobiprole in a clinical study of this the novel broad-spectrum cephalosporin (Baum *et al.*, 2009) and mutants expressing *mexXY* are readily selected by this β-lactam *in vitro* (Queenan *et al.*, 2010). Despite its ability to accommodate FQs, MexXY-OprM has seldom been linked to FQ resistance in clinical isolates (Wolter *et al.*, 2004; Niga *et al.*, 2005) although this agent is capable of selecting MexXY-expressing FQ-resistant mutants in the lab (Hocquet *et al.*, 2008). This efflux system was also seen upregulated in clinical isolates in conjunction with other RND type efflux systems, most commonly *mexAB-oprM* (Quale *et al.*, 2006; Hocquet *et al.*, 2007a; Ratkai *et al.*, 2010; Campo Esquisabel *et al.*, 2011), although its contribution to resistance in these instances was unclear.

Regulation

The *mexXY* operon is under the control of the product of the divergently transcribed *mexZ* repressor gene (Westbrock-Wadman *et al.*, 1999; Matsuo *et al.*, 2004). Not surprisingly, then, mutations in *mexZ* are common in pan-aminoglycoside resistant CF isolates of *P. aeruginosa* expressing *mexXY* (Islam *et al.*, 2004, 2009; Llanes *et al.*, 2004; Vogne *et al.*, 2004; Hocquet *et al.*, 2006; Henrichfreise *et al.*, 2007; Feliziani *et al.*, 2010; Mulcahy *et al.*, 2010). Intriguingly, *mexZ* has been identified as the most commonly mutated gene in CF isolates (Smith *et al.*, 2006; Feliziani *et al.*, 2010), suggesting that MexXY expression may be commonplace in CF lung isolates. Consistent with this, *mexX* is induced *in vitro* upon exposure of *P. aeruginosa* to human airway epithelial cells (Frisk *et al.*, 2004) and *mexY* shows enhanced expression in this organism in the CF lung (DNA array was performed on RNA isolated from sputum) (Son *et al.*, 2007). A number of studies highlight the absence of mutations in *mexZ* or the *mexXY* promoter region in *mexXY*-expressing pan-aminoglycoside-resistant CF isolates (Sobel *et al.*, 2003; Islam *et al.*, 2004, 2009; Llanes *et al.*, 2004; Hocquet *et al.*, 2006), indicating that additional genes/mutations are linked to expression of this efflux locus in *P. aeruginosa*. The observation, too, that sequential isolates from the same patient and bearing the same *mexZ* mutations displayed differences in levels of *mexXY* expression and resistance to aminoglycosides (Islam *et al.*, 2009) also argues that additional genes/mutations contribute to *mexXY* expression in aminoglycoside-resistant CF isolates. A recent report of an *in vitro*-selected *mexXY*-expressing aminoglycoside-resistant mutant lacking a *mexZ* mutation identified a novel gene, *parR*, as the site of mutation (Muller *et al.*, 2010). *parR* forms part of a two-gene operon, *parRS*, encoding a two-component regulatory system that impacts expression of several antimicrobial resistance determinants in *P. aeruginosa*, including *mexXY* (Muller *et al.*, 2010). This two-component systems also mediates induction of the *mexXY* operon in response to polycationic antimicrobials such as polymyxin B and colistin (Muller *et al.*, 2010). Significantly, mutations in *parR* are present in clinical isolates expressing *mexXY* but lacking mutations in *mexZ* (Muller *et al.*, 2010). While MexXY-OprM-mediated aminoglycoside resistance typically results from mutational upregulation of *mexXY*, there is a report of a point mutation in *mexY* in a CF isolate providing a modest (twofold) increase in resistance to aminoglycosides (and other MexXY-OprM substrates, including cefepime and fluoroquinolones) (Vettoretti *et al.*, 2009). Lab-selected FQ-resistant and cefepime-resistant *mexXY*-expressing mutants of *P. aeruginosa* with and without mutations in *mexZ* have also been described (Hocquet *et al.*, 2008) and mutations in *mexZ* are responsible for *mexXY*-expression in ceftobiprole-resistant isolates recovered during a clinical trial of this drug (Baum *et al.*, 2009).

The *mexXY* operon is inducible by many of the antimicrobials that it exports, specifically those that target the ribosome (i.e. aminoglycosides, tetracycline, macrolides, chloramphenicol, tigecycline) (Masuda *et al.*, 2000a; Dean *et al.*, 2003) (Table 9.1). The induction of *mexXY* by aminoglycosides likely explains the well-known phenomenon of adaptive (i.e. reversible) pan-aminoglycoside resistance in *P. aeruginosa* whereby resistance is enhanced upon exposure of the organism to an aminoglycoside and lost once the aminoglycoside is removed (Gilleland *et al.*, 1989; Barclay *et al.*, 1996). Antimicrobial induction of this efflux system is compromised by so-called ribosome protection mechanisms (Jeannot *et al.*, 2005), suggesting that the MexXY efflux system is recruited in response to ribosome disruption and not antibiotics per se. A gene, PA5471, encoding a conserved hypothetical protein has been shown to be induced by the same ribosome-disrupting antimicrobials as *mexXY* and to be required for antibiotic-inducible *mexXY* expression (Morita *et al.*, 2006b). Interestingly, PA5471 is strongly expressed in tobramycin-exposed *P. aeruginosa* biofilms growing on CF airway epithelial cells (Anderson *et al.*, 2008). Antimicrobial induction of PA5471 is governed by a transcriptional attenuation mechanism that is ultimately responsive to ribosome disruption by antibiotics (or mutation) (Morita *et al.*, 2009) (Fig. 9.3) with antimicrobial-induced PA5471 promoting *mexXY* expression, in part, via interaction with MexZ and modulation of its repressor activity (Yamamoto *et al.*, 2009). The recent crystal structure of MexZ has revealed that many of the *mexZ* mutations responsible for *mexXY* expression and aminoglycoside resistance map to the DNA-binding domain of this repressor as well as to a C-terminal domain that is implicated in effector (PA5471?) binding (Alguel *et al.*, 2010). Consistent with the link between translation disruption and induction of PA5471 and *mexXY*, mutations in *fmt* (encoding a methionyl-tRNA-formyltransferase) and *folD* (involved in folate biosynthesis and production of the formyl group added to initiator methionine), both of which will negatively impact protein synthesis, increase expression of PA5471 and *mexXY* (Caughlan *et al.*, 2009). Moreover, transposon disruption of the *rplY* gene encoding a probable ribosomal protein also promoted increased PA5471 and *mexXY* expression, although the resulting modest (twofold) increase in aminoglycoside resistance was deemed to be a product of the primary *rplY* disruption and not the secondary effects of *mexXY* expression (El'Garch *et al.*, 2007). In contrast, *mexXY* expression in *parR* mutants, or ParRS-dependent induction of *mexXY* by polycationic antimicrobials is independent of PA5471 and MexZ (Muller *et al.*, 2010).

Stress inducibility

Transcriptome studies have revealed that PA5471 is substantially upregulated in *P. aeruginosa* cells subjected to oxidative stress imposed by antiseptics such as peroxide (H_2O_2) (Chang *et al.*, 2005a) and peracetic acid (Chang *et al.*, 2005b). These

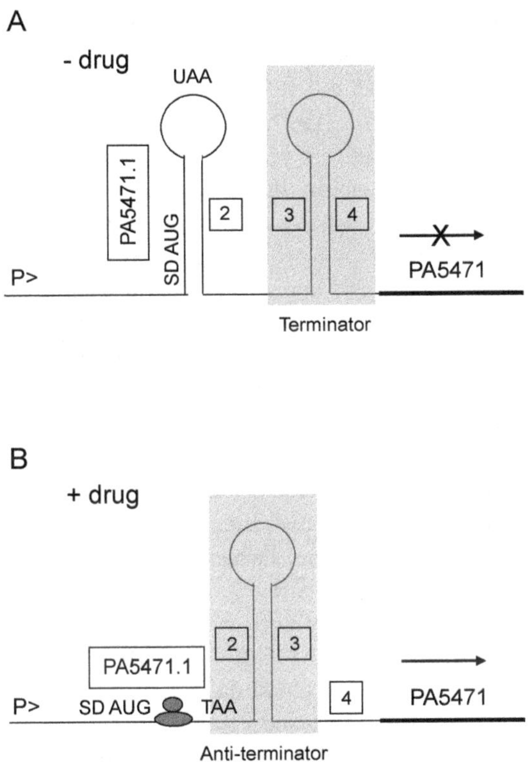

Figure 9.3 Transcriptional attenuation explains drug-inducible PA5471 expression. Transcription of the PA5471 upstream region (thin line) begins at the sole prompter (P>) that occurs upstream of PA5471 and an open-reading-frame encoding a small leader peptide, PA5471.1. This region carries sequences capable of adopting three stem–loop structures, with the stems formed by regions identified as PA5471.1–2, 2–3 and 3–4, where PA5471.1 is the coding sequence for this leader peptide, the PA5471.1–2 pair is the more energetically favourable pair, followed by 2–3 and 3–4, and the 3–4 stem–loop structure is a transcriptional terminator. In the absence of a ribosome-targeting antimicrobial (A), ribosomes that bind at the Shine–Dalgarno (SD) site of PA5471.1 and initiate translation at the AUG start codon proceed until the UAA stop codon, thereby not interfering with the pairing of the PA5471.1–2 regions, thus permitting pairing of the 3–4 regions. With the formation of the 3–4 terminator structure, transcription into the PA5471 coding region is blocked. In the presence of a ribosome-targeting antimicrobial (B) ribosomes stall during translation of the PA5471.1 leader peptide-encoding region preventing formation of the PA5471.1–2 pair. The now available region 2 pairs with region 3, rendering region 3 unavailable to pair with region 4. The formation of the 2–3 pair and the corresponding stem–loop structure thus blocks formation of the 3–4 terminator structure (and, so, is referred to as the anti-terminator) permitting transcription to proceed into PA5471.

agents also induce *mexXY* expression (Fraud and Poole, 2010) and such induction is, like *mexXY* induction by ribose-disrupting agents, PA5471-dependent (Fraud and Poole, 2010). Intriguingly, the CF lung is rich in oxidative stress-promoting reactive oxygen species (ROS) (Ciofu *et al.*, 2005) owing to the chronic inflammation apparently resultant from the CFTR defect that characterizes this disease and to chronic *P. aeruginosa* infection (Jacquot *et al.*, 2008; Rottner *et al.*, 2009). Given that MexXY-OprM-mediated efflux is the most common mechanism of pan-aminoglycoside resistance in *P. aeruginosa* CF isolates (Poole, 2005a), the implication here is that ROS may be promoting the development of aminoglycoside resistance in CF lung isolates, mediated by PA5471 and MexXY-OprM. Consistent with this, long-term *in vitro* exposure of *P. aeruginosa* to an ROS-generating agent (peroxide), while not enhancing pan-aminoglycoside resistance did increase the frequency with which pan-aminoglycoside-resistant mutants were recovered (Fraud and Poole, 2010). Moreover, this increased frequency of pan-aminoglycoside-resistant mutants in peroxide-exposed cell populations was dependent on PA5471 and MexXY-OprM and not explainable by ROS simply enhancing mutation rates as a result of its ability to damage DNA (Fraud and Poole, 2010). Thus, ROS in the CF lung appears to be providing a selective pressure for MexXY-dependent pan-aminoglycoside-resistant mutants, once again highlighting the potential for natural environments to promote efflux-mediated antimicrobial resistance independent of antimicrobial exposure.

Although the positive influence of ROS on *mexXY* expression and MexXY-OprM-dependent aminoglycoside resistance is known, why both ROS and ribosome-targeting antimicrobials induce *mexXY* expression in *P. aeruginosa* and do so via PA5471 is uncertain. A possible explanation lies in the link between translational (in) fidelity and protein oxidation. Abnormal polypeptides/proteins accumulate in non-growing (stationary phase), senescent bacteria owing to reduced translational fidelity (i.e. increased error rate) and are prone to cell-mediated oxidation/oxidative damage, with oxidation somehow identifying these as candidates for destruction and/or removal (Davies and Lin, 1988a,b; Lee *et al.*, 1988; Dean *et al.*, 1997; Ballesteros *et al.*, 2001). Consistent with this, protein oxidation in *E. coli* has been shown to increase with the age of stationary-phase cultures (Dukan and Nystrom, 1998). Ribosome disruption with antibiotics also leads to accumulation of abnormal polypeptides in bacteria (Thompson *et al.*, 2002; Harms *et al.*, 2003; Hermann, 2005), which may similarly be subjected to natural oxidative processes in the cell (which target abnormal over normal proteins) that target them for destruction or removal (by MexXY?). Indeed, using antibiotics (e.g. the aminoglycoside streptomycin) and specific mutations to compromise ribosome function, the production of aberrant proteins that are subsequently prone to oxidation has been seen in *E. coli* (Dukan *et al.*, 2000). Application of an exogenous oxidative stress (with antiseptics or ROS in the CF lung) will also lead to oxidation of normal proteins/polypeptides (Davies and Lin, 1988a; Dean *et al.*, 1997; Nishida *et al.*, 2006), possibly targeting them for destruction and removal via the same mechanism (hence the common recruitment of PA5471 and, possibly, MexXY by ribosome-targeting antibiotics and oxidative disinfectants). As such, PA5471/MexXY may contribute to a natural process for removal of abnormal proteins that accumulate in response to ageing and/or environmental stresses (and antibiotics). Alternatively, given that oxidative stress has been linked to frameshifting, ribosome dissociation (Kthiri *et al.*, 2010) and mistranslation (Ling and Soll, 2010) in *E. coli*, ROS may simply induce PA5471 and *mexXY*, just as ribosome-targeting antimicrobials do, via perturbation of ribosome function.

MexCD-OprJ

Quiescent in wild type cells grown under standard laboratory conditions, MexCD-OprJ does not contribute to intrinsic antimicrobial resistance in *P. aeruginosa* (Srikumar *et al.*, 1997), although its hyperexpression in so-called *nfxB* mutants is responsible for acquired multidrug resistance in this organism (Poole *et al.*, 1996a). The pump accommodates a broad range of antimicrobials (Table 9.2), including FQs, β-lactams, macrolides, tetracyclines, chloramphenicol (Poole *et al.*, 1996a), tigecycline (Dean *et al.*, 2003), biocides [e.g. triclosan (Chuanchuen *et al.*, 2001) and chlorhexidine (Fraud *et al.*, 2008)], organic solvents (Li *et al.*, 1998) dyes and detergents (Srikumar *et al.*, 1997; Srikumar *et al.*, 1998), with macrolides (e.g. azithromycin) (Mulet *et al.*, 2009) and triclosan (Chuanchuen *et al.*, 2001) readily selecting *mexCD-oprJ*-hyperexpressing mutants resistant to these agents. The selection of a *mexCD-oprJ*-hyperexpressing *nfxB* mutant on minimal medium supplemented with the iron chelator 2,2′-dipyridyl (Poole *et al.*, 1996a) and the subsequent demonstration that efflux pump inhibitors enhance the activity of such chelators (Liu *et al.*, 2010) suggests that 2,2′-dipyridyl is a substrate for MexCD-OprJ as well. An intriguing feature of *nfxB* strains is their hypersusceptibility to β-lactams such as carbenicillin (Hirai *et al.*, 1987; Srikumar *et al.*, 1997; Gotoh *et al.*, 1998), which appears to result from the reduced expression of MexAB-OprM (Gotoh *et al.*, 1998) and the AmpC β-lactamase (Masuda *et al.*, 2001) in these mutants. The observation that *nfxB* mutants are also hypersusceptible to aminoglycosides (Hirai *et al.*, 1987; Poole *et al.*, 1996a), a major substrate for the MexXY-OprM multidrug efflux system (see above), suggests that this efflux system, too, may be down regulated in *nfxB* mutants. The coordinated expression of multidrug efflux systems in *P. aeruginosa*, with increases in one compensated for by decreases in another also extends to additional multidrug efflux systems in this organism (Li *et al.*, 2000a), suggesting that there is some global control of net efflux gene expression in *P. aeruginosa*.

Clinical significance

FQs readily select MDR mutants expressing *mexCD-oprJ in vitro* (Köhler *et al.*, 1997b; Join-Lambert *et al.*, 2001) and this efflux system has been implicated in FQ resistance in clinical isolates (Poole, 2000; Zhanel *et al.*, 2004; Reinhardt *et al.*, 2007). Still, *mexCD-oprJ*-expressing mutants appear to be rare in a clinical setting (Jeannot *et al.*, 2008; Kiser *et al.*, 2010). This may be related to which FQs are used clinically, inasmuch as some FQs (e.g. trovafloxacin) have been shown to readily select *mexCD-oprJ*-hyperexpressing mutants in animal models of infection although others (e.g. ciprofloxacin) do not (Join-Lambert *et al.*, 2001).

Although MexCD-OprJ accommodates cefepime (Masuda *et al.*, 2000b) it is seldom linked to resistance to this agent in clinical isolates (Jeannot *et al.*, 2008). *mexC* was identified as one of the genes the most strongly induced by azithromycin (9 hour exposure) in one study (Kai *et al.*, 2009), contrasting with another that showed azithromycin induction of *mexCD-oprJ* expression in biofilms but not planktonic cells (Gillis *et al.*, 2005). Mutants lacking both MexCD-OprJ and MexAB-OprM are unable to form biofilms in the presence of azithromycin, suggesting that MexCD-OprJ is involved in azithromycin tolerance in biofilm cells (Gillis *et al.*, 2005). Consistent with this, azithromycin readily selects MexCD-OprJ-expressing *nfxB* mutants in biofilm-grown *P. aeruginosa* (Mulet *et al.*, 2009).

Hypermutable (or mutator) *P. aeruginosa* exhibiting increased mutation rates are common in chronic infections such as those that occur in the lungs of CF (Oliver, 2010; Oliver and Mena, 2010). The hypermutation phenotype of mutator stains results from defects in DNA repair, predominantly in the mismatch repair (MMR) system [encoded by *mutS*, *mutL* and *uvrD* (also known as *mutU*)] (Oliver, 2010) but also in the DNA oxidative repair (GO) system (encoded by *mutT*, *mutY* and *mutM*) (Oliver and Mena, 2010). Significantly from an antimicrobial resistance standpoint, mutator strains show higher rates of antimicrobial resistance development than non-mutator strains (Oliver *et al.*, 2000; Mandsberg *et al.*, 2009; Morero and Argarana, 2009; Ferroni *et al.*, 2009). Interestingly, and for reasons that are as yet unclear, ciprofloxacin-resistant mutants selected *in vitro* from mutants defective in the GO system were predominantly *mexCD-oprJ*-hyperexpressing *nfxB* mutants (Mandsberg *et al.*, 2009; Morero and Argarana, 2009) while ciprofloxacin-resistant derivatives of a *mutS* strain defective in MMR were predominantly target site *gyrA* and *parC* mutants (Morero and Argarana, 2009). Strikingly, this tendency away from *mexCD-oprJ*-expressing *nfxB* mutants in FQ-resistant *mutS* strains seemed to be specific to the *in vitro* situation inasmuch as ciprofloxacin-resistant mutants derived from a *mutS* strain passaged *in vivo* (i.e. in mice treated with this FQ) were invariably *mexCD-orpJ*-hyperexpressors (Macia *et al.*, 2006).

Regulation

mexCD-oprJ expression is controlled by a single known regulator, the NfxB repressor (Shiba *et al.*, 1995; Poole *et al.*, 1996a), and *in vitro*-selected (Hirai *et al.*, 1987; Masuda *et al.*, 1995; Poole *et al.*, 1996a) and clinical (Jakics *et al.*, 1992, 2000; Yoshida *et al.*, 1994; Jalal and Wretlind, 1998; Higgins *et al.*, 2003; Henrichfreise *et al.*, 2007) mutants expressing this efflux system and resistant to FQs invariably contain mutations in *nfxB* (Poole *et al.*, 1996a; Jalal and Wretlind, 1998; Jalal *et al.*, 2000; Higgins *et al.*, 2003; Henrichfreise *et al.*, 2007). MexCD-OprJ-hyperexpressing mutants selected by azithromycin (Mulet *et al.*, 2009) and triclosan (Chuanchuen *et al.*, 2001) also invariably contain mutations in *nfxB*. NfxB is a multimeric (A. Purssell and K. Poole, unpublished), presumed dimeric repressor that binds to the *nfxB-mexC* intergenic region (Shiba *et al.*, 1995; Poole *et al.*, 1996a) where it controls expression of *nfxB* and the *mexCD-oprJ* operon. Azithromycin-selected *mexCD-orpJ*-hyperexpressing mutants carry mutations throughout the *nfxB* gene (Mulet *et al.*, 2009) including in regions implicated in DNA-binding and NfxB dimerization (A. Purssell and K. Poole, unpublished).

Stress inducibility

An earlier observation that MexCD-OprJ-hyperexpressing FQ-resistant mutants of *P. aeruginosa* could be readily selected from a rat model of acute pneumonia in the absence of antimicrobial treatment (Join-Lambert *et al.*, 2001) was an indication that this efflux system was responding to a selective pressure in the animals that was unrelated to antimicrobials – i.e. this system had other than antimicrobial efflux as an intended function. That *mexCD-oprJ* hyperexpression in *nfxB* mutants is associated with impairment of *P. aeruginosa*'s type III secretion system (TTSS) (owing to impaired expression of the TTSS activator, ExsA and its target TTSS genes) (Linares *et al.*, 2005) and reduced fitness and virulence (Sanchez *et al.*, 2002a) also speak to this efflux system functioning as other than an antimicrobial export system. The subsequent demonstration that exposure of the organism to a variety of membrane-damaging biocides (chlorhexidine and benzalkonium chloride) and dyes (ethidium bromide and rhodamine) induced *mexCD-oprJ* expression (Morita *et al.*, 2003) was the first indication that envelope stress might be a 'natural' trigger for expression of this efflux system. More recently, additional biocides and membrane damaging agents (MDAs), including detergents, organic solvents, and cationic antimicrobial peptides have been shown to induce expression of *mexCD-oprJ* (Fraud *et al.*, 2008). Consistent with envelope stress being a signal for MexCD-OprJ recruitment, this induction was mediated by the *algU*-encoded envelope stress response sigma factor (Fraud *et al.*, 2008), a functionally equivalent homologue of *E. coli* RpoE (Yu *et al.*, 1995) that was first described as a regulator of alginate biosynthesis in *P. aeruginosa* (Martin *et al.*, 1993; Hershberger *et al.*, 1995). Not only is AlgU required for chlorhexidine induction of this efflux system, but hyperexpression of this efflux system in *nfxB* mutants is also substantially dependent upon AlgU (Fraud *et al.*, 2008). While MexCD-OprJ would appear to be a component of an envelope stress response in this organism its role in this is unclear. The exometabolome of an *nfxB* mutant relative to wild type does, however, show elevated levels of long-chain fatty aids, which have been proposed as possible MeCD-OprJ substrates (Stickland *et al.*, 2010). Perhaps this efflux system plays a role in fatty acid export as part of a system for exchanging these components of membrane lipids as the cell responds to envelope stress and restructures its membranes accordingly. How AlgU responds to MDAs in promoting *mexCD-oprJ* expression is, however, as yet unknown.

MexEF-OprN

Also quiescent in wild type cells grown under standard laboratory conditions (Köhler *et al.*, 1997a; Srikumar *et al.*, 1997) the MexEF-oprN multidrug efflux system is expressed and, so, contributes to multidrug resistance in so-called *nfxC* mutants (Fukuda *et al.*, 1995; Jalal *et al.*, 2000; Maseda *et al.*, 2000) and in mutants defective in the *mexS* gene [also known as *qrh* (Köhler *et al.*, 1999a; Ramsey and Whiteley, 2004) and PA2491 (Sobel *et al.*, 2005b)] encoding a putative oxidoreductase of as yet unknown function (Sobel *et al.*, 2005b). MexEF-OprN accommodates a rather narrow range of antimicrobial substrates that include FQs, trimethoprim and chloramphenicol (Table 9.2), the latter of which readily selects MexEF-oprN-expressing multidrug-resistant *mexS* mutants *in vitro* (Sobel *et al.*, 2005b). The MexEF-OprN pump also accommodates organic solvents (Li *et al.*, 1998), dyes (Germ *et al.*, 1999) and biocides such as triclosan (Chuanchuen *et al.*, 2001). MexEF-OprN-hyperproducing *nfxC* (Köhler *et al.*, 1997a, 2001) and *mexS* (*qrh*) (Ramsey and Whiteley, 2004) mutants also show reduced production of QS-dependent gene products and AHLs, consistent with an ability to accommodate QS signalling molecules and reminiscent of the AHL-exporting MexAB-OprM multidrug efflux system. Enhanced resistance to imipenem that is also seen in *nfxC* (Köhler *et al.*, 1997a) and *mexS* (Sobel *et al.*, 2005b) strains results not from MexEF-OprN expression but the concomitant decrease in outer membrane protein OprD in these mutants (Fukuda *et al.*, 1990; Masuda *et al.*, 1995; Sobel *et al.*, 2005b). OprD is an imipenem channel and a primary route of entry of this antibiotic in *P. aeruginosa* (Trias and Nikaido, 1990) whose absence is often seen in imipenem-resistant strains of *P. aeruginosa* (Ballestero *et al.*, 1996; Köhler *et al.*, 1999b; Livermore, 2001).

Clinical significance

While FQs do readily select *mexEF-oprN*-expressing *nfxC* mutants *in vitro* (Köhler *et al.*, 1997a,b; Join-Lambert *et al.*, 2001) there are few reports of FQ-resistant clinical *nfxC* mutants (Fukuda *et al.*, 1995; Jalal *et al.*, 2000) and no reports of clinical *mexS* mutants. As well, despite the ready selection of *mexEF-oprN*-expressing *mexS* mutants by chloramphenicol (Sobel *et al.*, 2005b) this drug is generally not used in the clinic.

Regulation

Unlike the other FQ-exporting RND-type efflux systems, expression of *mexEF-oprN* is regulated by a transcriptional activator, MexT (Köhler *et al.*, 1999a; Ochs *et al.*, 1999), a LysR family regulator that is required for efflux gene expression and, so, MexEF-OprN-mediated FQ/multidrug resistance (Fetar, H., unpublished observation). *mexT* occurs upstream of *mexEF-oprN* and downstream of *mexS* (Köhler *et al.*, 1999a), the latter gene also positively regulated by MexT (Köhler *et al.*, 1999a;

Tian *et al.*, 2009a). Unusually, many so-called wild type stains carry inactivating mutations in *mexT* (Maseda *et al.*, 2000), with *mexEF-oprN* expression and resistance in *nfxC* mutants resulting from reversion of these mutations (Maseda *et al.*, 2000). The coordinate reduction in *oprD* expression seen in *nfxC* mutants also appears to be MexT-dependent – cloned *mexT* in a wild type background reduces *oprD* expression to levels seen in an *nfxC* mutant (Köhler *et al.*, 1999a). MexT's negative influence on *oprD* expression occurs at the level of transcription (Ochs *et al.*, 1999) although whether this regulator binds to the *oprD* promoter region and impacts expression directly has not been assessed. Hyperexpression of *mexCD-oprJ* (and reduction in OprD production) is also seen in lab isolates disrupted in the *mexS* gene (Sobel *et al.*, 2005b; Fetar *et al.*, 2011), and *mexEF-oprN* hyperexpression in such mutants is also MexT-dependent (Sobel *et al.*, 2005b). *mexEF-oprN* expression and modest multidrug resistance has also been reported in a mutant disrupted in the *mvaT* gene (Westfall *et al.*, 2006) encoding a global regulator of virulence gene expression (Diggle *et al.*, 2002). There are, however, no reports of *mexEF-oprN*-expressing clinical resistant isolates harbouring mutations in *mexS* or *mvaT*. Both *mexS* and *mexEF-oprN* are downregulated in a mutant defective in the global regulatory gene *rsaL* gene that encodes a negative regulator of QS (Rampioni *et al.*, 2009) although how these genes might be linked operationally to RsaL and QS is unknown. MexT regulates a number of virulence traits in *P. aeruginosa* independent of its influence on *mexEF-orpN* and quorum-sensing (Tian *et al.*, 2009b), indicating that this regulator is a more global regulator in this organism with additional targets beyond efflux. Indeed, a transcriptomic study of genes positively influenced by cloned *mexT* identified a number of genes directly regulated by MexT including a xenobiotic reductase gene, *xenB*, an ABC transporter operon (which included a glutathione-S-transferase gene), and several genes encoding small, apparently exported proteins of unknown function (Tian *et al.*, 2009a). Interestingly, *xenB* is also upregulated in a *mexS* mutant but only in the absence of *mexEF-oprN* and this is MexT dependent (Fetar *et al.*, 2011).

The MexT protein shows a high degree of similarity with LysR family regulator NodD, first found in *Azorhizobium* sp. and shown to bind to a *nod* box having the nucleotide sequence ATC-N9-GAT (Goethals *et al.*, 1992). MexT binding to the *mexT*-mexE intergenic region, which contains two *nod* boxes, has been demonstrated (Tian *et al.*, 2009a; Maseda *et al.*, 2010) with only the more *mexT* proximal *nod* box ultimately shown to bind MexT *in vitro* (Maseda *et al.*, 2010) (H. Fetar and K. Poole, unpublished). This contrasted with a recent study that used *in silico* and microarray analyses in proposing a consensus (and likely MexT-binding) sequence of ATCA(N5)GTCGAT(N4)ACYAT upstream of *mexEF-oprN* and other MexT-regulated genes (Tian *et al.*, 2009a). This sequence contains the *mexT*-distal *nod* box which, while required for *mexEF-oprN* expression (Maseda *et al.*, 2010) did not bind MexT (Maseda *et al.*, 2010) (H. Fetar and K. Poole, unpublished). MexT binding in these *in vitro* studies occurred in the absence of a signal/coinducer, consistent with the known mechanism of operation of many LysR family regulators, which often bind to a promoter distal site upstream of target genes in the absence of coinducer, with coinducer then promoting regulator binding to a more promoter-proximal site that serves to stimulate target gene expression (Schell, 1993; Maddocks and Oyston, 2008). The exact nature of the signal(s)/inducer(s) to which MexT responds in promoting expression of *mexEF-oprN* and other MexT-regulated genes is unclear although a number of *in vitro* and *in vivo* treatments have been shown to induce expression of this efflux system (see below).

Stress inducibility

Despite the identification of regulatory genes impacting expression of *mexEF-oprN*, little is known about the signals to which the regulators respond in promoting efflux gene expression and, in fact, it is far from clear that antimicrobial efflux and resistance is the intended function of this pump. Mutants overexpressing MexEF-OprN were, for example, readily recovered from an experimental model of rat pneumonia in the absence of antibiotic selection (Join-Lambert *et al.*, 2001) indicating some advantage of MexEF-OprN expression *in vivo*, independent of antimicrobial export. Transcriptome analysis of *P. aeruginosa* exposed to airway epithelial cells has revealed substantial increases in expression *mexEF-oprN* and *mexS* (then called PA2491) expression, apparently in parallel with increased damage of the epithelial cells (and release of cell contents?) (Frisk *et al.*, 2004), although the efflux-recruiting signal(s) released by these cells is unknown. Possibly, epithelial cell contents generate intracellular pools of metabolite substrates for MexAB-OprM/PA2491, either following uptake of epithelial cell contents as, for example, nutrients or as a result of physiological changes promoted by exposure to these cell contents (a stress response?). In any case, these data suggest that *P. aeruginosa* may well encounter circumstances *in vivo* in which MexS and MexEF-OprN are needed.

The observation that mutational loss of the VsqR QS and virulence regulator compromises *mexEF-oprN* expression in cells under oxidative stress suggests that this system may normally be induced in response to this stressor (Juhas *et al.*, 2004). A synthetic derivate of a natural furanone compound that functions as an antagonist of QS in *P. aeruginosa* was also shown to induce *mexEF-oprN* expression (Hentzer *et al.*, 2003). Hyperexpression of MexEF-OprN has been shown to compromise virulence (Cosson *et al.*, 2002) possibly as a result of its negative impact on the TTSS charged with delivery of *P. aeruginosa* toxins into cells of infected tissues (Linares *et al.*, 2005). That MexEF-OprN hyperexpression compromises expression of TTSS genes was explained by its export of intracellular signalling molecules needed to activate these genes, although given the apparent connection between expression of certain metabolic genes and type III secretion, it could not be ruled out that hyperexpression of this efflux system somehow impacted cellular metabolism (Linares *et al.*, 2005). Clearly, however, these studies highlight a role for MexEF-OprN independent of its drug-exporting capability.

Most recently, *mexEF-oprN* was shown to be induced in response to nitrosative stress [e.g. in the presence of the nitrosating agent *S*-nitrosoglutathione (GSNO) or the NO·-generating agent diethylamine triamine NONOate (DETA)] and this was dependent on MexT (Fetar *et al.*, 2011). Moreover, several of the MexT targets identified in an array study of the MexT regulon (Tian *et al.*, 2009a) were also shown to be induced in response to nitrosative stress, including *xenB*, the ABC-GST operon and representatives of the 'exported small proteins of unknown function' (Fetar *et al.*, 2011) suggesting that MexT controls expression of a regulon with some function in a nitrosative stress response. Of note, however, the nitrosative stress-inducible nitric oxide-detoxifying proteins flavohemoglobin (Fhp) and nitric oxide reductase (NorBC) (Firoved *et al.*, 2004) that are linked to survival of nitrosative stress in other bacteria (Poole, 2005c) are not MexT regulated (Tian *et al.*, 2009a).

Interestingly, chloramphenicol but not the related compound, florfenicol, which lacks a nitro group, induces *mexEF-oprN* expression, again dependent upon MexT (Fetar *et al.*, 2011). This highlights the importance of the nitro moiety of chloramphenicol for this induction, an interesting observation given that some common products of nitrosative stress in bacteria are nitrated amino acids (i.e. chloramphenicol may resemble a nitrated nitrosative stress product that is an intended signal for MexT and substrate MexS and/or MexEF-OprN). The observation that XenB, implicated in removal of nitro groups from nitroglycerine and trinitrotoluene in *Pseudomonas* (Blehert *et al.*, 1999; Pak *et al.*, 2000), is co-regulated with *mexEF-oprN* by nitrosative stress is consistent with these playing a role in 'detoxifying' nitrated products of nitrosative stress. Still, neither chloramphenicol nor a model nitrated amino acid, nitrotyrosine, interacts with MexT *in vitro* (Fetar, H., and Poole, K., unpublished). As such, chloramphenicol treatment of *P. aeruginosa* may stimulate the production *in vivo* of the MexT/MexS/MexAB-OprM substrates much as GSNO treatment does (N.B. GSNO also fails to bind MexT *in vitro*; Fetar, H., and Poole, K., unpublished). What those product(s) may be remains a mystery.

The observed co-regulation of genes for a presumed oxidoreductase (MexS) and efflux system (MexEF-OprN) both inversely regulated with an outer membrane amino acid/carbapenem channel gene (*oprD*) raises a number of questions, particularity given that neither *mexS* nor *oprD* seem to be regulated by nitrosative stress (*mexS* may be induced but only in mutants lacking *mexEF-oprN* and data for *oprD* are inconclusive; H. Fetar and K. Poole, unpublished). Given that *mexS* (*qrh*) is moderately constitutively expressed (Köhler *et al.*, 1999a) and its loss is needed for *mexEF-oprM* upregulation (Sobel *et al.*, 2005b; Fetar *et al.*, 2011), it seems likely that MexS/MexEF-OprM substrates are present in cells under 'normal' laboratory growth conditions and 'dealt' with primarily by MexS. Only under circumstances where the enzyme is unable to keep up with 'substrate' production, due either to mutational loss of *mexS* or, perhaps, an excess of 'substrate' production would MexEF-OprN be recruited. That enhanced *mexEF-oprN* expression in *mexS* (and earlier *nfxC*) mutants is coupled with reduced production of OprD, suggests that such substrates might well be derived from/dependent on amino acids/peptides, whose reduced uptake in such mutants would clearly limit the production of downstream substrates for MexEF-OprN and MexS. How these substrates relate (if at all) to those presumably generated by chloramphenicol treatment or nitrosative stress is unclear.

Additional MexX efflux systems

The remaining Mex efflux systems in *P. aeruginosa* (Table 9.1) provide only very modest resistance to a limited number of agents, typically only in mutants lacking the aforementioned four RND-type efflux systems and often only when overexpressed from a plasmid (Table 9.1). Not surprisingly, these have not been linked to antimicrobial resistance in clinical isolates. With few exceptions, their natural functions remain a mystery. *mexPQ-opmE* is noteworthy for its induction by copper (Cu) (Teitzel *et al.*, 2006; Thaden *et al.*, 2010) mediated by the CueR transcriptional regulator whose binding to the *mexPQ* promoter has been demonstrated. This and the observation that knockouts in *mexP* or *mexQ* increase susceptibility of *P. aeruginosa* to Cu (Thaden *et al.*, 2010) is consistent with MexPQ-OpmE functioning as a metal ion exporter. *muxABC-opmB*, unique amongst RND-type efflux systems in *P. aeruginosa*, encodes two RND components both of which are required for pump function (Mima *et al.*, 2009). This is reminiscent of the MdtABC-TolC efflux system in *Escherichia coli* (Nagakubo *et al.*, 2002), which shows significant sequence similarity to MuxABC (Mima *et al.*, 2009). Intriguingly, knockouts in *muxA* render cells more resistant to β-lactams and attenuated for virulence (Yang *et al.*, 2011) although the nature of resistance and virulence changes or the genes involved are unknown. Perhaps the best studied of the 'non-clinical' RND type pumps in *P. aeruginosa* is that encoded by the *mexGHI-opmD* operon. Originally linked to vanadium susceptibility (Aendekerk *et al.*, 2002) this pump is most intimately linked to QS – the efflux operon is positively regulated by QS regulators (Deziel *et al.*, 2005) and QS signalling molecules (Whiteley *et al.*, 1999; Wagner *et al.*, 2003; Dietrich *et al.*, 2006), is upregulated in a mutant defective in the QS negative regulator RsaL (Rampioni *et al.*, 2009), and mutants lacking the pump show defects in the production of QS signalling molecules and, thus, QS-dependent phenotypes, including virulence (Aendekerk *et al.*, 2002, 2005). The recent observation that indole and its oxidized form (7-hydroxyindole) downregulate *mexGHI-opmD* expression (Lee *et al.*, 2009) likely reflects the anti-QS properties of these molecules (they repress production of the PQS signalling molecule that normally promotes *mexGHI-opmD* expression) (Tashiro *et al.*, 2010). Finally, uracil (presumably a uracil-containing signalling nucleotide) has been shown to promote expression of QS-regulated genes, including *mexGHI-opmD*, in *P. aeruginosa* (Ueda *et al.*, 2009). The observation that indole represses a variety of genes involved in uracil/uracil nucleotide synthesis in *E. coli* (Lee *et al.*, 2008) suggests that indole's impact on QS and, so, *mexGHI-opmD* occurs via a uracil-derived signalling molecule.

Efflux inhibition

Given the significant contribution of RND-type efflux systems to antimicrobial resistance in *P. aeruginosa*, some effort has been devoted to developing strategies for overcoming efflux (Table 9.3). The development of RND efflux pump inhibitors (EPIs) began more than a decade ago with a number of broad-spectrum as well as pump-specific (MexAB-OprM) inhibitors identified. These early broad-spectrum EPIs were dipeptides, the first of which, L-Phe-L-Arg-β-naphthylamine (PAβN) has been shown to be active against

MexAB-OprM, MexCD-OprJ and MexEF-OprN in *P. aeruginosa* (Lomovskaya *et al.*, 2001) as well RND type pumps in a variety of other organisms (Lomovskaya and Watkins, 2001). As such, this inhibitor now enjoys general usage *in vitro* as an EPI and diagnostic agent for assessing an efflux contribution to resistance in *P. aeruginosa* (Kriengkauykiat *et al.*, 2005; Tohidpour *et al.*, 2009) as well as other organisms (Mamelli *et al.*, 2003; Hasdemir *et al.*, 2004; Kurincic *et al.*, 2007; Mamelli *et al.*, 2007; Tran *et al.*, 2009; Hirata *et al.*, 2010). More recently described pump inhibitors with activity against *P. aeruginosa* include alkylaminoquinazolines (Chevalier *et al.*, 2010; Mahamoud *et al.*, 2011) and the anti-malarial compound mefloquine (Vidal-Aroca *et al.*, 2009), although alkylaminoquinazolines are less effective than PAβN and the target pump(s) of these inhibitors have not been identified. Despite these many reports of pump inhibitors, however, their clinical utility, if any, remains to be established.

In light of the earlier suggestion that RND pump-mediated resistance to antimicrobials is dependent on an outer membrane barrier (Nikaido, 1998) an alternative approach to countering efflux-mediated antimicrobial resistance is barrier destruction using membrane permeabilizers. Indeed, known membrane permeabilizers such as EDTA and sodium hexametaphosphate (Li *et al.*, 2000b) as well as recently described membrane-permeabilizing lactoferrin-derived small peptides (Sanchez-Gomez *et al.*, 2011) were effective at reversing antimicrobial resistance resultant from efflux pump hyperexpression in *P. aeruginosa*. Finally, preliminary efforts at down-regulating efflux gene expression (*mexAB-oprM*) as a way of countering efflux-mediated antimicrobial resistance have been described, with both liposome-encapsulated anti-sense *oprM* oligonucleotides (produces an antiparallel RNA duplex with *oprM* mRNA) (Wang *et al.*, 2010a) and parallel complementary *mexA* RNA (produces

Table 9.3 Approaches to counter efflux-mediated antimicrobial resistance in *P. aeruginosa*

Method	Reference
Efflux pump inhibition	
Broad-spectrum RND pump inhibitors	
L-Phe-L-Arg-β-naphthylamine (PAβN, MC-207,110)	Renau *et al.* (1999), Lomovskaya *et al.* (2001)
D-Orn-D-hPhe-3-aminoquinoline (MC-02,595) and analogues, including L-Phe-L-Orn-β-naphthylamine (MC-04,124)	Renau *et al.* (2001, 2002, 2003), Watkins *et al.* (2003)
Alkylaminoquinazolines	Chevalier *et al.* (2010), Mahamoud *et al.* (2011)
Mefloquine[a]	Vidal-Aroca *et al.* (2009)
Pump-specific inhibitors	
MexAB-OprM-specific inhibitor and its derivatives, including a pre-clinical candidate, D13–9001	Nakayama *et al.* (2003a,b, 2004a,b), Yoshida *et al.* (2006a,b, 2007)
Membrane permeabilization	
Lactoferrin-derived peptides[b]	Sanchez-Gomez *et al.* (2011)
EDTA, sodium hexametaphosphate[b]	Li *et al.* (2000)
Gene silencing/antisense	
oprM-directed antisense phosphorothionate oligonucleotide[c]	Wang *et al.* (2010)
mexA-specific parallel complementary RNA[d]	Liu *et al.* (2009)

[a]Mefloquine was shown to interfere with the efflux activity of MexAB-OpM, the only efflux system tested in *P. aeruginosa*, although an ability to potentiate antimicrobial agents was not assessed.
[b]These agents countered much of the antimicrobial resistance afford by efflux pump-hyperexpression in mutant strains.
[c]Reduced expression of *oprM* and enhanced susceptibility to MexAB-OprM substrate antimicrobials was seen in antisense-treated cells.
[d]Increased ethidium bromide accumulation (a sign of reduced efflux activity) and enhanced susceptibility to MexAB-OprM substrate antimicrobials was seen in cells expressing the indicated complementary RNA.

a parallel RNA duplex with *mexA* mRNA) (Liu *et al.*, 2009) both demonstrating an ability to reduce efflux and resistance to MexAB-OpM antimicrobial substrates.

Concluding remarks

Antimicrobial export mediated by RND-family multidrug efflux systems is a significant determinant of intrinsic insusceptibility of *P. aeruginosa* to a range of antimicrobials, and hyperexpression of these systems in mutants strains contributes to acquired resistance to a number of clinically relevant antimicrobials (Poole, 2004). In addition, the broad substrate specificity of these efflux system often undermines the utility of novel agents [e.g. tigecycline (Dean *et al.*, 2003)] and therapeutic approaches [e.g. antimicrobial photodynamic inactivation (Tegos *et al.*, 2008)] in treating *P. aeruginosa* as they are all too often substrates for RND-type multidrug efflux pumps. This is seen, too, in recent studies of plant natural products as antimicrobials and their ineffectiveness against *P. aeruginosa* owing to efflux-mediated resistance (Kuete *et al.*, 2010, 2011). Complicating matters is the recognition that antimicrobial export is not the intended, natural function of these efflux systems (Poole, 2008) and so unknown or poorly understood processes govern their expression and unknown or poorly understood environmental circumstances or physiological conditions inside the cell provide a stimulus for this expression and, possibly, selective pressure for efflux-expressing antimicrobial-resistant mutants. With few exceptions antimicrobials do not induce or influence efflux gene expression, and when they do it is as a result of stresses imposed on the cell or a mimicking of a probably natural substrate or inducer. It is, for example, increasingly appreciated that RND type efflux systems in *P. aeruginosa* respond to environmental stresses and, so, play some role in stress responses in this organism (see Poole, 2008, 2012, and above). As such, stress in the environment governs efflux gene expression and provides selective pressure for efflux-expressing antimicrobial-resistant mutants. This is seen in *in vitro* studies showing that oxidative stress enhances the frequency of recovery of pan-aminoglycoside-resistant mutants (Fraud and Poole, 2010), an observation with a clinically relevant connection in *P. aeruginosa* infection of the CF lung, the latter an environment rich in oxidative stress-promoting ROS (Ciofu *et al.*, 2005). Environmental stress control of efflux gene expression may also explain the enhanced recovery of MexEF-OpN-expressing FQ-resistant mutants *in vivo* versus *in vitro* (Join-Lambert *et al.*, 2001), inasmuch as this efflux system is induced by nitrosative stress (Fetar *et al.*, 2011) and infecting organisms experience nitrosative stress as a result of the elucidation of antimicrobial reactive nitrogen species by immune cells in response to infection (Zaki *et al.*, 2005). Finally, the mutational upregulation of multiple RND multidrug efflux systems in individual clinical isolates where the link to/correlation with antimicrobial resistance is not always clear (Quale *et al.*, 2006; Hocquet *et al.*, 2007a; Campo Esquisabel *et al.*, 2011) may also be explained by *in vivo* circumstances other than antimicrobials (possibly stress) providing the pressure for their selection.

Clearly, a better understanding of the 'intended' function of RND type multidrug efflux systems of clinical relevance, the environmental conditions and signals to which they respond, is necessary in order to appreciate when and where they are likely to be a confounding factor for chemotherapy. A fuller appreciation of the details of their regulation, their broad substrate-specificity and mode(s) of operation will be invaluable in informing strategies for countering their operation (using inhibitors or, possibly by modulating their expression, either directly or by nullifying environmental signals needed for their expression) and for designing more efficacious antimicrobials (i.e. those poorly accommodated by these pumps) (Poole and Lomovskaya, 2006). Knowing the how and why behind the broad substrate specificity of RND type pumps, which sees accommodation of a broad range of antimicrobials by some pumps (MexAB-OprM, MexCD-OprJ, MexXY-OprM) and a more restricted range in others (MexEF-OprN), and the identity of so-called intended substrates, will enhance our appreciation of the roles of these pumps in the cell and the intricacies of multisubstrate recognition by bacterial exporters. Ultimately, a minority of efflux systems in *P. aeruginosa*, including a minority of RND type efflux systems, accommodate clinically relevant antimicrobials in a meaningful way and so the study of these pumps is also a study of the basic biology of *P. aeruginosa*. While RND pumps are common in Gram-negative organisms, their numbers vary from organism to organism and *P. aeruginosa* harbours amongst the largest numbers of these pumps of any organism (eleven). Clearly, then, they speak of either functions/needs not present in many other organisms or alternative approaches to common functions/needs.

References

Aendekerk, S., Ghysels, B., Cornelis, P., and Baysse, C. (2002). Characterization of a new efflux pump, MexGHI-OpmD, from *Pseudomonas aeruginosa* that confers resistance to vanadium. Microbiology *148*, 2371–2381.

Aendekerk, S., Diggle, S.P., Song, Z., Hoiby, N., Cornelis, P., Williams, P., and Camara, M. (2005). The MexGHI-OpmD multidrug efflux pump controls growth, antibiotic susceptibility and virulence in *Pseudomonas aeruginosa* via 4-quinolone-dependent cell-to-cell communication. Microbiology *151*, 1113–1125.

Aires, J.R., Köhler, T., Nikaido, H., and Plesiat, P. (1999). Involvement of an active efflux system in the natural resistance of *Pseudomonas aeruginosa* to aminoglycosides. Antimicrob. Agents Chemother. *43*, 2624–2628.

Akama, H., Kanemaki, M., Yoshimura, M., Tsukihara, T., Kashiwagi, T., Yoneyama, H., Narita, S.I., Nakagawa, A., and Nakae, T. (2004a). Crystal structure of the drug-discharge outer membrane protein, OprM, of *Pseudomonas aeruginosa*: dual modes of membrane anchoring and occluded cavity end. J. Biol. Chem. *17*, 52816–52819.

Akama, H., Matsuura, T., Kashiwagi, S., Yoneyama, H., Narita, S., Tsukihara, T., Nakagawa, A., and Nakae, T. (2004b). Crystal structure of the membrane fusion protein, MexA, of the multidrug transporter in *Pseudomonas aeruginosa*. J. Biol. Chem. *279*, 25939–25942.

Alguel, Y., Lu, D., Quade, N., Sauter, S., and Zhang, X. (2010). Crystal structure of MexZ, a key repressor responsible for antibiotic resistance in *Pseudomonas aeruginosa*. J. Struct. Biol. *172*, 305–310.

Alonso, A., Campanario, E., and Martinez, J.L. (1999). Emergence of multidrug-resistant mutants is increased under antibiotic selective pressure in *Pseudomonas aeruginosa*. Microbiology *145*, 2857–2862.

Anderson, G.G., Moreau-Marquis, S., Stanton, B.A., and O'Toole, G.A. (2008). *In vitro* analysis of tobramycin-treated *Pseudomonas aeruginosa* biofilms on cystic fibrosis-derived airway epithelial cells. Infect. Immun. *76*, 1423–1433.

Ball, P.R., Shales, S.W., and Chopra, I. (1980). Plasmid-mediated tetracycline resistance in *Escherichia coli* involves increased efflux of the antibiotic. Biochem. Biophys. Res. Commun. *93*, 74–81.

Ballestero, S., Fernandez-Rodriguez, A., Villaverde, R., Escobar, H., Perez-Diaz, J.C., and Baquero, F. (1996). Carbapenem resistance in *Pseudomonas aeruginosa* from cystic fibrosis patients. J. Antimicrob. Chemother. *38*, 39–45.

Ballesteros, M., Fredriksson, A., Henriksson, J., and Nystrom, T. (2001). Bacterial senescence: protein oxidation in non-proliferating cells is dictated by the accuracy of the ribosomes. EMBO J. *20*, 5280–5289.

Barclay, M.L., Begg, E.J., Chambers, S.T., Thornley, P.E., Pattemore, P.K., and Grimwood, K. (1996). Adaptive resistance to tobramycin in *Pseudomonas aeruginosa* lung infection in cystic fibrosis. J. Antimicrob. Chemother. *37*, 1155–1164.

Baum, E.Z., Crespo-Carbone, S.M., Morrow, B.J., Davies, T.A., Foleno, B.D., He, W., Queenan, A.M., and Bush, K. (2009). Effect of MexXY overexpression on ceftobiprole susceptibility in *Pseudomonas aeruginosa*. Antimicrob. Agents Chemother *53*, 2785–2790.

Bissonnette, L., Champetier, S., Buisson, J.P., and Roy, P.H. (1991). Characterization of the nonenzymatic chloramphenicol resistance (*cmlA*) gene of the In4 integron of Tn*1696*: similarity of the product to transmembrane transport proteins. J. Bacteriol. *173*, 4493–4502.

Blair, J.M., and Piddock, L.J. (2009). Structure, function and inhibition of RND efflux pumps in Gram-negative bacteria: an update. Curr. Opin. Microbiol. *12*, 512–519.

Blehert, D.S., Fox, B.G., and Chambliss, G.H. (1999). Cloning and sequence analysis of two *Pseudomonas* flavoprotein xenobiotic reductases. J. Bacteriol. *181*, 6254–6263.

Boutoille, D., Corvec, S., Caroff, N., Giraudeau, C., Espaze, E., Caillon, J., Plesiat, P., and Reynaud, A. (2004). Detection of an IS*21* insertion sequence in the *mexR* gene of *Pseudomonas aeruginosa* increasing β-lactam resistance. FEMS Microbiol. Lett. *230*, 143–146.

Bryan, L.E., Haraphongse, R., and Van den Elzen, H.M. (1976). Gentamicin resistance in clinical-isolates of *Pseudomonas aeruginosa* associated with diminished gentamicin accumulation and no detectable enzymatic modification. Antibiot, J.. (Tokyo) *29*, 743–753.

Campo Esquisabel, A.B., Rodriguez, M.C., Campo-Sosa, A.O., Rodriguez, C., and Martinez-Martinez, L. (2011). Mechanisms of resistance in clinical isolates of *Pseudomonas aeruginosa* less susceptible to cefepime than to ceftazidime. Clin. Microbiol. Infect. *17*, 1817–1822.

Cao, L., Srikumar, R., and Poole, K. (2004). MexAB-OprM hyperexpression in NalC type multidrug resistant *Pseudomonas aeruginosa*: identification and characterization of the *nalC* gene encoding a repressor of PA3720-PA3719. Mol. Microbiol. *53*, 1423–1436.

Caughlan, R.E., Sriram, S., Daigle, D.M., Woods, A.L., Buco, J., Peterson, R.L., Dzink-Fox, J., Walker, S., and Dean, C.R. (2009). Fmt bypass in *Pseudomonas aeruginosa* causes induction of MexXY efflux pump expression. Antimicrob. Agents Chemother. *53*, 5015–5021.

Cavallo, J.D., Plesiat, P., Couetdic, G., Leblanc, F., and Fabre, R. (2002). Mechanisms of β-lactam resistance in *Pseudomonas aeruginosa*: prevalence of OprM.-overproducing strains in a French multicentre study (1997). J. Antimicrob. Chemother. *50*, 1039–1043.

Cavallo, J.D., Hocquet, D., Plesiat, P., Fabre, R., and Roussel-Delvallez, M. (2007). Susceptibility of *Pseudomonas aeruginosa* to antimicrobials: a 2004 French multicentre hospital study. J. Antimicrob. Chemother. *59*, 1021–1024.

Chang, W., Small, D.A., Toghrol, F., and Bentley, W.E. (2005a). Microarray analysis of *Pseudomonas aeruginosa* reveals induction of pyocin genes in response to hydrogen peroxide. BMC. Genomics *6*, 115.

Chang, W., Small, D.A., Toghrol, F., and Bentley, W.E. (2005b). Microarray analysis of toxicogenomic effects of peracetic acid on *Pseudomonas aeruginosa*. Environ. Sci. Technol. *39*, 5893–5899.

Chen, H., Hu, J., Chen, P.R., Lan, L., Li, Z., Hicks, L.M., Dinner, A.R., and He, C. (2008). The *Pseudomonas aeruginosa* multidrug efflux regulator MexR uses an oxidation-sensing mechanism. Proc. Natl. Acad. Sci. U.S.A. *105*, 13586–13591.

Chen, H., Yi, C., Zhang, J., Zhang, W., Ge, Z., Yang, C.G., and He, C. (2010). Structural insight into the oxidation-sensing mechanism of the antibiotic resistance of regulator MexR. EMBO Rep. *11*, 685–690.

Chevalier, J., Mahamoud, A., Baitiche, M., Adam, E., Viveiros, M., Smarandache, A., Militaru, A., Pascu, M.L., Amaral, L., and Pages, J.M. (2010). Quinazoline derivatives are efficient chemosensitizers of antibiotic activity in *Enterobacter aerogenes, Klebsiella pneumoniae* and *Pseudomonas aeruginosa* resistant strains. Int. J. Antimicrob. Agents *36*, 164–168.

Chevalier, J., Mahamoud, A., Baitiche, M., Adam, E., Viveiros, M., Smarandache, A., Militaru, A., Pascu, M.L., Amaral, L., and Pages, J.M. (2010). Quinazoline derivatives are efficient chemosensitizers of antibiotic activity in *Enterobacter aerogenes, Klebsiella pneumoniae* and *Pseudomonas aeruginosa* resistant strains. Int. J. Antimicrob. Agents *36*, 164–168.

Chuanchuen, R., Beinlich, K., Hoang, T.T., Becher, A., Karkhoff-Schweizer, R.R., and Schweizer, H.P. (2001). Cross-resistance between triclosan and antibiotics in *Pseudomonas aeruginosa* is mediated by multidrug efflux pumps: exposure of a susceptible mutant strain to triclosan selects *nfxB* mutants overexpressing MexCD-OprJ. Antimicrob. Agents Chemother. *45*, 428–432.

Chuanchuen, R., Murata, T., Gotoh, N., and Schweizer, H.P. (2005). Substrate-dependent utilization of OprM or OpmH by the *Pseudomonas aeruginosa* MexJK efflux pump. Antimicrob. Agents Chemother. *49*, 2133–2136.

Ciofu, O., Riis, B., Pressler, T., Poulsen, H.E., and Hoiby, N. (2005). Occurrence of hypermutable *Pseudomonas aeruginosa* in cystic fibrosis patients is associated with the oxidative stress caused by chronic lung inflammation. Antimicrob. Agents Chemother. *49*, 2276–2282.

Cosson, P., Zulianello, L., Join-Lambert, O., Faurisson, F., Gebbie, L., Benghezal, M., Van Delden, C., Curty, L.K., and Kohler, T. (2002). *Pseudomonas aeruginosa* virulence analyzed in a *Dictyostelium discoideum* host system. J. Bacteriol. *184*, 3027–3033.

Daigle, D.M., Cao, L., Fraud, S., Wilke, M.S., Pacey, A., Klinoski, R., Strynadka, N.C., Dean, C.R., and Poole, K. (2007). Protein modulator of multidrug efflux gene expression in *Pseudomonas aeruginosa*. J. Bacteriol. *189*, 5441–5451.

Davies, J.P., Chen, F.W., and Ioannou, Y.A. (2000). Transmembrane molecular pump activity of Niemann-Pick C1 protein. Science *290*, 2295–2298.

Davies, K.J., and Lin, S.W. (1988a). Degradation of oxidatively denatured proteins in *Escherichia coli*. Free Radic. Biol. Med. *5*, 215–223.

Davies, K.J., and Lin, S.W. (1988b). Oxidatively denatured proteins are degraded by an ATP-independent proteolytic pathway in *Escherichia coli*. Free Radic. Biol. Med. *5*, 225–236.

Dean, C.R., Visalli, M.A., Projan, S.J., Sum, P.-E., and Bradford, P.A. (2003). Efflux-mediated resistance to tigecycline (GAR-936) in *Pseudomonas aeruginosa* PAO1. Antimicrob. Agents Chemother. *47*, 972–978.

Dean, R.T., Fu, S., Stocker, R., and Davies, M.J. (1997). Biochemistry and pathology of radical-mediated protein oxidation. Biochem. J. *324*, 1–18.

Deziel, E., Gopalan, S., Tampakaki, A.P., Lepine, F., Padfield, K.E., Saucier, M., Xiao, G., and Rahme, L.G. (2005). The contribution of MvfR to *Pseudomonas aeruginosa* pathogenesis and quorum sensing circuitry regulation: multiple quorum sensing-regulated genes are modulated without affecting *lasRI, rhlRI* or the production of N-acyl-L-homoserine lactones. Mol. Microbiol. *55*, 998–1014.

Dietrich, L.E., Price-Whelan, A., Petersen, A., Whiteley, M., and Newman, D.K. (2006). The phenazine pyocyanin is a terminal signalling factor in the quorum sensing network of *Pseudomonas aeruginosa*. Mol. Microbiol. *61*, 1308–1321.

Diggle, S.P., Winzer, K., Lazdunski, A., Williams, P., and Camara, M. (2002). Advancing the quorum in *Pseudomonas aeruginosa*: MvaT and the regulation of *N*-acylhomoserine lactone production and virulence gene expression. J. Bacteriol. *184*, 2576–2586.

Dukan, S., and Nystrom, T. (1998). Bacterial senescence: stasis results in increased and differential oxidation of cytoplasmic proteins leading to developmental induction of the heat shock regulon. Genes Dev. *12*, 3431–3441.

Dukan, S., Farewell, A., Ballesteros, M., Taddei, F., Radman, M., and Nystrom, T. (2000). Protein oxidation in response to increased transcriptional or translational errors. Proc. Natl. Acad. Sci U.S.A. *97*, 5746–5749.

Dupont, P., Hocquet, D., Jeannot, K., Chavanet, P., and Plesiat, P. (2005). Bacteriostatic and bactericidal activities of eight fluoroquinolones against MexAB-OprM-overproducing clinical strains of *Pseudomonas aeruginosa*. J. Antimicrob. Chemother. *55*, 518–522.

El'Garch, F., Jeannot, K., Hocquet, D., Llanes-Barakat, C., and Plesiat, P. (2007). Cumulative effects of several nonenzymatic mechanisms on the resistance of *Pseudomonas aeruginosa* to aminoglycosides. Antimicrob. Agents Chemother. *51*, 1016–1021.

Evans, H.L., Lefrak, S.N., Lyman, J., Smith, R.L., Chong, T.W., McElearney, S.T., Schulman, A.R., Hughes, M.G., Raymond, D.P., Pruett, T.L., *et al.* (2007). Cost of Gram-negative resistance. Crit. Care Med. *35*, 89–95.

Evans, K., Passador, L., Srikumar, R., Tsang, E., Nezezon, J., and Poole, K. (1998). Influence of the MexAB-OprM multidrug efflux system on quorum-sensing in *Pseudmonas aeruginosa*. J. Bacteriol. *180*, 5443–5447.

Evans, K., Adewoye, L., and Poole, K. (2001). MexR repressor of the *mexAB-oprM* multidrug efflux operon of *Pseudomonas aeruginosa*: identification of MexR binding sites in the *mexA-mexR* intergenic region. J. Bacteriol. *183*, 807–812.

Feliziani, S., Lujan, A.M., Moyano, A.J., Sola, C., Bocco, J.L., Montanaro, P., Canigia, L.F., Argarana, C.E., and Smania, A.M. (2010). Mucoidy, quorum sensing, mismatch repair and antibiotic resistance in *Pseudomonas aeruginosa* from cystic fibrosis chronic airways infections. PLoS. ONE. *5*, pii, e12669.

Ferroni, A., Guillemot, D., Moumile, K., Bernede, C., Le Bourgeois, M., Waernessyckle, S., Descamps, P., Sermet-Gaudelus, I., Lenoir, G., Berche, P., *et al.* (2009). Effect of mutator *P. aeruginosa* on antibiotic resistance acquisition and respiratory function in cystic fibrosis. Pediatr. Pulmonol. *44*, 820–825.

Fetar, H., Gilmour, C., Klinoski, R., Daigle, D.M., Dean, C.R., and Poole, K. (2011). *mexEF-oprN* multidrug efflux operon of *Pseudomonas aeruginosa*: regulation by the MexT activator in response to nitrosative stress and chloramphenicol. Antimicrob. Agents Chemother. *55*, 508–514.

Firoved, A.M., Wood, S.R., Ornatowski, W., Deretic, V., and Timmins, G.S. (2004). Microarray analysis and functional characterization of the nitrosative stress response in nonmucoid and mucoid *Pseudomonas aeruginosa*. J. Bacteriol. *186*, 4046–4050.

Fraud, S., and Poole, K. (2010). Oxidative stress induction of the *mexXY* multidrug efflux genes and promotion of aminoglycoside resistance development in *Pseudomonas aeruginosa*. Antimicrob. Agents Chemother. *55*, 1068–1074.

Fraud, S., Campigotto, A.J., Chen, Z., and Poole, K. (2008). The MexCD-OprJ multidrug efflux system of *Pseudomonas aeruginosa*: involvement in chlorhexidine resistance and induction by membrane damaging agents dependent upon the AlgU stress-response sigma factor. Antimicrob. Agents Chemother. *52*, 4478–4482.

Frisk, A., Schurr, J.R., Wang, G., Bertucci, D.C., Marrero, L., Hwang, S.H., Hassett, D.J., and Schurr, M.J. (2004). Transcriptome analysis of *Pseudomonas aeruginosa* after interaction with human airway epithelial cells. Infect. Immun. *72*, 5433–5438.

Fukuda, H., Hosaka, M., Hirai, K., and Iyobe, S. (1990). New norfloxacin resistance gene in *Pseudomonas aeruginosa* PAO. Antimicrob. Agents Chemother. *34*, 1757–1761.

Fukuda, H., Hosaka, M., Iyobe, S., Gotoh, N., Nishino, T., and Hirai, K. (1995). *nfxC*-type quinolone resistance in a clinical isolate of *Pseudomonas aeruginosa*. Antimicrob. Agents Chemother. *39*, 790–792.

Germ, M., Yoshihara, E., Yoneyama, H., and Nakae, T. (1999). Interplay between the efflux pump and the outer membrane permeability barrier in fluorescent dye accumulation in *Pseudomonas aeruginosa*. Biochem. Biophys. Res. Commun. *261*, 452–455.

Ghosh, S., Cremers, C.M., Jakob, U., and Love, N.G. (2011). Chlorinated phenols control the expression of the multi-drug resistance efflux pump MexAB-OprM in *Pseudomonas aeruginosa* by activating NalC. Mol. Microbiol. *79*, 1547–1556.

Giamarellos-Bourboulis, E.J., Papadimitriou, E., Galanakis, N., Antonopoulou, A., Tsaganos, T., Kanellakopoulou, K., and Giamarellou, H. (2006). Multidrug resistance to antimicrobials as a predominant factor influencing patient survival. Int. J. Antimicrob. Agents *27*, 476–481.

Gilleland, L.B., Gilleland, H.E., Gibson, J.A., and Champlin, F.R. (1989). Adaptive resistance to aminoglycoside antibiotics in *Pseudomonas aeruginosa*. J. Med. Microbiol. *29*, 41–50.

Gillis, R.J., White, K.G., Choi, K.H., Wagner, V.E., Schweizer, H.P., and Iglewski, B.H. (2005). Molecular basis of azithromycin-resistant *Pseudomonas aeruginosa* biofilms. Antimicrob. Agents Chemother. *49*, 3858–3867.

Goethals, K., Mergaert, P., Gao, M., Geelen, D., Van Montagu, M., and Holsters, M. (1992). Identification of a new inducible nodulation gene in *Azorhizobium caulinodans*. Mol. Plant Microbe Interact. *5*, 405–411.

Gotoh, N., Tsujimoto, H., Poole, K., Yamagishi, J.-I., and Nishino, T. (1995). The outer membrane protein OprM of *Pseudomonas aeruginosa* is encoded by *oprK* of the *mexA-mexB-oprK* multidrug resistance operon. Antimicrob. Agents Chemother. *39*, 2567–2569.

Gotoh, N., Tsujimoto, H., Tsuda, M., Okamoto, K., Nomura, A., Wada, T., Nakahashi, M., and Nishino, T. (1998). Charaterization of the MexC-MexD-OprJ multidrug efflux system in Δ*mexA-mexB-oprM* mutants of *Pseudomonas aeruginosa*. Antimicrob. Agents Chemother. *42*, 1938–1943.

Gutierrez, O., Juan, C., Cercenado, E., Navarro, F., Bouza, E., Coll, P., Perez, J.L., and Oliver, A. (2007). Molecular epidemiology and mechanisms of carbapenem resistance in *Pseudomonas aeruginosa* isolates from Spanish hospitals. Antimicrob. Agents Chemother. *51*, 4329–4335.

Hammami, S., Ghozzi, R., Burghoffer, B., Arlet, G., and Redjeb, S. (2009). Mechanisms of carbapenem resistance in non-metallo-β-lactamase-producing clinical isolates of *Pseudomonas aeruginosa* from a Tunisian hospital. Pathol. Biol. (Paris) *57*, 530–535.

Hamzehpour, M.M., Pechere, J.-C., Plesiat, P., and Koehler, T. (1995). OprK and OprM define two genetically distinct multidrug efflux systems in *Pseudomonas aeruginosa*. Antimicrob. Agents Chemother. *39*, 2392–2396.

Harms, J.M., Bartels, H., Schlunzen, F., and Yonath, A. (2003). Antibiotics acting on the translational machinery. J. Cell. Sci. *116*, 1391–1393.

Hasdemir, U.O., Chevalier, J., Nordmann, P., and Pages, J.M. (2004). Detection and prevalence of active drug efflux mechanism in various multidrug-resistant *Klebsiella pneumoniae* strains from Turkey. J. Clin. Microbiol. *42*, 2701–2706.

He, G.X., Kuroda, T., Mima, T., Morita, Y., Mizushima, T., and Tsuchiya, T. (2004). An H^+-coupled multidrug efflux pump, PmpM, a member of the MATE family of transporters, from *Pseudomonas aeruginosa*. J. Bacteriol. *186*, 262–265.

Henrichfreise, B., Wiegand, I., Pfister, W., and Wiedemann, B. (2007). Resistance mechanisms of multiresistant *Pseudomonas aeruginosa* strains from Germany and correlation with hypermutation. Antimicrob. Agents Chemother. *51*, 4062–4070.

Hentzer, M., Wu, H., Andersen, J.B., Riedel, K., Rasmussen, T.B., Bagge, N., Kumar, N., Schembri, M.A., Song, Z., Kristoffersen, P., *et al.* (2003). Attenuation of *Pseudomonas aeruginosa* virulence by quorum sensing inhibitors. EMBO J. *22*, 3803–3815.

Hermann, T. (2005). Drugs targeting the ribosome. Curr. Opin. Struct. Biol *15*, 355–366.

Hershberger, C.D., Ye, R.W., Parsek, M.R., Xie, Z.D., and Chakrabarty, A.M. (1995). The *algT* (*algU*) gene of *Pseudomonas aeruginosa*, a key regulator involved in alginate biosynthesis, encodes an alternative sigma factor (σE). Proc. Natl. Acad. Sci U.S.A. *92*, 7941–7945.

Hidron, A.I., Edwards, J.R., Patel, J., Horan, T.C., Sievert, D.M., Pollock, D.A., and Fridkin, S.K. (2008). NHSN annual update: antimicrobial-resistant pathogens associated with healthcare-associated infections: annual summary of data reported to the National Healthcare Safety Network at the Centers for Disease Control and Prevention, 2006–2007. Infect. Control Hosp. Epidemiol. *29*, 996–1011.

Higgins, P.G., Fluit, A.C., Milatovic, D., Verhoef, J., and Schmitz, F.J. (2003). Mutations in GyrA, ParC, MexR and NfxB in clinical isolates of *Pseudomonas aeruginosa*. Int. J. Antimicrob. Agents *21*, 409–413.

Higgins, M.K., Bokma, E., Koronakis, E., Hughes, C., and Koronakis, V. (2004). Structure of the periplasmic component of a bacterial drug efflux pump. Proc. Natl. Acad. Sci U.S.A. *101*, 9994–9999.

Hirai, K., Suzue, S., Irikura, T., Iyobe, S., and Mitsuhashi,s. (1987). Mutations producing resistance to norfloxacin in *Pseudomonas aeruginosa*. Antimicrob. Agents Chemother. *31*, 582–586.

Hirakata, Y., Srikumar, R., Poole, K., Gotoh, N., Suematsu, T., Kohno, S., Kamihira, S., Hancock, R.E., and Speert, D.P. (2002). Multidrug efflux systems play an important role in the invasiveness of *Pseudomonas aeruginosa*. J. Exp. Med. *196*, 109–118.

Hirakata, Y., Kondo, A., Hoshino, K., Yano, H., Arai, K., Hirotani, A., Kunishima, H., Yamamoto, N., Hatta, M., Kitagawa, M., *et al.* (2009). Efflux pump inhibitors reduce the invasiveness of *Pseudomonas aeruginosa*. Int. J. Antimicrob. Agents *34*, 343–346.

Hirata, K., Suzuki, H., Nishizawa, T., Tsugawa, H., Muraoka, H., Saito, Y., Matsuzaki, J., and Hibi, T. (2010). Contribution of efflux pumps to clarithromycin resistance in *Helicobacter pylori*. Gastroenterol, J. . Hepatol. *25(Suppl. 1)*, S75–S79.

Hirsch, E.B., and Tam, V.H. (2010). Impact of multidrug-resistant *Pseudomonas aeruginosa* infection on patient outcomes. Expert. Rev. Pharmacoecon. Outcomes. Res. *10*, 441–451.

Hocquet, D., Nordmann, P., El Garch, F., Cabanne, L., and Plesiat, P. (2006). Involvement of the MexXY-OprM efflux system in emergence of cefepime resistance in clinical strains of *Pseudomonas aeruginosa*. Antimicrob. Agents Chemother. *50*, 1347–1351.

Hocquet, D., Berthelot, P., Roussel-Delvallez, M., Favre, R., Jeannot, K., Bajolet, O., Marty, N., Grattard, F., Mariani-Kurkdjian, P., Bingen, E., *et al.* (2007a). *Pseudomonas aeruginosa* may accumulate drug resistance mechanisms without losing its ability to cause bloodstream infections. Antimicrob. Agents Chemother. *51*, 3531–3536.

Hocquet, D., Roussel-Delvallez, M., Cavallo, J.D., and Plesiat, P. (2007b). MexAB-OprM – and MexXY-overproducing mutants are very prevalent among clinical strains of *Pseudomonas aeruginosa* with reduced susceptibility to ticarcillin. Antimicrob. Agents Chemother. *51*, 1582–1583.

Hocquet, D., Muller, A., Blanc, K., Plesiat, P., Talon, D., Monnet, D.L., and Bertrand, X. (2008). Relationship between antibiotic use and incidence of MexXY-OprM overproducers among clinical isolates of *Pseudomonas aeruginosa*. Antimicrob. Agents Chemother. *52*, 1173–1175.

Islam, S., Jalal, S., and Wretlind, B. (2004). Expression of the MexXY efflux pump in amikacin-resistant isolates of *Pseudomonas aeruginosa*. Clin. Microbiol. Infect. *10*, 877–883.

Islam, S., Oh, H., Jalal, S., Karpati, F., Ciofu, O., Hoiby, N., and Wretlind, B. (2009). Chromosomal mechanisms of aminoglycoside resistance in *Pseudomonas aeruginosa* isolates from cystic fibrosis patients. Clin. Microbiol. Infect. *15*, 60–66.

Jacquot, J., Tabary, O., Le Rouzic, P., and Clement, A. (2008). Airway epithelial cell inflammatory signalling in cystic fibrosis. Int. J. Biochem. Cell Biol. *40*, 1703–1715.

Jakics, E.B., Iyobe, S., Hirai, K., Fukuda, H., and Hashimoto, H. (1992). Occurrence of the *nfxB* type mutation in clinical isolates of *Pseudomonas aeruginosa*. Antimicrob. Agents Chemother. *36*, 2562–2565.

Jalal, S., and Wretlind, B. (1998). Mechanisms of quinolone resistance in clinical strains of *Pseudomonas aeruginosa*. Microbiol. Drug Resist. *4*, 257–261.

Jalal, S., Wretlind, G., Gotoh, N., and Wretlind, B. (1999). Rapid identification of mutations in a multidrug efflux pump in *Pseudomonas aeruginosa*. APMIS *107*, 1109–1116.

Jalal, S., Ciofu, O., Hoiby, N., Gotoh, N., and Wretlind, B. (2000). Molecular mechanisms of fluoroquinolone resistance in *Pseudomonas aeruginosa* isolates from cystic fibrosis. Antimicrob. Agents Chemother. *44*, 710–712.

Jeannot, K., Sobel, M.L., El Garch, F., Poole, K., and Plesiat, P. (2005). Induction of the MexXY efflux pump in *Pseudomonas aeruginosa* is dependent on drug–ribosome interaction. J. Bacteriol. *187*, 5341–5346.

Jeannot, K., Elsen, S., Kohler, T., Attree, I., Van Delden, C., and Plesiat, P. (2008). Resistance and virulence of *Pseudomonas aeruginosa* clinical strains overproducing the MexCD-OprJ efflux pump. Antimicrob. Agents Chemother. *52*, 2455–2462.

Jeong, J.H., Shin, K.S., Lee, J.W., Park, E.J., and Son, S.Y. (2009). Analysis of a novel class 1 integron containing metallo-β-lactamase gene VIM-2 in *Pseudomonas aeruginosa*. J. Microbiol. *47*, 753–759.

Join-Lambert, O.F., Michea-Hamzehpour, M., Kohler, T., Chau, F., Faurisson, F., Dautrey, S., Vissuzaine, C., Carbon, C., and Pechere, J.C. (2001). Differential selection of multidrug efflux mutants by trovafloxacin and ciprofloxacin in an experimental model of *Pseudomonas aeruginosa* acute pneumonia in rats. Antimicrob. Agents Chemother. *45*, 571–576.

Jones, C.S., Osborne, D.J., and Stanley, J. (1992). Cloning of a probe for a previously undescribed enterobacterial tetracycline resistance gene. Lett. Appl. Microbiol. *15*, 106–108.

Jones, R.N., Stilwell, M.G., Rhomberg, P.R., and Sader, H.S. (2009). Antipseudomonal activity of piperacillin/tazobactam: more than a decade of experience from the S.E.N.T.R.Y Antimicrobial Surveillance Program (1997–2007). Diagn. Microbiol. Infect. Dis. *65*, 331–334.

Juhas, M., Wiehlmann, L., Huber, B., Jordan, D., Lauber, J., Salunkhe, P., Limpert, A.S., von Gotz, F., Steinmetz, I., Eberl, L., *et al.* (2004). Global regulation of quorum sensing and virulence by VqsR in *Pseudomonas aeruginosa*. Microbiology *150*, 831–841.

Kai, T., Tateda, K., Kimura, S., Ishii, Y., Ito, H., Yoshida, H., Kimura, T., and Yamaguchi, K. (2009). A low concentration of azithromycin inhibits the mRNA expression of N-acyl homoserine lactone synthesis enzymes, upstream of *lasI* or *rhlI*, in *Pseudomonas aeruginosa*. Pulm. Pharmacol. Ther. *22*, 483–486.

Kallen, A.J., Hidron, A.I., Patel, J., and Srinivasan, A. (2010). Multidrug resistance among Gram-negative pathogens that caused healthcare-associated infections reported to the National Healthcare Safety Network, 2006–2008. Infect. Control Hosp. Epidemiol. *31*, 528–531.

Kazama, H., Hamashima, H., Sasatsu, M., and Arai, T. (1998). Distribution of the antiseptic-resistance genes *qacE* and *qacE*Δ1 in Gram-negative bacteria. FEMS Microbiol. Lett. *159*, 173–178.

Keen, E.F., I.I.I., Robinson, B.J., Hospenthal, D.R., Aldous, W.K., Wolf, S.E., Chung, K.K., and Murray, C.K. (2010). Prevalence of multidrug-resistant organisms recovered at a military burn center. Burns *36*, 819–825.

Kerr, K.G., and Snelling, A.M. (2009). *Pseudomonas aeruginosa*: a formidable and ever-present adversary. J. Hosp. Infect. *73*, 338–344.

Kiser, T.H., Obritsch, M.D., Jung, R., Maclaren, R., and Fish, D.N. (2010). Efflux pump contribution to multidrug resistance in clinical isolates of *Pseudomonas aeruginosa*. Pharmacotherapy *30*, 632–638.

Köhler, T., Epp, S.F., Curty, L.K., and Pechére, J.-C. (1999a). Characterization of MexT, the Regulator of the MexE-MexF-OprN Multidrug Efflux System of *Pseudomonas aeruginosa*. J. Bacteriol. *181*, 6300–6305.

Köhler, T., Michea-Hamzehpour, M., Epp, S.F., and Pechere, J.C. (1999b). Carbapenem activities against *Pseudomonas aeruginosa*: respective contributions of OprD and efflux systems. Antimicrob. Agents Chemother. *43*, 424–427.

Köhler, T., Michea-Hamzehpour, M., Henze, U., Gotoh, N., Curty, L.K., and Pechere, J.-C. (1997a). Characterization of MexE-MexF-OprN, a positively regulated multidrug efflux system of *Pseudomonas aeruginosa*. Mol. Microbiol. *23*, 345–354.

Köhler, T., Michea-Hamzehpour, M., Plesiat, P., Kahr, A.-L., and Pechere, J.-C. (1997b). Differential selection of multidrug efflux systems by quinolones in *Pseudomonas aeruginosa*. Antimicrob. Agents Chemother. *41*, 2540–2543.

Köhler, T., Van Delden, C., Curty, L.K., Hamzehpour, M.M., and Pechere, J.C. (2001). Overexpression of the MexEF-OprN multidrug efflux system affects cell-to-cell signaling in *Pseudomonas aeruginosa*. J. Bacteriol. *183*, 5213–5222.

Koronakis, V., Sharff, A., Koronakis, E., Luisi, B., and Hughes, C. (2000). Crystal structure of the bacterial membrane protein TolC central to multidrug efflux and protein export. Nature *405*, 914–919.

Kriengkauykiat, J., Porter, E., Lomovskaya, O., and Wong-Beringer, A. (2005). Use of an efflux pump inhibitor to determine the prevalence of efflux pump-mediated fluoroquinolone resistance and multidrug resistance in *Pseudomonas aeruginosa*. Antimicrob. Agents Chemother. *49*, 565–570.

Kthiri, F., Gautier, V., Le, H.T., Prere, M.F., Fayet, O., Malki, A., Landoulsi, A., and Richarme, G. (2010). Translational defects in a mutant deficient in YajL, the bacterial homolog of the parkinsonism-associated protein DJ-1. J. Bacteriol. *192*, 6302–6306.

Kuete, V., Ngameni, B., Tangmouo, J.G., Bolla, J.M., Alibert-Franco, S., Ngadjui, B.T., and Pages, J.M. (2010). Efflux pumps are involved in the defense of Gram-negative bacteria against the natural products isobavachalcone and diospyrone. Antimicrob. Agents Chemother. *54*, 1749–1752.

Kuete, V., Alibert-Franco, S., Eyong, K.O., Ngameni, B., Folefoc, G.N., Nguemeving, J.R., Tangmouo, J.G., Fotso, G.W., Komguem, J., Ouahouo, B.M., *et al.* (2011). Antibacterial activity of some natural products against bacteria expressing a multidrug-resistant phenotype. Int. J. Antimicrob. Agents *37*, 156–161.

Kurincic, M., Botteldoorn, N., Herman, L., and Smole, M.S. (2007). Mechanisms of erythromycin resistance of *Campylobacter* spp. isolated from food, animals and humans. Int. J. Food Microbiol. *120*, 186–190.

Lambert, M.L., Suetens, C., Savey, A., Palomar, M., Hiesmayr, M., Morales, I., Agodi, A., Frank, U., Mertens, K., Schumacher, M., *et al.* (2011). Clinical outcomes of health-care-associated infections and antimicrobial resistance in patients admitted to European intensive-care units: a cohort study. Lancet Infect. Dis. *11*, 30–38.

Laraki, N., Galleni, M., Thamm, I., Riccio, M.L., Amicosante, G., Frere, J.M., and Rossolini, G.M.

(1999). Structure of In31, a bla_{IMP}-containing *Pseudomonas aeruginosa* integron phyletically related to In5, which carries an unusual array of gene cassettes. Antimicrob. Agents Chemother. *43*, 890–901.

Lautenbach, E., Synnestvedt, M., Weiner, M.G., Bilker, W.B., Vo, L., Schein, J., and Kim, M. (2010). Imipenem resistance in *Pseudomonas aeruginosa*: emergence, epidemiology, and impact on clinical and economic outcomes. Infect. Control Hosp. Epidemiol. *31*, 47–53.

Lee, J., Zhang, X.S., Hegde, M., Bentley, W.E., Jayaraman, A., and Wood, T.K. (2008). Indole cell signaling occurs primarily at low temperatures in *Escherichia coli*. ISME. J. 2, 1007–1023.

Lee, J., Attila, C., Cirillo, S.L., Cirillo, J.D., and Wood, T.K. (2009). Indole and 7-hydroxyindole diminish *Pseudomonas aeruginosa* virulence. Microb. Biotechnol. 2, 75–90.

Lee, Y.S., Park, S.C., Goldberg, A.L., and Chung, C.H. (1988). Protease So from *Escherichia coli* preferentially degrades oxidatively damaged glutamine synthetase. J. Biol. Chem. *263*, 6643–6646.

Li, X.Z., and Poole, K. (1999). Organic solvent-tolerant mutants of *Pseudomonas aeruginosa* display multiple antibiotic resistance. Can. J. Microbiol. *45*, 18–22.

Li, X.Z., and Poole, K. (2001). Mutational analysis of the OprM outer membrane component of the MexA-MexB-OprM multidrug efflux system of *Pseudomonas aeruginosa*. J. Bacteriol. *183*, 12–27.

Li, X.-Z., Zhang, L., and Poole, K. (1998). Role of the multidrug efflux systems of *Pseudomonas aeruginosa* in organic solvent tolerance. J. Bacteriol. *180*, 2987–2991.

Li, X.Z., Barre, N., and Poole, K. (2000a). Influence of the MexA-MexB-oprM multidrug efflux system on expression of the MexC-MexD-OprJ and MexE-MexF-OprN multidrug efflux systems in *Pseudomonas aeruginosa*. J. Antimicrob. Chemother. *46*, 885–893.

Li, X.-Z., Zhang, L., and Poole, K. (2000b). Interplay between the MexAB-OprM multidrug efflux system and the outer membrane barrier in the multiple antibiotic resistance of *Pseudomonas aeruginosa*. J. Antimicrob. Chemother. *45*, 433–436.

Li, X.-Z., Poole, K., and Nikaido, H. (2003a). Contributions of MexAB-OprM and an EmrE homologue to intrinsic resistance of *Pseudomonas aeruginosa* to aminoglycosides and dyes. Antimicrob. Agents Chemother. *47*, 27–33.

Li, Y., Mima, T., Komori, Y., Morita, Y., Kuroda, T., Mizushima, T., and Tsuchiya, T. (2003b). A new member of the tripartite multidrug efflux pumps, MexVW-OprM, in *Pseudomonas aeruginosa*. J Antimicrob Chemother *52*, 572–575.

Libisch, B., Poirel, L., Lepsanovic, Z., Mirovic, V., Balogh, B., Paszti, J., Hunyadi, Z., Dobak, A., Fuzi, M., and Nordmann, P. (2008). Identification of PER-1 extended-spectrum β-lactamase producing *Pseudomonas aeruginosa* clinical isolates of the international clonal complex CC11 from Hungary and Serbia. FEMS Immunol. Med. Microbiol. *54*, 330–338.

Lim, D., Poole, K., and Strynadka, N.C. (2002). Crystal structureof the MexR repressor from the multidrug efflux operon in *Pseudomonas aeruginosa*. J. Biol. Chem. 277, 29253–29259.

Linares, J.F., Lopez, J.A., Camafeita, E., Albar, J.P., Rojo, F., and Martinez, J.L. (2005). Overexpression of the multidrug efflux pumps MexCD-OprJ and MexEF-OprN is associated with a reduction of type III secretion in *Pseudomonas aeruginosa*. J. Bacteriol. *187*, 1384–1391.

Ling, J., and Soll, D. (2010). Severe oxidative stress induces protein mistranslation through impairment of an aminoacyl-tRNA synthetase editing site. Proc. Natl. Acad. Sci. U.S.A. *107*, 4028–4033.

Liu, Y., Yang, L., and Molin, S. (2010). Synergistic activities of an efflux pump inhibitor and iron chelators against *Pseudomonas aeruginosa* growth and biofilm formation. Antimicrob. Agents Chemother. *54*, 3960–3963.

Liu, Y.S., Zhang, Y.Q., Yang, L.X., Yao, T.J., and Xiao, C.L. (2009). Gene-specific silencing induced by parallel complementary RNA in *Pseudomonas aeruginosa*. Biotechnol. Lett. *31*, 1571–1575.

Livermore, D.M. (2001). Of *Pseudomonas*, porins, pumps and carbapenems. J. Antimicrob. Chemother. *47*, 247–250.

Llanes, C., Hocquet, D., Vogne, C., Benali-Baitich, D., Neuwirth, C., and Plesiat, P. (2004). Clinical strains of *Pseudomonas aeruginosa* overproducing MexAB-OprM and MexXY efflux pumps simultaneously. Antimicrob. Agents Chemother. *48*, 1797–1802.

Lomovskaya, O., and Watkins, W. (2001). Inhibition of efflux pumps as a novel approach to combat drug resistance in bacteria. J. Mol. Microbiol. Biotechnol. 3, 225–236.

Lomovskaya, O., Warren, M.S., Lee, A., Galazzo, J., Fronko, R., Lee, M., Blais, J., Cho, D., Chamberland, S., Renau, T., *et al.* (2001). Identification and characterization of inhibitors of multidrug resistance efflux pumps in *Pseudomonas aeruginosa*: novel agents for combination therapy. Antimicrob. Agents Chemother. *45*, 105–116.

Macia, M.D., Borrell, N., Segura, M., Gomez, C., Perez, J.L., and Oliver, A. (2006). Efficacy and potential for resistance selection of antipseudomonal treatments in a mouse model of lung infection by hypermutable *Pseudomonas aeruginosa*. Antimicrob. Agents Chemother. *50*, 975–983.

McMurry, L.M., Petrucci, R.E., Jr., and Levy, S.B. (1980). Active efflux of tetracycline encoded by four genetically different tetracycline resistance determinants in *Escherichia coli*. Proc. Natl. Acad. Sci. U.S.A. *77*, 3974–3977.

Maddocks, S.E., and Oyston, P.C. (2008). Structure and function of the LysR-type transcriptional regulator (LTTR) family proteins. Microbiology *154*, 3609–3623.

Mahamoud, A., Chevalier, J., Baitiche, M., Adam, E., and Pages, J.M. (2011). An alkylaminoquinazoline restores antibiotic activity in Gram-negative resistant isolates. Microbiology *157*, 566–571.

Mahar, P., Padiglione, A.A., Cleland, H., Paul, E., Hinrichs, M., and Wasiak, J. (2010). *Pseudomonas aeruginosa* bacteraemia in burns patients: Risk factors and outcomes. Burns *36*, 1228–1233.

Maloney, J., Rimland, D., Stephens, D.S., Terry, P., and Whitney, A.M. (1989). Analysis of amikacin-resistant

Pseudomonas aeruginosa developing in patients receiving amikacin. Arch. Intern. Med. *149*, 630–634.

Mamelli, L., Amoros, J.P., Pages, J.M., and Bolla, J.M. (2003). A phenylalanine-arginine β-naphthylamide sensitive multidrug efflux pump involved in intrinsic and acquired resistance of *Campylobacter* to macrolides. Int. J. Antimicrob. Agents *22*, 237–241.

Mamelli, L., Demoulin, E., Prouzet-Mauleon, V., Megraud, F., Pages, J.M., and Bolla, J.M. (2007). Prevalence of efflux activity in low-level macrolide-resistant *Campylobacter* species. J. Antimicrob. Chemother. *59*, 327–328.

Mandsberg, L.F., Ciofu, O., Kirkby, N., Christiansen, L.E., Poulsen, H.E., and Hoiby, N. (2009). Antibiotic resistance in *Pseudomonas aeruginosa* strains with increased mutation frequency due to inactivation of the DNA oxidative repair system. Antimicrob. Agents Chemother. *53*, 2483–2491.

Maniati, M., Ikonomidis, A., Mantzana, P., Daponte, A., Maniatis, A.N., and Pournaras, S. (2007). A highly carbapenem-resistant *Pseudomonas aeruginosa* isolate with a novel bla_{VIM-4}/bla_{P1b} integron overexpresses two efflux pumps and lacks OprD. J. Antimicrob. Chemother. *60*, 132–135.

Martin, D.W., Holloway, B.W., and Deretic, V. (1993). Characterization of a locus determining the mucoid status of *Pseudomonas aeruginosa*: AlgU shows sequence similarities with a *Bacillus* sigma factor. J. Bacteriol. *175*, 1153–1164.

Maseda, H., Saito, K., Nakajima, A., and Nakae, T. (2000). Variation of the *mexT* gene, a regulator of the MexEF-oprN efflux pump expression in wild-type strains of *Pseudomonas aeruginosa*. FEMS Microbiol. Lett. *192*, 107–112.

Maseda, H., Uwate, M., and Nakae, T. (2010). Transcriptional regulation of the *mexEF-oprN* multidrug efflux pump operon by MexT and an unidentified repressor in *nfxC*-type mutant of *Pseudomonas aeruginosa*. FEMS Microbiol. Lett. *311*, 36–43.

Masuda, N., Sakagawa, E., and Ohya, S. (1995). Outer membrane proteins responsible for multiple drug resistance in *Pseudomonas aeruginosa*. Antimicrob. Agents Chemother. *39*, 645–649.

Masuda, N., Sakagawa, E., Ohya, S., Gotoh, N., Tsujimoto, H., and Nishino, T. (2000a). Contribution of the MexX-MexY-OprM efflux system to intrinsic resistance in *Pseudomonas aeruginosa*. Antimicrob. Agents Chemother. *44*, 2242–2246.

Masuda, N., Sakagawa, E., Ohya, S., Gotoh, N., Tsujimoto, H., and Nishino, T. (2000b). Substrate specificities of MexAB-OprM, MexCD-OprJ, and MexXY-OprM efflux pumps in *Pseudomonas aeruginosa*. Antimicrob. Agents Chemother. *44*, 3322–3327.

Masuda, N., Sakagawa, E., Ohya, S., Gotoh, N., and Nishino, T. (2001). Hypersusceptibility of the *Pseudomonas aeruginosa nfxB* mutant to β-actams due to reduced expression of the AmpC β-lactamase. Antimicrob Agents. Chemother. *45*, 1284–1286.

Matsuo, Y., Eda, S., Gotoh, N., Yoshihara, E., and Nakae, T. (2004). MexZ-mediated regulation of *mexXY* multidrug efflux pump expression in *Pseudomonas aeruginosa* by binding on the *mexZ-mexX* intergenic DNA. FEMS Microbiol. Lett. *238*, 23–28.

Mauldin, P.D., Salgado, C.D., Hansen, I.S., Durup, D.T., and Bosso, J.A. (2010). Attributable hospital cost and length of stay associated with health care-associated infections caused by antibiotic-resistant Gram-negative bacteria. Antimicrob. Agents Chemother. *54*, 109–115.

Mendez, B., Tachibana, C., and Levy, S.B. (1980). Heterogeneity of tetracycline resistance determinants. Plasmid 3, 99–108.

Mikolosko, J., Bobyk, K., Zgurskaya, H.I., and Ghosh, P. (2006). Conformational flexibility in the multidrug efflux system protein AcrA. Structure. *14*, 577–587.

Miller, P.F., and Sulavik, M.C. (1996). Overlaps and parallels in the regulation of intrinsic multiple-antibiotic resistance in *Esherichia coli*. Mol. Microbiol. *21*, 441–448.

Mima, T., Joshi, S., Gomez-Escalada, M., and Schweizer, H.P. (2007). Identification and characterization of TriABC-OpmH, a triclosan efflux pump of *Pseudomonas aeruginosa* requiring two membrane fusion proteins. J. Bacteriol. *189*, 7600–7609.

Mima, T., Kohira, N., Li, Y., Sekiya, H., Ogawa, W., Kuroda, T., and Tsuchiya, T. (2009). Gene cloning and characteristics of the RND-type multidrug efflux pump MuxABC-OpmB possessing two RND components in *Pseudomonas aeruginosa*. Microbiology *155*, 3509–3517.

Mine, T., Morita, Y., Kataoka, A., Mitzushima, T., and Tsuchiya, T. (1999). Expression in *Escherichia coli* of a new multidrug efflux pump, MexXY, from *Pseudomonas aeruginosa*. Antimicrob. Agents Chemother. *43*, 415–417.

Misra, R., and Bavro, V.N. (2009). Assembly and transport mechanism of tripartite drug efflux systems. Biochim. Biophys. Acta *1794*, 817–825.

Morero, N.R., and Argarana, C.E. (2009). *Pseudomonas aeruginosa* deficient in 8-oxodeoxyguanine repair system shows a high frequency of resistance to ciprofloxacin. FEMS Microbiol. Lett. *290*, 217–226.

Morita, Y., Kodama, K., Shiota, S., Mine, T., Kataoka, A., Mizushima, T., and Tsuchiya, T. (1998). NorM, a putative multidrug efflux protein, of *Vibrio parahaemolyticus* and its homolog in *Escherichia coli*. Antimicrob. Agents Chemother. *42*, 1778–1782.

Morita, Y., Kataoka, A., Shiota, S., Mizushima, T., and Tsuchiya, T. (2000). NorM of *Vibrio parahaemolyticus* is an Na^+-driven multidrug efflux pump. J. Bacteriol. *182*, 6694–6697.

Morita, Y., Murata, T., Mima, T., Shiota, S., Kuroda, T., Mizushima, T., Gotoh, N., Nishino, T., and Tsuchiya, T. (2003). Induction of *mexCD-oprJ* operon for a multidrug efflux pump by disinfectants in wild-type *Pseudomonas aeruginosa* PAO1. J. Antimicrob. Chemother. *5*, 991–994.

Morita, Y., Cao, L., Gould, G., Avison, M.B., and Poole, K. (2006a). *nalD* encodes a second repressor of the *mexAB-oprM* multidrug efflux operon of *Pseudomonas aeruginosa*. J. Bacteriol. *188*, 8649–8654.

Morita, Y., Sobel, M.L., and Poole, K. (2006b). Antibiotic inducibility of the MexXY multidrug efflux system of

Pseudomonas aeruginosa: involvement of the antibiotic-inducible PA5471 gene product. J. Bacteriol. *188*, 1847–1855.

Morita, Y., Gilmour, C., Metcalf, D., and Poole, K. (2009). Translational control of the antibiotic inducibility of the PA5471 gene required for *mexXY* multidrug efflux gene expression in *Pseudomonas aeruginosa*. J. Bacteriol. *191*, 4966–4975.

Mulcahy, L.R., Burns, J.L., Lory, S., and Lewis, K. (2010). Emergence of *Pseudomonas aeruginosa* strains producing high levels of persister cells in patients with cystic fibrosis. J. Bacteriol. *192*, 6191–6199.

Mulet, X., Macia, M.D., Mena, A., Juan, C., Perez, J.L., and Oliver, A. (2009). Azithromycin in *Pseudomonas aeruginosa* biofilms: bactericidal activity and selection of *nfxB* mutants. Antimicrob. Agents Chemother. *53*, 1552–1560.

Muller, C., Plesiat, P., and Jeannot, K. (2010). A two-component regulatory system interconnects resistance to polymyxins, aminoglycosides, fluoroquinolones, and β-lactams in *Pseudomonas aeruginosa*. Antimicrob. Agents Chemother. *55*, 1211–1221.

Muller, J.F., Stevens, A.M., Craig, J., and Love, N.G. (2007). Transcriptome analysis reveals that multidrug efflux genes are upregulated to protect *Pseudomonas aeruginosa* from pentachlorophenol stress. Appl. Environ. Microbiol. *73*, 4550–4558.

Murakami, S., Nakashima, R., Yamashita, E., Matsumoto, T., and Yamaguchi, A. (2006). Crystal structures of a multidrug transporter reveal a functionally rotating mechanism. Nature *443*, 173–179.

Mutlu, G.M., and Wunderink, R.G. (2006). Severe pseudomonal infections. Curr. Opin. Crit. Care *12*, 458–463.

Nagakubo, S., Nishino, K., Hirata, T., and Yamaguchi, A. (2002). The putative response regulator BaeR stimulates multidrug resistance of *Escherichia coli* via a novel multidrug exporter system, MdtABC. J. Bacteriol. *184*, 4161–4167.

Nakayama, K., Ishida, Y., Ohtsuka, M., Kawato, H., Yoshida, K., Yokomizo, Y., Hosono, S., Ohta, T., Hoshino, K., Ishida, H., *et al.* (2003a). MexAB-OprM-specific efflux pump inhibitors in *Pseudomonas aeruginosa*. Part 1: discovery and early strategies for lead optimization. Bioorg. Med. Chem. Lett. *13*, 4201–4204.

Nakayama, K., Ishida, Y., Ohtsuka, M., Kawato, H., Yoshida, K., Yokomizo, Y., Ohta, T., Hoshino, K., Otani, T., Kurosaka, Y., *et al.* (2003b). MexAB-OprM specific efflux pump inhibitors in *Pseudomonas aeruginosa*. Part 2: achieving activity *in vivo* through the use of alternative scaffolds. Bioorg. Med. Chem. Lett. *13*, 4205–4208.

Nakayama, K., Kawato, H., Watanabe, J., Ohtsuka, M., Yoshida, K., Yokomizo, Y., Sakamoto, A., Kuru, N., Ohta, T., Hoshino, K., *et al.* (2004a). MexAB-OprM specific efflux pump inhibitors in *Pseudomonas aeruginosa*. Part 3: Optimization of potency in the pyridopyrimidine series through the application of a pharmacophore model. Bioorg. Med. Chem. Lett. *14*, 475–479.

Nakayama, K., Kuru, N., Ohtsuka, M., Yokomizo, Y., Sakamoto, A., Kawato, H., Yoshida, K., Ohta, T., Hoshino, K., Akimoto, K., *et al.* (2004b). MexAB-OprM specific efflux pump inhibitors in *Pseudomonas aeruginosa*. Part 4: Addressing the problem of poor stability due to photoisomerization of an acrylic acid moiety. Bioorg. Med. Chem. Lett. *14*, 2493–2497.

Nehme, D., Li, X.Z., Elliot, R., and Poole, K. (2004). Assembly of the MexAB-OprM multidrug efflux system of *Pseudomonas aeruginosa*: identification and characterization of mutations in *mexA* compromising MexA multimerization and interaction with MexB. J. Bacteriol. *186*, 2973–2983.

Niga, T., Ito, H., Oyamada, Y., Yamagishi, J., Kadono, M., Nishino, T., Gotoh, N., and Inoue, M. (2005). Cooperation between alteration of DNA gyrase genes and over-expression of MexB and MexX confers high-level fluoroquinolone resistance in *Pseudomonas aeruginosa* strains isolated from a patient who received a liver transplant followed by treatment with fluoroquinolones. Microbiol. Immunol. *49*, 443–446.

Nikaido, H. (1998). Multiple antibiotic resistance and efflux. Curr. Opin. Microbiol. *1*, 516–523.

Nikaido, H., and Takatsuka, Y. (2009). Mechanisms of RND multidrug efflux pumps. Biochim. Biophys. Acta *1794*, 769–781.

Nishida, T., Orikasa, Y., Ito, Y., Yu, R., Yamada, A., Watanabe, K., and Okuyama, H. (2006). *Escherichia coli* engineered to produce eicosapentaenoic acid becomes resistant against oxidative damages. FEBS Lett. *580*, 2731–2735.

Ochs, M.M., McCusker, M.P., Bains, M., and Hancock, R.E. (1999). Negative regulation of the *Pseudomonas aeruginosa* outer membrane porin OprD selective for imipenem and basic amino acids. Antimicrob. Agents Chemother. *43*, 1085–1090.

Okamoto, K., Gotoh, N., and Nishino, T. (2002). Alterations of susceptibility of *Pseudomonas aeruginosa* by overproduction of multidrug efflux systems, MexAB-OprM, MexCD-OprJ, and MexXY/OprM to carbapenems: substrate specificities of the efflux systems. J. Infect. Chemother. *8*, 371–373.

Oliver, A. (2010). Mutators in cystic fibrosis chronic lung infection: Prevalence, mechanisms, and consequences for antimicrobial therapy. Int. J. Med. Microbiol. *300*, 563–572.

Oliver, A., and Mena, A. (2010). Bacterial hypermutation in cystic fibrosis, not only for antibiotic resistance. Clin. Microbiol. Infect. *16*, 798–808.

Oliver, A., Canton, R., Campo, P., Baquero, F., and Blazquez, J. (2000). High frequency of hypermutable *Pseudomonas aeruginosa* in cystic fibrosis lung infection. Science *288*, 1251–1254.

Pak, J.W., Knoke, K.L., Noguera, D.R., Fox, B.G., and Chambliss, G.H. (2000). Transformation of 2,4,6-trinitrotoluene by purified xenobiotic reductase B from *Pseudomonas fluorescens* I-C. Appl. Environ. Microbiol. *66*, 4742–4750.

Palma, M., Worgall, S., and Quadri, L.E. (2003). Transcriptome analysis of the *Pseudomonas aeruginosa* response to iron. Arch. Microbiol. *180*, 374–379.

Pamp, S.J., Gjermansen, M., Johansen, H.K., and Tolker-Nielsen, T. (2008). Tolerance to the antimicrobial peptide colistin in *Pseudomonas aeruginosa* biofilms is linked to metabolically active cells, and depends on the *pmr* and *mexAB-oprM* genes. Mol. Microbiol. *68*, 223–240.

Pearson, J.P., Van Delden, C., and Iglewski, B.H. (1999). Active efflux and diffusion are involved in transport of *Pseudomonas aeruginosa* cell-to-cell signals. J. Bacteriol. *181*, 1203–1210.

Pena, C., Suarez, C., Tubau, F., Juan, C., Moya, B., Dominguez, M.A., Oliver, A., Pujol, M., and Ariza, J. (2009). Nosocomial outbreak of a non-cefepime-susceptible ceftazidime-susceptible *Pseudomonas aeruginosa* strain overexpressing MexXY-OprM and producing an integron-borne PSE-1 β-lactamase. J. Clin. Microbiol. *47*, 2381–2387.

Phan, G., Benabdelhak, H., Lascombe, M.B., Benas, P., Rety, S., Picard, M., Ducruix, A., Etchebest, C., and Broutin, I. (2010). Structural and dynamical insights into the opening mechanism of *P. aeruginosa* OprM channel. Structure *18*, 507–517.

Poole, K. (2000). Efflux-mediated resistance to fluoroquinolones in Gram-negative bacteria. Antimicrob. Agents Chemother. *44*, 2233–2241.

Poole, K. (2004). Efflux pumps. In *Pseudomonas*, Vol. I. Genomics, Life Style and Molecular Architecture, Ramos, J.-L., ed. (Kluwer Academic/Plenum Publishers, New York), pp. 635–674.

Poole, K. (2005a). Aminoglycoside resistance in *Pseudomonas aeruginosa*. Antimicrob. Agents Chemother. *49*, 479–487.

Poole, K. (2005b). Efflux-mediated antimicrobial resistance. J. Antimicrob. Chemother. *56*, 20–51.

Poole, K. (2007). Efflux pumps as antimicrobial resistance mechanisms. Ann. Med. *39*, 162–176.

Poole, K. (2008). Bacteria multidrug efflux pumps serve other functions. Microbes *3*, 179–185.

Poole, K. (2011). *Pseudomonas aeruginosa*: resistance to the max. Front. Microbio. *2*, 65.

Poole, K. (2012). Stress responses as determinants of antimicrobial resistance in Gram-negative bacteria. Trends Microbiol. *20*, 227–234.

Poole, K., and Lomovskaya, O. (2006). Can efflux inhibitors really counter resistance? Drug Discov. Today Therapeut. Strat. *3*, 145–152.

Poole, K., Krebes, K., McNally, C., and Neshat, S. (1993). Multiple antibiotic resistance in *Pseudomonas aeruginosa*: evidence for involvement of an efflux operon. J. Bacteriol. *175*, 7363–7372.

Poole, K., Gotoh, N., Tsujimoto, H., Zhao, Q., Wada, A., Yamasaki, T., Neshat, S., Yamagishi, J.-I., Li, X.-Z., and Nishino, T. (1996a). Overexpression of the *mexC-mexD-oprJ* efflux operon in *nfxB* multidrug resistant strains of *Pseudomonas aeruginosa*. Mol. Microbiol. *21*, 713–724.

Poole, K., Tetro, K., Zhao, Q., Neshat, S., Heinrichs, D., and Bianco, N. (1996b). Expression of the multidrug resistance operon *mexA-mexB-oprM* in *Pseudomonas aeruginosa*: *mexR* encodes a regulator of operon expression. Antimicrob. Agents Chemother. *40*, 2021–2028.

Poole, R.K. (2005c). Nitric oxide and nitrosative stress tolerance in bacteria. Biochem. Soc. Trans. *33*, 176–180.

Pos, K.M. (2009). Drug transport mechanism of the AcrB efflux pump. Biochim. Biophys. Acta *1794*, 782–793.

Pournaras, S., Maniati, M., Spanakis, N., Ikonomidis, A., Tassios, P.T., Tsakris, A., Legakis, N.J., and Maniatis, A.N. (2005). Spread of efflux pump-overexpressing, non-metallo-β-lactamase-producing, meropenem-resistant but ceftazidime-susceptible *Pseudomonas aeruginosa* in a region with bla_{VIM} endemicity. J. Antimicrob. Chemother. *56*, 761–764.

Quale, J., Bratu, S., Gupta, J., and Landman, D. (2006). Interplay of efflux system, *ampC*, and *oprD* expression in carbapenem resistance of *Pseudomonas aeruginosa* clinical isolates. Antimicrob. Agents Chemother. *50*, 1633–1641.

Queenan, A.M., Shang, W., Bush, K., and Flamm, R.K. (2010). Differential selection of single-step AmpC or efflux mutants of *Pseudomonas aeruginosa* by using cefepime, ceftazidime, or ceftobiprole. Antimicrob. Agents Chemother. *54*, 4092–4097.

Rampioni, G., Schuster, M., Greenberg, E.P., Zennaro, E., and Leoni, L. (2009). Contribution of the RsaL global regulator to *Pseudomonas aeruginosa* virulence and biofilm formation. FEMS Microbiol. Lett. *301*, 210–217.

Ramsey, M.M., and Whiteley, M. (2004). *Pseudomonas aeruginosa* attachment and biofilm development in dynamic environments. Mol. Microbiol. *53*, 1075–1087.

Ratkai, C., Nagy, E., Peixe, L., Bertalan, V., and Hajdu, E. (2010). Isolation and characterization of an imported extremely-resistant *Pseudomonas aeruginosa* producing three different extended-spectrum β-lactamases and hyperproducing two multidrug-efflux pumps. J. Infect. *61*, 511–512.

Reinhardt, A., Kohler, T., Wood, P., Rohner, P., Dumas, J.L., Ricou, B., and Van Delden, C. (2007). Development and persistence of antimicrobial resistance in *Pseudomonas aeruginosa*: a longitudinal observation in mechanically ventilated patients. Antimicrob. Agents Chemother. *51*, 1341–1350.

Renau, T.E., Leger, R., Flamme, E.M., Sangalang, J., She, M.W., Yen, R., Gannon, C.L., Griffith, D., Chamberland, S., Lomovskaya, O., *et al.* (1999). Inhibitors of efflux pumps in *Pseudomonas aeruginosa* potentiate the activity of the fluoroquinolone antibacterial levofloxacin. J. Med. Chem. *42*, 4928–4931.

Renau, T.E., Leger, R., Flamme, E.M., She, M.W., Gannon, C.L., Mathias, K.M., Lomovskaya, O., Chamberland, S., Lee, V.J., Ohta, T., *et al.* (2001). Addressing the stability of C-capped dipeptide efflux pump inhibitors that potentiate the activity of levofloxacin in *Pseudomonas aeruginosa*. Bioorg. Med. Chem. Lett. *11*, 663–667.

Renau, T.E., Leger, R., Yen, R., She, M.W., Flamme, E.M., Sangalang, J., Gannon, C.L., Chamberland, S., Lomovskaya, O., and Lee, V.J. (2002). Peptidomimetics of efflux pump inhibitors potentiate the activity of levofloxacin in *Pseudomonas aeruginosa*. Bioorg. Med. Chem. Lett. *12*, 763–766.

Renau, T.E., Leger, R., Filonova, L., Flamme, E.M., Wang, M., Yen, R., Madsen, D., Griffith, D., Chamberland, S., Dudley, M.N., *et al.* (2003).

Conformationally-restricted analogues of efflux pump inhibitors that potentiate the activity of levofloxacin in *Pseudomonas aeruginosa*. Bioorg. Med. Chem. Lett. *13*, 2755–2758.

Roberts, M.C. (1996). Tetracycline resistance determinants: mechanisms of action, regulation of expression, genetic mobility, and distribution. FEMS Microbiol. Lett. *19*, 1–24.

Rodriguez-Martinez, J.M., Poirel, L., and Nordmann, P. (2009). Molecular epidemiology and mechanisms of carbapenem resistance in *Pseudomonas aeruginosa*. Antimicrob. Agents Chemother. *53*, 4783–4788.

Rottner, M., Freyssinet, J.M., and Martinez, M.C. (2009). Mechanisms of the noxious inflammatory cycle in cystic fibrosis. Respir. Res. *10*, 23.

Rubens, C.E., McNeill, W.F., and Farrar, W.E., Jr. (1979). Transposable plasmid deoxyribonucleic acid sequence in *Pseudomonas aeruginosa* which mediates resistance to gentamicin and four other antimicrobial agents. J. Bacteriol. *139*, 877–882.

Saito, K., Yoneyama, H., and Nakae, T. (1999). *nalB*-type mutations causing the overexpression of the MexAB-OprM efflux pump are located in the *mexR* gene of the *Pseudomonas aeruginosa* chromosome. FEMS Microbiol. Lett. *179*, 67–72.

Saito, K., Eda, S., Maseda, H., and Nakae, T. (2001). Molecular mechanism of MexR-mediated regulation of MexAB-OprM efflux pump expression in *Pseudomonas aeruginosa*. FEMS Microbiol. Lett. *195*, 23–28.

Sanchez, P., Linares, J.F., Ruiz-Diez, B., Campanario, E., Navas, A., Baquero, F., and Martinez, J.L. (2002a). Fitness of *in vitro* selected *Pseudomonas aeruginosa nalB* and *nfxB* multidrug resistant mutants. J. Antimicrob. Chemother. *50*, 657–664.

Sanchez, P., Rojo, F., and Martinez, J.L. (2002b). Transcriptional regulation of *mexR*, the repressor of *Pseudomonas aeruginosa mexAB-oprM* multidrug efflux pump. FEMS Microbiol. Lett. *207*, 63–68.

Sanchez-Gomez, S., Japelj, B., Jerala, R., Moriyon, I., Alonso, M.F., Leiva, J., Blondelle, S.E., Andra, J., Brandenburg, K., Lohner, K., *et al.* (2011). Structural features governing the activity of lactoferricin-derived peptides that act in synergy with antibiotics against *Pseudomonas aeruginosa in vitro* and *in vivo*. Antimicrob. Agents Chemother. *55*, 218–228.

Schell, M.A. (1993). Molecular biology of the LysR family of transcriptional regulators. Annu. Rev. Microbiol. *47*, 597–626.

Schluter, A., Heuer, H., Szczepanowski, R., Poler, S.M., Schneiker, S., Puhler, A., and Top, E.M. (2005). Plasmid pB8 is closely related to the prototype IncP-1beta plasmid R751 but transfers poorly to *Escherichia coli* and carries a new transposon encoding a small multidrug resistance efflux protein. Plasmid *54*, 135–148.

Schulz, R., Vargiu, A.V., Collu, F., Kleinekathofer, U., and Ruggerone, P. (2010). Functional rotation of the transporter AcrB: insights into drug extrusion from simulations. PLoS. Comput. Biol. *6*, e1000806.

Schwarz, S., Kehrenberg, C., Doublet, B., and Cloeckaert, A. (2004). Molecular basis of bacterial resistance to chloramphenicol and florfenicol. FEMS Microbiol. Rev. *28*, 519–542.

Schweizer, H.P. (1998). Intrinsic resistance to inhibitors of fatty acid biosynthesis in *Pseudomonas aeruginosa* is due to efflux: application of a novel technique for generation of unmarked chromosomal mutations for the study of efflux systems. Antimicrob. Agents Chemother. *42*, 394–398.

Seeger, M.A., Schiefner, A., Eicher, T., Verrey, F., Diederichs, K., and Pos, K.M. (2006). Structural asymmetry of AcrB trimer suggests a peristaltic pump mechanism. Science *313*, 1295–1298.

Sekiya, H., Mima, T., Morita, Y., Kuroda, T., Mizushima, T., and Tsuchiya, T. (2003). Functional cloning and characterization of a multidrug efflux pump, MexHI-OpmD, from a *Pseudomonas aeruginosa* mutant. Antimicrob. Agents Chemother. *47*, 2990–2992.

Sennhauser, G., Bukowska, M.A., Briand, C., and Grutter, M.G. (2009). Crystal structure of the multidrug exporter MexB from *Pseudomonas aeruginosa*. J. Mol. Biol. *389*, 134–145.

Shawar, R.M., MacLeod, D.L., Garber, R.L., Burns, J.L., Stapp, J.R., Clausen, C.R., and Tanaka, S.K. (1999). Activities of tobramycin and six other antibiotics against *Pseudomonas aeruginosa* isolates from patients with cystic fibrosis. Antimicrob. Agents Chemother. *43*, 2877–2880.

Shiba, T., Ishiguro, K., Takemoto, N., Koibuchi, H., and Sugimoto, K. (1995). Purification and characterization of the *Pseudomonas aeruginosa* NfxB protein, the negative regulator of the *nfxB* gene. J. Bacteriol. *177*, 5872–5877.

Shorr, A.F. (2009). Review of studies of the impact on Gram-negative bacterial resistance on outcomes in the intensive care unit. Crit. Care Med. 37, 1463–1469.

Sivaneson, M., Mikkelsen, H., Ventre, I., Bordi, C., and Filloux, A. (2010). Two component regulatory systems in *Pseudomonas aeruginosa*: an intricate network mediating fimbrial and efflux pump gene expression. Mol. Microbiol. *79*, 1353–1366.

Slama, T.G. (2008). Gram-negative antibiotic resistance: there is a price to pay. Crit. Care *12(Suppl. 4)*, S4.

Smith, E.E., Buckley, D.G., Wu, Z., Saenphimmachak, C., Hoffman, L.R., D'Argenio, D.A., Miller, S.I., Ramsey, B.W., Speert, D.P., Moskowitz, S.M., *et al.* (2006). Genetic adaptation by *Pseudomonas aeruginosa* to the airways of cystic fibrosis patients. Proc. Natl. Acad. Sci. U.S.A. *103*, 8487–8492.

Sobel, M.L., McKay, G.A., and Poole, K. (2003). Contribution of the MexXY multidrug transporter to aminoglycoside resistance in *Pseudomonas aeruginosa* clinical isolates. Antimicrob. Agents Chemother. *47*, 3202–3207.

Sobel, M.L., Hocquet, D., Cao, L., Plesiat, P., and Poole, K. (2005a). Mutations in PA3574 (*nalD*) lead to increased MexAB-OprM expression and multidrug resistance in lab and clinical isolates of *Pseudomonas aeruginosa*. Antimicrob. Agents Chemother. *49*, 1782–1786.

Sobel, M.L., Poole, K., and Neshat, S. (2005b). Mutations in PA2491 (*mexS*) promote MexT-dependent *mexEF-oprN* expression and multidrug resistance in a clinical strain of *Pseudomonas aeruginosa*. J. Bacteriol. *187*, 1246–1253.

Son, M.S., Matthews, W.J., Jr., Kang, Y., Nguyen, D.T., and Hoang, T.T. (2007). *In vivo* evidence of *Pseudomonas aeruginosa* nutrient acquisition and pathogenesis in the lungs of cystic fibrosis patients. Infect. Immun. *75*, 5313–5324.

Srikumar, R., and Poole, K. (1999). Demonstration of ethidium bromide efflux by multiresistant pumps of *Pseudomonas aeruginosa*. Clin. Microbiol. Infect. *5*, 5S58–5S59.

Srikumar, R., Li, X.-Z., and Poole, K. (1997). Inner membrane efflux components are responsible for the β-lactam specificity of multidrug efflux pumps in *Pseudomonas aeruginosa*. J. Bacteriol. *179*, 7875–7881.

Srikumar, R., Kon, T., Gotoh, N., and Poole, K. (1998). Expression of *Pseudomonas aeruginosa* multidrug efflux pumps MexA-MexB-OprM and MexC-MexD-OprJ in a multidrug-sensitive *Escherichia coli* strain. Antimicrob. Agents Chemother. *42*, 65–71.

Srikumar, R., Paul, C.J., and Poole, K. (2000). Influence of mutations in the *mexR* repressor gene on expression of the MexA-MexB-OprM multidrug efflux system of *Pseudomonas aeruginosa*. J. Bacteriol. *182*, 1410–1414.

Starr, L., Fruci, M., and Poole, K. (2012). Pentachlorophenol induction of the *Pseudomonas aeruginosa mexAB-oprM* efflux operon: involvement of repressors NalC and MexR and the antirepressor ArmR. PLoS One 7, e32684.

Stickland, H.G., Davenport, P.W., Lilley, K.S., Griffin, J.L., and Welch, M. (2010). Mutation of *nfxB* causes global changes in the physiology and metabolism of *Pseudomonas aeruginosa*. J. Proteome. Res. *9*, 2957–2967.

Symmons, M.F., Bokma, E., Koronakis, E., Hughes, C., and Koronakis, V. (2009). The assembled structure of a complete tripartite bacterial multidrug efflux pump. Proc. Natl. Acad. Sci. U.S.A. *106*, 7173–7178.

Tashiro, Y., Toyofuku, M., Nakajima-Kambe, T., Uchiyama, H., and Nomura, N. (2010). Bicyclic compounds repress membrane vesicle production and *Pseudomonas* quinolone signal synthesis in *Pseudomonas aeruginosa*. FEMS Microbiol. Lett. *304*, 123–130.

Tegos, G.P., Masago, K., Aziz, F., Higginbotham, A., Stermitz, F.R., and Hamblin, M.R. (2008). Inhibitors of bacterial multidrug efflux pumps potentiate antimicrobial photoinactivation. Antimicrob. Agents Chemother. *52*, 3202–3209.

Teitzel, G.M., Geddie, A., De Long, S.K., Kirisits, M.J., Whiteley, M., and Parsek, M.R. (2006). Survival and growth in the presence of elevated copper: transcriptional profiling of copper-stressed *Pseudomonas aeruginosa*. J. Bacteriol. *188*, 7242–7256.

Thaden, J.T., Lory, S., and Gardner, T.S. (2010). Quorum-sensing regulation of a copper toxicity system in *Pseudomonas aeruginosa*. J. Bacteriol. *192*, 2557–2568.

Thompson, J., O'Connor, M., Mills, J.A., and Dahlberg, A.E. (2002). The protein synthesis inhibitors, oxazolidinones and chloramphenicol, cause extensive translational inaccuracy *in vivo*. J. Mol. Biol. *322*, 273–279.

Tian, Z.X., Fargier, E., Mac, A.M., Adams, C., Wang, Y.P., and O'Gara, F. (2009a). Transcriptome profiling defines a novel regulon modulated by the LysR-type transcriptional regulator MexT in *Pseudomonas aeruginosa*. Nucleic Acids Res. *37*, 7546–7559.

Tian, Z.X., Mac, A.M., O'Connor, H.F., Fargier, E., Mooij, M.J., Adams, C., Wang, Y.P., and O'Gara, F. (2009b). MexT modulates virulence determinants in *Pseudomonas aeruginosa* independent of the MexEF-OprN efflux pump. Microb. Pathog. *47*, 237–241.

Tohidpour, A., Najar, P.S., Mehrabadi, J.F., and Rezaei, Y.H. (2009). Determination of the efflux pump-mediated resistance prevalence in *Pseudomonas aeruginosa*, using an efflux pump inhibitor. Curr. Microbiol. *59*, 352–355.

Tomas, M., Doumith, M., Warner, M., Turton, J.F., Beceiro, A., Bou, G., Livermore, D.M., and Woodford, N. (2010). Efflux pumps, OprD porin, AmpC β-lactamase, and multiresistance in *Pseudomonas aeruginosa* isolates from cystic fibrosis patients. Antimicrob. Agents Chemother. *54*, 2219–2224.

Tran, Q.T., Dupont, M., Lavigne, J.P., Chevalier, J., Pages, J.M., Sotto, A., and Davin-Regli, A. (2009). Occurrence of efflux mechanism and cephalosporinase variant in a population of *Enterobacter aerogenes* and *Klebsiella pneumoniae* isolates producing extended-spectrum β-lactamases. Antimicrob. Agents Chemother. *53*, 1652–1656.

Trias, J., and Nikaido, H. (1990). Outer membrane protein D2 catalyzes facilitated diffusion of carbapenems and penems through the outer membrane of *Pseudomonas aeruginosa*. Antimicrob. Agents Chemother. *34*, 52–57.

Tseng, J.T., Bryan, L.E., and Van den Elzen, H.M. (1973). Mechanisms and spectrum of streptomycin resistance in a natural population of *Pseudomonas aeruginosa*. Antimicrob. Agents Chemother. 2, 136–141.

Tumbarello, M., Repetto, E., Trecarichi, E.M., Bernardini, C., De Pascale, G., Parisini, A., Rossi, M., Molinari, M.P., Spanu, T., Viscoli, C., *et al.* (2011). Multidrug-resistant *Pseudomonas aeruginosa* bloodstream infections: risk factors and mortality. Epidemiol. Infect. *139*, 1740–1749.

Ueda, A., Attila, C., Whiteley, M., and Wood, T.K. (2009). Uracil influences quorum sensing and biofilm formation in *Pseudomonas aeruginosa* and fluorouracil is an antagonist. Microb. Biotechnol. *2*, 62–74.

Vettoretti, L., Floret, N., Hocquet, D., Dehecq, B., Plesiat, P., Talon, D., and Bertrand, X. (2009). Emergence of extensive-drug-resistant *Pseudomonas aeruginosa* in a French university hospital. Eur. J. Clin. Microbiol. Infect Dis. *28*, 1217–1222.

Vidal-Aroca, F., Meng, A., Minz, T., Page, M.G., and Dreier, J. (2009). Use of resazurin to detect mefloquine as an efflux-pump inhibitor in *Pseudomonas aeruginosa* and *Escherichia coli*. J. Microbiol. Methods *79*, 232–237.

Vogne, C., Aires, J.R., Bailly, C., Hocquet, D., and Plesiat, P. (2004). Role of the multidrug efflux system MexXY in the emergence of moderate resistance to aminoglycosides among *Pseudomonas aeruginosa* isolates from patients with cystic fibrosis. Antimicrob. Agents Chemother. *48*, 1676–1680.

Wagner, V.E., Bushnell, D., Passador, L., Brooks, A.I., and Iglewski, B.H. (2003). Microarray analysis of *Pseudomonas aeruginosa* quorum-sensing regulons: effects of growth phase and environment. J. Bacteriol. *185*, 2080–2095.

Wang, H., Meng, J., Jia, M., Ma, X., He, G., Yu, J., Wang, R., Bai, H., Hou, Z., and Luo, X. (2010a). *oprM* as a new target for reversion of multidrug resistance in *Pseudomonas aeruginosa* by antisense phosphorothioate oligodeoxynucleotides. FEMS Immunol. Med. Microbiol. *60*, 275–282.

Wang, J., Zhou, J.Y., Qu, T.T., Shen, P., Wei, Z.Q., Yu, Y.S., and Li, L.J. (2010b). Molecular epidemiology and mechanisms of carbapenem resistance in *Pseudomonas aeruginosa* isolates from Chinese hospitals. Int. J. Antimicrob. Agents *35*, 486–491.

Watkins, W.J., Landaverry, Y., Leger, R., Litman, R., Renau, T.E., Williams, N., Yen, R., Zhang, J.Z., Chamberland, S., Madsen, D., *et al.* (2003). The relationship between physicochemical properties, *in vitro* activity and pharmacokinetic profiles of analogues of diamine-containing efflux pump inhibitors. Bioorg. Med. Chem. Lett. *13*, 4241–4244.

Weir, T.L., Stull, V.J., Badri, D., Trunck, L.A., Schweizer, H.P., and Vivanco, J. (2008). Global gene expression profiles suggest an important role for nutrient acquisition in early pathogenesis in a plant model of *Pseudomonas aeruginosa* infection. Appl. Environ. Microbiol. *74*, 5784–5791.

Welch, A., Awah, C.U., Jing, S., van Veen, H.W., and Venter, H. (2010). Promiscuous partnering and independent activity of MexB, the multidrug transporter protein from *Pseudomonas aeruginosa*. Biochem. J. *430*, 355–364.

Westbrock-Wadman, S., Sherman, D.R., Hickey, M.J., Coulter, S.N., Zhu, Y.Q., Warrener, P., Nguyen, L.Y., Shawar, R.M., Folger, K.R., and Stover, C.K. (1999). Characterization of a *Pseudomonas aeruginosa* efflux pump contributing to aminoglycoside impermeability. Antimicrob. Agents Chemother. *43*, 2975–2983.

Westfall, L.W., Carty, N.L., Layland, N., Kuan, P., Colmer-Hamood, J.A., and Hamood, A.N. (2006). *mvaT* mutation modifies the expression of the *Pseudomonas aeruginosa* multidrug efflux operon *mexEF-oprN*. FEMS Microbiol. Lett. *255*, 247–254.

Whiteley, M., Lee, K.M., and Greenberg, E.P. (1999). Identification of genes controlled by quorum sensing in *Pseudomonas aeruginosa*. Proc. Natl. Acad. Sci. U.S.A. *96*, 13904–13909.

Wilke, M.S., Heller, M., Creagh, A.L., Haynes, C.A., McIntosh, L.P., Poole, K., and Strynadka, N.C. (2008). The crystal structure of MexR from *Pseudomonas aeruginosa* in complex with its antirepressor ArmR. Proc. Natl. Acad. Sci. U.S.A. *105*, 14832–14837.

Wolter, D.J., Smith-Moland, E., Goering, R.V., Hanson, N.D., and Lister, P.D. (2004). Multidrug resistance associated with *mexXY* expression in clinical isolates of *Pseudomonas aeruginosa* from a Texas hospital. Diagn. Microbiol. Infect. Dis. *50*, 43–50.

Wu, Y., Li, H., Li, J., and Huang, Z.H. (2008). Detection of *Pseudomonas aeruginosa* carried a new array of gene cassettes within class 1 integron isolated from a teaching hospital in Nanjing, China. J. Microbiol. *46*, 687–691.

Yamamoto, M., Ueda, A., Kudo, M., Matsuo, Y., Fukushima, J., Nakae, T., Kaneko, T., and Ishigatsubo, Y. (2009). Role of MexZ and PA5471 in transcriptional regulation of *mexXY* in *Pseudomonas aeruginosa*. Microbiology *155*, 3312–3321.

Yang, L., Chen, L., Shen, L., Surette, M., and Duan, K. (2011). Inactivation of MuxABC-OpmB transporter system in *Pseudomonas aeruginosa* leads to increased ampicillin and carbenicillin resistance and decreased virulence. J. Microbiol. *49*, 107–114.

Yerushalmi, H., Lebendiker, M., and Schuldiner, S. (1995). EmrE, an *Escherichia coli* 12-kDa multidrug transporter, exchanges toxic cations and H^+ and is soluble in organic solvents. J. Biol. Chem. *270*, 6856–6863.

Yoshida, T., Muratani, T., Iyobe, S., and Mitsuhashi, S. (1994). Mechanisms of high-level resistance to quinolones in urinary tract isolates of *Pseudomonas aeruginosa*. Antimicrob. Agents Chemother. *38*, 1466–1469.

Yoshida, K., Nakayama, K., Kuru, N., Kobayashi, S., Ohtsuka, M., Takemura, M., Hoshino, K., Kanda, H., Zhang, J.Z., Lee, V.J., *et al.* (2006a). MexAB-OprM specific efflux pump inhibitors in *Pseudomonas aeruginosa*. Part 5: Carbon-substituted analogues at the C-2 position. Bioorg. Med. Chem. *14*, 1993–2004.

Yoshida, K., Nakayama, K., Yokomizo, Y., Ohtsuka, M., Takemura, M., Hoshino, K., Kanda, H., Namba, K., Nitanai, H., Zhang, J.Z., *et al.* (2006b). MexAB-OprM specific efflux pump inhibitors in *Pseudomonas aeruginosa*. Part 6: exploration of aromatic substituents. Bioorg. Med. Chem. *14*, 8506–8518.

Yoshida, K., Nakayama, K., Ohtsuka, M., Kuru, N., Yokomizo, Y., Sakamoto, A., Takemura, M., Hoshino, K., Kanda, H., Nitanai, H., *et al.* (2007). MexAB-OprM specific efflux pump inhibitors in *Pseudomonas aeruginosa*. Part 7: highly soluble and *in vivo* active quaternary ammonium analogue D13-9001, a potential preclinical candidate. Bioorg. Med. Chem. *15*, 7087–7097.

Yu, H., Schurr, M.J., and Deretic, V. (1995). Functional equivalence of *Escherichia coli* σ^E and *Pseudomonas aeruginosa* AlgU: *E. coli rpoE* restores mucoidy and reduces sensitivity to reactive oxygen intermediates in *algU* mutants of *P. aeruginosa*. J. Bacteriol. *177*, 3259–3268.

Zaki, M.H., Akuta, T., and Akaike, T. (2005). Nitric oxide-induced nitrative stress involved in microbial pathogenesis. J. Pharmacol. Sci. *98*, 117–129.

Zhanel, G.G., Hoban, D.J., Schurek, K., and Karlowsky, J.A. (2004). Role of efflux mechanisms on fluoroquinolone resistance in *Streptococcus pneumoniae* and *Pseudomonas aeruginosa*. Int. J. Antimicrob. Agents *24*, 529–535.

Zhanel, G.G., Decorby, M., Laing, N., Weshnoweski, B., Vashisht, R., Tailor, F., Nichol, K.A., Wierzbowski, A., Baudry, P.J., Karlowsky, J.A., *et al.* (2008). Antimicrobial-resistant pathogens in intensive care units in Canada: results of the Canadian National Intensive Care Unit (CAN-ICU) study, 2005–2006. Antimicrob. Agents Chemother. *52*, 1430–1437.

Zhanel, G.G., Decorby, M., Adam, H., Mulvey, M.R., McCracken, M., Lagace-Wiens, P., Nichol, K.A., Wierzbowski, A., Baudry, P.J., Tailor, F., *et al.* (2010). Prevalence of antimicrobial-resistant pathogens in canadian hospitals: results of the Canadian Ward Surveillance Study (CANWARD 2008). Antimicrob. Agents Chemother. *54*, 4684–4693.

Zhang, L., and Mah, T.F. (2008). Involvement of a novel efflux system in biofilm-specific resistance to antibiotics. J. Bacteriol. *190*, 4447–4452.

Zhao, W.H., Hu, Z., Chen, G., Ito, R., and Hu, Z.Q. (2009). Contributions of IMP-10 metallo-β-lactamase, the outer membrane barrier and the MexAB-OprM efflux system to high-level carbapenem resistance in *Pseudomonas aeruginosa*. Chemotherapy (Basel) *55*, 168–174.

Ziha-Zarifi, I., Llanes, C., Koehler, T., Pechere, J.-C., and Plesiat, P. (1999). *In vivo* emergence of multidrug-resistant mutants of *Pseudomonas aeruginosa* overexpressing the active efflux system MexA-MexB-OprM. Antimicrob. Agents Chemother. *43*, 287–291.

Function and Regulation of *Neisseria gonorrhoeae* Efflux Pumps

10

Yaramah M. Zalucki, Alexandra D. Mercante, Jason M. Cloward, Elizabeth A. Ohneck, Justin L. Kandler, Maira Goytia, Paul J.T. Johnson and William M. Shafer

Abstract

The export action of efflux pumps is a nearly universal mechanism used by bacteria to escape the action of toxic compounds in their environment. Antimicrobials faced by bacteria include various biocides (natural or synthetic) and classical antibiotics used in therapy of infections. Certain efflux pumps also export antimicrobials produced by their hosts and this ability likely enhances the survival of the infecting pathogen, especially during early stages of infection when mediators of innate host defence normally function to reduce the microbial load. This review is concerned with the roles of efflux pumps produced by *Neisseria gonorrhoeae* in contributing to its resistance to antimicrobials used in therapy of infections or those that participate in innate host defence. Specific emphasis is placed on the genetic organization, transcriptional regulation, and function of gonococcal efflux pumps. The major theme of this review is that in addition to their role in enhancing bacterial resistance to classical antibiotics and biocides, certain efflux pumps, such as those harboured by strict human pathogens like gonococci, can also influence *in vivo* fitness and survival of bacteria since they provide a mechanism to resist natural antimicrobials produced by their host.

Introduction

N. gonorrhoeae is an important, strict human pathogen that is transmitted by sexual contact and causes the disease gonorrhoea, which can have many clinical manifestations. Historical writings suggest that the gonococcus (or GC) has been a human curse for thousands of years (Shafer *et al.*, 2010). Thus, the ancient nature of gonorrhoea is suggested by warnings in the Book of Leviticus in the Old Testament that women should avoid men with discharges. Later descriptions of the disease by the second century Greek physician Galen confirms that gonorrhoea has been a disease of mankind for centuries. In sharp contrast, the earliest description of disease caused by the related human pathogen *N. meningitidis* (the meningococcus or MC), which can be carried commensally in the nasopharynx of 10–30% of the population, can be found in the early nineteenth century writings by Vieusseaux (Stephens *et al.*, 2007). Owing to the lack of earlier accounts of disease, which can have dramatic presentations in the forms of meningitis or septicaemia and result in rapid death, it has been suggested that only in recent times did MC evolve into a pathogen.

As emphasized in other chapters of this volume and elsewhere (Levy, 1992; Nikaido, 1996; Shafer *et al.*, 2010), drug efflux pumps can help bacteria resist the action of clinically useful antibiotics. One can even imagine that in the pre-antibiotic era, soil microbes found efflux pumps beneficial to help them resist antibiotics produced by other microbes. At first glance, it is more difficult to understand how bacteria that principally live inside a host developed the need for efflux pumps unless one considers that the respective host and competing flora also produce antimicrobials. Thus, strict human pathogens such as GC are confronted with a wide range of host-derived antimicrobials (e.g. antimicrobial peptides, toxic free fatty acids, bile salts and progesterone, etc.) that can have potent activity against them. We now know that many of these antimicrobials can be recognized and exported by efflux pumps produced by GC (Shafer *et al.*, 2010). Moreover, there is very good evidence (see below) that production and levels of one particular pump, MtrC–MtrD–MtrE, can substantially enhance both the duration of a gonococcal infection (Jerse *et al.*, 2003) and *in vivo* fitness (Warner *et al.*, 2007, 2008) of GC in an experimental female mouse model of lower genital tract infection (Jerse, 1999). This same efflux pump is also required for high level, chromosomally mediated resistance to penicillin that can be expressed by GC (Veal *et al.*, 2002). Accordingly, studies on efflux pumps produced by GC provide a unique opportunity to understand the role of bacterial drug efflux pumps in the overall pathogenic mechanisms of bacteria during infection and how they also contribute to resistance to antibiotics used clinically today or in the past. Thus, this review will concentrate on these issues as well as provide insights as to how GC controls expression of efflux pump genes and how such regulatory systems modulate expression of other genes involved in fitness and pathogenicity.

GC efflux pumps and roles in antimicrobial resistance and survival *in vivo*

In contrast to other bacteria that express a wide number and type of drug efflux pumps (Paulsen *et al.*, 2002), most GC strains produce only four drug efflux pumps (Fig. 10.1). These pumps belong to the resistance-nodulation-division (RND) family (MtrC–MtrD–MtrE), the major facilitator (MF) family (FarA–FarB–MtrE), the ABC transporter family (MacA–MacB–MtrE) and the multidrug toxic compound extrusion (MATE) family (NorM); these pumps have been reviewed elsewhere (Shafer *et al.*, 2010). Additionally, some (rare) clinical isolates have been reported to harbour the *mef* gene (Luna *et al.*, 2000), which encodes a pump that exports macrolides. The MtrC–MtrD–MtrE efflux pump, like other RND drug efflux pumps, can export structurally diverse hydrophobic antimicrobial agents, including antibiotics (macrolides, beta-lactams, rifampin, etc.), non-ionic detergents (Triton X-100 [TX-100]) (Lucas *et al.*, 1995; Veal *et al.*, 1998), biocides (the spermicide nonoxynol-9) and host-derived antimicrobials (e.g. certain antimicrobial peptides, progesterone, bile salts) that participate in innate host defence (Hagman *et al.*, 1995; Shafer *et al.*, 1998; Rouquette *et al.*, 1999; Warner *et al.*, 2008). Export of human antimicrobial peptides, such as LL-37, by the MtrC–MtrD–MtrE pump by both GC (Shafer *et al.*, 1998) and MC (Tzeng *et al.*, 2005), provides a way for these pathogens to capture them in the periplasm after they have breached the barrier imposed by the outer membrane. This is important since LL-37 has been proposed to inhibit cell wall biogenesis (Sochacki

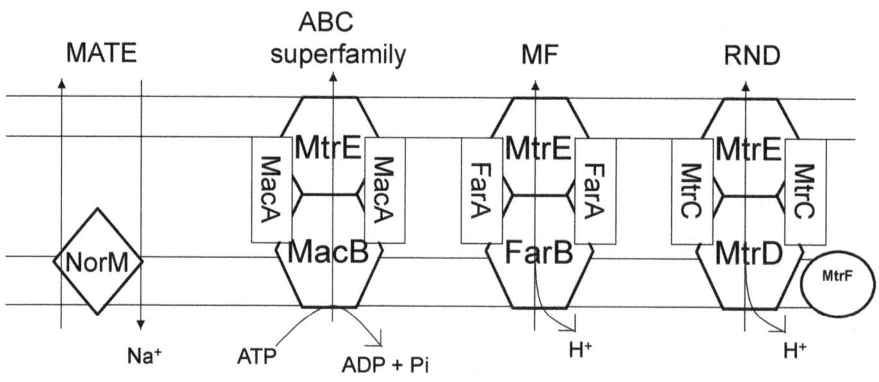

Figure 10.1 Membrane organization of antimicrobial efflux pumps of gonococci. The proposed membrane organization and class of the known antimicrobial efflux pumps common to gonococcal strains (NorM, MacA-MacB-MtrE, FarA-FarB-MtrE and MtrC-MtrD-MtrE) and how each pump is energized are depicted in this figure. MtrE acts as an outer membrane channel protein for three pumps. MtrF is shown as an accessory protein for the MtrC-MtrD-MtrE pump.

et al., 2011). Sublethal levels of LL-37 frequently bathe the mucosal surfaces, including the genital tract (Cole, 2007) and can transiently increase LL-37 resistance by a mechanism involving a two-component regulatory system that indirectly enhances transcription of *mtrCDE* (see below). Progesterone, also recognized by MtrC–MtrD–MtrE, behaves as a detergent on gonococcal membranes (Morse *et al.*, 1982) and its importance in protecting the female lower genital tract from infection is supported by the observation that surgical removal of ovaries from mice increased their susceptibility to an experimental GC infection (Jerse *et al.*, 2003). The FarA–FarB–MtrE efflux pump recognizes long-chain, unsaturated fatty acids (e.g. oleic and palmitic acid) (Lee and Shafer, 1999), which can have detergent-like action against GC (Morse *et al.*, 1982). The MacA–MacB pump can export macrolides, but is often silent due to promoter mutations (Rouquette-Loughlin *et al.*, 2005). NorM has the ability to export quaternary ammonium compounds and quinolone antibiotics (Rouquette-Loughlin *et al.*, 2003).

Mutations that increase the level of bacterial efflux pump proteins can reduce the efficacy of antibiotic treatment regimens (Levy, 1992; Nikaido, 1996; Shafer *et al.*, 2010). As a specific example, overexpression of the *mtrCDE* efflux pump operon due to *trans* – or *cis*-acting mutations (see below) can decrease gonococcal susceptibility to penicillin (Veal *et al.*, 2002). Importantly, in strains that harbour other mutations that impact the structure of penicillin-binding proteins 1 (PBP1) and 2 (PBP2), or outer membrane porin proteins (Faruki and Sparling, 1986; Spratt, 1988; Ropp *et al.*, 2002; Veal *et al.*, 2002; Olesky, *et al.*, 2006; Shafer *et al.*, 2010) overexpression of *mtrCDE* conferred a level of penicillin resistance that resulted in clinical failure (Faruki *et al.*, 1985). Collectively, these mutations ultimately led to the removal of penicillin from the CDC-recommended treatment regimen. Importantly, genetic inactivation of the pump operon in a resistant strain resulted in a return to a level of penicillin-susceptibility well below the MIC breakpoint (Veal *et al.*, 2002). This observation supports the notion (Lomovskaya and Watkins, 2002) that efflux pump inhibitors (EPI) could restore previously used antibiotics that were lost due to the development of resistance. Moreover, inhibiting the action of efflux pumps that also recognize host antimicrobials may have the added benefit of making EPI-treated bacteria less fit during natural infection. Looking towards the future, the third generation cephalosporin ceftriaxone is now the mainstay antibiotic used for treating gonorrhoea, but strains expressing decreased susceptibility to this important antibiotic have emerged and a few strains have been documented (Ohnishi *et al.*, 2011; Camara *et al.*, 2012; Unemo *et al.*, 2012) to be clinically resistant to cure by ceftriaxone (M. Unemo, personal communication); hence, it will be important to follow the spread of such strains and determine the role that efflux pumps like MtrC–MtrD–MtrE play in resistance. As has been recently emphasized (Dionne-Odom, 2011), the public health problem of GC strains resistant to antibiotics is becoming more severe and strains expressing clinical resistance to ceftriaxone would leave little therapeutic options in controlling gonorrhoea. In the absence of new antibiotics that recognize novel targets, this problem will only worsen.

In the absence of classical antibiotics, does production of an efflux pump provide an advantage for gonococci during infection? Results from studies on the MtrC–MtrD–MtrE pump suggest that this is the case since its production enhanced gonococcal survival during experimental infection in the lower genital tract of female mice (Jerse *et al.*, 2003). Moreover, differential expression of the *mtrCDE* operon (see below) can modulate fitness levels in this infection model (Warner *et al.*, 2007, 2008). Thus, we proposed that the MtrC–MtrD–MtrE efflux pump is important in the ability of GC to survive *in vivo* when challenged by mediators of innate host defence (Shafer *et al.*, 1996, 1998; Warner *et al.*, 2008). This capacity to resist mediators of innate host defence may be important during human infection because GC elicit a strong Th17, pro-inflammatory response (Liu *et al.*, 2011) characterized by enhanced production of antimicrobial peptides. Other RND-type pumps similar to MtrC-MtrD-MtrE produced by Gram-negative enteric pathogens (e.g. AcrA–AcrB–TolC) are also likely to enhance bacterial survival during infection (Bailey *et al.*, 2010; see also Chapters 9 and 12). The capacity of GC to export free long-chained fatty acids (e.g. palmitic and oleic acid) via the FarA-FarB-MtrE pump may help it survive in the rectum where concentrations of such fatty acids are elevated (Morse *et al.*, 1982; Shafer *et al.*, 1996); however, direct experimental evidence is lacking since FarA–FarB–MtrE was not required for *in vivo* survival in the mouse infection model (Jerse *et al.*, 2003). Nevertheless, results from studies on the Mtr system suggest that prior to the clinical use of antibiotics, efflux pumps served important roles for bacteria during stages of infection when they encounter host-derived antimicrobials.

Organization of GC efflux pump genes

Of the four GC efflux pumps (Shafer *et al.*, 2010), three are encoded by genes organized in an operon (*mtrCDE, farAB* and *macAB*) while the fourth (*norM*) is encoded by a stand-alone gene (Fig. 10.2). The *mtr* (*m*ultiple *t*ransferable *r*esistance) system was first identified by Maness and Sparling (1973) when they isolated a spontaneous mutant that exhibited increased resistance to multiple, structurally diverse antimicrobial hydrophobic compounds. It was originally thought that *mtr* regulated outer membrane permeability by over-producing a membrane protein and increasing the degree of peptidoglycan cross-linking (Guymon *et al.*, 1978). However, subsequent cloning/sequencing experiments in the 1990s (Pan and Spratt, 1994; Hagman *et al.*, 1995) showed that the mutation was located within a gene encoding a transcriptional repressor (MtrR) of an upstream, but transcriptionally divergent operon (*mtrCDE*) that would encode the tripartite MtrC-MtrD-MtrE efflux pump similar to other RND-type pumps of Gram-negative bacteria (Pan and Spratt, 1994; Hagman *et al.*, 1995, 1997; Delahay *et al.*, 1997). Like other RND efflux pumps, the three proteins that form the pump are a cytoplasmic membrane transporter (MtrD), a membrane fusion protein (MtrC) and an outer membrane channel protein (MtrE) (Fig. 10.1). MtrD and MtrE function as trimers, and are connected by three homodimers of MtrC, making the stoichiometry of an assembled pump 3:6:3 (Janganan *et al.*, 2011). It is likely that MtrE also serves as the outer membrane channel protein for the FarA–FarB and MacA–MacB efflux pumps (Fig. 10.1). Directly or indirectly, other proteins also participate in efflux mediated by the Mtr pump. In this respect, Veal and Shafer (2003) identified an accessory protein (MtrF) which, for reasons that are not yet clear, is required when the host strain expresses high levels of antimicrobial resistance via the MtrC-MtrD-MtrE efflux system. Additionally, energy supplied by the TonB-ExbB-ExbD system is needed for inducible antimicrobial resistance mediated by MtrC–MtrD–MtrE (Rouquette *et al.*, 2002) that also requires the participation of a transcriptional activator termed MtrA (see below). Lipooligosaccharide (LOS) structure is also important in the function of the MtrC–MtrD–MtrE efflux pump in strains overexpressing *mtrCDE* because a deep rough LOS mutant expressing a core oligosaccharide that was severely truncated was unable to express high-level resistance to substrates of the pump (Lucas *et al.*, 1995).

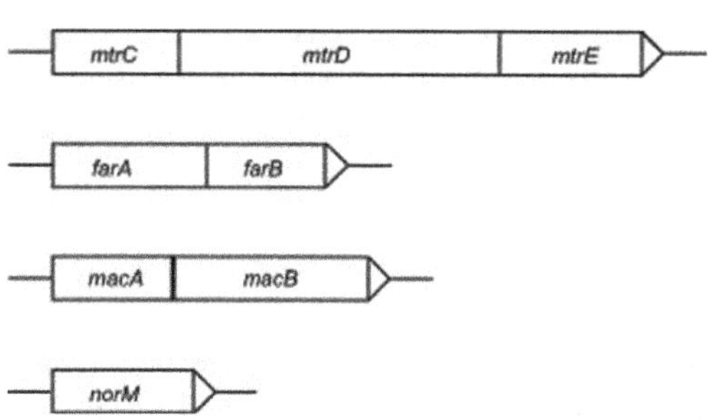

Figure 10.2 Genetic organization of the efflux components of gonococci. Shown are the genes that encode components of the four drug efflux pumps (MtrC-MtrD-MtrE, FarA-FarB-MtrE, MacA-MacB-MtrE and NorM) expressed by gonococci; all genes are drawn to scale. Please see text for details of genes and corresponding pumps.

Transcriptional regulation of GC efflux genes

Discoveries of bacterial efflux pumps were often made prior to the availability of genome sequence databases and in these instances were facilitated by the isolation of mutants that expressed decreased susceptibility to antimicrobials. (Nikaido, 1996; Shafer *et al.*, 2010). These resistance-conferring mutations frequently mapped to a gene that encoded a DNA-binding protein that would normally dampen expression of a closely linked gene or operon encoding efflux pump proteins. Pan and Spratt (1994) discovered the *mtr* locus (Fig. 10.2) in GC strain CH95 in this way when they identified a mutation in a gene (*mtrR*) that encodes a DNA-binding protein similar to TetR and QacR (reviewed in Grovic *et al.*, 2002) responsible for the Mtr phenotype described by Maness and Sparling (1973). CH95 expressed high-level resistance to a panel of structurally diverse hydrophobic agents (HA), but introduction of CH95 *mtrR* coding mutation (H105Y) into a sensitive recipient only resulted in a slight (2 – to 4-fold) increase in resistance. However, when the DNA sequence upstream of *mtrR*, which included promoters for *mtrR* and *mtrCDE* transcription, was introduced into a sensitive strain, higher levels of HA-resistance similar to the donor could be obtained in transformants. The reason for this difference was discovered by Hagman *et al.* (1995) to be due to the presence of a single base pair (bp) deletion in a 13-bp inverted repeat sequence within the *mtrR* promoter of the donor strains CH95 or FA171 (Fig. 10.3). Subsequent examination (Shafer *et al.*, 1996, 2010) of CH95 and other high-level resistant GC strains showed that this single bp deletion was frequently present in a 13 bp inverted repeat and that introduction of this mutation with or without *mtrR* coding region mutations into the sensitive strain FA19 would result in high-level HA-resistance. Taken together, GC has both *cis* – and *trans*-acting elements that impact transcription of efflux pump genes and these are discussed separately below.

Cis-acting factors that regulate efflux pump genes

Point mutations, deletions, or insertions in the nucleotide sequences between *mtrR* and *mtrCDE* can provide GC with higher levels of HA-resistance than mutations within the *mtrR*-coding region (Fig. 10.3). As mentioned above, a single bp deletion within the 13-bp inverted repeat element localized in the *mtrR* promoter can significantly enhance transcription of *mtrCDE* (Hagman and Shafer, 1995). This mutation is frequently found in GC clinical isolates that express the Mtr phenotype (Xia *et al.*, 2000, Liao *et al.*, 2011); some clinical isolates have a dinucleotide insertion within this inverted repeat (Zarantonelli *et al.*, 2001). In either case, the optimal 17 bp spacing between the −10 and −35 elements is disrupted and this significantly reduces *mtrR* transcription. This, however, cannot be the sole reason why such strains express high-level HA resistance since their level of *mtrCDE* expression is greater than strains with *mtrR* loss-of-function or null mutations. Instead, because the promoters for *mtrR* and *mtrCDE* transcription overlap and are divergent (Fig. 10.3), it is most likely that the mutations enhance RNA polymerase interactions with the *mtrCDE* promoter.

A limited number of GC strains [e.g. MS11 (Warner *et al.*, 2008)] have a point mutation (C→T) located 120 nucleotides upstream of the *mtrC* translational start codon (mtr_{120}), changing the sequence at this locus from TATAAC to TATAAT and thereby generating a consensus −10 element (Fig. 10.3) (Ohneck *et al.*, 2011). This new −10 element acts as a stronger promoter for *mtrCDE* transcription and results in high levels of *mtrCDE* expression and HA resistance, as well as enhanced *in vivo* fitness (Warner *et al.*, 2008). It appears not to be under the control of the transcriptional regulators MtrR or MtrA (see below), which act on the wild type promoter. Genome sequence analysis of 14 GC strains commonly used in the laboratory revealed that this mutation occurs in only one of these strains. However, this mutation was identified in one clinical isolate in a collection of 121 strains studied by M. Unemo and coworkers (Ohneck *et al.*, 2011), as well as in five recent clinical isolates from Pakistan and Vietnam (M. Unemo, personal communication). Thus, although rare, this mutation is clinically relevant. Since the mtr_{120} mutation significantly enhanced GC fitness in a murine model of lower genital tract infection (Warner *et al.*, 2008), it may provide the bacteria with a competitive advantage in the community. More detailed molecular surveillance will be needed to determine if such strains emerge at a higher frequency.

In 1999 a number of azithromycin-resistant GC clinical isolates were obtained from patients in Kansas City, MO (Johnson *et al.*, 2003). These strains contained a 153 bp insertion in the *mtrR-mtrCDE* intervening region and expressed

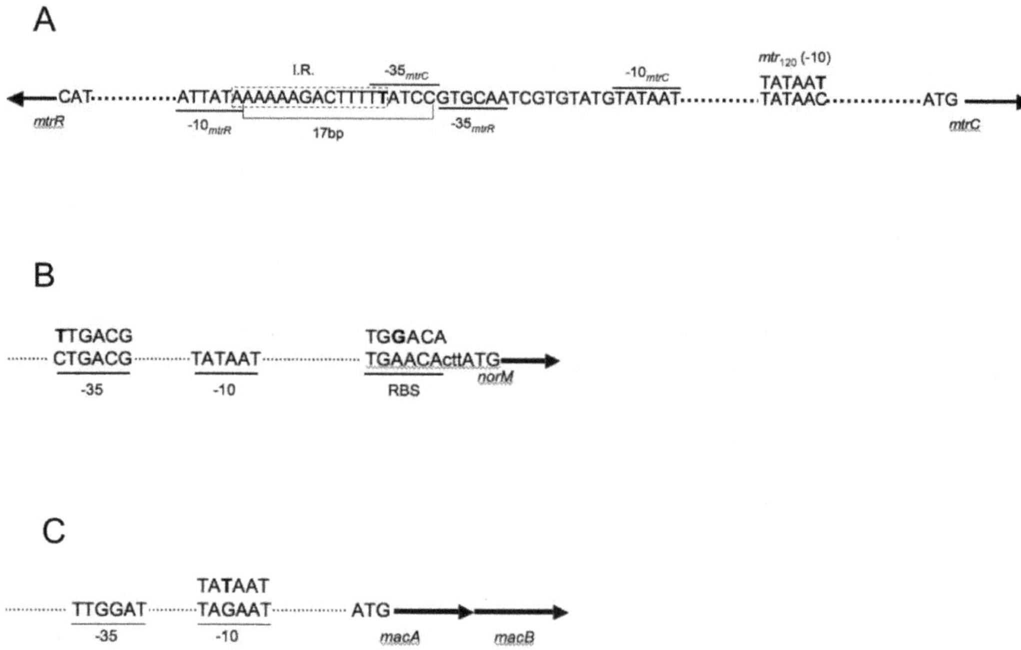

Figure 10.3 *Cis*-acting mutations regulate efflux pump expression in gonococci. A. The *mtrR-mtrC* intergenic region. The 13 bp inverted repeat is boxed. The site of the single bp deletion described by Hagman *et al.* (1995) is bolded. The consensus −10 element generated by the mtr_{120} mutation is shown, with the C→T mutation in bold text. B. The *norM* promoter region. Single base pair mutations resulting in increased *norM* expression are indicated by bold nucleotides. C. The *macAB* promoter region. The G→T mutation in the −10 element that results in increased *macAB* expression is bolded.

cross-resistance to structurally diverse HA that was mediated by the MtrC-MtrD-MtrE efflux pump. Our work (Rouquette-Loughlin *et al.*, 2004), showed that virtually all meningococci (MC) contain a 155–159 bp insertion within this region, identified as a Correia element (CE) (Correia *et al.*, 1988), with some serogroup Y strains also having a tandemly linked IS*1301* sequence; more recent studies have shown a low percentage of meningococcal strains lacking the CE (Enriquez *et al.*, 2010). In MC, the presence of the CE element dampened *mtrCDE* expression as a result of providing a binding site for integration host factor (IHF) and a new site for post-transcriptional processing of the *mtrC* transcript; IHF is also important in regulating the *farAB* efflux pump operon (see below). It is unclear, however, how the CE and perhaps more importantly the IHF-binding site and IHF functions in GC since HA-resistance is elevated in strains bearing the CE sequence. Nevertheless, the presence (albeit rare) of the CE at this site in GC suggests that horizontal gene transfer occurred between MC and GC and emphasizes the importance of recombination events in generating diversity in clinical strains of both pathogens.

The *norM* and *macAB* genes are also regulated by *cis*-acting control elements that were defined by mutations (Fig. 10.3). The presence of point mutations in the −35 hexamer of the *norM* promoter (C→T) or in the ribosome binding site (A→G) can enhance bacterial resistance to certain quaternary ammonium compounds as well as to norfloxacin and ciprofloxacin (Rouquette-Loughlin *et al.*, 2003). A point mutation in the −10 hexamer of the *macAB* promoter (G→T) has been identified that increases expression of *macAB* and levels of macrolide resistance in GC (Rouquette-Loughlin *et al.*, 2005). Unlike *cis*-acting mutations that influence *mtrCDE* expression, for both the *norM* and *macAB* systems, the aforementioned mutations are not likely by themselves to confer clinically significant levels of resistance to their antibiotic substrates. However, together with other mutations that impact antibiotic susceptibility levels, the presence of these *cis*-acting mutations could influence the efficacy of antibiotic therapy, particularly for strains with susceptibilities near the MIC breakpoint for the relevant antibiotics. Like the mtr_{120} mutation for *mtrCDE*, the point mutations found to regulate *norM* and *macAB* alter the sequence of their respective promoter elements to be closer to consensus (Shine and Dalgarno, 1975; Browning and Busby, 2004). Thus, it appears that point mutations within bacterial promoters can upregulate gene expression by enhancing promoter recognition by RNA polymerase outside of the control of transcriptional regulators, or, in the case of *norM*, may increase translation through improved ribosomal recognition of the *norM* transcript.

Trans-acting factors that regulate efflux pump genes

A host of DNA-binding proteins directly or indirectly control expression of genes encoding structural proteins of drug efflux pumps in gonococci (Grovic *et al.*, 2002; Shafer *et al.*, 2010). Many of these DNA-binding proteins also have the ability to directly modulate expression levels of other genes important in gonococcal physiology and metabolism, as well as regulating genes encoding transcriptional regulators of efflux pump determinants. In this respect, the DNA-binding proteins that control *mtrCDEF* and *farAB* expression are interconnected (Fig. 10.4) and specific examples are discussed in detail below.

MtrR control of gene expression

MtrR, first described by Pan and Spratt (1994), performs a central role in regulating *mtrCDE* directly and *farAB* indirectly (Lee *et al.*, 2003). It is a member of the TetR/QacR family and binds as two homodimers to the *mtrCDE* promoter region to dampen expression of this operon (Fig. 10.5) (Lucas *et al.*, 1997; Hoffmann *et al.*, 2005). Studies on clinical isolates that express elevated levels of resistance to hydrophobic agents recognized by the MtrC-MtrD-MtrE efflux pump have shown that such strains often bear mutations that cause radical amino acid replacements in the helix–turn–helix (HTH) motif (residues 32–53) that significantly reduce MtrR-binding to target DNA sequences (Shafer *et al.*, 1996). Some mutations that can confer a phenotype are outside of the HTH region and they might alter dimer formation or drug interactions (Shafer *et al.*, 1996). These loss-of-function mutations can enhance *mtrCDE* transcription 2 – to 3-fold and, depending on the substrate, increase resistance to antimicrobials by 4 – to 10-fold (Zarantonelli *et al.*, 2001; Veal *et al.*, 2002). It is important to emphasize that because loss-of-function mutations of *mtrR* have been observed in clinical isolates, their occurrence is relevant for considering antibiotic treatment regimens used worldwide. In addition to contributing to antibiotic resistance, the loss of MtrR, leading to increased levels of the MtrC-MtrD-MtrE efflux

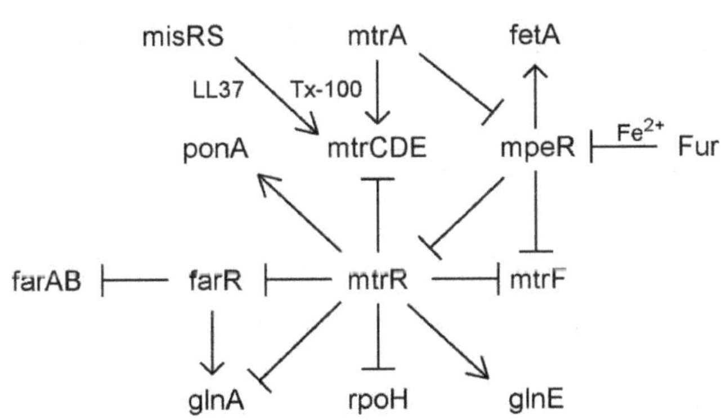

Figure 10.4 Schematic of the regulatory circuit mediated by the DNA-binding proteins that modulate *mtr* efflux expression. The details of the genes and regulation are discussed in the text; arrows indicate gene activation and barred lines indicate gene repression.

pump, conferred a fitness advantage *in vivo* in the female mouse model of lower genital tract infection (Warner *et al.*, 2007). These results confirmed previous findings showing that bacterial strains expressing the MtrC-MtrD-MtrE efflux pump had greater and longer survival in the genital tract of BALB/c mice than isogenic strains that did not express an active pump (Jerse *et al.*, 2003).

Using classical genetic and microarray technologies (Lee *et al.*, 2003; Folster *et al.*, 2007, 2009), we have learned that MtrR directly or indirectly controls over 65 genes at different phases of growth. We performed a microarray analysis using RNA extracted from mid-logarithmic phase cultures of strain FA19 and an isogenic derivative bearing an *mtrR* deletion to identify MtrR-repressed and – activated genes (Folster *et al.*, 2009). The cumulative results from these studies have led us to conclude that MtrR is a global regulatory protein, and its capacity to control genes involved in stress response (*rpoH*) (Folster *et al.*, 2009), peptidoglycan biosynthesis (*ponA*) (Folster *et al.*, 2007), amino acid biosynthesis (*glnA*) (Johnson *et al.*, 2011), polyamine uptake (*potF* and *potH*) and regulation of a regulator (*farR*) of the *farAB* efflux operon (Lee *et al.*, 2003) may be as important as its regulation of *mtrCDE*. We have termed these genes as being 'off-target', since MtrR is acting as a *trans*-regulator to distinguish them from the adjacent *mtrCDE* genes (Johnson *et al.*, 2011).

Examples of specific off-target genes and their role in GC physiology are described below. MtrR can directly repress *rpoH*, which encodes an essential sigma factor involved in environmental stress response (Gunesekere *et al.*, 2006). This regulation results in expression changes of a panel of RpoH-RNA polymerase transcribed

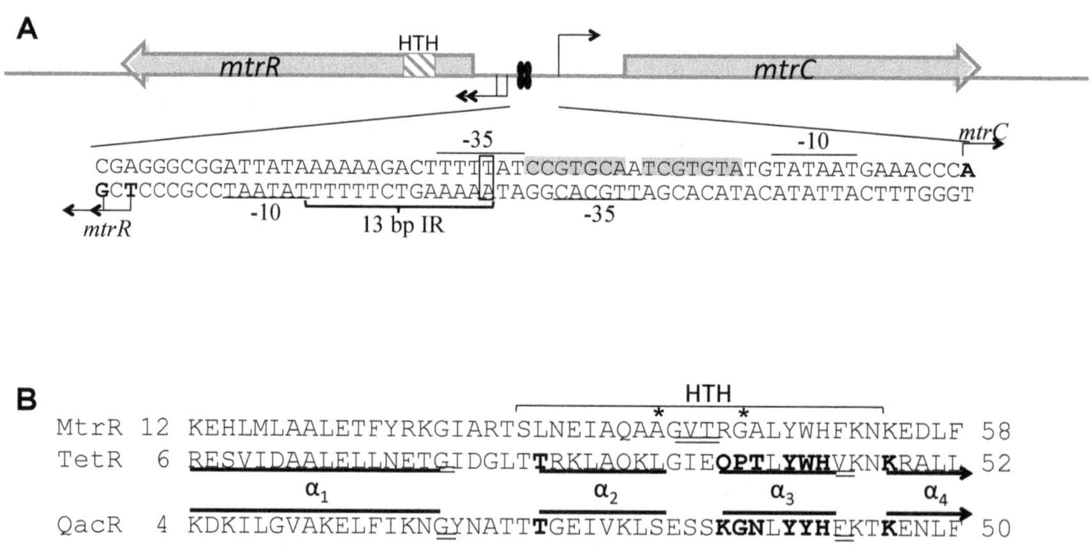

Figure 10.5 MtrR binding site within the *mtrR-mtrC* intergenic region and helix–turn–helix motif of MtrR. (A) Schematic representation of the *mtrR* and *mtrC* genes in *N. gonorrhoeae*. The double-stranded nucleotide sequence pictures the start sites (+1) of *mtrR* and *mtrCDE* (in boldface) and the −10 and − 35 sequences for *mtrR* and *mtrC* (underlined and overlined, respectively). The binding site for the two MtrR homodimers (Hoffman *et al.*, 2005) is highlighted in grey. The *cis*-acting 13-bp inverted repeat element within the *mtrR* promoter is shown and the site of the single bp deletion frequently found in clinical isolates is indicated by a box. In the diagram, bent arrows indicate the starting sites for each gene. The *mtrR* gene region encoding the helix–turn–helix (HTH) motif, characteristic of TetR/QacR transcriptional regulators, is indicated by the hatched pattern in the *mtrR* coding region. (B) Amino acid sequence alignment of the helix–turn–helix motif of MtrR involved in DNA binding is over lined. Residues involved in turns are indicated by double lines. Missense mutations frequently found in clinical isolated are indicated by an asterisk (A39T, G45D). Residues involved in DNA binding identified by crystal structures in TetR and QacR are indicated in boldface (for review see Ramos, 2005). α-helices in TetR and QacR are shaded. Highly conserved residues overlap with helix and turns.

genes (Folster *et al.*, 2009). RpoH expression is involved in the ability of GC to resist oxidative stress and loss of MtrR production enhances bacterial resistance to hydrogen peroxide (Folster *et al.*, 2009). Thus, MtrR can modulate levels of GC susceptibility to both oxidative (e.g. H_2O_2) and non-oxidative (e.g. antimicrobial peptides) killing systems of neutrophils. Furthermore, MtrR can regulate expression of efflux genes other than *mtrCDE* (Lee *et al.*, 2003). It can directly repress *farR*, which encodes the transcriptional repressor of the *farAB* efflux pump operon (Fig. 10.4). In opposing manners, MtrR and FarR can regulate expression of *glnA*, which encodes glutamine synthetase. FarR binds upstream of the *glnA* translational start and activates its expression (Johnson *et al.*, 2011). However, MtrR can bind upstream of the FarR-binding site and dampen transcription of *glnA* by reducing FarR binding to the *glnA* promoter and FarR expression. Curiously, MtrR can activate *glnE* (Folster *et al.*, 2009), which reveals that it can differentially control genes encoding enzymes that function in the same metabolic pathway. In addition to the mentioned regulatory roles of MtrR, preliminary studies suggest that MtrR also influences the extent of biofilm formation in *N. gonorrhoeae*. However, further research is needed to determine the MtrR-regulated genes involved in this process since none of the genes known to be involved in biofilm formation, such as *norB*, *estD*, or *aniA* (Falsetta *et al.*, 2011), were identified as MtrR-regulated by microarray analysis (Folster *et al.*, 2009).

An important evolutionary question is why did GC not dispense with MtrR since its loss increases antibiotic resistance and *in vivo* fitness? It is likely that genes activated by MtrR hold the key. In the mouse infection experiments performed by Warner *et al.* (2007), loss of MtrR afforded a fitness advantage for the first 5 days, but this advantage waned, suggesting that possession of MtrR may be advantageous at later stages of infection. One of these MtrR-activated genes that may be important *in vivo* is *glnE* (Folster *et al.*, 2009), which is of significant importance in *Salmonella typhimurium* growth and fitness *in vivo* (Klose and Mekalanos, 1997), is essential in mycobacteria (Parish and Stoker, 2000) and apparently is essential in GC as we have been unable to construct a *glnE* null mutation. GlnE encodes the enzymatic regulator of glutamine synthetase and can activate or de-activate the enzyme. Since levels of glutamine are low at mucosal surfaces and within phagocytes, the ability to synthesize glutamine *in vivo* may be important for survival and maximal fitness. Alternatively, or in conjunction with this necessity to produce glutamine at mucosal surfaces, the importance of MtrR regulation of off-target genes may be more integral in the long-term maintenance of gonococcal infections than previously understood due to its involvement in the regulation of biosynthetic pathways. Indeed, we found (Folster *et al.*, 2009) that the largest subpopulation of MtrR-regulated genes of known function were those involved in biosynthetic pathways and it is not without precedent that TetR/QacR transcriptional regulators can play a role in regulating expression of genes involved in such pathways (Ramos *et al.*, 2005). Certainly, other examples of MtrR-activated genes exist, but this subset of the MtrR regulon may explain why *mtrR* has not been discarded over the millennia.

MtrA regulation of efflux gene expression

MtrA is a member of the AraC family of transcriptional regulators and was discovered during experiments that evaluated whether growth of the gonococci in sublethal level of substrates recognized by MtrC-MtrD-MtrE could enhance bacterial resistance to such antimicrobials (Rouquette *et al.*, 1999). A panel of strains was tested and two urogenital isolates (strains FA19, FA889) and an invasive bloodstream isolate (strain UUI) were able to increase resistance to TX-100 when incubated overnight in sublethal levels of TX-100. For five other strains tested, including strain FA1090, no increase in resistance was observed (Rouquette *et al.*, 1999). A database search revealed an open reading frame that was likely to encode a transcriptional activator (MtrA) similar to those that positively regulate the expression of other bacterial efflux operons (Gambino *et al.*, 1993; Kaniga *et al.*, 1994; Ariza *et al.*, 1995). Comparing the sequence of *mtrA* from both strain FA19 and FA1090 revealed the presence of an 11 bp deletion in the coding region of FA1090, which would result in the production of a non-functional protein. Subsequent sequencing of PCR products from many gonococcal strains and whole genome sequencing efforts have shown that the 11 bp deletion in FA1090 is common and present in nearly 40% of strains. Introducing a null mutation into *mtrA* in strain FA19 resulted in the loss of the induction phenotype, suggesting that MtrA induces expression of the *mtrCDE* efflux pump when grown in sublethal concentrations of TX-100 (Rouquette *et al.*, 1999). This inducible resistance was later found to be dependent on the TonB-ExbD-ExbD system for energy (Rouquette-Loughlin *et al.*, 2002).

MtrA is not only a positive regulator of *mtrCDE*, it also controls (directly or indirectly) other genes involved in efflux or regulation of efflux genes (e.g. *mpeR*). Importantly, loss of MtrA was found to negatively impact the *in vivo* fitness of gonococci in the mouse infection model (Warner *et al.*, 2007). This could be due to the loss of induction of the *mtrCDE* pump operon by MtrA when GC are in the presence of antimicrobials recognized by the efflux pump. Supporting this conclusion, two second site suppressor mutations in the *mtrR* gene were isolated that gave increased fitness of the *mtrA* mutant in the mouse model (Warner *et al.*, 2007). This suggests that loss of the repressor (MtrR) compensates for the loss of the activator (MtrA) of the *mtrCDE* efflux pump operon. However, as MtrA is a global regulator of genes, further experiments are needed to determine the cause of the loss of fitness of MtrA mutants in the mouse model.

Recently, our group (Zalucki *et al.*, 2012) has purified MtrA as a maltose-binding protein fusion protein and found that it can bind in a specific manner upstream of the MtrR-binding site near *mtrCDE* (Fig. 10.5). The binding of MtrA to this site is enhanced by the presence of an efflux pump substrate (e.g. TX-100) and in this instance, endows MtrA with a greater affinity for the target DNA than that of MtrR. Taken together, the earlier observations of Rouquette *et al.* (1999) regarding the induction of *mtrCDE* and TX-100 resistance can now be explained by the ability of MtrA to bind upstream of *mtrCDE* with a greater affinity than the natural repressor (MtrR) and promote transcription of the efflux pump operon.

MpeR regulation of GC genes

MpeR was discovered by Folster and Shafer (2005) during a search of the FA1090 genome sequence for regulators, in addition to MtrR, that might control expression of *mtrF*. As described above, MtrF is an apparent accessory protein of the MtrC-MtrD-MtrE efflux pump and is needed for high-level resistance of gonococci to antimicrobials recognized by this pump (Veal and Shafer, 2003; Folster and Shafer, 2005). MpeR also negatively regulates expression of *mtrR* (Fig. 10.4; Mercante *et al.*, 2012). Interestingly, *mpeR* was found to be under the negative control of Fur complexed with iron (Jackson *et al.*, 2010). Additionally, *mpeR* is maximally expressed at the late log phase of growth when available free iron levels begin to diminish (Mercante *et al.*, 2012). The capacity of Fur + iron to repress *mpeR* and the ability of MpeR to repress *mtrR*, likely explains why *mtrCDE* is maximally expressed late in growth when levels of free iron would be depleted. Recent work (Hollander *et al.*, 2011) has shown that MpeR activates expression of *fetA*, which encodes a single-component TonB-dependent receptor that allows the gonococcus to acquire iron from enterobactin-like siderophores produced by enteric bacteria (Carson *et al.*, 1999, Hollander *et al.*, 2011). This regulation was observed in strain FA1090, but not FA19, because MtrA is a negative regulator of *mpeR* (Fig. 10.4) and, as mentioned above, FA1090 is a natural *mtrA* null mutant, thus removing this regulatory control mechanism. Taken together, these findings show that MpeR plays important roles in both drug efflux and iron acquisition by gonococci, emphasizing the need to consider regulators of efflux pump genes in a larger context with respect to bacterial physiology and pathogenesis.

Involvement of a two-component regulatory system in controlling *mtrCDE*

The MisR response regulator and MisS sensor kinase proteins form a classical two-component regulatory system (TCS) that is similar to but distinct from the PhoP/PhoQ system of *Salmonella enterica* serovar Typhimurium (Miller *et al.*, 1989). The *misRS* operon has been extensively studied in MC through both classical genetics and microarray technology (Tzeng *et al.*, 2008), and was named for its ability to influence meningococcal LOS inner core structure (Tzeng *et al.*, 2004). It also controls expression of *dsbD*, which is required for the maintenance of disulphide bonds in periplasmic/envelope proteins that may become damaged by the host redox defences (Kumar *et al.*, 2011). MisR/MisS also enhances expression of haemoglobin uptake and utilization proteins by serogroup B MC, suggesting that this TCS plays a role in iron scavenging (Zhao *et al.*, 2010). Other studies have shown that MisR/MisS is crucial for meningococcal pathogenesis. Serogroup C MC that lacked a functional *misR* gene were deficient in their ability to colonize human endothelial cells *in vitro* (Jamet *et al.*, 2009) and were avirulent in a murine intraperitoneal infection model (Newcombe *et al.*, 2004).

In GC, a link between *misRS* expression and the *mtrCDE* efflux pump operon was established when growth in sublethal levels of the antimicrobial peptide LL-37 increased transcription of the *misR* and *misS* genes as analysed by microarray and qRT-PCR (W.M. Shafer *et al.*, unpublished observations), resulting in inducible, high-level LL-37 resistance, which required this TCS. Because LL-37 is a known substrate of the MtrC-MtrD-MtrE efflux pump (W.M. Shafer *et al.*, 1998), we tested if MisR/MisS could regulate expression of *mtrCDE* and found that the MisR/MisS TCS is necessary for activation of the *mtrCDE* operon. This is true in both the presence and absence of the MtrR master repressor (J.L. Kandler and W.M. Shafer, unpublished). Thus, strong inducible activation of *mtrCDE* is most likely to occur when 1) levels of MpeR, which represses MtrR (Fig. 10.4), are high during iron starvation; 2) when MtrA binds to pump substrates (such as TX-100) and competes for binding at *mtrCDE* with MtrR; and 3) when the MisR/MisS TCS is activated by sublethal concentrations of host defence peptides. Since the human host restricts the availability of free iron, and mucosal surfaces often contain low levels of LL-37 during inflammation resulting from infection (Cole, 2007), it is likely that *mtrCDE* expression is higher during infection than when GC are cultivated in the laboratory in iron-rich culture media. Furthermore, just as MpeR allows GC to adjust gene expression to deal with falling iron levels (Mercante *et al.*, 2012), the MisR/MisS TCS may provide a mechanism for GC to sense and adapt to increasing concentrations of antimicrobial peptides as they progress through the different stages of infection. MisR/MisS also modulates levels of antimicrobial peptide susceptibility in MC (Johnson *et al.*, 2001; Newcombe *et al.*, 2004; Tzeng *et al.*, 2004). Although the MtrC-MtrD-MtrE pump produced by MC can export antimicrobial peptides, it is unclear if meningococcal MisR/MisS acts to modulate levels of peptide resistance by an efflux-dependent process.

The FarR DNA-binding protein and gene expression

FarR is the local repressor of the *farAB* efflux operon (Lee and Shafer, 2003). FarR-mediated repression of *farAB* also requires IHF, and a model for this joint repression has been published (Lee *et al.*, 2006). Interestingly, expression of *farR* is negatively regulated by MtrR (Lee *et al.*, 2003). Thus, MtrR can directly and indirectly regulate levels of two efflux systems in gonococci. In contrast, *farAB* is not negatively regulated by FarR in the naturally fatty-acid resistant MC (Schielke *et al.*, 2011). Instead, meningococcal FarR appears to control expression of *nadA* (Schielke *et al.*, 2009), which encodes a surface-exposed outer membrane protein produced by 40% of capsular serogroup B strains involved in attachment to and invasion of host cells (Capecchi *et al.*, 2005). NadA is currently under investigation as a vaccine candidate that could be used to help protect against capsular serogroup B MC strains, for which no vaccine is available (Comanducci *et al.*, 2002).

Conclusions

We propose that for strict human pathogens, such as GC, bacterial efflux pumps perform important roles in promoting survival when they are confronted by host antimicrobials or antibiotics. Unlike other Gram-negative pathogens, GC possesses a limited number of pumps (Fig. 10.1), yet uses these few pumps effectively to avoid the action of antimicrobials. From a clinical perspective, overexpression of the *mtrCDE*-encoded efflux pump system was a key component in the downfall of penicillin as a useful (and cheap) antibiotic in treating gonorrhoea (Veal *et al.*, 2002). We further propose that the ability of GC to transcriptionally modulate expression of efflux pump genes promotes survival and determines fitness during infection. There is clear evidence for this in an animal model system, but the frequent isolation of strains bearing mutations in regulatory genes that control efflux pump levels (especially those negative regulatory proteins) from patients suggests that the same is true for human infections (Shafer *et al.*, 1996, 2010). It is important to recognize that many of these regulatory proteins are interconnected and also control other important genes involved in basic metabolism and pathogenesis. Thus, by using efflux pump gene regulators to control a larger set of genes, GC has economically tailored itself to both respond to adverse conditions (e.g. presence of antimicrobials) as well as perform vital physiological functions, both of which are important during infection. Finally, it is essential that we further consider how host factors impact efflux pump gene expression during infection and how these changes influence survival of GC *in vivo*.

References

Ariza, R.R., Li, Z., Ringstad, N., and Demple, B. (1995). Activation of multiple antibiotic resistance and binding of stress-inducible promoters by *Escherichia coli* Rob protein. J. Bacteriol. *177*, 1655–1661.

Bailey, A.M., Ivens, A., Kingsley, R., Cottel, J.L., Wain, J., and Piddock, L.J.. (2010). RamA, a member of the AraC/XylS family, influences both virulence and efflux in *Salmonella enteritica* serovar Typhimurium. J. Bacteriol. *192*, 1607–1616.

Browning, D.F., and Busby, S.J.W. (2004). The regulation of bacterial transcription initiation. Nat. Rev. Microbiol. 2, 57–65.

Camara, J., Serra, J., Ayats, J., Bastida, T., Carnicer-pont, D., Andreu, A., and Ardanuy, C. (2012). Molecular characterization of two high-level ceftriaxone-resistant *Neisseria gonorrhoeae* isolates detected in Catalonia, Spain. J. Antimicrob. Chemother. 67, 1858-1860.

Capecchi, B., Adu-Bobie, J., Di Marcello, F., Ciucchi, L., Masignani, V., Taddei, A., Rappuoli, R., Pizza, M., and Arico, B. (2005). *Neisseria meningitidis* NadA is a new invasin which promotes bacterial adhesion to and penetration into human epithelial cells. Mol. Microbiol. 55, 687–698.

Carson, S.D.B., Klebba, P.E., Newton, S.M.C., and Sparling, P.F. (1999). Ferric enterobactin binding and utilization by *Neisseria gonorrhoeae*. J. Bacteriol. *181*, 2895–2901.

Cole, A.M. (2007). Innate host defense of human vaginal and cervical mucosae. In Antimicrobial Peptides and Human Disease, Shafer, W., ed. (Springer-Verlag, Heidelberg, Germany), pp. 201–230.

Comanducci, M., Bambini, S., Brunelli, B., Adu-Bobie, J., Arico, B., Capecchi, B., Giuliani, M.M., Masignani, V., Santini, L., Savino, S., *et al.* (2002). NadA, a novel vaccine candidate of *Neisseria meningitidis*. J. Exp. Med. *195*,1445–1454.

Correia, F.F., Inouye, S., and Inouye, M. (1988). A family of small repeated elements with some transposon-like properties in the genome of *Neisseria gonorrhoeae*. J. Biol. Chem. *263*, 12914–12916.

Delahay, R.M., Robertson, B.D., Balthazar, J.T., Shafer, W.M., and Ison, C.A. (1997). Involvement of the gonococcal MtrE protein in the resistance of *Neisseria gonorrhoeae* to toxic hydrophobic agents. Microbiol. *143*, 2127–2133.

Dionne-Odom, J., Tambe, P., Yee, E., Weinstock, H., and del Rio, C. (2011). Antimicrobial resistant gonorrhea in Atlanta: 1988–2006. Sex. Trans. Infect. *38*, 1–3.

Enriquez, R., Abad, R., Chanto, G., Corso, A., Cruces, R., Gabastou, J.M., Goria, M.V., Maldonado, A., Moreno, J., Muros-Le-Rouzic, Sorhouet, C., *et al.* (2010). Deletion of the Correia element in the *mtr* gene complex of *Neisseria meningitidis*. J. Med. Microbiol. *59*, 1055–1060.

Falsetta, M.L., Steichen, C.T., McEwan, A.G., Cho, C., Ketterer, M., Shao, J., Hunt, J., Jennings, M.P., and Apicella, M.A. (2011). The composition and metabolic phenotype of *Neisseria gonorrhoeae* biofilms. Front. Microbiol. 2, 1–11.

Faruki, H., and Sparling, P.F. (1986). Genetics of resistance in a non-beta-lactamase-producing gonococcus with relatively high level penicillin resistance. Antimicrob. Agents Chemother. *30*, 856–860.

Faruki, H., Kohmescher, R.N., McKinney, W.P., and Sparling, P.F. (1985). A community-based outbreak of infection with penicillin-resistant *Neisseria gonorrhoeae* not producing penicillinase (chromosomally-mediated resistance). N. Engl. J. Med. *313*, 607–611.

Folster, J.P., and Shafer, W.M. (2005). Regulation of *mtrF* expression in *Neisseria gonorrhoeae* and its role in high-level antimicrobial resistance. J. Bacteriol. *187*,3713–3720.

Folster J.P., Dhulipala, V., Nicholas, R.A., and Shafer, W.M. (2007). Differential regulation of *ponA* and *pilMNOPQ* expression by the *Neisseria gonorrhoeae* MtrR transcriptional regulatory protein. J. Bacteriol. *189*, 4569–4577.

Folster, J.P., Johnson, P.J.T., Jackson, L., Dhulipala, V., Dyer, D.W., and Shafer, W.M. (2009). MtrR modulates *rpoH* expression and levels of antimicrobial resistance in *Neisseria gonorrhoeae*. J.Bacteriol. *191*, 287–297.

Gambino, L., Gracheck, S.J., and Miller, P.F. (1993). Overexpression of the MarA positive regulator is sufficient to confer multiple antibiotic resistance in *Escherichia coli*. J. Bacteriol. *175*, 2888–2894.

Grkovic, S., Brown, M.H., and Skurray, R.A. (2002). Regulation of bacterial drug export systems. Microbiol. Mol. Bio. Rev. *66*, 671–701.

Gunesekere, I.C., Kahler, C.M., Powell, D.R., Snyder, L.A., Saunders, N.J., Rood, J.I., and Davies, J.K. (2006). Comparison of the RpoH-dependent regulon and stress response in *Neisseria gonorrhoeae*. J. Bacteriol. *188*, 4769–4776.

Guymon, L.F., Walstad, D.L., and Sparling, P.F. (1978). Cell envelope alterations in antibiotic-sensitive and – resistant strains of *Neisseria gonorrhoeae*. J. Bacteriol. *136*, 391–401.

Hagman, K.E., and Shafer, W.M. (1995). Transcriptional control of the *mtr* efflux system of *Neisseria gonorrhoeae*. J. Bacteriol. *171*, 4162–4165.

Hagman, K.E., Pan, W., Spratt, B.G., Balthazar, J.T., Judd, R.C., and Shafer, W.M. (1995). Resistance of *Neisseria gonorrhoeae* to antimicrobial hydrophoobic agents is modulated by the *mtrCDE* efflux system. Microbiol. *141*, 611–622.

Hagman, K.E., Lucas, C.E., Balthazar, J.T., Snyder, L., Nilles, M., Judd, R.C., and Shafer, W.M. (1997). The MtrD protein of *Neisseria gonorrhoeae* is a member of the resistance/nodulation/division protein family constituting part of an efflux system. Microbiol. *143*, 2117–2123.

Hoffmann, K.M., Williams, D.C., Shafer, W.M., and Brennan, R.G. (2005). Characterization of the multiple transferrable repressor, MtrR, from *Neisseria gonorrhoeae*. J. Bacteriol. *187*, 5008–5012.

Hollander, A., Mercante, A.D., Shafer, W.M., and Cornelissen, C.N. (2011). The iron-repressed, AraC-like regulator MpeR activates expression of *fetA* in *Neisseria gonorrhoeae*. Infect. Immun. *79*, 4764-4775.

Jackson, L.A., Ducey, T.F., Day, M.W., Zaitshik, J.B., Orvis, J., and Dyer, D.W. (2010). Transcriptional and functional analysis of the *Neisseria gonorrhoeae* Fur regulon. J. Bacteriol. *192*, 77–85.

Jamet, A., Rousseau, C., Monfort, J.B., Frapy, E., Nassif, X., and Martin, P. (2009). A two-component system is required for colonization of host cells by meningococcus. Microbiology *155*, 2288–2295.

Janganan, T.K., Bavro, V.N., Zhang, L., Matak-Vinkovic, D., Barrera, N.P., Venien-Bryan, C., Robinson, C.V., Borges-Walmsley, M.I., and Walmsley, A.R. (2011). Evidence for the assembly of a bacterial tripartite multidrug pump with a stoichiometry of 3:6:3. J. Biol. Chem. *286*, 26900-26912.

Jerse, A.E. (1999). Experimental gonococcal genital tract infection and opacity protein expression in estradiol-treated mice. Infect. Immun. *67*, 5699–5708.

Jerse, A.E., Sharma, N.D., Bodner, A.N.B., Snyder, L.A., and Shafer, W.M. (2003). A gonococcal efflux pump enhances bacterial survival in a female mouse model of genital tract infection. Infect. Immun. *71*, 5576–5582.

Johnson, C.R., Newcombe, J., Thorne, S., Borde, H.A., Eales-Reynolds, L.J., Gorringe, A.R., Funnell, S.G., and McFadden, J.J. (2001). Generation and characterization of a PhoP homologue mutant of *Neisseria meninigitidis*. Mol. Microbiol. *39*, 1345–1355.

Johnson, P.J., Stringer, V.A., and Shafer, W.M. (2011). Off-target gene regulation mediated by transcriptional repressors of antimicrobial efflux pump genes in *Neisseria gonorrhoeae*. Antimicrob. Agents Chemother. *55*, 2559-2565.

Johnson, S.R., Sandul, A.L., Parekh, M., Wang, S.A., Knapp, J.S., and Trees, D.L. (2003). Mutations causing *in vitro* resistance to azithromycin in *Neisseria gonorrhoeae*. Int. J. Antimicrob. Agents *21*, 414–419.

Kaniga, K., Bossio, J.C., and Galan, J.E. (1994). The *Salmonella typhimurium* invasion genes *invF* and *invG* encode homologues of the AraC and PulD family of proteins. Mol. Microbiol. *13*, 555–568.

Klose, K.E., and Mekalanos, J.J. (1997). Simultaneous prevention of glutamine synthesis and high-affinity transport attenuates *Salmonella typhimurium* virulence. Infect. Immun. *65*, 587–596.

Kumar, P., Sannigrahi, S., Scoullar, J., Kahler, C., and Tzeng, Y. (2011). Characterization of DsbD in *Neisseria meningitidis*. Mol. Microbio. *79*, 1557–1573.

Lee, E.-H., and Shafer, W.M. (1999). The *farAB*-encoded efflux pump medicates resistance of gonococci to long-chained fatty acids. Mol. Microbiol. *33*, 839–845.

Lee, E.-H., Rouquette-Loughlin, C., Folster, J.P., and Shafer, W.M. (2003). FarR regulates the farAB-encoded efflux pump in *Neisseria gonorrhoeae* by an MtrR regulatory mechanism. J. Bacteriol. *185*, 7145–7152.

Lee E.-H., Hill, S.A., Napier, R., and Shafer, W.M. (2006). Integration host factor is required for FarR repression of the *farAB*-encoded efflux pump of *Neisseria gonorrhoeae*. Mol. Microbiol. *60*, 1381–1400.

Levy, S.B. (1992). Active efflux mechanisms for antimicrobial resistance. Antimicrob. Agents Chemother. *36*, 695–703.

Liao, M., Gu, W.-M., Yang, Y., and Dillon, J.-A., R. (2011). Analysis of multiple loci of *Neisseria gonorrhoeae* isolates reveals of PIB, PBP2 and MtrR on reduced susceptibility to ceftriaxone. J. Antimicrob. Chemother.. *66*, 1016–1023.

Liu, Y., Feinen, B., and Russell, M.W. (2011). New concepts in immunity to *Neisseria gonorrhoeae*: innate responses and suppression of adaptive immunity favor the pathogen, not the host. Front. Microbiol. *2(article 52)*, 1–8.

Lomovskaya, O., and Wilkins, W. (2002). Inhibition of efflux pumps as a novel approach to combat drug resistance in bacteria. In Microbial Multidrug Efflux, Paulsen, I.T., and Lewis, K., eds. (Horizon Press, Norfolk, UK), pp. 201–232.

Lucas, C.E., Hagman, K.E., Levin, J.C., Stein, D.C., and Shafer, W.M. (1995). Importance of lipooligosaccharide structure in determining gonococcal resistance to hydrophobic antimicrobial agents resulting from the *mtr* efflux system. Mol. Microbiol. *16*, 1001–1009.

Lucas, C.E., Balthazar, J.T., and Shafer, W.M. (1997). The MtrR repressor binds the DNA sequence between the *mtrR* and *mtrC* genes of *Neisseria gonorrhoeae*. J.Bacteriol. *179*, 4123–4128.

Luna, V.A., Cousin, S., Jr., Whittington, W.L.H., and Roberts, M.C. (2000). Identification of the conjugative *mef* gene in clinical isolates of *Acinetobacter junii* and *Neisseria gonorrhoeae* isolates. Antimicrob. Agents Chemother. *44*, 2503–2506.

Maness, M.J., and Sparling, P.F. (1973). Multiple antibiotic resistance due to a single mutation in *Neisseria gonorrhoeae*. J. Infect. Dis. *128*, 321–330.

Mercante, A.D., Jackson, L., Johnson, P.J., Stringer, V.A., Dyer, D.W., and Shafer, W.M. (2012). MpeR regulates the *mtr* efflux locus in *Neisseria gonorrhoeae* and modulates antimicrobial resistance by an iron-represponsive mechanism. Antimicrob. Agents Chemother. *56*, 1491-1501.

Miller, S.I., Kukral, A.M., and Mekalanos, J.J. (1989). A two component regulatory system (*phoP phoQ*) controls *Salmonella typhimurium* virulence. Proc. Natl. Acad. Sci. U.S.A. *86*, 5054–5058.

Morse, S.A., Lysko, P.C., McFarland, L., Knapp, J.S., Sandstrom, E., Critchlow, C., and Holmes, K.K. (1982). Gonococcal strains from homosexual men have outer membranes with reduced permeability to hydrophobic molecules. Infect. Immun. *37*, 432–438.

Newcombe J., Eales-Reynolds, L., Wootton, L., Gorringe, A.R., Funnell, S.G., Taylor, S.C., and McFadden, J.J. (2004). Infection with an avirulent *phoP* mutant of *Neisseria meningitidis* confers broad cross-reactive immunity. Infect. Immun. 72, 338–344.

Nikaido, H. (1996). Multidrug efflux pumps of Gram-negative bacteria. J. Bacteriol. *178*, 5853–5859.

Ohneck, E.A., Zalucki, Y.M., Johnson, P.J., Dhulipala, V., Golparian, D., Unemo, M., Jerse, A.E., and Shafer, W.M. (2011). A novel mechanism of high-level, broad-spectrum antibiotic resistance caused by a single base pair change in *Neisseria gonorrhoeae*. MBio 2, pii:300187-11.

Ohnishi, M., Saika, T., Hoshina, S., Iwasaku, K., Nakayama, S., Watanabe, H., Kitawaki, J. 92011). Ceftriaxone-resistant *Neisseria gonorrhoeae*, Japan. Emerg. Infect. Dis. *17*, 148-149.

Olesky, M., Zhao, S., Rosenberg, R.I., and Nicholas, R.A. (2006). Porin-mediated antibiotic resistance in *Neisseria gonorrhoeae*: ion, solute, and antibiotic permeation through PIB proteins with *penB* mutations. J. Bacteriol. *188*, 2300–2308.

Pan, W., and Spratt, B.G. (1994). Regulation of the permeability of the gonococcal cell envelope by the *mtr* system. Mol. Microbiol. *11*, 769–775.

Parish, T., and Stoker, N.G. (2000). *glnE* is an essential gene in *Mycobacterium tuberculosis*. J. Bacteriol. *182*, 5715–5720.

Paulsen, I.T., Chen, J., Nelson, K.E., and Saier, M.H., Jr. (2002). Comparative genomics of multidrug efflux systems. In Microbial Multidrug Efflux, Paulsen, I.T., and Lewis, K., eds. (Horizon Press, Norfolk, UK), pp. 5–20.

Ramos, J.L., Martinez-Bueno, M., Molina-Henares, A.J., Teran, W., Watanabe, K., Zhang, X., Gallegos, M.T., Brennan, R., and Tobes, R. (2005). The TetR family of transcriptional repressors. Microbiol. Mol. Biol. Rev. *69*, 326–356.

Ropp, P.A., Hu, M., Olesky, M., and Nicholas, R.A. (2002). Mutations in *ponA*, the gene encoding penicillin binding-protein 1, and a novel locus, *penC*, are required for high-level penicillin resistance in *Neisseria gonorrhoeae*. Antimicrob. Agents Chemother. *46*, 769–777.

Rouquette, C., Harmon, J.B., and Shafer, W.M. (1999). Induction of the *mtrCDE*-encoded efflux pump system of *Neisseria gonorrhoeae* requires MtrA, an AraC-like protein. Mol. Microbiol. *33*, 651–658.

Rouquette,-Loughlin, C., Stojiljkovic, I., Hrobowski, T., Balthazar, J.T., and Shafer, W.M. (2002). Inducible, but not constitutive, resistance of gonococci to hydrophobic agents due to the MtrC-MtrD-MtrE efflux pump requires the TonB-ExbB-ExbD proteins. Antimicrob. Agents Chemother. *46*, 561–565.

Rouquette-Loughlin, C.E., Dunham, S.A., Kuhn, M., Balthazar, J.T., and Shafer, W.M. (2003). The NorM efflux pump of *Neisseria gonorrhoeae* and *Neisseria meningitidis* recognizes antimicrobial cationic compounds. J. Bacteriol. *185*, 1101 - 1106.

Rouquette-Loughlin C.E., Balthazar, J.T., Hill, S.A., and Shafer, W.M. (2004). Modulation of the *mtrCDE*-encoded efflux pump gene complex of *Neisseria meningitidis* due to a Correia element insertion sequence. Mol. Microbiol. *54*, 731–741.

Rouquette-Loughlin, C.E., Balthazar, J.T., and Shafer, W.M. (2005). Characterization of the MacA-MacB efflux system in *Neisseria gonorrhoeae*. J. Antimicrob. Chemother. *56*, 856–860.

Schielke, S., Huebner, C., Spatz, C., Nagele, V., Ackermann, N., Frosch, M., Kurzai, O., and Schubert-Unkmeir, A. (2009). Expression of the meningococcal adhesin NadA is controlled by a transcriptional regulator of the MarR family. Mol. Microbiol. 72,1054–1067.

Schielke, S., Spatz, C., Schwarz, R.F., Joseph, B., Schoen, C., Schulz, S.M., Hubert, K., Frosch, M., Schubert-Unkmeir, A., and Kurzai, O. (2011). Characterization of FarR as a highly specialized, growth phase-dependent transcriptional regulator in *Neisseria meningitidis*. Int. J. Med. Microbiol. *301*, 325–333.

Shafer, W.M., Balthazar, J.T., Hagman, K.E., and Morse, S.A. (1995). Missense mutations that alter the DNA-binding domain of the MtrR protein occur frequently in rectal isolates of *Neisseria gonorrhoeae* that are resistant to faecal lipids. Microbiol. *141*, 907–911.

Shafer, W.M., Qu, X., Waring, A.J., and Lehrer, R.I. (1998). Modulation of *Neisseria gonorrhoeae* susceptibility to vertebrate antibacterial peptides due to a member of the resistance/nodulation/division efflux pump family. Proc. Natl. Acad. Sci. U.S.A. 95, 1829–1833.

Shafer, W.M., Folster, J.P., and Nicholas, R.A. (2010). Molecular mechanisms of antibiotic resistance expressed by the pathogenic Neisseriae. In *Neisseria*: Molecular Mechanisms of Pathogenesis, Genco, C.A., and Wetzler, L., eds. (Horizon Press, Norfolk, UK), pp. 245–268.

Shine, J., and Dalgarno, L. (1975). Determinant of cistron specificity in bacterial ribosomes. Nature *254*, 34–38.

Sochacki, K.A., Barns, K.J., Bucki, R., and Weisshaar, J. (2011). Real-time attack on single *Escherichia coli* cells by the human antimicrobial peptide LL-37. Proc. Natl. Acad. Sci. U.S.A. *108*, 877–881.

Spratt, B.G. (1988). Hybrid penicillin-binding proteins in penicillin-resistant strains of *Neisseria gonorrhoeae*. Nature 332, 173–176.

Stephens, D.S., Greenwood, B., and Brandtzaeg, P. (2007). Epidemic meningitis, meningococcemia, and *Neisseria meningitidis*. Lancet *369,* 2196–2210.

Tzeng Y., Datta, A., Ambrose, K., Lo, M., Davies, J.K., Carlson, R.W., Stephens, D.S., and Kahler, C.M. (2004). The MisR/MisS two-component regulatory system influences inner core structure and immunotype of lipooligosaccharide in *Neisseria meningitidis*. J. Biol. Chem. *279,* 35053–35062.

Tzeng, Y.-L., Ambrose, K.D., Zughaier, S., Zhou, X., Miller, Y.K., Shafer, W.M., and Stephens, D.S. (2005). Cationic antimicrobial peptide resistance in *Neisseria meningitidis*. J. Bacteriol. *187,* 5387–5396.

Tzeng Y., Kahler, C.M., Zhang, X., and Stephens, D.S. (2008). MisR/MisS Two-Component Regulon in *Neisseria meningitidis*. Infect. Immun. *76,* 704–716.

Unemo, M., Golparian, D., Nicholas, R., Ohnishi, M., Gallay, A., Sednaoui, P. (2012). High-level cefixime- and ceftriaxone-resistant *Neisseria gonorrhoeae* in France: novel *penA* mosaic allele in a successful international clone causes treatment failure. Antimicrob. Agents Chemother. *56,* 1273-1280.

Veal. W.L., and Shafer, W. Shafer, W.M. (2003). Identification of a cell envelope protein (MtrF) involved in hydrophobic antimicrobial resistance in *Neisseria gonorrhoeae*. J. Antimicrob. Chemother. *51,* 27–37.

Veal, W.L., Yellen, A., Balthazar, J.T., Pan, W., Spratt, B.G., and Shafer, W.M. (1998). Loss of function mutations in the *mtr* efflux system of *Neisseria gonorrhoeae*. Microbiology *144,* 621–627.

Veal, W.L., Nicholas, R.A., and Shafer, W.M. (2002). Overexpression of the MtrC-MtrD-MtrE efflux pump due to an *mtrR* mutation is required for chromosomally mediated penicillin resistance in *Neisseria gonorrhoeae*. J. Bacteriol. *184,* 5619–5624.

Warner, D.M., Folster, J.P., Shafer, W.M., and Jerse, A.E. (2007). Regulation of the MtrC-MtrD-MtrE efflux pump system modulates the *in vivo* fitness of *Neisseria gonorrhoeae*. J. Infect. Dis. *196,* 1804–1812.

Warner, D.M., Shafer, W.M., and Jerse, A.E. (2008). Clinically relevant mutations that cause derepression of the *Neisseria gonorrhoeae* MtrC-MtrD-MtrE efflux pump system confer different levels of antimicrobial resistance and *in vivo* fitness. Mol. Microbiol. *70,* 462 – 478.

Xia, M., Whittington, W.I.H., Shafer, W.M., and Holmes, K.K., (2000). Gonorhea among men who have sex with men: outbreak caused by a single genotype of erythromycin-resistant *Neisseria gonorrhoeae* with a single basepair deletion in the *mtrR* promoter region. J. Infect. Dis. *181,* 2080–2082.

Zalucki, Y.M., Dhulipala, V., and Shafer, W.M. (2012) Dueling regulatory properties of a transcriptional activator (MtrA) and repressor (MtrR) that control efflux pump gene expression in *Neisseria gonorrhoeae*. MBio. *3*, pii: e00446-12.

Zarantonelli, L., Borthagary, G., Lee, E.-H., Veal, W.L., and Shafer, W.M. (2001). Decreased susceptibility to azithromycin and erythromycin mediated by a novel *mtrR* promoter mutation in *Neisseria gonorrhoeae*. Antimicrob. Agents Chemother. *47,* 651–654.

Zhao, S., Montanez, G.E., Kumar, P., Sannagrahi, S., and Tzeng, Y.-L. (2010). Regulatory role of the MisR/S two-component system in hemoglobin utilization in *Neisseria meningitidis*. Infect. Immun. *78,* 1109–1122.

Multidrug Efflux Transporters in *Campylobacter*

11

Zhangqi Shen, Chih-Chia Su, Edward W. Yu and Qijing Zhang

Abstract

As a major food-borne pathogen, *Campylobacter* is frequently exposed to antibiotics used for both animal production and human medicine. The increasing prevalence of antibiotic resistant *Campylobacter* has become a significant concern for public health. Among all known antibiotic resistance mechanisms, multidrug efflux systems play essential roles in the intrinsic and acquired resistance to structurally diverse antimicrobials. In *Campylobacter*, several multidrug efflux pumps, such as CmeABC, CmeDEF, CmeG, and Acr3, have been functionally characterized, which revealed that these efflux systems not only contribute to the resistance of antimicrobials, but also play important roles in facilitating the adaptation of *Campylobacter* to various environments, including the intestinal tract of animal hosts. The expression of these efflux transporters are controlled by transcriptional regulators, which sense the presence of toxic substrates and modulates the transcription of these efflux genes. Inhibiting the production or function of these multidrug efflux transporters, especially CmeABC, has been evaluated using efflux pump inhibitors and antisense peptide nucleic acid (PNA), demonstrating the potential of this approach for controlling antibiotic resistance in *Campylobacter*. In this Chapter, we will review the recent advance in understanding multidrug efflux systems and discuss the development of potential intervention strategies by targeting antimicrobial efflux pumps in *Campylobacter*.

Introduction

Campylobacter spp. are Gram-negative organisms that grow the best in microaerobic environments. Morphologically, *Campylobacter* cells are in the form of spirally curved rods 0.2–0.8 µm wide and 0.5–5 µm long. *Campylobacter* is recognized as a leading bacterial cause of food-borne diseases in the United States and other developed countries (Slutsker *et al.*, 1998) and is accounted for 400–500 million infected cases each year worldwide (Ruiz-Palacios, 2007). A recent CDC report indicated that campylobacteriosis is estimated to affect over 0.84 million people every year in the United States (Scallan *et al.*, 2011). At present, there are 18 validly named *Campylobacter* species, including *C. canadensis, C. coli, C. concisus, C. curvus, C. fetus, C. gracilis, C. helveticus, C. hominis, C. hyointestinalis, C. insulaenigrae, C. jejuni, C. lanienae, C. lari, C. mucosalis, C. rectus, C. showae, C. sputorum*, and *C. upsaliensis*. According to the published data, *C. jejuni* and *C. coli* are the most common species associated with *Campylobacter* enteritis in human (Gillespie *et al.*, 2002). Typical symptoms of campylobacteriosis include diarrhoea (bloody), fever, headache, myalgia, nausea, vomiting, and abdominal pain. Typically, campylobacteriosis is self-limited. Hospitalization and antibiotic treatments are only required in severe cases. However, *C. jejuni* is the predominant preceding infection for Guillain-Barré syndrome (GBS), which is the most common form of acute neuromuscular paralysis (Jacobs *et al.*, 2008).

C. jejuni and *C. coli* are increasingly resistant to antimicrobials, which has become a significant threat to public health (Gupta *et al.*, 2004; Luangtongkum *et al.*, 2006; Luangtongkum *et al.*, 2008; Chatre *et al.*, 2010; Chen *et al.*, 2010; Cody *et al.*, 2010; Kim *et al.*, 2010; Smith and Fratamico, 2010; Zhao *et al.*, 2010; Qin *et al.*, 2011; Schweitzer *et al.*, 2011). To counteract the selection pressure from antimicrobials, *Campylobacter* has evolved multiple mechanisms (Zhang and Plummer, 2008), which include (i) synthesis of enzymes to modify/inactivate antibiotics (e.g. β-lactamase); (ii) alternation of antibiotic targets (e.g. mutations in *gyrA* or 23S rRNA genes); (iii) protection of antibiotic targets [e.g. synthesis of protective protein Tet(O)]; (iv) reduced permeability of cellular membranes to antibiotics; and (v) active extrusion of antimicrobials out of *Campylobacter* cells through efflux transporters (e.g. CmeABC). *Campylobacter* harbours multiple antibiotic efflux transporters and several of them have been characterized. In this chapter, we will focus on reviewing the new information concerning the function and regulation of these efflux systems.

CmeABC

CmeABC is the predominant antibiotic efflux system in *C. jejuni* and belongs to the resistance-nodulation-cell division (RND) superfamily of multidrug efflux transporters. This efflux system is encoded by a three-gene operon including *cmeA*, *cmeB*, and *cmeC* (Lin *et al.*, 2002). CmeA is a membrane fusion protein, and its amino acid sequence shows similarity to the membrane fusion component in other bacterial efflux systems, including MtrC of *Neisseria gonorrhoeae*, MexA and MexC of *Pseudomonas aeruginosa*, and AcrA of *Escherichia coli*. CmeB is an inner membrane transporter and is homologous to AcrB, AcrD, and AcrF of *E. coli*, MexB and MexF of *P. aeruginosa*, and AcrB of *Salmonella typhi*. CmeC is an outer membrane protein and is predicted to be similar to the OprM and OprN of *P. aeruginosa* and TolC of *E. coli* (Lin *et al.*, 2002).

Insertional mutation of *cmeB* in *Campylobacter* resulted in increased susceptibility to several structurally diverse antimicrobial compounds, including ciprofloxacin, norfloxacin, nalidixic acid, cefotaxime, ampicillin, erythromycin, rifampin, tetracycline, ethidium bromide, heavy metals($CoCl_2$ and $CuCl_2$), acridine orange, SDS, tetracycline, and bile salts (chenodeoxycholate, deoxycholate, cholate, and taurocholate) (Lin *et al.*, 2002; Pumbwe and Piddock, 2002). Similar results were also observed with inactivation of *cmeC* (Lin *et al.*, 2003), suggesting that the entire CmeABC system is required for conferring antibiotic resistance. CmeABC functions synergistically with other mechanisms in mediating high-level resistance to antibiotics. For example, mutation of *cmeB* resulted in a 8-fold decrease in the minimum inhibitory concentration (MIC) of tetracycline in the *Campylobacter* strains carrying tetracycline resistance gene *tet(O)* (Lin *et al.*, 2002; Piddock *et al.*, 2008); inactivation of *cmeABC* in fluoroquinolone-resistant mutants (carrying specific GyrA mutations) rendered the resistant mutants susceptible to ciprofloxacin and enrofloxacin (Luo *et al.*, 2003); and the MICs of erythromycin decreased 8- to 1024-fold upon inactivation of the *cmeB* gene in the macrolide-resistant strains that carried a A2075G, A2074G, or A2074C mutation in the 23S rRNA genes (Cagliero *et al.*, 2005; Gibreel *et al.*, 2007; Lin *et al.*, 2007; Caldwell *et al.*, 2008) or modifications in the L4 and L22 ribosomal proteins (Cagliero *et al.*, 2006b). These examples illustrate the important role of CmeABC in the acquired resistance to clinically important antibiotics such as macrolides and fluoroquinolones. Recent studies also revealed that CmeABC is responsible for both the intrinsic and acquired resistance to bacteriocins and antimicrobial peptides produced by bacteria (Hoang *et al.*, 2011a,b).

In addition, CmeABC contributes to the adaptation of *Campylobacter* in the intestinal tract of animal hosts. As an enteric organism, *Campylobacter* must have the ability to deal with bile compounds, which are normally present in the digestive system. Bile contains a group of detergent-like bile salts, which exhibit potent bactericidal activity (Gunn, 2000). *In vitro* susceptibility tests indicate that inactivation of *cmeABC* resulted in a dramatically decreased resistance in *Campylobacter* to various bile salts (Lin *et al.*, 2002, 2003). Additionally, Lin *et al.* demonstrated that CmeABC plays a critical role in the intestinal colonization of *C. jejuni* (Lin *et al.*, 2003, 2005c). In the chicken model, the *cmeABC* mutants failed to colonize the inoculated chickens, while complementation of the *cmeABC* mutants *in trans* fully restored the colonization to the level of the wild-type strain (Lin *et al.*, 2003). These findings strongly suggest that facilitating adaptation in the intestinal tract is an important function of CmeABC.

The *cmeABC* operon is conserved among different *Campylobacter* species and is widely distributed in *Campylobacter* isolates (Lin *et al.*, 2002; Cagliero *et al.*, 2005; Ge *et al.*, 2005; Olah *et al.*, 2006; Guo *et al.*, 2010). This efflux system has been functionally characterized in *C. jejuni* (Lin *et al.*, 2002, 2003; Pumbwe and Piddock, 2002), *C. coli* (Cagliero *et al.*, 2005; Ge *et al.*, 2005), *C. lari, C. fetus,* and *C. hyointestinalis* (Guo *et al.*, 2010), and contributed to antibiotic resistance in all examined species. In general, the sequences of *cmeABC* are highly conserved within a species, but significant sequence polymorphisms are observed in the *cmeABC* genes among different *Campylobacter* species (Cagliero *et al.*, 2006a; Olah *et al.*, 2006; Fakhr and Logue, 2007; Guo *et al.*, 2010). In addition, analysis of the eight published whole genomes of *C. jejuni* indicated that all of them contain the CmeABC efflux systems. BLAST (Basic Local Alignment Search Tool) search of *cmeABC* against the gene database indicated that this efflux system is widely present in *C. jejuni* and *C. coli* as well as other *Campylobacter* species, such as *C. upsaliensis, C. fetus, C. lari, C. venerealis, C. gracilis, C. showae,* and *C. rectus*. Together, these observations indicate that the CmeABC efflux pump system is conserved at both the genomic and functional levels in different *Campylobacter* Spp.

The expression of *cmeABC* is controlled by a transcriptional regulator named CmeR (Lin *et al.*, 2005a). The *cmeR* gene (*cj0368c*) is located immediately upstream of the *cmeABC* operon and encodes a 210-amino acid (aa) protein. The N-terminal sequence of CmeR shows homology with the members of the TetR family of transcriptional repressors, including QacR of *S. aureus*, AcrR of *E. coli*, and MtrR of *N. gonorrhoeae* (Lin *et al.*, 2005a). CmeR functions as a repressor for *cmeABC* and inactivation of *cmeR* leads to the overexpression of *cmeABC,* which has been shown by immunoblotting, transcriptional fusion, and DNA microarray (Lin *et al.*, 2005a; Guo *et al.*, 2008). Electrophoretic mobility shift assays (EMSA) further demonstrated that CmeR binds specifically to the intergenic region (IT) between *cmeR* and *cmeABC* (Lin *et al.*, 2005a), specifically to the 16bp inverted repeat (IR) sequence (5′TGTAATAAATATTACA3′) located between the predicted −10 and −35 sequences of the *cmeABC* promoter (Lin *et al.*, 2005a). Mutations in the IR or nearby the IR that affect the binding of CmeR to the promoter of *cmeABC* would result in the overexpression of this efflux system (Lin *et al.*, 2005a; Cagliero *et al.*, 2007).

The *cmeABC* operon is inducible by bile salts and salicylate (Lin *et al.*, 2005c; Shen *et al.*, 2011). Transcriptional fusion assays indicated that presence of various bile salts in culture media induced the expression of *cmeABC* by 6- to 16-fold over the basal level of *cmeABC* transcription (Lin *et al.*, 2005c). Moreover, a recent study revealed that salicylate functions as an inducer for *cmeABC* as well. Presence of salicylate in Mueller–Hinton (MH) culture resulted in increased expression of *cmeABC* as demonstrated by immunoblotting, transcriptional fusion assay, and real-time PCR (Shen *et al.*, 2011). The presence of bile salts or salicylate did not alter the expression level of CmeR, suggesting that the induction was via altering the function of CmeR. Indeed, EMSA assay and surface plasmon resonance showed that the presence of bile salts or salicylate inhibited the binding of CmeR to the DNA of *cmeABC* promoter (Lin *et al.*, 2005c; Shen *et al.*, 2011). Recently, Lei *et al.* further confirmed the interaction of CmeR with bile acids using isothermal titration calorimetry and fluorescence polarization (Lei *et al.*, 2011). These findings indicate that certain substrates of CmeABC interact with CmeR and induce the expression of the CmeABC efflux system.

CmeABC not only contributes to antibiotic resistance, but also affects the frequency of emergence of fluoroquinolone (FQ) resistant *Campylobacter* mutants (Yan *et al.*, 2006; Shen *et al.*, 2011). Inactivation of *cmeB* reduced the frequency of emergence, while overexpression of *cmeABC* increased the frequency of emergence of FQ-resistant mutants under antibiotic selection (Yan *et al.*, 2006; Shen *et al.*, 2011). This is due to the fact that CmeABC functions synergistically with GyrA mutations in conferring FQ resistance. Without the function of CmeABC, GyrA mutations alone are unable to confer the level of resistance that is required for the mutants to survive the selection pressure (Yan *et al.*, 2006; Luangtongkum *et al.*, 2009).

CmeR

In addition to regulating *cmeABC*, CmeR functions as a pleiotropic regulator and modulates the expression of multiple genes in *Campylobacter* (Guo *et al.*, 2008). Microarray data revealed that inactivation of *cmeR* affected the transcription of at least 28 genes in *C. jejuni*. Particularly, CmeR also represses the expression of Cj0035c (a transporter of the major facilitator superfamily) and Cj0561c (encoding a periplasmic membrane fusion protein). Two CmeR binding sites (IRs) are identified in the promoter region of *cj0561c* and CmeR directly controls the expression of Cj0561c (Guo *et al.*, 2008). Both CmeR and Cj0561c are required for optimal colonization *in vivo* as inactivation of either gene reduced the fitness of *C. jejuni* in the intestinal tract of chickens (Guo *et al.*, 2008).

Recently, the crystal structure of CmeR has been determined by Gu *et al.* (Gu *et al.*, 2007). Similar to other members of the TetR family, CmeR forms a dimeric structure with an entirely helical architecture (Fig. 11.1). Each subunit of CmeR is composed of nine α helices and can be divided into two domains: an N-terminal DNA binding domain and a C-terminal multi-ligand-binding domain. Distinct from the other members of the TetR family, including TetR (Hinrichs *et al.*, 1994; Orth *et al.*, 2000), QacR (Schumacher *et al.*, 2001, 2002), CprB (Natsume *et al.*, 2004), and EthR (Dover *et al.*, 2004; Frenois *et al.*, 2004), the crystal structure of CmeR revealed that the α3 helix, involved in DNA recognition, was replaced by a random coil. To facilitate the comparison with the structures of other TetR members, helices of CmeR are numbered from the N terminus as α1 (7–29), α2 (36–43), α4 (57–81), α5 (88–104), α6 (106–118), α7 [a(125–136) and b(138–148)], α8 (152–170), α9 (172–180), and α10 (187–203), in which helix α3 has been omitted. Therefore, the N-terminal DNA-binding domain contains helices α1, α2, and a random loop (residues 47–53), and the C-terminal ligand-binding domain includes helices from α4 to α10 (Gu *et al.*, 2007; Routh *et al.*, 2009).

The overall structure of the N-terminal DNA-binding domain of CmeR is very distinct from to those of other TetR family members (Gu *et al.*, 2007). The first helix of CmeR, consisting of 23 amino acids, is relatively long among all structurally known members of the TetR family. For instance, the lengths of helices α1 in QacR, TetR, and EthR are 16, 13, and 17 residues, respectively. As mentioned above, the lack of α3 helix in CmeR is the most striking feature compared

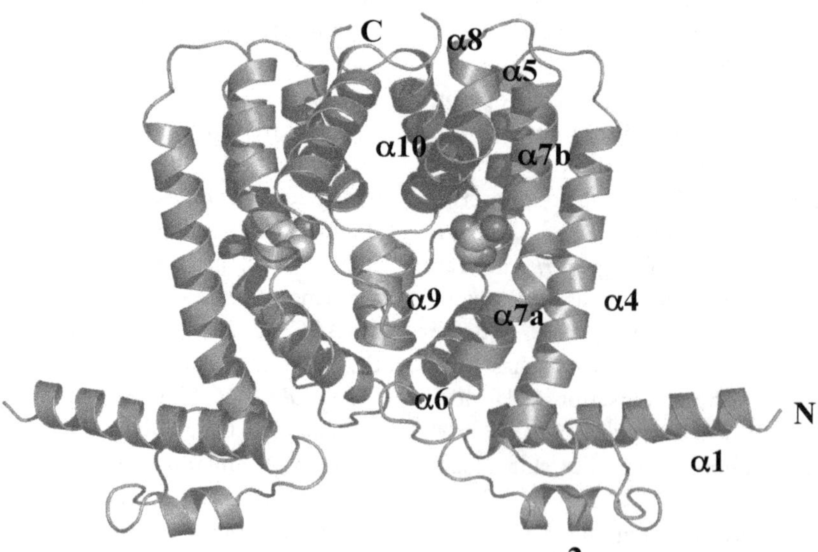

Figure 11.1 Ribbon diagram of the glycerol-bound CmeR homodimer generated by crystallographic symmetry. The Figure was prepared using PyMOL (http://www.pymol.sourceforge.net).

with other TetR family members. So far, CmeR is the only regulator missing helix α3 among the TetR family. The helix α3 together with α2 in TetR regulators was considered to form a helix–turn–helix (HTH) DNA-binding motif, which plays an important role in recognizing the target DNA (Ramos *et al.*, 2005). It is likely that CmeR might transform the flexible coil into a helix when bound to the target DNA or induced by certain conditions. Furthermore, CmeR functions as a pleiotropic regulator of a large set of genes and the flexibility of the DNA-binding domain might permit CmeR to recognize multiple DNA sites (Routh *et al.*, 2009).

The C-terminal domain of CmeR consists of helices α4 through α10 and five of them (α4, α5, α7, α8, and α10) form an anti-parallel five-helix bundle. The crystal structure of CmeR indicated that helices α6, α8, α9, and α10 are involved in the dimer formation (Gu *et al.*, 2007; Routh *et al.*, 2009). The overall structure of the C-terminal domain of CmeR is closer to that of QacR among members of the TetR family. The distinct feature of CmeR in the C-terminal domain is that the helix α9, between the two anti-parallel helices α8 and α10, deviates from the direction of α8 by 40°. Therefore, the bending of helix α9 towards the next subunit of the dimer ensures the secure interaction between the dimer (Gu *et al.*, 2007). One unique feature, not found in other regulators of TetR family, is the large tunnel-like cavity in each subunit of CmeR (Fig. 11.2). This tunnel, surrounded by mostly hydrophobic residues of helices α4–α9 in the C-terminal domain of CmeR, opens horizontally from the front to the back of each protomer. Helices α7 and α8 from one subunit, and α9′ from the other subunit of the regulator make the entrance of the tunnel, while helices α4–α6 form the end of this hydrophobic tunnel (Routh *et al.*, 2009).

This tunnel is approximately 20 Å in length and occupies a volume of about 1000 Å, which is distinctly larger than the binding pockets of many other members of the TetR family (Gu *et al.*, 2007). Unexpectedly, a fortuitous glycerol molecule was found to bind in the binding tunnel of each monomer (Fig. 11.1). Residues F99, F103, F137, S138, Y139, V163, C166, T167, and K170

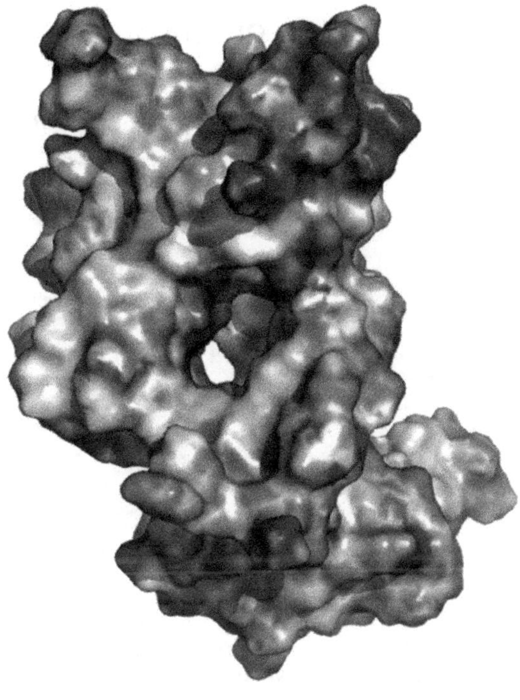

Figure 11.2 Views of the tunnel-like cavity in the ligand-binding domain of CmeR. This is an electrostatic surface potential of one subunit of CmeR. The view shows the long tunnel spanning through the C-terminal domain of CmeR. Blue (+15 k_BT) and red (–15 k_BT) indicate the positively and negatively charged areas, respectively, of the protein. A colour version of this figure is located in the plate section at the back of the book.

are responsible for forming this glycerol-binding site. The volume of the ligand-binding tunnel of CmeR is large enough to accommodate a few of the ligand molecules. The crystal structures of CmeR in complexes with taurocholate and cholate were also elucidated, which indicated that these two ligands bind distinctly in the binding tunnel (Fig. 11.3). Residues I68, C69, H72, F103, A108, F111, I115, W129, Q134, F137, V163, K170, H174, H175′, L176′, and L179′ are involved in taurocholate binding, whereas residues L65, C69, F103, A108, F111, G112, I115, W129, F137, Y139, C166, K170, P172′, H174, H175′, and L179′ interact with cholate in the tunnel. These quite distinct binding manners highlighted the plasticity and promiscuity of the ligand-binding tunnel of CmeR (Lei *et al.*, 2011).

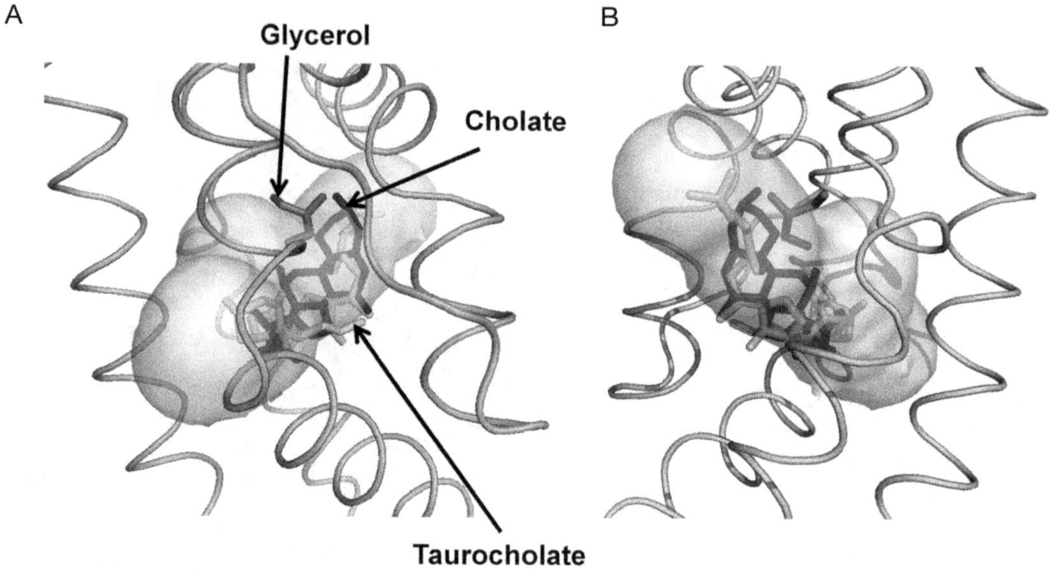

Figure 11.3 The bile acid and glycerol-binding sites of CmeR. This is a composite figure showing the locations of the bound ligands in the ligand binding tunnel of the left (a) and right (b) subunit of the CmeR dimer. The ligands shown in stick models are taurocholate, cholate and glycerol. The hydrophobic binding tunnel is coloured grey. The surrounding secondary structural elements, based on the structure of the CmeR–glycerol complex, are shown as ribbons.

CmeDEF

CmeDEF is another RND-type efflux pump identified in *C. jejuni*. CmeD (Cj1031) is an outer membrane protein of 424 aa, which shares low, but significant sequence homology to HefA of *H. pylori* and TolC of *E. coli*, the outer membrane components of the tripartite antibiotic efflux systems (Lin *et al.*, 2005b). CmeE (Cj1032) is a membrane fusion protein composed of 246 aa, which shares significant homology with the membrane fusion protein of HefB in *H. pylori*. CmeF, similar to CmeB, is an inner membrane transporter and is predicted to contain 12-transmembrane (TM) helical domain structure. The sequence of CmeF (1005 aa) indicates that this membrane protein shares a moderate homology with many other RND-type efflux transporters, such as HefC of *H. pylori*, and AcrB, AcrD, and AcrF of *E. coli* (Lin *et al.*, 2005b; Pumbwe *et al.*, 2005). The low sequence identity between CmeDEF and CmeABC suggests these two efflux systems may have different functions and abilities to extrude antibiotics and other toxic compounds.

Several studies have been found to investigate the contribution of *cmeDEF* to antimicrobial resistance. Pumbwe *et al.* reported that insertional mutation of *cmeF* in *Campylobacter* resulted in increased susceptibility to structurally unrelated antimicrobial compounds, including ampicillin, ethidium bromide, acridine orange, SDS, sodium deoxycholate, bile, detrimide, and triclosan (Pumbwe *et al.*, 2005). Akiba *et al.* also reported that the *cmeF* mutant of 11168 showed a twofold decrease in the resistance to ampicillin and ethidium bromide, but did not observe any changes in the susceptibility to the other tested antimicrobials including bile salts (Akiba *et al.*, 2006). Another study by Ge *et al.* found that inactivation of *cmeF* in 81–176 had no effect on the susceptibility to ciprofloxacin, erythromycin, tetracycline, and chloramphenicol. Interestingly, inactivation of *cmeF* in some *Campylobacter* strains (81–176, 21190, and 164) resulted in a 2- to 4-fold increase in the resistance to multiple antimicrobials and toxic compounds, including ciprofloxacin, cefotaxime, rifampicin, tetracycline, novobiocin, fusidic acid, ethidium bromide, acriflavine, SDS, triton X-100, tween 20, empigen, and different bile salts (Akiba *et al.*, 2006). In order to examine if the expression of CmeABC pump potentially masked the function of CmeDEF on antimicrobial resistance, the 81–176 and 21190 *cmeB/cmeF* double mutants were generated. In comparison with the 81–176 *cmeB* mutant, the 81–176 *cmeB/cmeF* double mutant showed a twofold decrease in the resistance to ciprofloxacin, tetracycline, fusidic acid, acriflavine, and a 2- to 16-fold decrease in the resistance to various detergents and bile salts. Likewise, in comparison with the 21190 *cmeB* mutant, the 21190 *cmeB/cmeF* double mutant showed a twofold increase in the susceptibility to fusidic acid, acriflavine, polymyxin B, novobiocin, and a 2- to 128-fold increase in the susceptibility to detergents and bile salts (Akiba *et al.*, 2006). Together, these results suggest that the CmeDEF tripartite efflux pump plays a modest role in antibiotic resistance in a strain-dependent manner and its function is normally masked by that of CmeABC.

One interesting finding is that at least one of the two efflux systems (CmeABC or CmeDEF) is required for maintaining cell viability in certain *Campylobacter* strains (Akiba *et al.*, 2006). Mutation of either *cmeB* or *cmeF* did not affect the viability or growth characteristics of *Campylobacter* (Lin *et al.*, 2003; Akiba *et al.*, 2006). However, the 81–176 *cmeB/cmeF* double mutant of *C. jejuni* showed a decreased in viability at the stationary phase (Akiba *et al.*, 2006). Additionally, repeated experiments failed to generate the *cmeB/cmeF* double mutation in NCTC 11168, suggesting that the double mutation is lethal to NCTC 11168 (Pumbwe *et al.*, 2005; Akiba *et al.*, 2006). These results suggest that CmeDEF has an uncharacterized physiological function in *Campylobacter*.

The baseline expression (in culture media) of *cmeDEF* in *Campylobacter* is much lower than that of *cmeABC* as determined by transcriptional fusion and immunoblotting assays (Akiba *et al.*, 2006). The regulatory mechanism for *cmeDEF* remains unknown and the conditions for inducing the expression of *cmeDEF* have not been determined. Nonetheless, it has been observed that CmeR, the transcriptional repressor for *cmeABC*, does not control the expression of *cmeDEF*.

CmeG

CmeG (Cj1375) is one of the MFS transporters in *Campylobacter* and is present in all *C. jejuni* strains sequenced to date. Amino acid sequence analysis revealed that CmeG shows sequence homology to Bmr of *B. subtilis* and NorA of *S. aureus* (Parkhill *et al.*, 2000; Jeon *et al.*, 2011). Both Bmr and NorA were demonstrated to contribute to multidrug resistance in other bacteria (Ubukata *et al.*, 1989; Neyfakh *et al.*, 1991, 1993; Jeon *et al.*, 2011). In addition, CmeG is predicated to be an inner membrane protein and possesses 12 transmembrane helices (TMs). Inactivation of *cmeG* significantly reduced the resistance to a variety of antimicrobials, including ciprofloxacin, erythromycin, tetracycline, gentamicin, ethidium bromide, and cholic acid. Overexpression of *cmeG* has been shown to enhance the resistance to various fluoroquinolones, including ciprofloxacin, enrofloxacin, norfloxacin, and moxifloxacin, but not to the other classes of antibiotics (Jeon *et al.*, 2011). Accumulation assays showed that the *cmeG* mutant accumulated more ethidium bromide and ciprofloxacin than the wild-type strain (Jeon *et al.*, 2011). These results indicate that CmeG functions as an efflux transporter in *Campylobacter*.

An important function of CmeG is its contribution to oxidative stress resistance in *Campylobacter*. Mutation of *cmeG* increased the susceptibility of *C. jejuni* to hydrogen peroxide and complementation restored the resistance to the wild-type level (Jeon *et al.*, 2011). This finding clearly indicates the significant role of CmeG in oxidative stress response, but how CmeG is involved in this process is unknown. Additionally, CmeG appears to be important for bacterial viability as inactivation of *cmeG* in *C. jejuni* 11168 resulted in a significant growth defect in conventional culture media and the growth defect was completely restored by *in trans* complementation (Jeon *et al.*, 2011). Furthermore, a *cmeG* mutant could not be generated in *C. jejuni* 81–176 and 81116 (Ge *et al.*, 2005; Jeon *et al.*, 2011), suggesting that *cmeG* is an essential gene in certain *Campylobacter* strains. These findings also imply that CmeG has an important physiological function in *Campylobacter*. The expression of *cmeG* appears to be regulated by Fur protein and iron concentrations as inactivation of Fur or depleting iron resulted in upregulated expression of *cmeG* (Palyada *et al.*, 2004, 2009; Holmes *et al.*, 2005). The detailed mechanism for *cmeG* regulation remains unknown.

Acr3

Recently, an arsenic detoxification system has been identified in *C. jejuni* (Wang *et al.*, 2009). The organic form of arsenic compounds is commonly used as a feed additive in poultry production to promote healthier and bigger chicken. Thus, the ability to resist the action of arsenic compounds is necessary for *Campylobacter* adaptation in the guts of these poultries. Indeed, *Campylobacter* isolates from conventional poultry products have significantly higher incidence of arsenic resistance than those isolates from antimicrobial-free poultry products (Sapkota *et al.*, 2006). The arsenic resistance and arsenic-sensing system identified in *C. jejuni* is encoded by a four-gene operon, which includes a putative membrane permease (ArsP), a transcriptional repressor (ArsR), an arsenate reductase (ArsC), and an arsenite efflux protein (Acr3) (Wang *et al.*, 2009). There are two families of arsenite transporters: the ArsB family and the Acr3 family (Rosen, 1999). The ArsB family has been found only in bacteria and archaea, while the Acr3 family exists in prokaryotes and fungi, as well as in plant genomes (Rosen, 1999, 2002; Fu *et al.*, 2009; Indriolo *et al.*, 2010).

Acr3 in *C. jejuni* is a 347 amino-acid membrane protein that contains ten predicted transmembrane helices. The presence of the *acr3*-containing operon is significantly associated with elevated resistance to arsenite and arsenate in *Campylobacter*. Furthermore, it has been found that inactivation of *acr3* led to 8- and 4-fold reductions in the MICs of arsenite and arsenate, respectively, but mutation of *acr3* did not affect the susceptibility to different classes of antibiotics, including erythromycin, tilmicosin, ciprofloxacin, enrofloxacin, oxytetracycline, ceftiofur, and polymyxin B (Wang *et al.*, 2009).

The expression of *acr3* is directly regulated by a transcriptional regulator ArsR, encoded by the second gene of the *ars* operon (Wang *et al.*, 2009). ArsR is a small regulatory protein containing a helix–turn–helix (HTH) DNA-binding motif. In-frame deletion of *arsR* greatly increased the expression of the other three genes in the operon including *acr3*, suggesting that ArsR functions as a repressor. Analysis of the *ars* promoter region indicated that there is an 18-bp inverted repeat forming a dyad structure, suggesting a binding site for ArsR. Indeed, EMSA showed the direct binding of ArsR to the promoter of the *ars* operon, specifically to the region containing the inverted repeat (Wang *et al.*, 2009). The expression of the *ars* operon is inducible by arsenite and arsenate compounds. In addition, it has been found that the ArsR protein contains a conserved metal binding motif (ELCVCDL). Thus, it is possible that arsenic compounds induce the expression of the *ars* operon by inhibiting the interaction of ArsR with the promoter. Indeed, EMSA demonstrated that arsenite, but not arsenate, inhibited the binding of ArsR to the promoter DNA, providing an explanation for arsenite-induced expression of the operon. Additionally, arsenate is reduced by ArsC to arsenite in bacterial cells and thus is also able to induce the expression of the *ars* operon in *Campylobacter* (Wang *et al.*, 2009).

In addition to the Acr3 efflux pump, *C. jejuni* strains also contain putative ArsB and ArsP transporters, which are associated with arsenic resistance. Our recent studies showed that mutation of the *arsB* gene resulted in 8- and 4-fold reduction in the MICs of arsenite and arsenate, respectively, compared to the wild-type strain, while inactivation of *arsP* led to a 4-fold reduction in the MIC of roxarsone (an organic arsenic compound) compared to the wild-type strains (Shen, Z., unpublished). Additionally, overexpression of *arsB* and cloning of *arsP* to an *arsP*-negative strain resulted in 8-fold increase in the MICs of arsenite and roxarsone, respectively (Shen, Z., unpublished). These findings indicate that *Campylobacter* harbours multiple arsenic efflux transporters that confer resistance to both organic and inorganic arsenic compounds.

Targeting efflux mechanisms to control *Campylobacter*

As discussed above, antibiotic efflux transporters are not only key players in the resistance to antibiotics and toxic compounds, but also have important physiological functions in *Campylobacter*, providing promising targets for the control of antibiotic resistant *Campylobacter*. This is especially true with CmeABC, which is essential for *Campylobacter* adaptation in the intestinal tract (Lin *et al.*, 2002, 2003). Thus, inhibition of CmeABC will not only reduce antibiotic resistance, but also potentially prevent *in vivo* colonization. Inhibition of antibiotic efflux in bacterial cells can be achieved by two different approaches. The first one is to inhibit the function of multidrug efflux transporters by using efflux pump inhibitors (EPIs) (Lomovskaya *et al.*, 2001; Lin and Martinez, 2006; Quinn *et al.*, 2007). Recently, a promising EPI, Phe-Arg β-naphthyl-amide dihydrochloride (MC-207,110; PAβN), was found to target the inner membrane drug transporter of RND-type efflux pumps and potentiate the activity of antimicrobial agents in *Pseudomonas aeruginosa* (Lomovskaya *et al.*, 2001). So far, PAβN has been demonstrated to be effective in potentiating antibiotics against a variety of Gram-negative bacteria (Ribera *et al.*, 2002; Mamelli *et al.*, 2003; Chan *et al.*, 2004; Chollet *et al.*, 2004; Coban *et al.*, 2004; Hasdemir *et al.*, 2004; Cebrian *et al.*, 2005; Martinez and Lin, 2006; Bina *et al.*, 2009; Lister *et al.*, 2012). Several studies have tried to inhibit CmeABC and antibiotic efflux in *Campylobacter* by using PAβN (Corcoran *et al.*, 2005; Mamelli *et al.*, 2005; Martinez and Lin, 2006; Quinn *et al.*, 2007). The general findings are that PAβN has a significant impact on erythromycin (Ery) MICs (8 to >64-fold reduction) and that this effect is largely due to the inhibition of CmeABC and is especially prominent in Ery^R mutants that do not harbour known target mutations in the 23S RNA genes (Quinn *et al.*, 2007). This EPI also showed good potentiating activities for other antimicrobials including rifampin, novobiocin, fusidic acid, and bile salts. In contrast to the findings in other bacteria, multiple studies have found that PAβN has limited or no effect on the MICs of FQ antimicrobials in *Campylobacter* (Quinn *et al.*, 2007; Hannula and Hanninen, 2008). Thus PAβN appears to be an effective potentiating agent for macrolide antibiotics, but not effective for potentiating FQ antimicrobials in *Campylobacter*.

The second approach for targeting antibiotic efflux systems is to inhibit their production. Recently, Jeon and Zhang reported the feasibility of using antisense peptide nucleic acid (PNA) to inhibit the production of CmeABC in *Campylobacter* (Jeon and Zhang, 2009). PNA agents are DNA-mimic synthetic polymers, carrying a pseudo-peptide backbone with nucleic acid bases (Nielsen *et al.*, 1991). PNA is neutrally charged due to the lack of a phosphodiester backbone, giving high affinity to nucleic acids. In addition, the extreme resistance of PNA to proteases and nucleases, and its stability in acidic pH have made PNA an ideal candidate for various antisense applications (Nielsen *et al.*, 1991). An anti-cmeA PNA was used to specifically inhibit the expression of CmeABC in *C. jejuni* (Jeon and Zhang, 2009). The specific inhibition was confirmed by immunoblotting using anti-CmeABC antibodies (Jeon and Zhang, 2009). Importantly, the anti-cmeA PNA potentiated the anti-*Campylobacter* activity of erythromycin and ciprofloxacin and resulted in 2- to > 64-fold reduction in their MICs (Jeon and Zhang, 2009). These results clearly indicated that the feasibility and potential of the PNA antisense technology in controlling antibiotic-resistant *Campylobacter*.

Conclusions and future directions

The importance of multidrug efflux pumps in antibiotic resistance and pathobiology is increasingly recognized in *Campylobacter*. It is well documented that the antibiotic efflux systems have functions beyond extrusion of antibiotics, and evidence is accumulating that they are intertwined with physiological pathways in *Campylobacter*. Thus, it is likely that these efflux systems have additional functions yet to be defined. Future studies should focus on understanding how these antibiotic efflux systems interact with various physiological pathways and how they facilitate the adaptation of *Campylobacter* to various unfavourable niches. Additionally, targeting efflux systems has become a promising approach for controlling antimicrobial resistance and preventing *Campylobacter* colonization in the intestinal tract of animal hosts. The inhibitor PAβN has been shown to be effective in potentiating macrolide antibiotics such as erythromycin, while anti-*cmeA* PNA sensitizes *Campylobacter* to both macrolides and fluoroquinolones. To enhance the usefulness of the approach, additional EPIs (both synthetic compounds and natural products) should be evaluated to inhibit antibiotic efflux systems in *Campylobacter*. Combination of EPIs with PNAs may also increase the potentiating effects on various antibiotics. Additionally, enhanced efforts should be directed towards elucidating the crystal structures and structure–function relationship of antibiotic efflux transporters, which will eventually facilitate the design of small molecules to block the function of these efflux pumps.

Acknowledgements

This work was supported by NIH grants GM086431 (to E.W.Y.) and DK063008 (to Q.Z.).

References

Akiba, M., Lin, J., Barton, Y.W., and Zhang, Q. (2006). Interaction of CmeABC and CmeDEF in conferring antimicrobial resistance and maintaining cell viability in *Campylobacter jejuni*. J. Antimicrob. Chemother. *57*, 52–60.

Bina, X.R., Philippart, J.A., and Bina, J.E. (2009). Effect of the efflux inhibitors 1-(1-naphthylmethyl)-piperazine and phenyl-arginine-beta-naphthylamide on antimicrobial susceptibility and virulence factor production in *Vibrio cholerae*. J. Antimicrob. Chemother. *63*, 103–108.

Cagliero, C., Mouline, C., Payot, S., and Cloeckaert, A. (2005). Involvement of the CmeABC efflux pump in the macrolide resistance of *Campylobacter coli*. J. Antimicrob. Chemother. *56*, 948–950.

Cagliero, C., Cloix, L., Cloeckaert, A., and Payot, S. (2006a). High genetic variation in the multidrug transporter *cmeB* gene in *Campylobacter jejuni* and *Campylobacter coli*. J. Antimicrob. Chemother. *58*, 168–172.

Cagliero, C., Mouline, C., Cloeckaert, A., and Payot, S. (2006b). Synergy between efflux pump CmeABC and modifications in ribosomal proteins L4 and L22 in conferring macrolide resistance in *Campylobacter jejuni* and *Campylobacter coli*. Antimicrob. Agents Chemother. *50*, 3893–3896.

Cagliero, C., Maurel, M.C., Cloeckaert, A., and Payot, S. (2007). Regulation of the expression of the CmeABC efflux pump in *Campylobacter jejuni*: identification of a point mutation abolishing the binding of the CmeR repressor in an *in vitro*-selected multidrug-resistant mutant. FEMS Microbiol. Lett. *267*, 89–94.

Caldwell, D.B., Wang, Y., and Lin, J. (2008). Development, stability, and molecular mechanisms of macrolide resistance in *Campylobacter jejuni*. Antimicrob. Agents Chemother. *52*, 3947–3954.

Cebrian, L., Rodriguez, J.C., Escribano, I., and Royo, G. (2005). Characterization of *Salmonella* spp. mutants with reduced fluoroquinolone susceptibility: importance of efflux pump mechanisms. Chemotherapy *51*, 40–43.

Chan, Y.Y., Tan, T.M., Ong, Y.M., and Chua, K.L. (2004). BpeAB-OprB, a multidrug efflux pump in *Burkholderia pseudomallei*. Antimicrob. Agents Chemother. *48*, 1128–1135.

Chatre, P., Haenni, M., Meunier, D., Botrel, M.A., Calavas, D., and Madec, J.Y. (2010). Prevalence and antimicrobial resistance of *Campylobacter jejuni* and *Campylobacter coli* isolated from cattle between 2002 and 2006 in France. J. Food Prot. *73*, 825–831.

Chen, X., Naren, G.W., Wu, C.M., Wang, Y., Dai, L., Xia, L.N., Luo, P.J., Zhang, Q., and Shen, J.Z. (2010). Prevalence and antimicrobial resistance of *Campylobacter* isolates in broilers from China. Vet. Microbiol. *144*, 133–139.

Chollet, R., Chevalier, J., Bryskier, A., and Pages, J.M. (2004). The AcrAB-TolC pump is involved in macrolide resistance but not in telithromycin efflux in *Enterobacter aerogenes* and *Escherichia coli*. Antimicrob. Agents Chemother. *48*, 3621–3624.

Coban, A.Y., Birinci, A., Ekinci, B., and Durupinar, B. (2004). Investigation of the effects of efflux pump inhibitors on ciprofloxacin MIC values of high level fluoroquinolone resistant *Escherichia coli* clinical isolates. Mikrobiyol. Bul. *38*, 21–25.

Cody, A.J., Clarke, L., Bowler, I.C., and Dingle, K.E. (2010). Ciprofloxacin-resistant campylobacteriosis in the UK. Lancet *376*, 1987.

Corcoran, D., Quinn, T., Cotter, L., and Fanning, S. (2005). Relative contribution of target gene mutation and efflux to varying quinolone resistance in Irish *Campylobacter* isolates. FEMS Microbiol. Lett. *253*, 39–46.

Dover, L.G., Corsino, P.E., Daniels, I.R., Cocklin, S.L., Tatituri, V., Besra, G.S., and Futterer, K. (2004). Crystal structure of the TetR/CamR family repressor *Mycobacterium tuberculosis* EthR implicated in ethionamide resistance. J. Mol. Biol. *340*, 1095–1105.

Fakhr, M.K., and Logue, C.M. (2007). Sequence variation in the outer membrane protein-encoding gene *cmeC*, conferring multidrug resistance among *Campylobacter jejuni* and *Campylobacter coli* strains isolated from different hosts. J. Clin. Microbiol. *45*, 3381–3383.

Frenois, F., Engohang-Ndong, J., Locht, C., Baulard, A.R., and Villeret, V. (2004). Structure of EthR in a ligand bound conformation reveals therapeutic perspectives against tuberculosis. Mol. Cell *16*, 301–307.

Fu, H.L., Meng, Y., Ordonez, E., Villadangos, A.F., Bhattacharjee, H., Gil, J.A., Mateos, L.M., and Rosen, B.P. (2009). Properties of arsenite efflux permeases (Acr3) from *Alkaliphilus metalliredigens* and *Corynebacterium glutamicum*. J. Biol. Chem. *284*, 19887–19895.

Ge, B., McDermott, P.F., White, D.G., and Meng, J. (2005). Role of efflux pumps and topoisomerase mutations in fluoroquinolone resistance in *Campylobacter jejuni* and

Campylobacter coli. Antimicrob. Agents Chemother. *49*, 3347–3354.

Gibreel, A., Wetsch, N.M., and Taylor, D.E. (2007). Contribution of the CmeABC efflux pump to macrolide and tetracycline resistance in *Campylobacter jejuni*. Antimicrob. Agents Chemother. *51*, 3212–3216.

Gillespie, I.A., O'Brien, S.J., Frost, J.A., Adak, G.K., Horby, P., Swan, A.V., Painter, M.J., and Neal, K.R. (2002). A case-case comparison of *Campylobacter coli* and *Campylobacter jejuni* infection: a tool for generating hypotheses. Emerg. Infect. Dis. *8*, 937–942.

Gu, R., Su, C.C., Shi, F., Li, M., McDermott, G., Zhang, Q., and Yu, E.W. (2007). Crystal structure of the transcriptional regulator CmeR from *Campylobacter jejuni*. J. Mol. Biol. *372*, 583–593.

Gunn, J.S. (2000). Mechanisms of bacterial resistance and response to bile. Microbes Infect. *2*, 907–913.

Guo, B., Wang, Y., Shi, F., Barton, Y.W., Plummer, P., Reynolds, D.L., Nettleton, D., Grinnage-Pulley, T., Lin, J., and Zhang, Q. (2008). CmeR functions as a pleiotropic regulator and is required for optimal colonization of *Campylobacter jejuni in vivo*. J. Bacteriol. *190*, 1879–1890.

Guo, B., Lin, J., Reynolds, D.L., and Zhang, Q. (2010). Contribution of the multidrug efflux transporter CmeABC to antibiotic resistance in different *Campylobacter* species. Foodborne Pathog. Dis. *7*, 77–83.

Gupta, A., Nelson, J.M., Barrett, T.J., Tauxe, R.V., Rossiter, S.P., Friedman, C.R., Joyce, K.W., Smith, K.E., Jones, T.F., Hawkins, M.A., *et al.* (2004). Antimicrobial resistance among *Campylobacter* strains, United States, 1997–2001. Emerg. Infect. Dis. *10*, 1102–1109.

Hannula, M., and Hanninen, M.L. (2008). Effect of putative efflux pump inhibitors and inducers on the antimicrobial susceptibility of *Campylobacter jejuni* and *Campylobacter coli*. J. Med. Microbiol. *57*, 851–855.

Hasdemir, U.O., Chevalier, J., Nordmann, P., and Pages, J.M. (2004). Detection and prevalence of active drug efflux mechanism in various multidrug-resistant *Klebsiella pneumoniae* strains from Turkey. J. Clin. Microbiol. *42*, 2701–2706.

Hinrichs, W., Kisker, C., Duvel, M., Muller, A., Tovar, K., Hillen, W., and Saenger, W. (1994). Structure of the Tet repressor–tetracycline complex and regulation of antibiotic resistance. Science *264*, 418–420.

Hoang, K.V., Stern, N.J., and Lin, J. (2011a). Development and stability of bacteriocin resistance in *Campylobacter* spp. J. Appl. Microbiol. *111*, 1544–1550.

Hoang, K.V., Stern, N.J., Saxton, A.M., Xu, F., Zeng, X., and Lin, J. (2011b). Prevalence, development, and molecular mechanisms of bacteriocin resistance in *Campylobacter*. Appl. Environ. Microbiol. *77*, 2309–2316.

Holmes, K., Mulholland, F., Pearson, B.M., Pin, C., McNicholl-Kennedy, J., Ketley, J.M., and Wells, J.M. (2005). *Campylobacter jejuni* gene expression in response to iron limitation and the role of Fur. Microbiology *151*, 243–257.

Indriolo, E., Na, G., Ellis, D., Salt, D.E., and Banks, J.A. (2010). A vacuolar arsenite transporter necessary for arsenic tolerance in the arsenic hyperaccumulating fern Pteris vittata is missing in flowering plants. Plant Cell *22*, 2045–2057.

Jacobs, B.C., Belkum, A.V., and Endtz, H.P. (2008). Guillain-Barré Syndrome and *Campylobacter* infection. In *Campylobacter*, Nachamkin, I., Szymanski, C.M., and Blaser, M.J., eds. (ASM Press, Washington DC), pp. 245–261.

Jeon, B., and Zhang, Q. (2009). Sensitization of *Campylobacter jejuni* to fluoroquinolone and macrolide antibiotics by antisense inhibition of the CmeABC multidrug efflux transporter. J. Antimicrob. Chemother. *63*, 946–948.

Jeon, B., Wang, Y., Hao, H., Barton, Y.W., and Zhang, Q. (2011). Contribution of CmeG to antibiotic and oxidative stress resistance in *Campylobacter jejuni*. J. Antimicrob. Chemother. *66*, 79–85.

Kim, H.J., Kim, J.H., Kim, Y.I., Choi, J.S., Park, M.Y., Nam, H.M., Jung, S.C., Kwon, J.W., Lee, C.H., Kim, Y.H., *et al.* (2010). Prevalence and characterization of *Campylobacter* spp. isolated from domestic and imported poultry meat in Korea, 2004–2008. Foodborne Pathog. Dis. *7*, 1203–1209.

Lei, H.T., Shen, Z., Surana, P., Routh, M.D., Su, C.C., Zhang, Q., and Yu, E.W. (2011). Crystal structures of CmeR-bile acid complexes from *Campylobacter jejuni*. Protein Sci. *20*, 712–723.

Lin, J., and Martinez, A. (2006). Effect of efflux pump inhibitors on bile resistance and *in vivo* colonization of *Campylobacter jejuni*. J. Antimicrob. Chemother. *58*, 966–972.

Lin, J., Michel, L.O., and Zhang, Q. (2002). CmeABC functions as a multidrug efflux system in *Campylobacter jejuni*. Antimicrob. Agents Chemother. *46*, 2124–2131.

Lin, J., Sahin, O., Michel, L.O., and Zhang, Q. (2003). Critical role of multidrug efflux pump CmeABC in bile resistance and *in vivo* colonization of *Campylobacter jejuni*. Infect. Immun. *71*, 4250–4259.

Lin, J., Akiba, M., Sahin, O., and Zhang, Q. (2005a). CmeR functions as a transcriptional repressor for the multidrug efflux pump CmeABC in *Campylobacter jejuni*. Antimicrob. Agents Chemother. *49*, 1067–1075.

Lin, J., Akiba, M., and Zhang, Q. (2005b). Multidrug efflux systems in *Campylobacter*. In *Campylobacter*: Molecular and Cellular Biology, Ketley, J.M., and Konkel, M.E., eds. (Horizon Press, Norfolk, UK), pp. 205–218.

Lin, J., Cagliero, C., Guo, B., Barton, Y.W., Maurel, M.C., Payot, S., and Zhang, Q. (2005c). Bile salts modulate expression of the CmeABC multidrug efflux pump in *Campylobacter jejuni*. J. Bacteriol. *187*, 7417–7424.

Lin, J., Yan, M., Sahin, O., Pereira, S., Chang, Y.J., and Zhang, Q. (2007). Effect of macrolide usage on emergence of erythromycin-resistant *Campylobacter* isolates in chickens. Antimicrob. Agents Chemother. *51*, 1678–1686.

Lister, I.M., Raftery, C., Mecsas, J., and Levy, S.B. (2012). *Yersinia pestis acrAB-tolC* in antibiotic resistance and virulence. Antimicrob. Agents Chemother. *56*, 1120–1123.

Lomovskaya, O., Warren, M.S., Lee, A., Galazzo, J., Fronko, R., Lee, M., Blais, J., Cho, D., Chamberland, S., Renau, T., *et al.* (2001). Identification and characterization of inhibitors of multidrug resistance efflux pumps in

Pseudomonas aeruginosa: novel agents for combination therapy. Antimicrob. Agents Chemother. *45*, 105–116.

Luangtongkum, T., Morishita, T.Y., Ison, A.J., Huang, S., McDermott, P.F., and Zhang, Q. (2006). Effect of conventional and organic production practices on the prevalence and antimicrobial resistance of *Campylobacter* spp. in poultry. Appl. Environ. Microbiol. *72*, 3600–3607.

Luangtongkum, T., Morishita, T.Y., Martin, L., Choi, I., Sahin, O., and Zhang, Q. (2008). Prevalence of tetracycline-resistant *Campylobacter* in organic broilers during a production cycle. Avian Dis. *52*, 487–490.

Luangtongkum, T., Jeon, B., Han, J., Plummer, P., Logue, C.M., and Zhang, Q. (2009). Antibiotic resistance in *Campylobacter*: emergence, transmission and persistence. Future Microbiol. *4*, 189–200.

Luo, N., Sahin, O., Lin, J., Michel, L.O., and Zhang, Q. (2003). *In vivo* selection of *Campylobacter* isolates with high levels of fluoroquinolone resistance associated with *gyrA* mutations and the function of the CmeABC efflux pump. Antimicrob. Agents Chemother. *47*, 390–394.

Mamelli, L., Amoros, J.P., Pages, J.M., and Bolla, J.M. (2003). A phenylalanine-arginine beta-naphthylamide sensitive multidrug efflux pump involved in intrinsic and acquired resistance of *Campylobacter* to macrolides. Int. J. Antimicrob. Agents *22*, 237–241.

Mamelli, L., Prouzet-Mauleon, V., Pages, J.M., Megraud, F., and Bolla, J.M. (2005). Molecular basis of macrolide resistance in *Campylobacter*: role of efflux pumps and target mutations. J. Antimicrob. Chemother. *56*, 491–497.

Martinez, A., and Lin, J. (2006). Effect of an efflux pump inhibitor on the function of the multidrug efflux pump CmeABC and antimicrobial resistance in *Campylobacter*. Foodborne Pathog. Dis. *3*, 393–402.

Natsume, R., Ohnishi, Y., Senda, T., and Horinouchi, S. (2004). Crystal structure of a gamma-butyrolactone autoregulator receptor protein in *Streptomyces coelicolor* A3(2). J. Mol. Biol. *336*, 409–419.

Neyfakh, A.A., Bidnenko, V.E., and Chen, L.B. (1991). Efflux-mediated multidrug resistance in *Bacillus subtilis*: similarities and dissimilarities with the mammalian system. Proc. Natl. Acad. Sci. U.S.A. *88*, 4781–4785.

Neyfakh, A.A., Borsch, C.M., and Kaatz, G.W. (1993). Fluoroquinolone resistance protein NorA of *Staphylococcus aureus* is a multidrug efflux transporter. Antimicrob. Agents Chemother. *37*, 128–129.

Nielsen, P.E., Egholm, M., Berg, R.H., and Buchardt, O. (1991). Sequence-selective recognition of DNA by strand displacement with a thymine-substituted polyamide. Science *254*, 1497–1500.

Olah, P.A., Doetkott, C., Fakhr, M.K., and Logue, C.M. (2006). Prevalence of the Campylobacter multi-drug efflux pump (CmeABC) in *Campylobacter* spp. Isolated from freshly processed Turkeys. Food Microbiol. *23*, 453–460.

Orth, P., Schnappinger, D., Hillen, W., Saenger, W., and Hinrichs, W. (2000). Structural basis of gene regulation by the tetracycline inducible Tet repressor-operator system. Nat. Struct. Biol. *7*, 215–219.

Palyada, K., Threadgill, D., and Stintzi, A. (2004). Iron acquisition and regulation in *Campylobacter jejuni*. J. Bacteriol. *186*, 4714–4729.

Palyada, K., Sun, Y.Q., Flint, A., Butcher, J., Naikare, H., and Stintzi, A. (2009). Characterization of the oxidative stress stimulon and PerR regulon of *Campylobacter jejuni*. BMC Genomics *10*, 481.

Parkhill, J., Wren, B.W., Mungall, K., Ketley, J.M., Churcher, C., Basham, D., Chillingworth, T., Davies, R.M., Feltwell, T., Holroyd, S., *et al.* (2000). The genome sequence of the food-borne pathogen *Campylobacter jejuni* reveals hypervariable sequences. Nature *403*, 665–668.

Piddock, L.J., Griggs, D., Johnson, M.M., Ricci, V., Elviss, N.C., Williams, L.K., Jorgensen, F., Chisholm, S.A., Lawson, A.J., Swift, C., *et al.* (2008). Persistence of *Campylobacter* species, strain types, antibiotic resistance and mechanisms of tetracycline resistance in poultry flocks treated with chlortetracycline. J. Antimicrob. Chemother. *62*, 303–315.

Pumbwe, L., and Piddock, L.J. (2002). Identification and molecular characterisation of CmeB, a *Campylobacter jejuni* multidrug efflux pump. FEMS Microbiol. Lett. *206*, 185–189.

Pumbwe, L., Randall, L.P., Woodward, M.J., and Piddock, L.J. (2005). Evidence for multiple-antibiotic resistance in *Campylobacter jejuni* not mediated by CmeB or CmeF. Antimicrob. Agents Chemother. *49*, 1289–1293.

Qin, S.S., Wu, C.M., Wang, Y., Jeon, B., Shen, Z.Q., Zhang, Q., and Shen, J.Z. (2011). Antimicrobial resistance in *Campylobacter coli* isolated from pigs in two provinces of China. Int. J. Food Microbiol. *146*, 94–98.

Quinn, T., Bolla, J.M., Pages, J.M., and Fanning, S. (2007). Antibiotic-resistant *Campylobacter*: could efflux pump inhibitors control infection? J. Antimicrob. Chemother. *59*, 1230–1236.

Ramos, J.L., Martinez-Bueno, M., Molina-Henares, A.J., Teran, W., Watanabe, K., Zhang, X., Gallegos, M.T., Brennan, R., and Tobes, R. (2005). The TetR family of transcriptional repressors. Microbiol. Mol. Biol. Rev. *69*, 326–356.

Ribera, A., Ruiz, J., Jiminez de Anta, M.T., and Vila, J. (2002). Effect of an efflux pump inhibitor on the MIC of nalidixic acid for *Acinetobacter baumannii* and *Stenotrophomonas maltophilia* clinical isolates. J. Antimicrob. Chemother. *49*, 697–698.

Rosen, B.P. (1999). Families of arsenic transporters. Trends Microbiol. *7*, 207–212.

Rosen, B.P. (2002). Biochemistry of arsenic detoxification. FEBS Lett. *529*, 86–92.

Routh, M.D., Su, C.C., Zhang, Q., and Yu, E.W. (2009). Structures of AcrR and CmeR: insight into the mechanisms of transcriptional repression and multi-drug recognition in the TetR family of regulators. Biochim. Biophys. Acta *1794*, 844–851.

Ruiz-Palacios, G.M. (2007). The health burden of *Campylobacter* infection and the impact of antimicrobial resistance: playing chicken. Clin. Infect. Dis. *44*, 701–703.

Sapkota, A.R., Price, L.B., Silbergeld, E.K., and Schwab, K.J. (2006). Arsenic resistance in *Campylobacter* spp.

isolated from retail poultry products. Appl. Environ. Microbiol. *72*, 3069–3071.

Scallan, E., Hoekstra, R.M., Angulo, F.J., Tauxe, R.V., Widdowson, M.A., Roy, S.L., Jones, J.L., and Griffin, P.M. (2011). Foodborne illness acquired in the United States – major pathogens. Emerg. Infect. Dis. *17*, 7–15.

Schumacher, M.A., Miller, M.C., Grkovic, S., Brown, M.H., Skurray, R.A., and Brennan, R.G. (2001). Structural mechanisms of QacR induction and multidrug recognition. Science *294*, 2158–2163.

Schumacher, M.A., Miller, M.C., Grkovic, S., Brown, M.H., Skurray, R.A., and Brennan, R.G. (2002). Structural basis for cooperative DNA binding by two dimers of the multidrug-binding protein QacR. EMBO J. *21*, 1210–1218.

Schweitzer, N., Dan, A., Kaszanyitzky, E., Samu, P., Toth, A.G., Varga, J., and Damjanova, I. (2011). Molecular epidemiology and antimicrobial susceptibility of *Campylobacter jejuni* and *Campylobacter coli* isolates of poultry, swine, and cattle origin collected from slaughterhouses in Hungary. J. Food Prot. *74*, 905–911.

Shen, Z., Pu, X.Y., and Zhang, Q. (2011). Salicylate functions as an efflux pump inducer and promotes the emergence of fluoroquinolone-resistant *Campylobacter jejuni* mutants. Appl. Environ. Microbiol. *77*, 7128–7133.

Slutsker, L., Altekruse, S.F., and Swerdlow, D.L. (1998). Foodborne diseases. Emerging pathogens and trends. Infect. Dis. Clin. North Am. *12*, 199–216.

Smith, J.L., and Fratamico, P.M. (2010). Fluoroquinolone resistance in *Campylobacter*. J. Food Prot. *73*, 1141–1152.

Ubukata, K., Itoh-Yamashita, N., and Konno, M. (1989). Cloning and expression of the norA gene for fluoroquinolone resistance in *Staphylococcus aureus*. Antimicrob. Agents Chemother. *33*, 1535–1539.

Wang, L., Jeon, B., Sahin, O., and Zhang, Q. (2009). Identification of an arsenic resistance and arsenic-sensing system in *Campylobacter jejuni*. Appl. Environ. Microbiol. *75*, 5064–5073.

Yan, M., Sahin, O., Lin, J., and Zhang, Q. (2006). Role of the CmeABC efflux pump in the emergence of fluoroquinolone-resistant *Campylobacter* under selection pressure. J. Antimicrob. Chemother. *58*, 1154–1159.

Zhang, Q., and Plummer, P.J. (2008). Mechnaisms of Antibiotic Resistance in *Campylobacter*. In *Campylobacter*, Nachamkin, I., Szymanski, C.M., and Blaser, M.J., eds. (ASM Press, Washington DC), pp. 263–276.

Zhao, S., Young, S.R., Tong, E., Abbott, J.W., Womack, N., Friedman, S.L., and McDermott, P.F. (2010). Antimicrobial resistance of *Campylobacter* isolates from retail meat in the United States between 2002 and 2007. Appl. Environ. Microbiol. *76*, 7949–7956.

Index

A

CzcD 6, 84, 87, 94, 101, 104, 105, 114, 118

D

N

O

P

Q

R

U

V

W

X

Y

Z

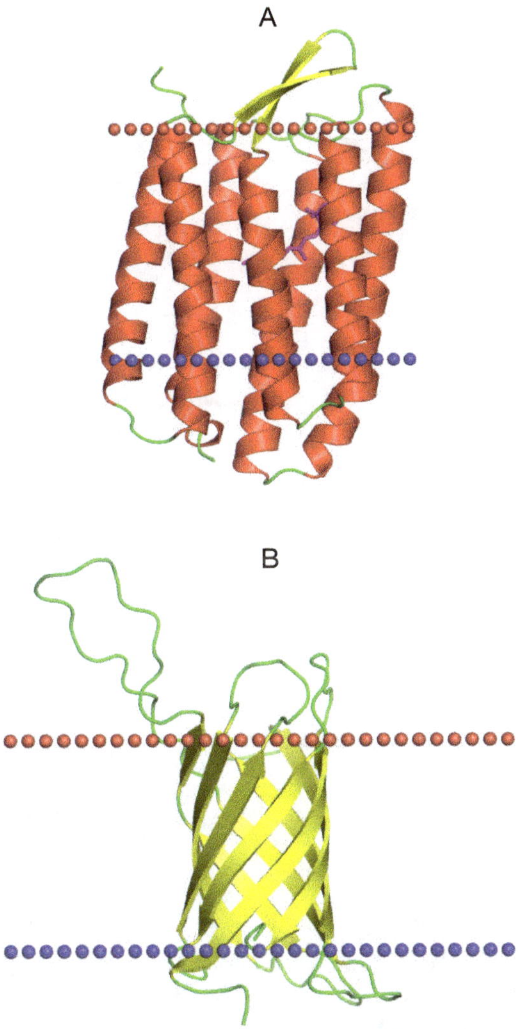

Figure 1.1 Representative structures of alpha (A) and beta (B) transmembrane proteins. The archaeal rhodopsin-1 (A; TC # 3.E.1) and beta-barrel porin (B; TC # 1.B.6) proteins are represented. Top and bottom of each figure show the extracellular and cytoplasmic sides, respectively. Both figures were obtained from the Orientations of Proteins in Membranes (OPM) database.

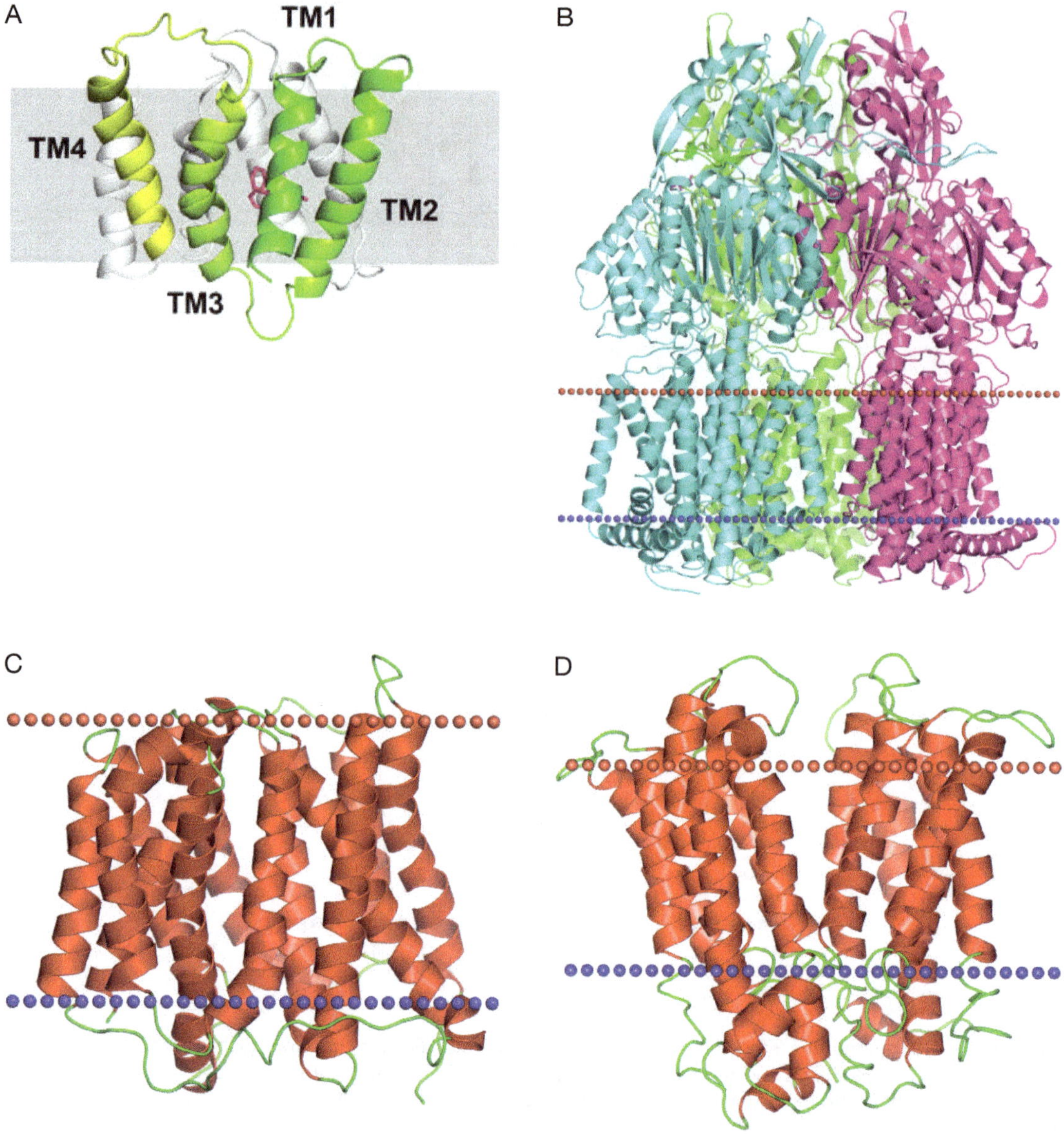

Figure 1.4 Structures of the dimeric EmrE (A; TC # 2.A.7.1.3; DMT superfamily), the trimeric AcrB (B; TC # 2.A.6.2.2; RND superfamily), EmrD (C; TC# 2.A.1.2; MFS) and NorM (D; TC # 2.A.66.1.2; MOP superfamily) proteins. Red and blue dots signify the periplasmic and cytoplasmic sides of the Gram-negative bacterial envelope, respectively. The image for EmrE was published by Chen *et al.* (2007). A–C were obtained from the OPM database.

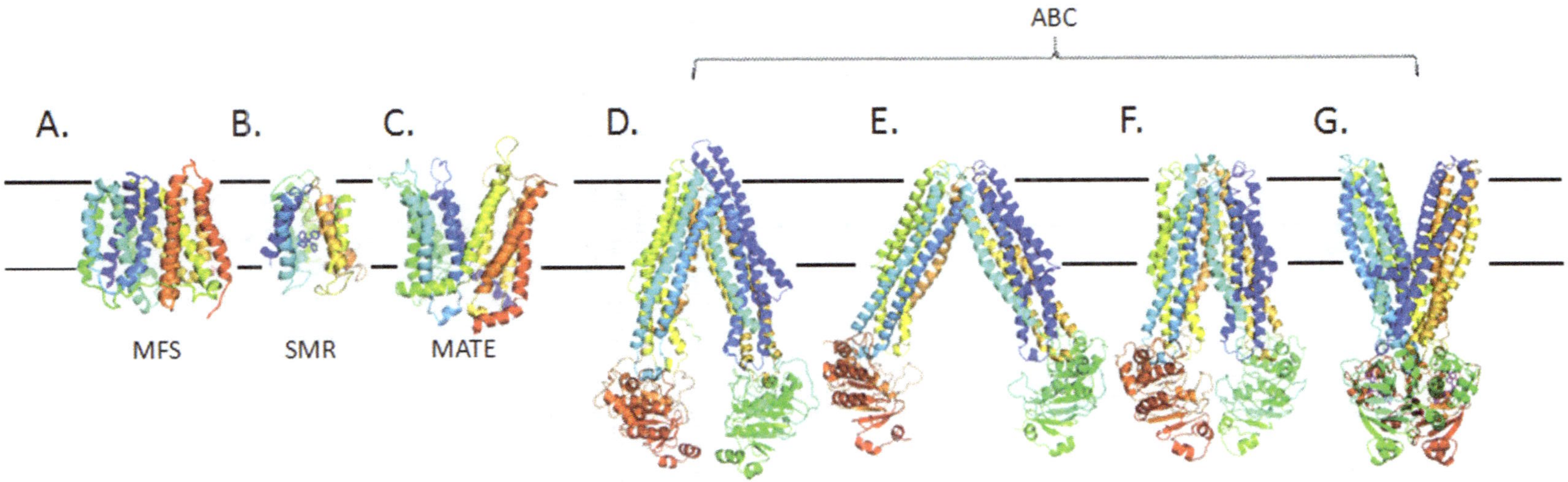

Figure 2.1 X-ray structures from the four MDR/lipid transporter families found in nature where the polyspecific binding site is located in the cell membrane (indicated by black lines). These crystal structures provide a molecular structural basis for drug/lipid transport and are from the major facilitator superfamily (MSF), small multidrug resistance (SMR), multidrug and toxin extrusion family (MATE), and ATP-binding cassette (ABC) transporter families. The X-ray structures, from left to right, are EmrD (Yin *et al.*, 2006), EmrE (Chen *et al.*, 2007), and NorM (He *et al.*, 2010), P-glycoprotein (Aller *et al.*, 2009), and the open conformation of MsbA, closed conformation of MsbA, outward-facing conformation (AMPPNP) of MsbA (Ward *et al.*, 2007).

A.

B.

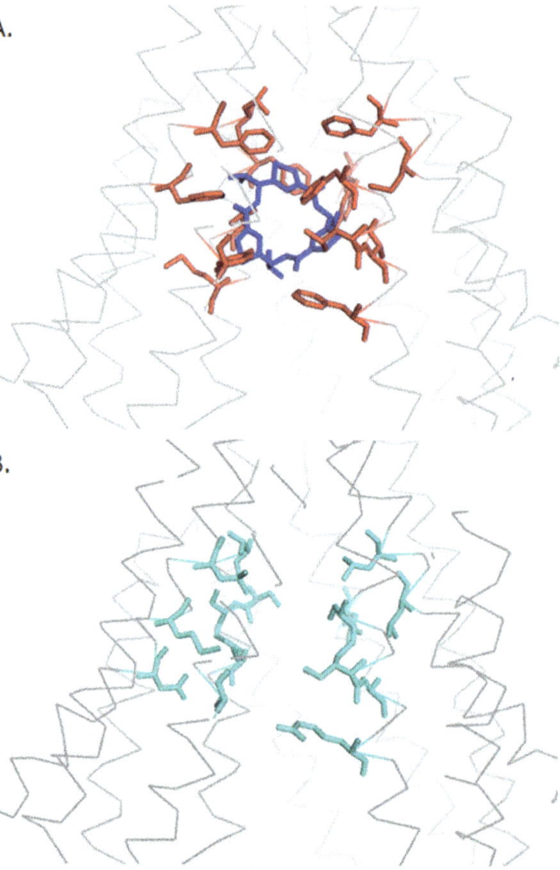

C.

Comparison of Pgp and MsbA

TM	Mouse Pgp	EC-MsbA
1	His 60	Asn 37
	Gly 61	Ala 38
	Ala 63	Ser 40
	Leu 64	Asp 41
	Pro 65	Thr 42
	Met 67	Met 44
	Met 68	Leu 45
	Phe 71	Leu 48
2	Tyr 110	Met 67
	Tyr 113	Val 70
	Val 121	Arg 78
3	Gly 187	Ile 144
	Phe 190	Val 147
	Gln 191	Arg 148
	Ala 194	Ala 151
4	Gly 222	Ser 179
	Ala 225	Ile 182
	Ala 229	Ser 186
	Ser 233	Arg 190
5	Gly 296	Pro 253
	Phe 299	Gln 256
	Leu 300	Leu 257
	Tyr 303	Ser 260
	Tyr 306	Leu 263
	Phe 310	Leu 267
	Trp 311	Tyr 268
6	Leu 328	Thr 285
	Thr 329	Val 286
	Phe 332	Ser 289
	Ser 333	Ser 290
	Leu 335	Ile 292
	Ile 336	Ala 293
	Ala 338	Met 295
	Phe 339	Arg 296
	Gly 342	Lys 299
	Gln 343	Ser 300
	Pro 346	Asn 303

TM	Mouse Pgp	EC-MsbA
7	Asn 717	Asn 37
	Gln 721	Asp 41
	Phe 724	Met 44
	Phe 728	Leu 48
8	Leu 758	Leu 74
	Ser 762	Arg 78
	Phe 766	Ser 82
9	Ala 830	Ile 144
	Phe 833	Val 147
10	Ile 860	Leu 174
	Val 861	Ala 175
	Ile 864	Val 178
	Ala 865	Ser 179
	Ala 867	Ala 181
	Gly 868	Ile 182
	Glu 871	Val 185
11	Phe 938	Asp 252
	Thr 941	Ile 255
	Gln 942	Gln 256
	Met 945	Ala 259
	Ser 948	Ala 262
	Tyr 949	Leu 263
	Ala 952	Phe 265
	Phe 953	Leu 267
12	Val 970	Ile 284
	Leu 971	Thr 285
	Phe 974	Phe 288
	Ser 975	Ser 289
	Val 978	Ile 292
	Gly 980	Leu 294
	Ala 981	Met 295
	Met 982	Arg 296
	Gly 985	Lys 299
	Gln 986	Ser 300
	Ser 988	Thr 302
	Ser 989	Asn 303

Internal Cavity Residues

Figure 2.2 Comparison of residues in the substrate-binding pocket between mouse Pgp (Aller *et al.*, 2009) and MsbA from *E. coli* (Ward *et al.*, 2007). (A) Pgp structure with QZ59-RRR (Aller *et al.*, 2009) bound. (B) Residues corresponding to MsbA mapped onto the Pgp structure as a reference. (C) Comparison of residues lining the substrate-binding pocket of mouse Pgp and MsbA from *E. coli*. Aromatic residues in the binding pocket of Pgp and MsbA are highlighted yellow and cyan, respectively.

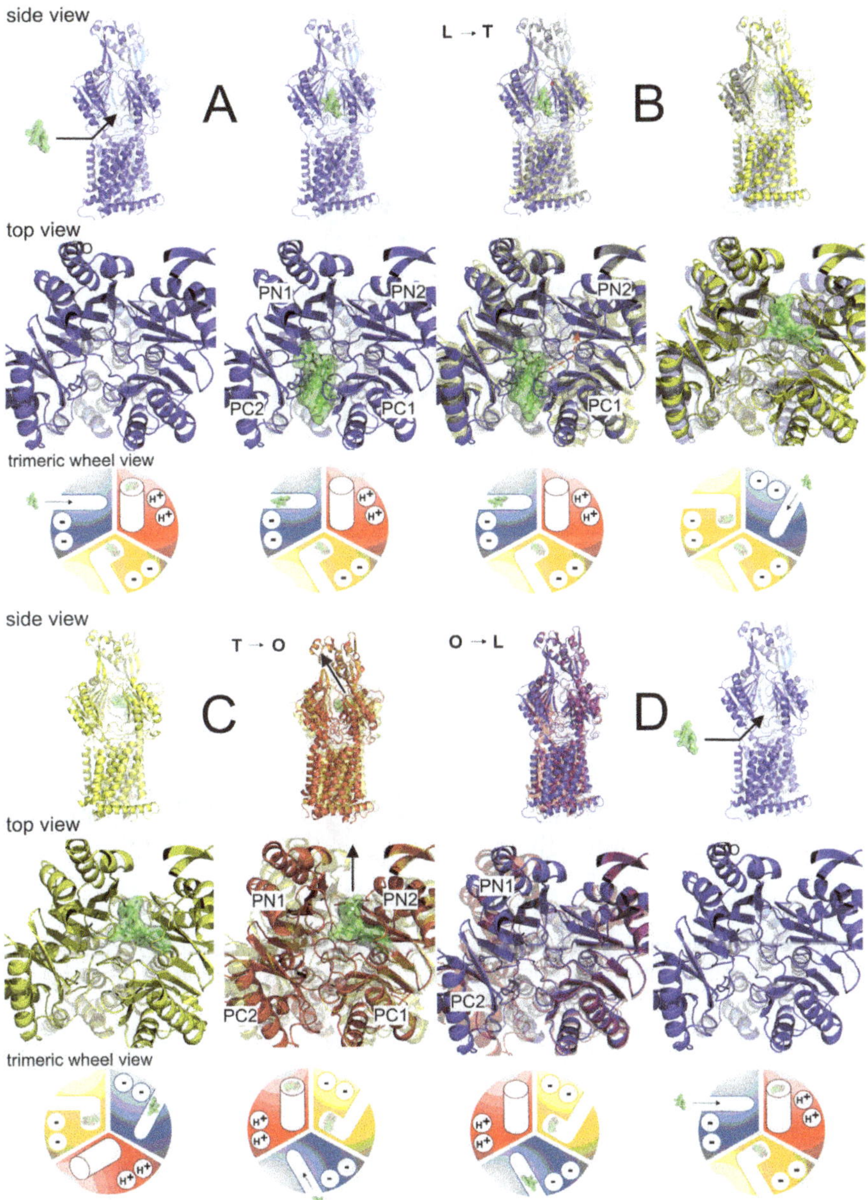

Figure 3.2 Stepwise AcrB-catalysed drug transport mediated by cyclic conformations loose (L, blue), tight (T, yellow) and open (O, red). Shown are four panels (A–D) indicating the sequential drug transport steps and conformational cycling of one AcrB monomer within the trimer. The panels show in the upper row a side view of the monomer orientated such that the membrane plane is horizontal, and the view is towards the PC1 and PC2 subdomains and the access pocket region. The second row is a view on the porter domain perpendicular to the membrane plane. The third row displays in a trimeric wheel view the state of the monomer from rows one and two (upper left monomer in the wheel, L in blue, T in yellow and O in red) shown in (A–D) in relation to the other monomers in the AcrB trimer. Substrate (doxorubicin) is depicted in green. The putative protonation state of D407 and D408 inside the transmembrane domain is indicated (L and T are deprotonated, O is protonated). (A) Initial drug uptake via the access pocket in the L monomer. (B) Transfer of the drug from the low affinity access pocket to the high affinity deep binding pocket during the L to T transition. (C) Directional drug transport during the T to O transition. Subdomains PC1 and PC2 close the lateral access and prevent backsliding of the drug, which is extruded through tunnel 3. (D) Resetting the monomer by the O to L transition to reinitialize drug uptake via the access pocket (Taken from Eicher *et al.*, 2012).

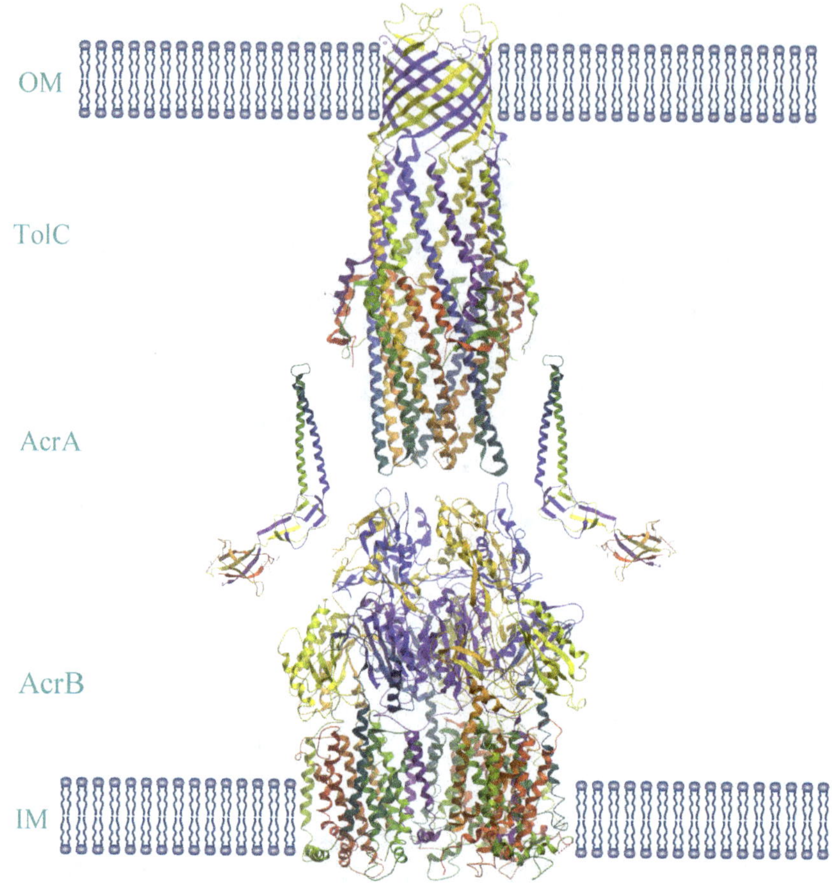

Figure 3.9 Schematic of the assembly. The subunit stoichiometry of the assembly has not been established unequivocally. The membrane-proximal domain is not included in the structure of the AcrA.

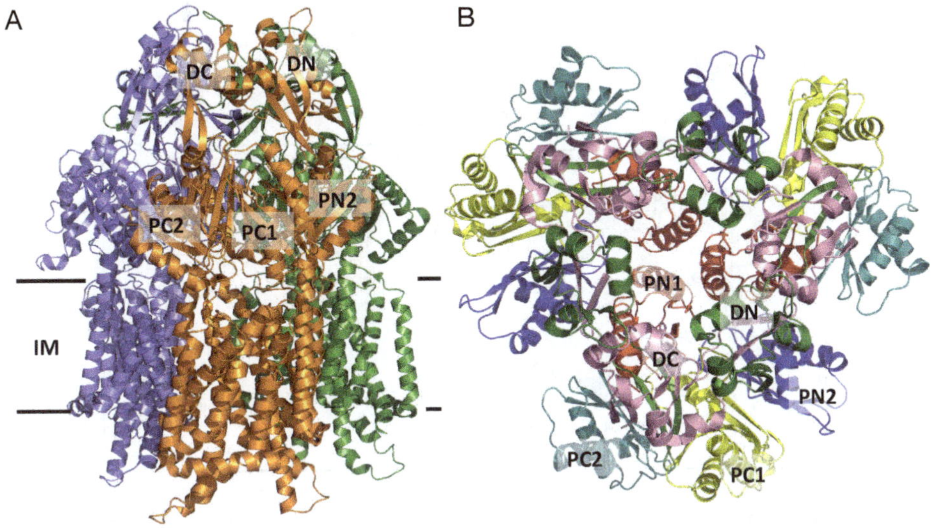

Figure 4.3 Crystal structures of the CusA efflux pump. (A) Ribbon diagram of the CusA homotrimer viewed in the membrane plane. Each molecule is labelled with a different colour: molecule 1 is purple, molecule 2 is orange, and molecule 3 is green. Each subunit of CusA consists of 12 transmembrane helices (TM1-TM12) and a large periplasmic domain formed by two periplasmic loops between TM1 and TM2, and TM7 and TM8, respectively. (B) Ribbon diagram of the CusA homotrimer subdomains as viewed from the top. The six subdomains are PN1 (red), PN2 (blue), PC1 (yellow), PC2 (teal), DN (green) and DC (pink).

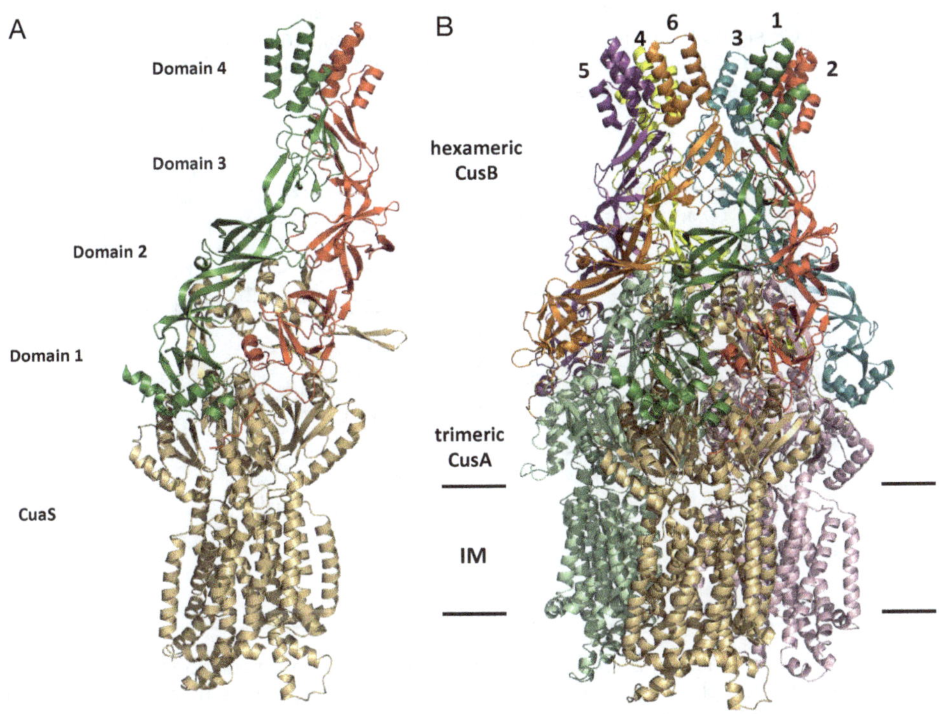

Figure 4.6 Crystal structures of the CusBA adaptor-transporter efflux complex. (A) The asymmetrical unit shows two CusB molecules (colours green and red) interacting with one CusA molecule (light orange). Domains 1–4 of CusB are labelled. (B) The full structure showing the hexameric CusB adaptor structure interacting with the trimeric CusA pump. Each CusB molecule is indicated in a different colour: green, red, light blue, yellow, purple, and orange. Each CusA molecule is indicated in different, lighter colours: light orange, light red, and light green.

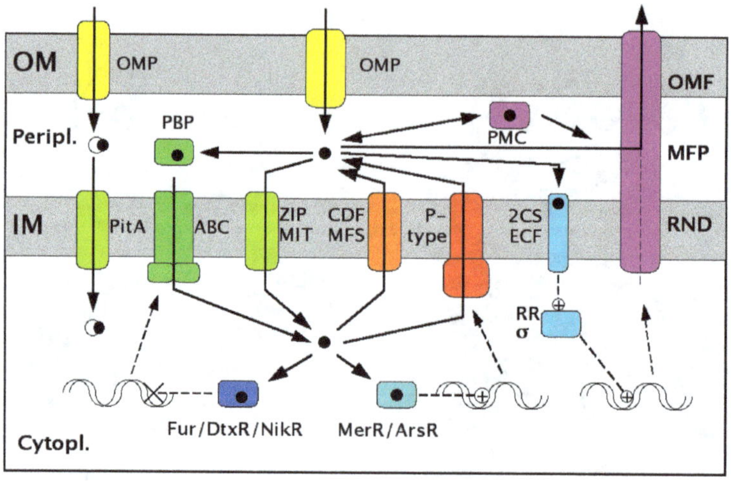

Figure 5.1 Basic components of transition metal cation homeostasis on bacteria: defining the physiological role of RND systems. Metal cations (black dots) cross the outer membrane (OM) via outer membrane porins (OMP, yellow). They are imported further on by fast and unspecific secondary uptake systems, e.g. from the ZIP or MIT protein families (green). If this is not sufficient, genes repressed by regulatory proteins of the Fur, DtxR, or NikR family (dark blue) are being de-repressed, which encode primary uptake systems like ABC transporters (medium green) that interact with periplasmic metal-binding proteins (PBP, medium green). If, on the other hand, a metal is in surplus, regulatory proteins of the MerR or ArsR protein families activate expression of genes for primary efflux pumps such as P-type ATPases (red) that cooperate with secondary efflux such as CDF and MFS members (orange). There might be also uptake of metal:phosphate complexes by OMPs and the PitA transporter (on the left hand). Periplasmic cations are being sensed by two-component or ECF sigma factor regulatory systems (light blue), which control expression of the genes for RND-driven systems and maybe periplasmic metal chaperones (purple). For the RND protein, PPE is indicated as thick line, TEE as thin and dashed line.

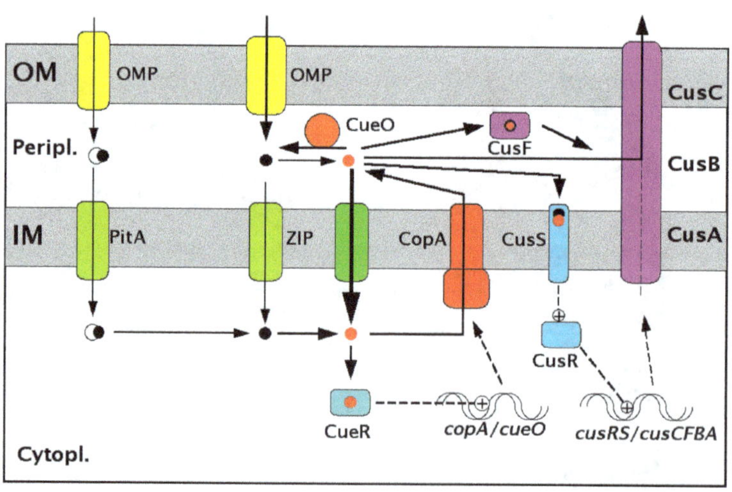

Figure 5.6 Copper homeostasis in *E. coli*. This figure uses the same colour code as Fig. 5.1. Cu(II) and maybe Cu(II):phosphate complexes are transported by OMPs into the periplasm and further on by PitA and ZIP into the cytoplasm but this seems to be a slow uptake process. In either compartment, Cu(II) (black dots) is reduced to Cu(I)(red dots). Periplasmic Cu(I) is rapidly taken up by (an) unknown system(s). Cytoplasmic Cu(I) is sensed by the MerR-type regulator CueR that controls *copA* and *cueR*. CopA is a P-type ATPase that transports Cu(I) back to the periplasm where is gets oxidized by CueO back to Cu(II) in a reaction that depends on molecular oxygen. Periplasmic copper is sensed by the two-component regulatory system CusSR in control of the *cus* determinant. CusF binds periplasmic copper. This copper and that not bound to CusF is exported back to the outside by the RND-driven CusCBA protein complex.

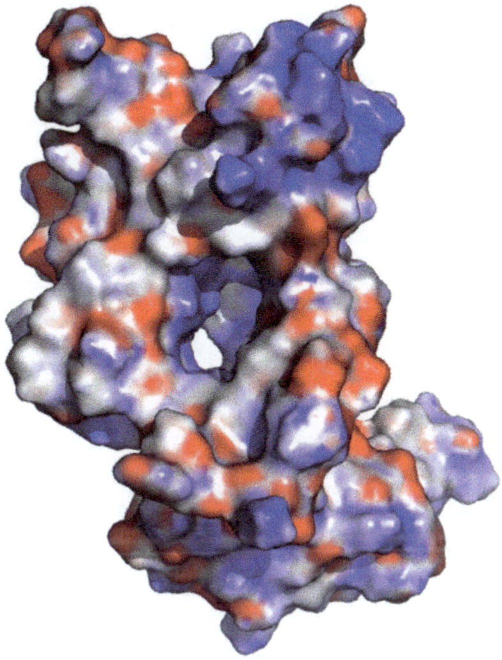

Figure 11.2 Views of the tunnel-like cavity in the ligand-binding domain of CmeR. This is an electrostatic surface potential of one subunit of CmeR. The view shows the long tunnel spanning through the C-terminal domain of CmeR. Blue (+15 k_BT) and red (–15 k_BT) indicate the positively and negatively charged areas, respectively, of the protein.

www.ingramcontent.com/pod-product-compliance
Lightning Source LLC
LaVergne TN
LVHW062229200926
848484LV00006B/1503